Polymers in Microlithography

ACS SYMPOSIUM SERIES **412**

Polymers in Microlithography
Materials and Processes

Elsa Reichmanis, EDITOR
AT&T Bell Laboratories

Scott A. MacDonald, EDITOR
IBM Almaden Research Center

Takao Iwayanagi, EDITOR
Hitachi Central Research Laboratory

Developed from a symposium sponsored
by the Division of Polymeric Materials: Science and Engineering
at the 197th National Meeting
of the American Chemical Society,
Dallas, Texas,
April 9–14, 1989

American Chemical Society, Washington, DC 1989

Library of Congress Cataloging-in-Publication Data

Polymers in microlithography: materials and processes / Elsa Reichmanis, Scott A. MacDonald, Takao Iwayanagi.

p. cm.—(ACS Symposium Series, 0097–6156; 412).

Includes bibliographical references.

Developed from a symposium sponsored by the Division of Polymeric Materials: Science and Engineering at the 197th National Meeting of the American Chemical Society, Dallas, Texas, April 9–14, 1989.

ISBN 0–8412–1701–7

1. Polymers—Congresses. 2. Photoresists—Congresses. 3. Microlithography—Materials—Congresses.

I. Reichmanis, Elsa, 1953– . II. MacDonald, Scott A. III. Iwayanagi, Takao, 1949– . IV. American Chemical Society. Division of Polymeric Materials: Science and Engineering. V. Series.

TK7871.15.P6P635 1989
621.381'531—dc20   89–17931
                                                    CIP

The paper used in this publication meets the minimum requirements of American National Standard for Information Sciences—Permanence of Paper for Printed Library Materials, ANSI Z39.48–1984.

∞

Copyright © 1989

American Chemical Society

All Rights Reserved. The appearance of the code at the bottom of the first page of each chapter in this volume indicates the copyright owner's consent that reprographic copies of the chapter may be made for personal or internal use or for the personal or internal use of specific clients. This consent is given on the condition, however, that the copier pay the stated per-copy fee through the Copyright Clearance Center, Inc., 27 Congress Street, Salem, MA 01970, for copying beyond that permitted by Sections 107 or 108 of the U.S. Copyright Law. This consent does not extend to copying or transmission by any means—graphic or electronic—for any other purpose, such as for general distribution, for advertising or promotional purposes, for creating a new collective work, for resale, or for information storage and retrieval systems. The copying fee for each chapter is indicated in the code at the bottom of the first page of the chapter.

The citation of trade names and/or names of manufacturers in this publication is not to be construed as an endorsement or as approval by ACS of the commercial products or services referenced herein; nor should the mere reference herein to any drawing, specification, chemical process, or other data be regarded as a license or as a conveyance of any right or permission to the holder, reader, or any other person or corporation, to manufacture, reproduce, use, or sell any patented invention or copyrighted work that may in any way be related thereto. Registered names, trademarks, etc., used in this publication, even without specific indication thereof, are not to be considered unprotected by law.

PRINTED IN THE UNITED STATES OF AMERICA

ACS Symposium Series

M. Joan Comstock, *Series Editor*

1989 ACS Books Advisory Board

Paul S. Anderson
Merck Sharp & Dohme Research Laboratories

Alexis T. Bell
University of California—Berkeley

Harvey W. Blanch
University of California—Berkeley

Malcolm H. Chisholm
Indiana University

Alan Elzerman
Clemson University

John W. Finley
Nabisco Brands, Inc.

Natalie Foster
Lehigh University

Marye Anne Fox
The University of Texas—Austin

G. Wayne Ivie
U.S. Department of Agriculture, Agricultural Research Service

Mary A. Kaiser
E. I. du Pont de Nemours and Company

Michael R. Ladisch
Purdue University

John L. Massingill
Dow Chemical Company

Daniel M. Quinn
University of Iowa

James C. Randall
Exxon Chemical Company

Elsa Reichmanis
AT&T Bell Laboratories

C. M. Roland
U.S. Naval Research Laboratory

Stephen A. Szabo
Conoco Inc.

Wendy A. Warr
Imperial Chemical Industries

Robert A. Weiss
University of Connecticut

Foreword

The ACS SYMPOSIUM SERIES was founded in 1974 to provide a medium for publishing symposia quickly in book form. The format of the Series parallels that of the continuing ADVANCES IN CHEMISTRY SERIES except that, in order to save time, the papers are not typeset but are reproduced as they are submitted by the authors in camera-ready form. Papers are reviewed under the supervision of the Editors with the assistance of the Series Advisory Board and are selected to maintain the integrity of the symposia; however, verbatim reproductions of previously published papers are not accepted. Both reviews and reports of research are acceptable, because symposia may embrace both types of presentation.

Contents

Preface xi

1. Polymers in Microlithography: An Overview 1
   Elsa Reichmanis and Larry F. Thompson

   CHEMICALLY AMPLIFIED RESIST CHEMISTRY

2. Brönsted Acid Generation from Triphenylsulfonium
   Salts in Acid-Catalyzed Photoresist Films 27
   D. R. McKean, U. Schaedeli, and Scott A. MacDonald

3. Chemically Amplified Resists: Effect of Polymer
   and Acid Generator Structure 39
   Francis M. Houlihan, Elsa Reichmanis, Larry F.
   Thompson, and Regine G. Tarascon

4. Copolymer Approach to Design of Sensitive
   Deep-UV Resist Systems with High Thermal Stability
   and Dry Etch Resistance 57
   Hiroshi Ito, Mitsuru Ueda, and Mayumi Ebina

5. Nonswelling Negative Resists Incorporating Chemical
   Amplification: The Electrophilic Aromatic Substitution
   Approach 74
   Jean M. J. Fréchet, Stephen Matuszczak, Harald
   D. H. Stöver, C. Grant Willson, and Berndt Reck

6. Acid-Catalyzed Cross-Linking in Phenolic-Resin-Based
   Negative Resists 86
   A. K. Berry, K. A. Graziano, L. E. Bogan, Jr., and
   J. W. Thackeray

7. New Design for Self-Developing Imaging Systems Based on Thermally Labile Polyformals 100
   Jean M. J. Fréchet, C. Grant Willson, T. Iizawa, T. Nishikubo, K. Igarashi, and J. Fahey

MULTILEVEL RESIST CHEMISTRY AND PROCESSING

8. Polysilanes: Solution Photochemistry and Deep-UV Lithography 115
   R. D. Miller, G. Wallraff, N. Clecak, R. Sooriyakumaran, J. Michl, T. Karatsu, A. J. McKinley, K. A. Klingensmith, and J. Downing

9. Syntheses of Base-Soluble Si Polymers and Their Application to Resists 133
   Shuzi Hayase, Rumiko Horiguchi, Yasunobu Onishi, and Toru Ushirogouchi

10. Lithographic Evaluation of Phenolic Resin–Dimethyl Siloxane Block Copolymers 158
    M. J. Jurek and Elsa Reichmanis

11. Preparation of a Novel Silicone-Based Positive Photoresist and Its Application to an Image Reversal Process 175
    Akinobu Tanaka, Hiroshi Ban, and Saburo Imamura

12. Photooxidation of Polymers: Application to Dry-Developed Single-Layer Deep-UV Resists 189
    Omkaram Nalamasu, Frank A. Baiocchi, and Gary N. Taylor

13. Kinetics of Polymer Etching in an Oxygen Glow Discharge 210
    Charles W. Jurgensen

14. Quantitative Analysis of a Laser Interferometer Waveform Obtained During Oxygen Reactive-Ion Etching of Thin Polymer Films 234
    B. C. Dems, P. D. Krasicky, and F. Rodriguez

15. Evaluation of Several Organic Materials as Planarizing Layers for Lithographic and Etchback Processing 252
    L. E. Stillwagon and Gary N. Taylor

    NOVEL CHEMISTRY AND PROCESSES FOR MICROLITHOGRAPHY

16. New Negative Deep-UV Resist for KrF Excimer Laser Lithography 269
    Masayuki Endo, Yoshiyuki Tani, Masaru Sasago, and Noboru Nomura

17. Characterization of a Thiosulfate Functionalized Polymer: A Water-Soluble Photosensitive Zwitterion 280
    C. E. Hoyle, D. E. Hutchens, and S. F. Thames

18. Pyrimidine Derivatives as Lithographic Materials 303
    Yoshiaki Inaki, Minoo Jalili Moghaddam, and Kiichi Takemoto

19. Synthesis of New Metal-Free Diazonium Salts and Their Applications to Microlithography 319
    Shou-ichi Uchino, Michiaki Hashimoto, and Takao Iwayanagi

20. Photobleaching Chemistry of Polymers Containing Anthracenes 332
    James R. Sheats

21. Lithography and Spectroscopy of Ultrathin Langmuir–Blodgett Polymer Films 349
    S. W. J. Kuan, P. S. Martin, L. L. Kosbar, C. W. Frank, and R. F. W. Pease

22. Dissolution of Phenolic Resins and Their Blends 364
    J. P. Huang, E. M. Pearce, A. Reiser, and T. K. Kwei

23. Solvent Concentration Profile of Poly(methyl methacrylate) Dissolving in Methyl Ethyl Ketone: A Fluorescence-Quenching Study 385
    William Limm, Mitchell A. Winnik, Barton A. Smith, and Deirdre T. Stanton

24. Molecular Studies on Laser Ablation Processes of Polymeric Materials by Time-Resolved Luminescence Spectroscopy...............400
    Hiroshi Masuhara, Akira Itaya, and Hiroshi Fukumura

25. Mechanism of Polymer Photoablation Explored with a Quartz Crystal Microbalance...............411
    Sylvain Lazare and Vincent Granier

26. Mechanism of UV- and VUV-Induced Etching of Poly(methyl methacrylate): Evidence for an Energy-Dependent Reaction...............424
    Nobuo Ueno, Tsuneo Mitsuhata, Kazuyuki Sugita, and Kenichiro Tanaka

INDEXES

Author Index 438

Affiliation Index 439

Subject Index 439

Preface

PROGRESS IN MICROELECTRONICS, and especially in microlithographic technology, is proceeding at an astonishing rate. Today, it is believed that conventional photolithography, which uses 365–405-nm radiation, will be able to print 0.5–0.6-μm features in production and that it will remain the dominant printing technology well into the 1990s. Diazonaphthoquinone–novolac materials will most likely remain the materials of choice for production of these devices. The costs of introducing new resist materials and new hardware are strong driving forces pushing photolithography to its absolute limit. The technological alternatives to conventional photolithography are largely the same as they were a decade ago, that is, deep-UV photolithography, scanning electron-beam lithography, and X-ray lithography. The leading candidate for the production of devices with features as small as 0.3 μm is deep-UV lithography.

No matter which technology eventually replaces photolithography, the new resists and processes that will be required will necessitate an enormous investment in research and process development. The polymer materials that are used as radiation-sensitive resist films must be carefully designed to meet the specific requirements of the lithographic technology and device process. Although these requirements vary according to the radiation source and device process, properties such as sensitivity, contrast, resolution, etching resistance, shelf life, and purity are ubiquitous.

This volume is not intended to be comprehensive, but the chapters found here should provide the reader with an appreciation for the diversity of chemical research efforts that are required for the development of new resist materials and processes. They span the range of novel synthetic reactions that may be applied to imaging processes, to new processing techniques to enhance image quality, to understanding the fundamental science behind processes such as polymer dissolution and photoablation. The contents have been divided into three sections: Chemically Amplified Resist Chemistry, Multilevel Resist Chemistry and Processing, and Novel Chemistry and Processes for Microlithography. Each section contains an introduction written by a recognized expert in the field.

Acknowledgments

We are indebted to many people and organizations for making the symposium and book possible, particularly the authors, for their efforts in providing manuscripts of their presentations. We are especially grateful to the Petroleum Research Fund and the Division of Polymeric Materials: Science and Engineering for financial support. Finally, our sincerest thanks are extended to Cheryl Shanks and the production staff of the ACS Books Department for their efforts in publishing this volume.

ELSA REICHMANIS
AT&T Bell Laboratories
Murray Hill, NJ 07974

SCOTT A. MACDONALD
IBM Almaden Research Center
San Jose, CA 95120–6099

TAKAO IWAYANAGI
Hitachi Central Research Laboratory
Kokubunji, Tokyo 185, Japan

July 31, 1989

Chapter 1

Polymers in Microlithography

An Overview

Elsa Reichmanis and Larry F. Thompson

AT&T Bell Laboratories, 600 Mountain Avenue, Murray Hill, NJ 07974

> The evolution in microelectronics technology has progressed at an astonishing rate during the past decade. This is particularly true of microlithographic technology which is the technology used to generate the high resolution circuit elements characteristic of today's integrated circuits. While almost all of these commercial devices are made by photolithographic techniques that utilize 365-436nm UV radiation, within the next 3-8 years, new lithographic strategies will be required. These technological alternatives, such as deep-UV, e-beam and x-ray lithography, will require new polymeric resist materials and processes. A brief overview of the current trends in microlithography is presented along with an examination of the varied chemistries that can be applied to this technology. The reader is referred to alternate sources for detailed reviews of the field.

The evolution in microelectronics technology, and microlithography in particular, has progressed at an astonishing rate during the past decade. The speed of integrated circuit devices has increased by several orders of magnitude, while the cost associated per bit has decreased at a still faster rate (Figure 1). These improvements are a direct result of the increase in the number of components per chip, a trend that has progressed at a rate of 10^2- 10^3 per decade. It is expected that this trend will continue, although perhaps at a slower rate (1).

This increase in circuit density has been made possible by decreasing the minimum feature size on the chip. In the mid 1970's, the state-of-the-art dynamic random access memory (DRAM) device was capable of storing 4000 bits of data and had features 5 μm in size. Today, 4 megabit DRAM's are in production with minimum features in the 0.8 - 1.0 μm range, while state-of-the-art devices with 0.6 - 0.8 μm features are in pilot production (2). As shown in Figure 2 and Table I, contact printing was the dominant lithographic technology used for device

0097-6156/89/0412-0001$07.00/0
© 1989 American Chemical Society

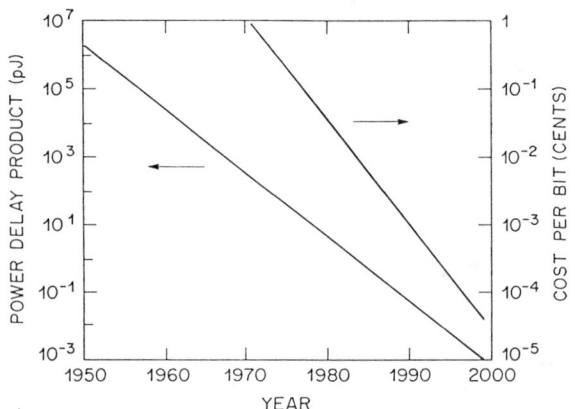

Figure 1: Plot of the power delay product vs. year and cost per bit vs. year for commercially available VLSI devices.

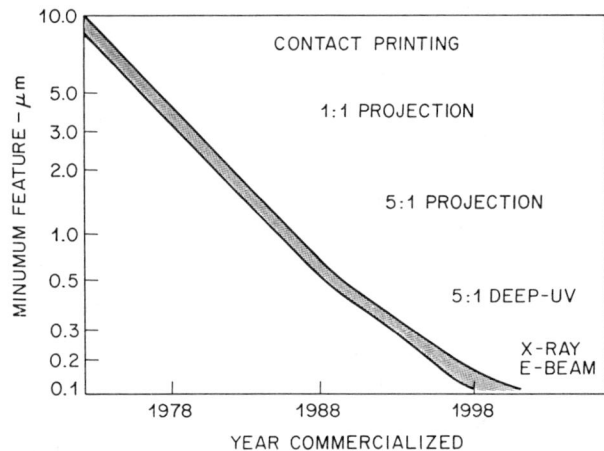

Figure 2: Graphical representation of the minimum feature size vs. year of commercialization for MOS devices.

production well into the 1970's. Although contact printing is ostensibly a high resolution technology, its utility is limited by mask and wafer defects which begin to affect yield and economics at geometries below about 3μm. Contact printing thus gave way to one-to-one projection printing which obviated the contact induced defects by separating the mask and wafer, and allowed a reduction in feature size to about 1.5 μm. At that time, this was believed to be the resolution limit for photolithography in a production environment. The development of reduction step-and-repeat exposure tools allowed further improvement of the resolution obtainable by optical techniques. High numerical aperture(NA) steppers operating at the conventinal 405 or 436nm (H or G) lines of the Hg arc lamp are used to produce today's state-of-the-art devices and it is generally believed that such tools will be capable of producing chips with features as small as 0.6 μm. Further reduction in feature size will require the introduction of a new lithographic technology the strategy for which is discussed below. Conventional G-line (436 nm) lithography employing 0.4 NA reduction lenses is currently used in manufacturing to produce today's devices with features of 0.8 μm. Higher NA G-line and I-line (365nm) lenses are available that will push the technology to the sub-0.5 μm regime. Concomitantly, deep-UV (230-260nm) systems are becoming available that effectively compete with the I-line technology. Work is progressing on development of higher NA (0.45) deep-UV systems which should be available by mid-to-late 1990 (3). This technology is expected to have a production resolution capability of at least 0.4 μm. E-beam and x-ray lithography will play an increasingly significant role in the production of devices by the turn of the century when minimum features are expected to reach the sub-0.25 μm level (4,5).

The technological alternatives to conventional photolithography are largely the same as they were a decade ago, viz., deep-UV photolithography, scanning electron-beam and x-ray lithography (1,6). The leading candidate for the production of devices with features perhaps as small as 0.3 μm is deep-UV lithography (Figure 2, Table I) (2,7). Major advances in this technology in the past decade relating to improved quartz lenses and high output light sources have

Table I: Photolithographic Trends

Lithographic Technology	Years in Use	Minimum Feature
Contact Printing	1960-1973	5 μm
1:1 Projection	1973-1982	1.5 μm
5:1 Projection Step and Repeat	1982-present	0.8 μm
Deep/Mid UV Step and Repeat	1988-	0.3 μm
X-ray E-beam	1995-	0.1 μm

occurred. Several step and repeat 5x and 10x reduction systems that use excimer laser sources have been designed and/or built (8,9). Systems using refractive optics require a very narrow bandwidth light source (less than 0.001 Å) since it is not practical to correct for chromatic aberrations in quartz lenses. Laser sources provide such narrow bandwidths with enough intensity to accommodate resists with ~50 mJ cm^{-2} sensitivities, enabling a rather wide choice of resist chemistries. Work is also being done on 4x reduction systems based on all reflective optical systems and wide bandwidth Hg arc sources in the 240 to 260 nm region (10). However, since the intensity of these sources is less than that of laser sources, more sensitive resists (<10 mJ cm^{-2}) will be required for high throughput.

No matter which technology eventually replaces photolithography, new resists and processes will be required, necessitating enormous investments in research and process development. The polymer materials that are used as radiation sensitive resist films must be carefully designed to meet the specific requirements of the lithographic technology and device process. Although these requirements vary according to the radiation source and device process, the following properties are common to all resists: sensitivity, contrast, resolution, etching resistance, shelf-life and purity.

The sensitivity of a resist must be sufficiently high to meet the throughput requirements of a given exposure tool, while the contrast must be adequate to ensure high differential solubility between the exposed and unexposed regions of the resist film. These parameters are determined by measuring the change in solubility as a function of the radiation dose received by the material (6,11). Although the sensitivity and contrast of a resist are dependent on many variables such as developer strength, photon or particle energy, and processing conditions, plotting the sensitivity curves as depicted in Figure 3 provides a standard means of comparing the sensitivity and contrast of different resist materials. The contrast (γ) for either a positive or negative resist is detemined from the slope of the linear portion of the curves shown in Figure 3. The sensitivity for a positive resist is the minimum dose required to clear a given feature and for a negative resist it is the dose required to effect 50-70 % film thickness remaining after development. The resolution of a resist is defined as the smallest feature that can be resolved in a densely patterned area. For a negative resist, resolution should be reported in terms of the smallest trench that is resolvable between two large pads; and for a positive resist, the smallest isolated line. An accepted alternative is to use an equal line-space pattern for both tones of resist. Resolution is a difficult property to quantify since many external variables, e.g., radiation type, developer selection, pattern density, resist thickness and substrate material, affect the size of the minimum feature that can be resolved. When reporting the resolution of a resist system it is critical to give complete experimental details to ensure that valid comparisons can be made.

The etching resistance of a resist is simply a function of how well the material withstands the pattern transfer processes used in device manufacturing. Its resistance must be sufficiently high to allow precise transfer of the resist image into the underlying substrate with < 10% linewidth change. This is not an easy

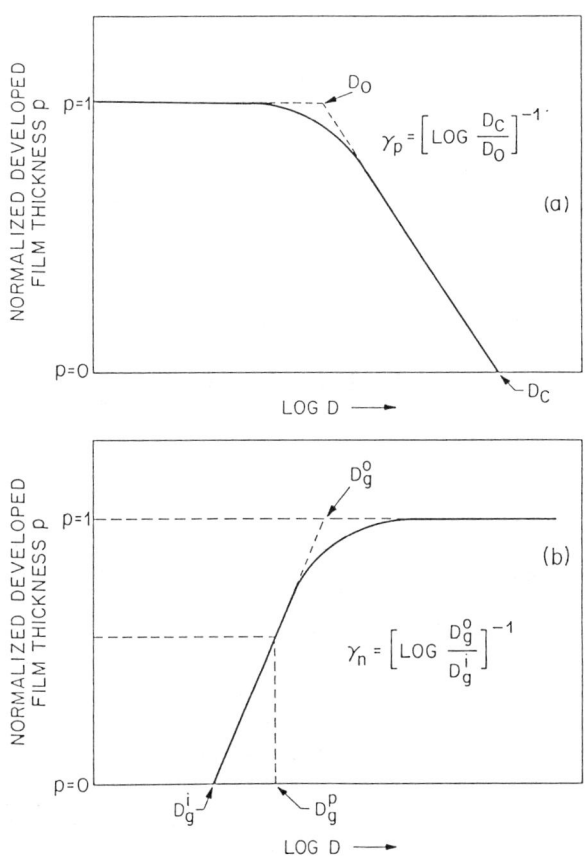

Figure 3: Representative contrast curves for a) positive resists and b) negative resists.

requirement to meet, especially for plasma and reactive-ion etching (RIE) processes which are extremely harsh. Etching resistance is often reported in relative terms by comparing the etching rate of a given resist with that of a conventional positive photoresist. As with resolution, many process and materials variables affect the etching rate of a polymer and complete experimental details should be reported to allow valid comparisons of different resists.

Generally, the desired resist properties can be achieved by careful manipulation of the structure and molecular properties of resist components. As mentioned earlier, new resists will be required that are both sensitive to different types of radiation and compatible with advanced processing requirements. Further, the necessity to accommodate substrate topography will likely demand some form of "multi-level" resist technology. This technology utilizes two or more discrete layers, each of which provides a specific function. Considerable work is underway to develop suitable multi-level resists and processes. In particular, there have been numerous reports on metal containing polymers suitable for "bilevel" applications which are the simplest of the multi-level schemes. Since all the alternatives to conventional wavelength photolithography employ rather high energy radiation, resists designed for one type of radiation, e.g. e-beam, will frequently be applicable to the other lithographic technologies. The reader is referred to references 1, 7, 11, 12, 13, and 14 for general, in depth reviews on the subject of polymers for microlithography. A brief overview is given below.

Single-Level Resist Chemistry

Negative resists are a class of materials that become less soluble in a developer after exposure to radiation. Generally, the chemistry of negative resists involves some form of radiation induced crosslinking. The parent polymers are usually soluble in organic solvents, which in turn are used as developers (crosslinking novolac and poly(hydroxystyrene) systems are an exception). Since polymer dissolution occurs first by swelling of the matrix followed by chain disentanglement, it is critical to select a developer that minimizes swelling of the crosslinked regions thereby facilitating high resolution. The ideal developer should be a kinetically good, but thermodynamically poor solvent for the resist. This ensures that the developer will dissolve the unexposed regions of the film while minimizing the swelling volume of the irradiated regions in a given development time. Novembre and co-workers (15) developed a method based upon the Hansen 3-dimensional solubility parameter model to screen potential organic-based resist developers. This methodology facilitates selection of an optimal developer without the tedious trial and error approach commonly used. The sensitivity of negative resists is generally high since only a few events per chain are required to achieve differential solubility. Some negative systems crosslink via a chain reaction yielding even higher sensitivities.

The first resist used to fabricate solid-state devices was a negative resist based on cyclized poly(*cis*-1,4-isoprene) which is crosslinked using a photoactive bisaryldiazide crosslinking agent (16,17). A major limitation of these resists is that they require organic solvent developers which cause image distortion due to swelling. An alternate, aqueous developable material that utilizes poly(p-vinylphenol), an aqueous-base soluble phenolic resin, as the matrix has been reported and overcomes this problem (18,19).

Three classes of inherently crosslinking polymers have been reported to be useful as negative-acting resists for deep-UV, e-beam and x-ray applications. These include epoxy (glycidyl), vinyl and halogen containing materials in which the radiation sensitive unit is an integral part of the polymer (Figure 4). Crosslinking of glycidyl and vinyl groups occurs *via* chain mechanisms leading to high crosslinking efficiency and high resist sensitivity. Unfortunately, these highly efficient reactions give rise to poor resist contrast and line-size control due to post-exposure curing reactions (20-25).

The incorporation of halogen groups into acrylate and styrene based polymers facilitates radiation-induced crosslinking with high crosslinking efficiency without the presence of a chain-propagation mechanism. Radiation induced cleavage of the carbon-halogen bond generates a radical that may then undergo rearrangement, abstraction or recombination reactions leading to the formation of a crosslinked network. The mechanism of crosslinking in chloromethylated polystyrene has been extensively studied by Tabata and Tagawa (26). The localized nature of the crosslinking reaction in these polymers, as opposed to the chain-propagation mechanism occurring in the epoxy and vinyl containing resists, eliminates the post-exposure curing effects seen in the vinyl and epoxy materials. Taylor and coworkers (27) found that incorporating halogen into methacrylate and acrylate polymers results in a significant increase in sensitivity to radiation-induced crosslinking. Poly(2,3-dichloropropyl acrylate) is nearly three orders of magnitude more sensitive to x-ray irradiation, for example, than the parent, poly(propyl acrylate). Some of this increase can be attributed to increased X-ray absorption; however, much of it is due to enhanced susceptibility to radiation-induced crosslinking. This observation was applied to styrene based electron resists by Thompson et al (28). Polystyrene is a high resolution, negative electron and x-ray resist with excellent plasma resistance. However, its sensitivity is too low for practical application. The sensitivity of polystyrenes can be improved through incorporation of Cl or chloromethyl groups into the polymer *via* substitution on the aromatic ring. Alternatively, sensitivity can be improved through copolymerization with known radiation-sensitive monomers such as glycidyl methacrylate (29). The copolymer of chlorostyrene and glycidyl methacrylate is an example of a resist that incorporates both of these sensitivity-enhancing features. The incorporation of chloromethyl groups into polystyrene can be accomplished by a variety of routes including polymerization of chloromethylstyrene (30), chlorination of poly(methylstyrene) (31) and chloromethylation of polystyrene (32). As little as 5 wt% of chlorine, substituted at the methyl moiety, results in over an order of magnitude improvement in sensitivity compared with polystyrene of similar molecular weight (2 μC cm^{-2} vs. 50 μC cm^{-2} at 10 kV). Many such resists have been described, all of which exhibit good lithographic performance.

Materials that exhibit enhanced solubility after exposure to radiation are defined as positive resists. The mechanism of positive resist action in most of these materials involves either main-chain scission or a polarity change. Positive photoresists that operate on the polarity change principle have been widely used for over three decades in the fabrication of VLSI devices and they exhibit high resolution and excellent dry etching resistance. Ordinarily, the chain scission mechanism is only operable at photon wavelengths below 300 nm where the energy is sufficient to break main chain bonds.

The "classic" positive resist that undergoes chain scission upon irradiation is poly(methyl methacrylate) (PMMA) (33). The radiation chemistry involves initial cleavage of the methacrylate side chain to generate a radical which may undergo β-scission. This results in a reduction in polymer molecular weight leading to enhanced solubility of the exposed regions (34). Choice of an appropriate developer such as methyl isobutyl ketone allows selective removal of the irradiated areas with minimum swelling of the remaining resist.

While PMMA exhibits high resolution, its sensitivity to radiation-induced degradation is low and dry etching pattern transfer characteristics are poor (33,35,36). However, the high-resolution characteristics of PMMA have prompted several investigators to examine substituted systems and copolymers to improve sensitivity and etching resistance. Examples include substitution of the backbone α-methyl group with electronegative groups such as Cl or CN, (37) introduction of bulky groups that provide steric hindrance to weaken the main chain of the polymer, (38) or improvement of the absorption characteristics of the polymer (39,40). Additionally, incorporation of fluorine into the ester groups of PMMA has been shown to be effective in improving the susceptibility of PMMA to radiation-induced degradation (41-43). Systems based on methacrylic acid that form inter- and intramolecular anhydrides also show improved sensitivity due to the enhanced radiation sensitivity of the anhydride linkages and the excess strain in the polymer due to cyclization (44-48).

Another class of "chain scission" positive resists is the poly(olefin sulfones). These polymers are alternating copolymers of an olefin and sulfur dioxide. The relatively weak C-S bond is readily cleaved upon irradiation and several sensitive resists have been developed based on this chemistry (49,50). One of these materials, poly(butene-1 sulfone) (PBS) has been made commercially available for mask making. PBS exhibits an e-beam sensitivity of 1.6 μC cm^{-2} at 20 kV and 0.25 μm resolution.

The most widely used positive resists are those that operate on the basis of a dissolution inhibition mechanism. Such resists are generally two-component materials consisting of an alkali soluble matrix resin that is rendered insoluble in aqueous alkaline solutions through addition of a hydrophobic, radiation-sensitive material. Upon irradiation, the hydrophobic moiety may be either removed or converted to an alkali soluble species, allowing selective removal of the irradiated portions of the resist by an alkaline developer.

The best known of these so called dissolution inhibition resists is "conventional positive photoresist"; a photosensitive material that uses a novolac (phenol-formaldehyde) resin with a diazonaphthoquinone photoactive compound (PAC) as a dissolution inhibitor. The novolac matrix resin is a condensation polymer of a substituted phenol (often cresol) and formaldehyde. These resins are soluble in organic solvents, facilitating spin coating of uniform, high quality, glassy films. They are also soluble in basic solutions such as aqueous, sodium hydroxide or tetramethylammonium hydroxide (TMAH). The novolac resin is rendered insoluble in aqueous base through the addition of 10-20 wt % of the PAC. Upon irradiation, the diazonaphthoquinone undergoes a Wolff rearrangement followed by hydrolysis to generate a base-soluble indene carboxylic acid (Figure 5) (51). The exposed regions of the film may then be removed by treatment with aqueous base. While the basic components of all conventional photoresists are the same, the precise performance characteristics depend on the substitution pattern on the novolac resin and/or the PAC (52-55).

Through creative chemistry and resist processing, schemes have been developed that produce negative-tone images in positive photoresist. One example of such an "image reversal" process requires addition of small amounts of base additives such as monazoline, imidazole or triethanolamine to diazoquinone-novolac resists (56-58). The chemistry and processes associated with such systems are shown in Figure 6. Thermally induced, base catalyzed decarboxylation of the indene carboxylic acid destroys the aqueous base solubility of the exposed resist. Subsequent flood exposure renders the previously masked regions soluble in aqueous base, allowing generation of negative tone patterns. It is not always necessary to add the base to the resist prior to exposure. Alternate image reversal processes have been developed involving treatment of exposed photoresist with a gaseous amine in a vacuum environment (59). Recently, it has been shown that select sulfonic acids generated from an aryl diazonaphthoquinone PAC can catalyze the crosslinking of the novolac matrix in the exposed regions, providing yet another image reversal scheme (60).

While "conventional positive photoresists" are sensitive, high-resolution materials, they are essentially opaque to radiation below 300 nm. This has led researchers to examine alternate chemistry for deep-UV applications. Examples of deep-UV sensitive dissolution inhibitors include aliphatic diazoketones (61-64) and nitrobenzyl esters (65). Certain onium salts have also recently been shown to be effective inhibitors for phenolic resins (66). A novel e-beam sensitive dissolution inhibition resist was designed by Bowden, et al a (67) based on the use of a novolac resin with a poly(olefin sulfone) dissolution inhibitor. The aqueous, base-soluble novolac is rendered less soluble via addition of ~10 wt % poly(2-methyl pentene-1 sulfone)(PMPS). Irradiation causes main chain scission of PMPS followed by depolymerization to volatile monomers (68). The dissolution inhibitor is thus effectively "vaporized", restoring solubility in aqueous base to the irradiated portions of the resist. Alternate resist systems based on this chemistry have also been reported (69,70).

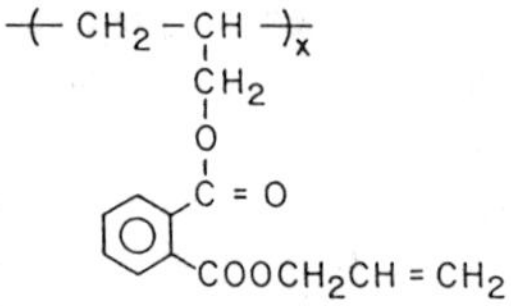

POLY (DIALLYL-ORTHO-
PHTHALATE)

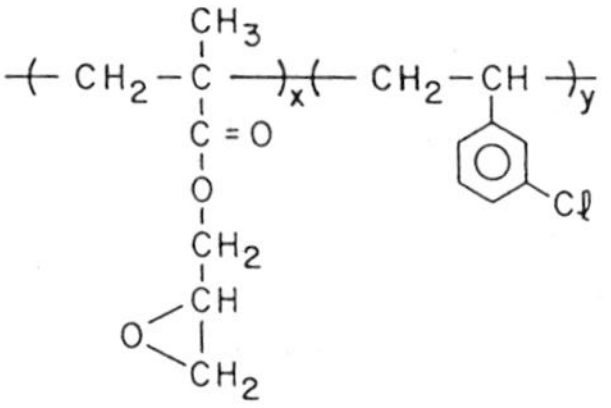

POLY (GLYCIDYL METHACRYLATE-
CO-CHLOROSTYRENE)

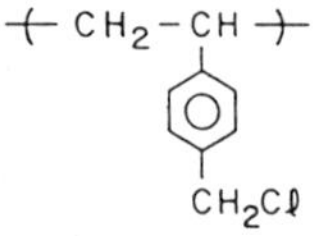

POLY (CHLOROMETHYLSTYRENE)

Figure 4: Select examples of resists incorporating radiation sensitive groups that undergo crosslinking.

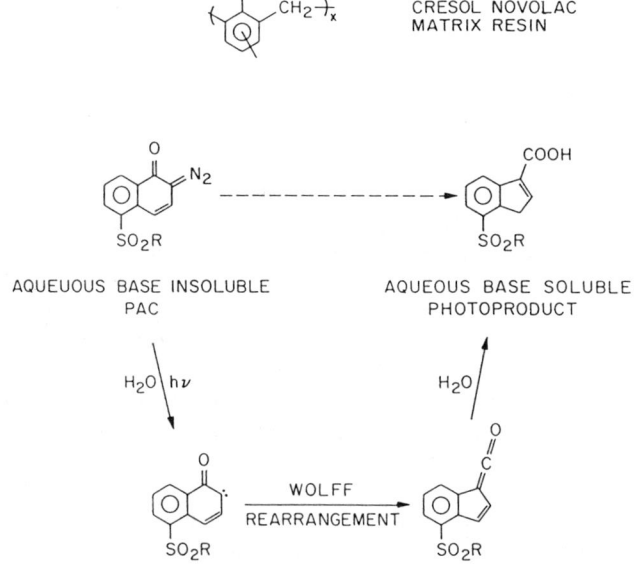

Figure 5: Chemistry associated with a typical conventional positive photoresist.

As device features move into the submicron regime, advanced processing techniques and new lithographic technologies will be required to accomodate high-resolution, high-aspect ratio imaging over device topography. This necessitates the development of new resist materials with improved etching resistance, resolution and sensitivity. One approach to improving sensitivity involves the concept of chemical amplification (71-73). Aryldiazonium, diaryliodonium and triarylsulfonium metal halides, for example, dissociate upon irradiation to produce an acid that can, in turn, catalyze a variety of bond-forming or bond-breaking reactions in a surrounding matrix polymer. The quantum efficiency of such reactions is thus effectively much higher than the quantum yield for initial onium salt dissociation. Many chemistries based on these processes have been developed and the reader is referred to reference 7 for an in-depth discussion of this topic. Recent developments are discussed in subsequent chapters of this book.

This chemical amplification principle has been used to design a number of negative resists based on acid catalyzed cationic polymerization of appropriate monomers, (72,74) or crosslinking of polymers (73,75). However, such materials generally exhibit poor contrast and resolution. A notable exception is the three component resist developed by Feely, et al (75) that consists of a blend of novolac resin, melamine crosslinking agent and acid generator. Irradiation generates an acid catalyst that induces formation of a crosslinked network between the novolac matrix resin and amine additive.

Ito, et al. proposed and applied the concept of chemical amplification to the development of high resolution, dual-tone resist materials. Their initial studies dealt with the catalytic deprotection of poly(t-butoxycarbonyloxystyrene) (PBOCS) (73,76,77). Here, the thermally stable, acid labile, t-butoxycarbonyl group is used to mask the hydroxy functionality of poly(vinylphenol). Irradiation of PBOCS films containing small amounts of an onium salt such as diphenyliodonium hexafluoroarsenate with UV light liberates an acid species that upon subsequent baking, catalyzes cleavage of the protecting group to generate poly(p-hydroxystyrene) and regeneration of the acid, making it available for subsequent deprotection reactions (Figure 7). Aqueous base developers selectively remove the irradiated regions affording high-resolution, high-aspect ratio images. Alternatively, organic solvent developers remove the unirradiated portions of the resist and high quality negative images obtain.

Recently, nonionic acid precursors based on nitrobenzyl ester photochemistry have been developed for chemically amplified resist processes (78-80). These ester based materials (Figure 8) exhibit a number of advantages over the onium salt systems. Specifically, the esters are easily synthesized, are soluble in a variety organic solvents, are nonionic in character, and contain no potential device contaminants such as arsenic or antimony. In addition, their absorption characteristics are well suited for deep-UV exposure.

Three component, aqueous-base developable, positive-tone resists utilizing the chemical amplification principle have also recently been reported (81,82). In these systems, irradiation of a phenolic resin/inhibitor/acid generator resist generates an acid which upon mild heating, catalyzes either depolymerization or deblocking of a

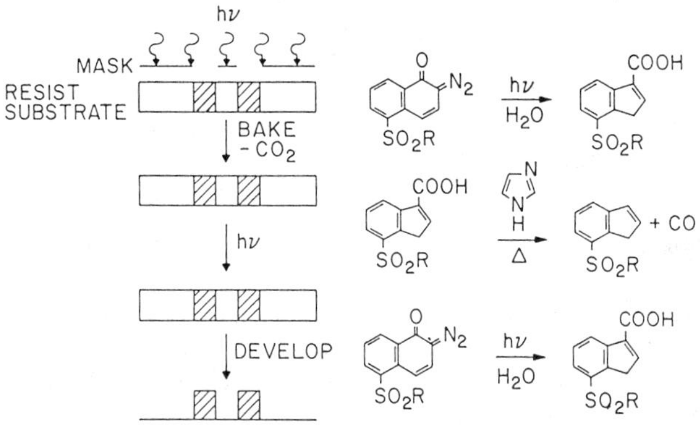

Figure 6: Image reversal process and related chemistry based on a conventional positive photoresist and monazoline.

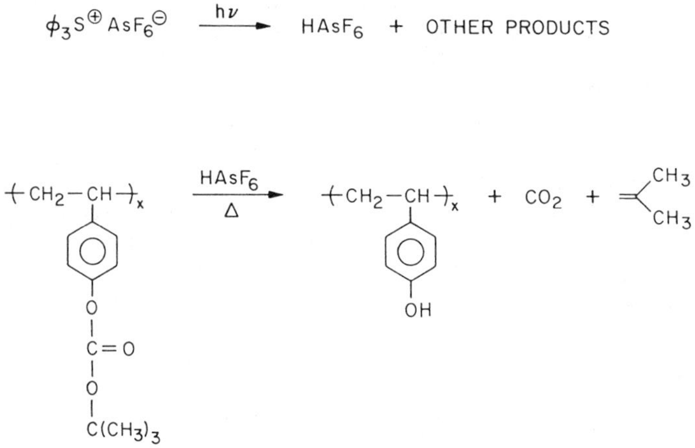

Figure 7: Chemical reactions associated with a typical positive acting chemically amplified resist system.

dissolution inhibitor, allowing aqueous base removal of the irradiated regions. Specifically, Ito found that polyphthalaldehyde, a polymer that undergoes acid-catalyzed depolymerization to volatile monomer, is an effective dissolution inhibitor for phenolic resins, (81) while a class of non-polymeric inhibitors containing acid labile blocking groups (Figure 9) was developed by O'Brien (82).

Multi-Level Resist Chemistry

The increasing complexity and miniaturization of integrated circuit technology are pushing conventional single-layer resist processes to their limit. The demand for improved resolution requires imaging features with increasingly higher aspect ratios and smaller linewidth variation over steep substrate topography. The decrease in feature size can lead to other problems associated with the particular lithographic technique employed. For instance, in photolithography, feature size can be affected by standing wave effects and reflections from the substrate surface which limit the resolution attained with optical techniques. A number of schemes have been proposed to address these problems including the use of polymeric planarizing layers, anti-reflection coatings, and contrast enhancement materials (83). The latter two schemes involve the use of thin organic films that have precisely designed light absorption characteristics. Antireflective coatings are designed to be highly absorbing at the exposure wave length and are applied to the substrate prior to spin-coating the resist. Contrast enhancement materials (CEM) are generally applied to the surface of the resist and through photochemical bleaching during exposure, improve the quality of the aerial light image (84-86). The process utilizes photobleachable materials that are opaque before exposure but bleach (i.e. become transparent) on exposure to radiation. The contrast enhancement layer is coated directly onto the surface of the resist. The dynamics of the bleaching process are such that the transmitted image is of much higher contrast than the incident, aerial image (85) thereby improving the resist's ability to discriminate mask features. The optical density of CEM films should be greater than 2 for thicknesses less than 0.5 μm.

The first CEM system described by Griffing and West (84) consisted of an organic dye dispersed in an inert polymer film that is spin coated onto the surface of a resist and subsequently removed following exposure but prior to resist development. The chemistry of this system is based on the photoisomerization of an aromatic dye to an oxaziridine (87) (Figure 10). Other workers have evaluated polysilanes (88) and diazonium salt chemistry (89,90) for CEM applications.

It is predicted that CEM techniques will extend resolution to feature sizes as small as $0.4\lambda/NA$, or ~0.6 μm for currently available exposure tools operating at 405 and 436 nm (91). The currently available materials afford improved resolution and yield, and increased process latitude (92). Further developments to achieve practical, water soluble systems plus a better match with the absorption characteristics of the resist would be desirable.

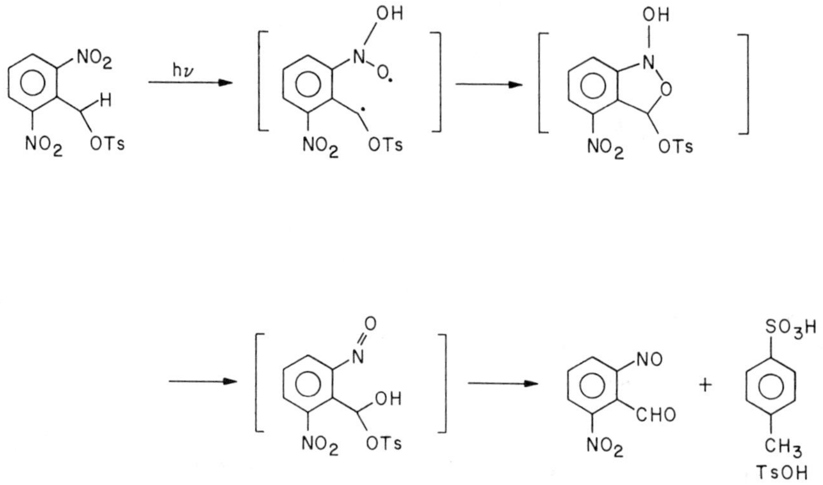

Figure 8: Structure and photochemical reaction of a dinitrobenzyl ester based acid generator.

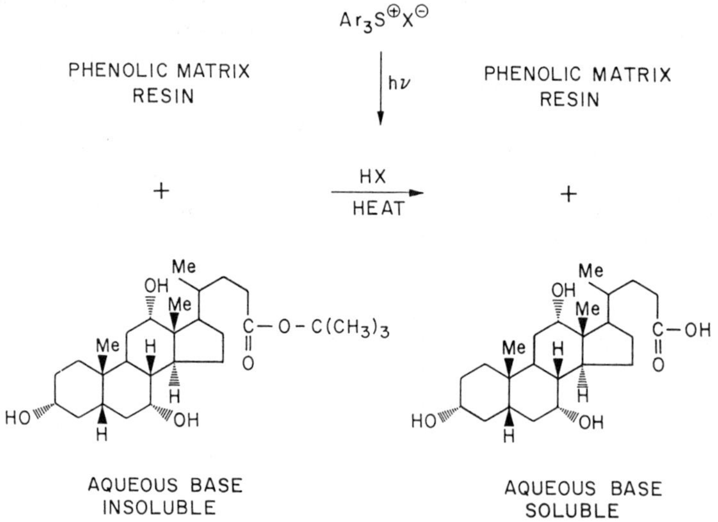

Figure 9: Chemistry associated with a three component aqueous-base developable, chemically amplified resist.

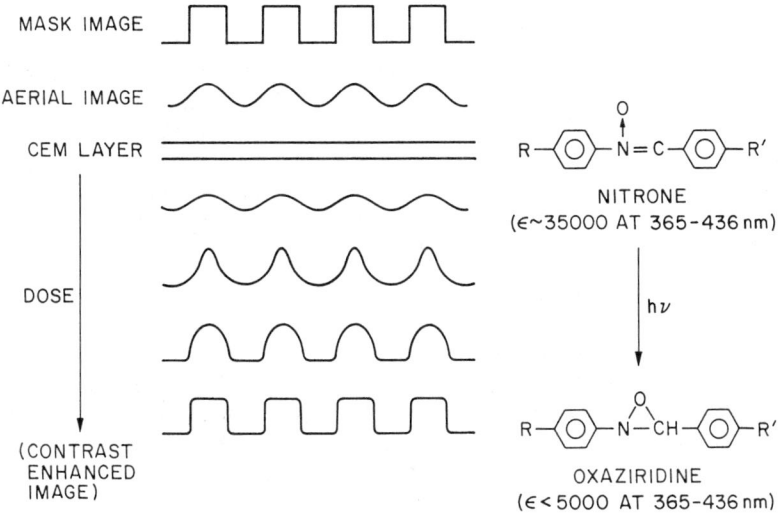

Figure 10: Schematic representation of the CEM process and oxaziridine CEM chemistry.

There are a number of multilevel resist schemes summarized in Figure 11, that are based on planarization of the topographic features on the substrate surface. The process involves spin coating thick (1-2μm) films of thermoplastic materials such as PMMA, polystyrene, polyimide or novolac resins (usually as conventional photoresist formulations) onto the surface of the wafer. This process readily provides local planarization of closely spaced features, but global, or large area (wafer), planarization where both closely spaced and isolated features are planarized, is rarely achieved by this process. Global planarization is important to produce a perfectly level surface that will minimize depth of focus problems found with photolithographic techniques and associated with variations in film thickness resulting from topographic features on the wafer surface, induced from previous processes (93). Mathematical models that predict the degrees of local and global planarization of different polymer coatings have recently been presented (94,95)

Two approaches have been used to provide good local and global planarization. First, planarizing layers consisting of relatively low T_g materials that can be spin coated from solution and flowed in a low viscosity state prior to curing have achieved 70-90% planarization over an entire wafer. Alternatively, planarizing layers of a low viscosity epoxy resin containing an onium salt curing agent that is spin coated without solvent, allowed to stabilize, and then cured to a high T_g film, have shown excellent planarization properties. A good review of the planarization process and materials has been recently published and the reader is referred to this review for more detailed information (96).

After deposition of the planarizing layer, the imaging resist is spin-coated either directly on top of the planarizing layer or onto an intermediate RIE barrier layer and subsequently exposed and developed by conventional methods. The problem then becomes one of transferring the defined image through the planarizing layer to the substrate. There are two generic types of processes used, viz. liquid developed and dry developed. The liquid developed processes which typically involve a second exposure step have been extensively discussed by Lin (83) and will not be covered here. The dry developed processes based on RIE for pattern transfer have received considerable attention and use materials with properties that depend on the specific strategy employed. There are two pattern transfer schemes that have been used: trilevel and bilevel and these are briefly discussed below.

Trilevel processing begins with planarization of the device topography with a thick layer of some organic polymer such as a polyimide, or positive photoresist that has been "hard-baked" (baked to induce crosslinking), or otherwise treated, to render it insoluble in most solvents. Next, an intermediate RIE barrier such as silicon dioxide is deposited, and finally the structure is coated with the desired resist material. A pattern is delineated in the top resist (imaging) layer and subsequently transferred into the planarizing layer by dry etching techniques (97,98).

Several variations of the generic process have been reported. The most common intermediate oxygen RIE barrier layer is SiO_2. While trilevel lithography is a time consuming process requiring precise contol of several processing steps, it improves the resolution capability of conventional resists by separating the imaging

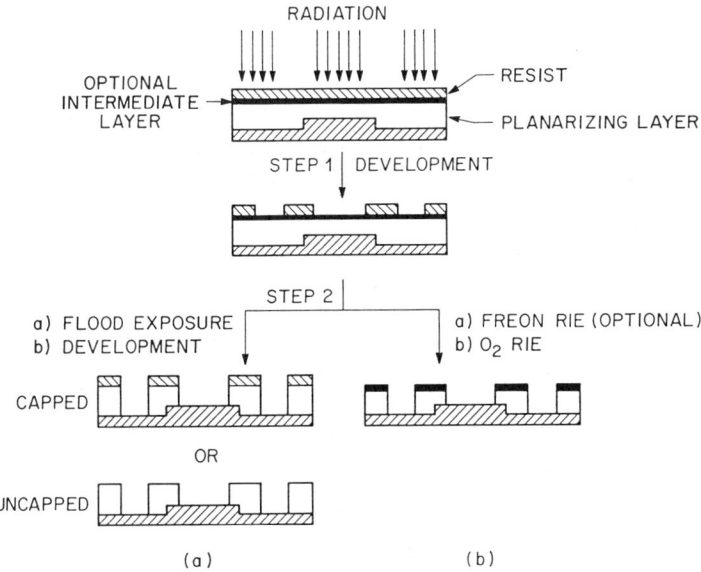

Figure 11: The process sequence for several multi-level resist schemes involving planarization techniques.

function of the resist from the subsequent etching mask function and permits imaging in a relatively thin resist film.

A simplification of trilevel lithography involves combining the properties of the top imaging layer with those of the oxygen RIE resistant intermediate layer into a single layer which is coated onto the surface of the thick planarizing layer resulting in a "bilevel" structure. Conventional processing allows pattern definition of this upper layer, and the pattern is then transferred to the substrate by oxygen RIE techniques. One of the first examples was described by Shaw, et al who demonstrated the utility of polysiloxanes as deep-UV and e-beam resists (99). Organometallic polymers are, in general, ideal candidates for bilevel lithography (100). Treatment of such compounds, particularly organosilicon materials, with an oxygen plasma leads to the formation of the corresponding metal oxide. Taylor and Wolf have shown that incorporation of silicon into organic polymers renders them resistant to erosion in oxygen plasmas (101). This surface passivation results from the formation of a protective coating of SiO_2 on the polymer surface. Modeling studies predict that the thickness of this layer should be about 50Å, a value that has been confirmed by surface analysis (102,103). The preliminary work of Taylor et al has led to development of numerous silicon containing resist systems for bilevel RIE pattern transfer processes (1,13). Selected examples of silicon containing resists that utilize standard exposure and development techniques are shown in Figure 12.

Gas phase functionalization is a new technique that is receiving increased attention as a method for combining advantages of multilevel processing and solventless development in a single layer. Many of these schemes involve the use of inorganic reagents to selectively funtionalize the resist film after exposure to enhance oxygen RIE resistance (104-108). In one example, the chemically amplified resist based on poly(t-butoxycarbonyloxystyrene) is converted to poly (vinylphenol) after irradiation and post-exposure-bake. The phenolic polymer thus generated is reactive towards silylating agents so that treatment with an organosilicon reagent such as hexamethyldisilazane or chlorotrimethylsilane results in selective incorporation of silicon into the irradiated parts of the film, allowing high-resolution oxygen RIE pattern transfer (107). Another approach utilizes conventional novolac-diazoquinone chemistry to induce differential reactivity towards an organometallic reagent (108). In these systems, the dissolution inhibitor blocks the diffusion of a silylating agent in the unexposed regions of the film, while diffusion and thus reactivity are enhanced in the exposed portions. RIE pattern transfer of the resultant images affords sub-micron resolution for 365-405nm exposure.

Conclusion

The continual progress in VLSI device development is placing increasing demands on the lithographic technologies used for their manufacture. At the present time, almost all commercial devices are made by photolithography utilizing

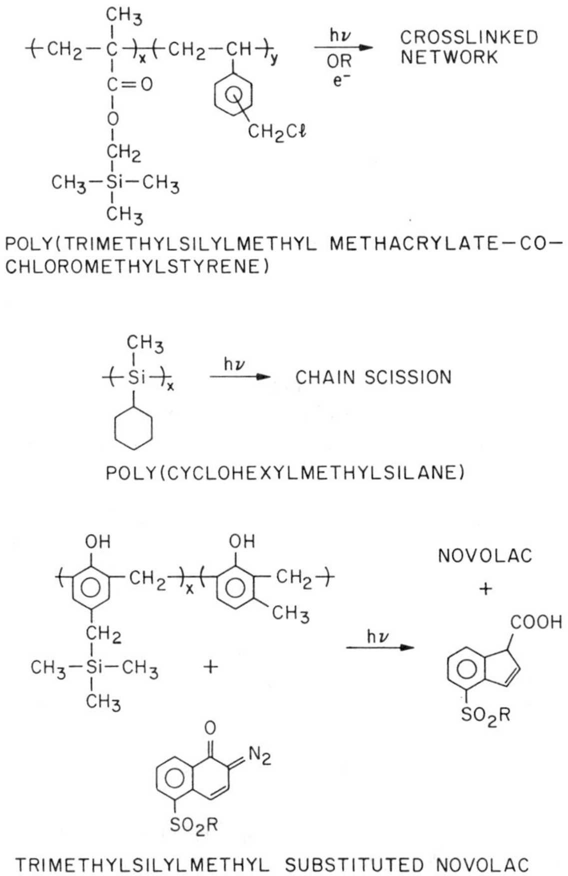

Figure 12: Select examples of organosilicon resist chemistries that undergo crosslinking, chain scission or solution inhibition.

365-436nm UV radiation. However, within the next 3-8 years, new lithgraphic strategies such as deep-UV, e-beam and x-ray lithography will be required to meet resolution needs extending below 0.5 µm. Each of these alternative technologies will require new polymeric resist materials and processes, whose development contains many challenges in the areas of both fundamental and applied polymer chemistry as well as process engineering.

Literature Cited

1. "Electronic and Photonic Applications of Polymers", *ACS Advances in Chemistry Series* **218**. Bowden, M. J., Turner, S. R., eds., ACS, Washington, D.C., 1988.
2. Powell, M. W., *Solid State Technology*, 1989, **32**(3), 66.
3. McCoy, J. H., Lee, W., Varnell, G. L., *Solid State Technology*, 1989, **32**(3), 87.
4. Peters, D. W., Frankel, R. D., *ibid.*, 77.
5. Clemens, J. T., *ibid.*, 69.
6. Thompson, L. F. in "Introduction to Microlithography", *ACS Symposium Series* **219**. Thompson, L. F., Willson, C. G., Bowden, M. J., eds, ACS, Washington, D.C., 1983.
7. Iwayanagi, T., Ueno, T., Nonogaki, S., Ito, H., Willson, C. G. "Materials and Processes for Deep-UV Lithography", In ref.1, pg. 109.
8. Pol, V., Bennewitz, J. H., Escher, G. C., Feldman, M., Firtion, V. A., Jewell, T. E., Wilcomb, B. E., Clemens, J. T., *Proc. SPIE*, 1986, **633**, 6.
9. Pol, V., *Solid State Technology*, 1987, **30**(1), 71.
10. Buckley, J. D., Karatzas, C., *Proc. SPIE*, 1989, **1088**.
11. Moreau, W. M., "Semiconductor Lithography, Principles, Practices, and Materials", Plenum, N. Y., 1988.
12. Thompson, L. F., Reichmanis, E., *Annual Rev. Mat. Sci.* 1987, **17**, 235.
13. Reichmanis, E., Thompson, L. F., *Chemical Reviews*, 1989, in press.
14. Bowden, M. J. In "Materials for Microlithography", *ACS Symposium Series* **266**, Thompson, L. F., Willson, C. G., Frechet, J. M. J.; eds, ACS, Washington, D.C., 1984, pg. 39.
15. Novembre, A. E., Masakowski, L. M., Hartney, M. A. *Poly. Eng. Sci.*, 1986, **26**(16), 1158.
16. Thompson, L. F., Kerwin, R. E. *Annual Review of Materials Science*, Huggins, R. A., Bube, R. H., Roberts, R. W., eds., 1976, **6**, 267.
17. Minsk, L. M. US Pat 2, 725, 372 (1955).
18. Iwayanagi, T., Kohashi, T., Nonogaki, S., Matsuzawa, T., Douta, K., Yanazawa, H., *IEEE Trans. Elec. Dev.*, 1981, ED **28**(11), 1306.
19. Hashimoto, M., Iwayanagi, T., Shiraishi, H. and Nonagaki, S., "Proc. Reg. Tech. Conf. on Photopolymers", Mid-Hudson Sect. SPE, Ellenville, NY, 1985, pg. 11.
20. Tan, Z. C., Petropoulos, C. C., Rauner, F. J *Vac. Sci. technol.*, 1981, **19**(4), 1348.

21. Bartelt, J. L., *appl. Poly. Symp.*, 1974, **23**, 139.
22. Hirai, T., Hatano, Y., Nonogaki, S. *J. Electrochem. Soc.*, 1971, **118**(4): 669.
23. Thompson, L. F., Feit, E. D., Heidenreich, R. D., *Poly. Eng. and Sci.*, 1974, **14**(7): 529.
24. Feit, E. D., Thompson, L. F., Heidenreich, R. D., *ACS Div. of Org. Coat. and Plast. Chem. Preprint.*, 1973, 383.
25. Taniguchi, T., Hatano, Y., Shiraishi, H., Horigome, S., Nonagoki, S., Naraoka, K., *Japan J. Appl. Phys.*, 1979, **28**; 1143.
26. Tabata, Y., Tagawa, S., Washio, M. in "Materials for Microlithography", Thompson, L. F., Willson, C. G., Frechet, J. M. J., eds. *ACS Symposium Series* **266**, ACS, Washington, D.C., 1984, pp. 151.
27. Taylor, G. N., Wolf, T. M., *J. Electrochem. Soc.*, 1980, **127**, 2665.
28. Thompson, L. F., Doerries, E. M., *J. Electrochem. Soc.*, 1979, **126**(10); 1699.
29. Thompson, L. F., Yau, L., Doerris, E. M., *J. Electrochem. Soc.*, 1979, **126** (10); 1703.
30. Feit, E. D., Thompson, L. F., Wilkins, C. W., Jr., Wurtz, M. E., Doerries, E. M., Stillwagon, L. E., *J. Vac. Sci. Technol.* 1979, **16**(6); 1997.
31. Hartney, M. A., Tarascon, R. G., Novembre, A. E., *J. Vac. Sci. Technol.*, 1985, **B3**, 360.
32. Imamura, S. *J. Electrochem. Soc.*, 1979, **126** (9), 1268.
33. Hatzakis, M., *J. Electrochem. Soc.*, 1969, **116**, 1033.
34. Ranby, B., Rabek, J. F., *Photodegradation, Photooxidation and Photostabilization of Polymers*, New York, NY, John Wiley & Sons, 1975, pp. 156.
35. Mimura, Y., Ohkubo, T., Takanichi, T., and Sekikawa, K., *Jpn. Appl. Phys.*, 1978, **17**, 541.
36. Lin, B. J., *J. Vac. Sci. Technol.*, 1975, **12**, 1317-20.
37. Helbert, J. N., Chen, C. Y., Pittman, C. U., Jr., Hagnauer, G. L., *Macromolecules*, 1978, **11**, 1104.
38. Moreau, W. M., *Proc. SPIE, Submicron Lithog.*, 1982, **333**, 2.
39. Chandross, E. A., Reichmanis, E., Wilkins, C. W., Jr., and Hartless, E. L., *Can. J. Chem.*, 1983, **61** (5), 817.
40. Reichmanis, E., Wilkins, C. W., Jr., "Polymer Materials for Electronic Applications", *ACS Symposium Series*, **184**, Feit, E. D., Wilkins, C. W., Jr., eds. American Chemical Soc., Wash., D.C., 1982, pp. 19.
41. Kakuchi, M., Sugawara, S., Murase, K., Matsuyama, K., *J. Electrochem. Soc.*, 1977, **124**, 1648.
42. Tada, T., *J. Elcetrochem. Soc.*, 1979, **126**, 1829.
43. Tada, T., *J. Electrochem. Soc.*, 1983, **130**, 912.
44. Roberts, E. D., *ACS Div. Org. Coat. And Plasitcs Chem. Preprints*, 1973, **33**(1), 359.
45. Roberts, E. D., *ACS Div. Org. Coat. and Plastics Chem. Preprints*, 1977, **37**(2), 36.
46. Moreau, W., Merritt, D., Moyer, W., Hatzakis, M., Johnson, D., Pederson, L. *J. Vac. Sci. Technol.*, 1979, **16**(16), 1989-91.
47. Namastse, Y. M. N., Obendorf, S. K., Anderson, C. C., Krasciky, P. D., Rodriguez, F., and Tiberio, R., *J. Vac. Sci. Technol. B.*, 1983, **1**(4), 1160.

48. Namastse, Y. M. N., Obendorf, S. K., Anderson, C. C., Roderiguez, F., *Proc. SPIE* 1984, **469**, 144.
49. Bowden, M. J., Thompson, L. F., *Solid State Technol.*, 1979, **22**, 72.
50. Bowden, M. J., Thompson, L. F., Ballantyne, J. P., *J. Vac. Sci. Technol.*, 1975, **12**(6), 1294.
51. Pacansky, J. and Lyerla, J. R., IBM J. Res. Develop., 1979, **23**(1), 42.
52. Hanabata, M., Furuta, A., Uermura, Y., *Proc. SPIE*, 1986, **681**, 76.
53. Templeton, M. K., Szmanda, C. R, Zampini, A., *Proc. SPIE*, 1987, **771**; 136.
54. Trefonas, P., III, Daniels, B. K., Fischer, R. L., *Solid State Technol.*, 1987, **30**, 131.
55. Miller, R. D., Willson, C. G., McKean, D. R., Tompkins, T., Clecak, N., Michl, J., Downing, J., *Proc. Reg. Tech. Conf.* on "Photopolymers, Principles, Processes and Materials", Mid-Hudson Section, SPE. Ellenville, N.Y., Nov. 8-10, 1982, p. 111.
56. Moritz, H., *IEEE Trans. Electron Devices*, 1985, **ED-32**(3), 672.
57. Takahashi, Y., Shinozaki, F., Ikeda, T., *Jpn. Kokai Tokkyo Koho*, 1980, **88**; 8032.
58. MacDonald, S. A., Ito, H., Willson, C. G., *Microelectron Eng.*, 1983, **1**, 269.
59. Alling, E., Stauffer, C., *Proc. SPIE*, 1985, **539**, 194.
60. Buhr, G., Lenz, H., Scheler, S., *Proc. SPIE*, 1989, **1086**, in press.
61. Grant, B. D., Clecak, N. J., Twieg, R. J., Willson, C. G., *IEEE Trans. Electron Dev.*, 1981, **ED-28**(11), 1300.
62. Willson, C. G., Miller, R. D., McKean, D. R., *Proc, SPIE*, 1987, **771**, 2.
63. Schwartzkopf, G. *Proc. SPIE*, 1988, **920**; 51-59.
64. Sugiyama, H., Ebata, K., Mizushima, A., Nate, K., *Proc. Reg. Tech. Conf.* on "Photopolymers, Principles, Processes and Materials", Mid-Hudson Section, SPE, Ellenville, NY, Oct. 30-Nov. 2, 1988, p.51.
65. Reichmanis, E., Wilkins, C. W., Jr., Chandross, E. A., *J. Vac. Sci. Technol.*, 1981, **19**(4), 1338.
66. Ito, H., Flores, E., *J. Electrochem. Soc.*, 1988, **135**, 2322.
67. Bowden, M. J., Thompson, L. F., Fahrenholtz, S. R., Doeries. E. M., *J. Electrochem. Soc.*, 1981, **128**, 1304.
68. Bowden, M. J., Allara, D. L., Vroom, W. I., Frackoviak, S., Kelley, L. C., Falcone, D. R. ACS Symposium Series, 242, "Polymers in Electronics", Davidson, T., ed. American Chemical Society, Washington, D.C., 1984. pp. 135.
69. Shiraishi, H., Isobe, A., Murai, F., Nongaki, S. *ACS Symposium Series 242*, "Polymers in Electronics", Davidson, T., ed., ACS, Washington, D.C., 1984 pp. 167.
70. Ito, H., Pederson, L. A., MacDonald, S. A., Cheng, Y. Y., Lyerla, J. R., Willson, C. G., *Proc. Reg. Tech. Conf.* on "Photopolymers, Principles, Processes and Materials", Mid-Hudson Sect. SPE, Oct. 28-30, 1985, Ellenville, NY, p. 127.
71. Ito, H., Willson, C. G., *Proc. Reg. Tech. Conf.* on "Photopolymers, Principles, Processes and Materials", Mid-Hudson Sect. SPE, Nov. 8-10, 1982, p. 331.

72. Crivello, J. V., *ACS Symp. Series 242*, "Polymers in Electronics", Davidson, T. ed., Amer. Chem. Soc., Wash. D.C., 1984, pp. 3.
73. Ito, H., Willson, C. G., ibid, pp. 11.
74. Crivello, J. V., *Proc. SPE Reg. Tech. Conf.* on Photopolymers: Princ. Proc. and Mat., Nov. 8-10, 1982, Ellenville, N.Y. pp. 267.
75. Feely, W. E., Inhof, J. C., Stein, C. M., *Polym. Eng. Sci.*, 1986, **26**, 1101.
76. Frechet, J. M., Eichler, E., Ito, H., Willson, C. G., *Polymer*, 1980, **24**; 995.
77. Ito, H., Willson, C. G., Frechet, J. M. J., Farrall, M. J., Eichler, E., *Macromolecules*, 1983, **16**, 510.
78. Houlihan, F. M., Shugard, A., Gooden, R., and Reichmanis, E., *Macromolecules*, 1988, **21**, 2001.
79. Neenan, T. X., Houlihan, F. M., Kometani, J. M., Tarascon, R. G., Reichmanis, E., Thompson, L. F., *Proc. SPIE*, 1989, **1086**, 2.
80. Tarascon, R. G., Reichmanis, E., Houlihan, F. M., Shugard, A., Thompson, L. F., *Proc. Reg. Tech. Conf. on Photopolymers, Princ., Proc. Materials*, Mid-Hudson Sec. SPE Oct. 30 - Nov. 2, 1988, Ellenville, NY, p. 11.
81. Ito, H., *Proc. SPIE*, 1988, **920**, 33.
82. O'Brien, M. J., *Proc. Reg. Tech. Conf.* on "Photopolymers, Principles, Processes and Materials", Mid-Hudson Sect. SPE, Oct 30-Nov. 2, 1988, Ellenville, NY, pg 1.
83. Lin, B. J., In "Introduction to Microlithography", Thompson, L. F., Bowden, M. J., Willson, C. G. eds., ACS Washington, D.C. *ACS Symposium Series 219*, 1983, p. 287.
84. Griffing, B. F., West, P. R., *Proc. Reg. Tech. Conf.*, "Photopolymers: Princ., Proc. and Mat.", Mid-Hudson Sect. SPE, Nov. 8-10, 1982, Ellenville, N.Y., p. 185.
85. Griffing, B. F., West, P. R., *Proc. SPIE*, 1984, **469**, 102.
86. Griffing, B. F., West, P. R., *Solid State Technol.*, 1985, **28**(5), 152.
87. West, P. R., Davis, G. C., Griffing, B. F., 1985. *Polymer Preprints*, 1985, **26**(2), 337.
88. Hofer, D. C., Miller, R. D., Willson, C. G., Neureuther, A. R., *Proc. SPIE*, 1985, **469**, 108.
89. Halle, L., *J. Vac. Sci. Technol. B.*, 1985, **3**(1), 323.
90. Uchino, S.-I., Ueno, T., Iwayanagi, T., Morishita, H., Nonogaki, S., Shirai, S.-I., Moriuchi, N., *Proc. Poly. Mat. Sci. and Eng.*, 1986, **55**, 604.
91. Neureuther, A. R., Hofer, D. C., Willson, C. G., *Microcircuit Engineering 84*. Heuberger, A., Beneking, H., eds. Academic Press, London, 1985, p. 53.
92. Strom, D. R., *Semiconductor Internat.'l*, 1986, **9**(5), 162.
93. Wilson, C. G., Bowden, M. J., In "*Electronic and Photonic Applications of Polymers*", Bowden, M. J. and Turner, S. R., eds., ACS Advances in Chemistry Series, **218**, ACS, Washington, D.C. 1988, pp. 90-92.
94. LaVergne, D. B., Hofer, D. C., *Proc. SPIE*, 1985, **539**, 115.
95. Stillwagon, L. E., and Larson, R. G. *Proc. SPIE*, 1988, **920**, 312.
96. Stillwagon, L. E., *Solid State Technol.*, 1987, **30**(6), 67.
97. Havas, J., *Electrochem. Soc. Extended Abstracts*, 1976, **76**(2), 743.
98. Moran, J. M., Maydan D., *J. Vac. Sci. Technol.*, 1979, **16**(6), 162.

99. Shaw, J. M., Hatzakis, M., Paraszczak, J., Liutkus, J., Babich, E., *Proc. Reg. Tech. Conf.* on "Photopolymers, Principles, Processes and Materials", Mid-Hudson Sect. SPE, Nov. 8-10, 1982, Ellenville, NY, p. 285.
100. Reichmanis, E., Smolinsky, G. and Wilkins, C. W., Jr., *Solid State Technology*, 1985, **28**(8), 130.
101. Taylor, G. N., Wolf, T. M., *Polym. Eng. Sci.*, 1980, **20**, 1087.
102. Watanabe, F., Ohnishi, Y., *J. Vac. Sci. Technol. B*, 1986, **4**(1), 422.
103. Jurgensen, C. W., Shugard, A., Dudash, N., Reichmanis, E., Vasile, M. J., *Vac. Sci. Technol.*, 1988, **A6**, 2938.
104. Taylor, G. N., Stillwagon, L. E., Venkatesan, T., *J. Electrochem. Soc.*, 1984, **131**, 1658.
105. Wolf, T. M., Taylor, G. N., Venkatesan T., Kraetsch, R. T., *J. Electrochem. Soc.*, 1984, **131**, 1664.
106. Stillwagon, L. E., Silverman, P. J., Taylor, G. N., *Proc. Reg. Tech. Conf.* on "Photopolymers; Principles, Processes and Materials", Mid-Hudson Section SPE, Ellenville, N.Y., Oct. 28-30, 1985, p. 87.
107. MacDonald, S. A., Ito, H., Hiraoka, H., Willson, C. G., *ibid.* p. 177.
108. Coopmans, F., Roland, B., *Solid State Technology*, 1987, **30** (6), 93.

RECEIVED August 28, 1989

CHEMICALLY AMPLIFIED RESIST CHEMISTRY

CHEMICALLY AMPLIFIED RESIST CHEMISTRY

It is gratifying to see the tremendous growth that has occurred in the area of high resolution, chemically amplified resists. Certainly one reason for growth, is that this approach affords both the sensitivity and resolution, required to make deep-UV lithography commercially viable. It is also interesting that the introduction of Kr-F excimer lasers, as intense deep-UV sources, has not reduced the requirement for a sensitive deep-UV photoresist. For while the excimer lasers illuminating present day step-and-repeat systems are extremely bright, only a small percentage of the light passes through the optical system and onto the photosensitive resist film.

The papers presented in this section cover several aspects of acid catalyzed resist chemistry. Investigators have studied basic issues such as, acid diffusion within the matrix, and the influence of polymeric structure on resist performance. Also, the chemistry associated with the original work on poly(tert-butyloxycarbonlyoxystyrene) has been extended to several other systems.

The use of these acid catalyzed resist systems represents a major break with the conventional diazonaphthoquinone/novolac resists, currently dominating high resolution photolithographic processes. One issue remaining, for chemically amplified resist systems, is to demonstrate the ability to perform on a semiconductor manufacturing line. Given the number of groups studying these systems, I am confident this demonstration will come to pass.

Scott A. MacDonald
IBM Almaden Research Center
650 Harry Road
San Jose, CA 95120–6099

Chapter 2

Brönsted Acid Generation from Triphenylsulfonium Salts in Acid-Catalyzed Photoresist Films

D. R. McKean, U. Schaedeli[1], and Scott A. MacDonald

IBM Research Division, Almaden Research Center, 650 Harry Road, San Jose, CA 95120-6099

Acid-catalyzed photoresists have emerged as the most sensitive resists for deep-uv application. We have studied one such system containing poly(t-4-butoxycarbonyloxystyrene) and triphenylsulfonium hexafluoroantimonate and analyzed both the photochemical and thermal resist chemistry. The amount of acid generated in the resist film has been measured directly using a merocyanine dye technique. The quantum efficiencies for acid generation have been determined and compared with the quantum yield in solution. The catalytic chain length for the acid-catalyzed deprotection step was measured and then used to determine the range of acid migration during the postbake step.

Semiconductor microelectronic device manufacturers have been remarkably successful in continuing to increase circuit densities while reducing the cost per circuit element. The increased density has been achieved by improvements in lithographic processing techniques, materials, and exposure tools which have permitted large reductions in the size of circuit elements. The theoretical resolution obtained from an optical exposure system is proportional to λ/NA where λ is the wavelength and NA is the numerical aperature of the lens system. Improvements in resolution and corresponding circuit element size reduction can be achieved by either decreasing wavelength or increasing numerical aperature. However, an increase in numerical aperature is accompanied by a decrease in field size and depth of focus. Resolution improvements obtained by means of decreased exposure radiation wavelength can be obtained without the loss of field size or depth of focus.

Of the several competing strategies for obtaining 0.5 micrometer resolution necessary for 4 megabit static RAM devices (1), use of deep-uv exposure tools is the

[1]Current address: Ciba-Geigy AG, Fribourg, Switzerland

0097-6156/89/0412-0027$06.00/0
© 1989 American Chemical Society

only means of achieving desired resolution without undesirable loss of field size and depth of focus or introduction of significant processing complexity. The major drawback associated with deep-uv photolithography is that conventional positive photoresists cannot be utilized due to low transparency in the wavelength region of interest (235-255 nm).

While many new deep-uv photoresist strategies have been considered, much of the attention has focused on systems which have incorporated acid-catalyzed chemistry (2,3). These resists are advantageous because of high sensitivity to deep-uv irradiation which permits high wafer throughput. A number of acid-catalyzed schemes have been developed including cationic polymerization, depolymerization, polymer cross-linking reactions, and acid-catalyzed functional group interconversions or rearrangements.

The development of new classes of cationic photoinitiators has played a critical role in the production of highly sensitive, acid-catalyzed deep-uv photoresists. Sulfonium salts have been widely used in this respect (4). These materials are relatively easy to prepare and structural modifications can be used to produce desired wavelength sensitivity. Triphenylsulfonium salts are particularly well suited for deep-uv application and in addition can be photosensitized for longer wavelength. These salts are quite stable thermally and certain ones such as the hexafluoroantimonate salt are soluble in casting solvents and thus easily incorporated within resist materials.

The generally accepted mechanism for the generation of acid from irradiation of triphenylsulfonium salt (1) is depicted in Scheme 1. The sulfonium salt excited state undergoes homolytic cleavage of the carbon-sulfur bond to give an intermediate sulfur-centered radical cation along with phenyl radical. Brönsted acid is believed to arise from hydrogen atom abstraction by the radical cation followed by dissociation. Phenylthiobiphenyl rearrangement products have also been observed (5), suggesting that acid may arise by photorearrangement followed by dissociation. Some evidence has been presented which suggests that phenyl cation is produced by heterolytic cleavage of the excited state of the sulfonium salt (6). Nucleophiles such as ethers, epoxides and sulfides, if present, react rapidly with the radical cation (7). The quantum efficiency for production of Brönsted acid from triphenylsulfonium hexafluoroarsenate is known (8). Acetonitrile solutions were irradiated at 254 nm. Acid was analyzed by either aqueous or potentiometric titration. A quantum yield of 0.7 was observed.

A number of acid-catalyzed resist schemes have used the t-butoxycarbonyl (t-BOC) functionality as protecting groups. These groups have been widely employed in peptide chemistry and are well known for their lability toward strong acids (9). The t-BOC groups have been employed as pendant groups on acid-sensitive polymers (10-12) and also on small molecule dissolution inhibitors (13). Sulfonium salts or other acid-photogenerators are dissolved in the polymer solution and films are prepared. After irradiation the films are heated and the t-BOC groups are converted to hydroxyl groups (Scheme 2). The reaction proceeds by an A_{AL}-1 mechanism in which the acid is regenerated in the final step of the mechanism. While the generation of the acid is limited by the dose of radiation and the quantum yield, the overall chemistry is amplified because of the catalytic nature of the deprotection step.

The characterization of t-BOC containing resists has generally been done by infrared analysis of the carbonyl group before and after exposure and postbake. This analysis provides a measure of overall chemistry within the resist film but doesn't

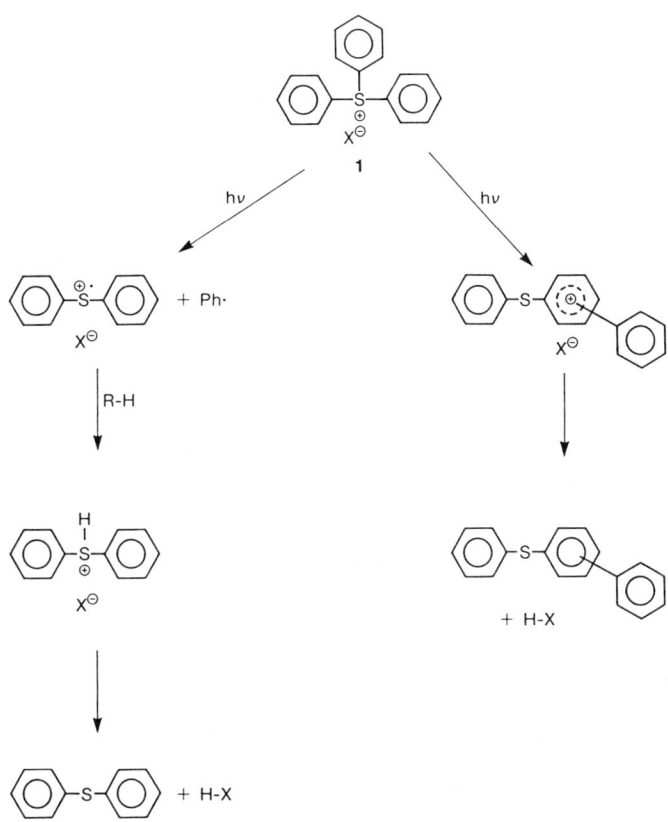

Scheme 1. Mechanisms for acid generation from triphenylsulfonium salts.

give any information about the amount of acid generated or the extent of the catalytic chain. We have carried out a detailed characterization of a resist film containing triphenylsulfonium hexafluoroantimonate in poly(4-t-butoxycarbonyloxystyrene) and have measured quantum yields, the amount of acid generated, and the extent of the catalytic chain for this resist.

Results and Discussion

Merocyanine Dye Method for Acid Analysis. Resist photochemistry can often be monitored by the changes in ultraviolet absorption spectra associated with a bleaching of the sensitizer absorbance. In the case of resist systems with triphenylsulfonium salts, no change in the film absorption is observed on irradiation. In order to determine the amount of acid produced, a direct method for acid analysis was required. A highly sensitive method was desirable since the amount of acid produced is approximately 10^{-6} mmol for a 1 micrometer thick film on a 2 inch wafer. Furthermore a nonaqueous technique is preferred in order to avoid hydrolysis of the hexafluoroantimonate salt. Hydrolysis gives hydrogen fluoride (14) which makes accurate acid determination more difficult.

Gaines (15) has previously described the use of merocyanine dyes as a nonaqueous means of determining Brönsted acid concentration. Merocyanine dyes are protonated by strong acids to produce protonated dye which has a distinct visible absorption (Figure 1). The unprotonated dye form (3) has a solvent dependent visible absorption maxima. The present studies were performed in acetonitrile or dichloromethane solvent where absorption maxima were at 576 nm and 610 nm respectively. The absorbance of the protonated form (4) is relatively unaffected by choice of solvent and is clearly separable from the absorbance of the free dye. The extinction coefficient of the free dye is quite large (71,000 in dichloromethane) which allows determination of small amounts of acid such as 10^{-6} mmol with an average error of less than 10%.

The equilibrium reaction depicted in Figure 1 is dependent on acid strength. Accurate determination of acid concentration is dependent on shifting the equilibrium to favor product formation. Experiments were performed to determine the dependence of the equilibrium on acid strength. Known quantities of trifluoroacetic acid were added to the dye solution and the visible absorption spectra were recorded (Figure 2). A direct correlation was observed between the number of equivalents of acid added and the degree of bleaching of the dye which indicates complete product formation. A similar result was observed by Gaines (15) for addition of methanesulfonic acid. Since the superacids such as hexafluoroantimonic acid are stronger than either trifluoroacetic acid or methanesulfonic acid (16), the bleaching of dye absorbance with superacids will correlate well with acid concentration.

Acid Generation in Photoresist Films. A resist composition was chosen to give incomplete t-BOC removal after irradiation with 0.5-2.0 mJ/cm^2 dose at 254 nm followed by postbake at 100°C for one minute. These conditions were satisfied for films consisting of 1% triphenylsulfonium hexafluoroantimonate (1) in poly(4-t-butoxycarbonyloxystyrene) (2). The amount of acid produced on irradiation was determined by performing acid analyses before and after irradiation and taking the difference between these two values. All analyses were performed on four different wafers and the results were taken as the average of the four replicate experiments. Average error for the four experiments was less than 10%.

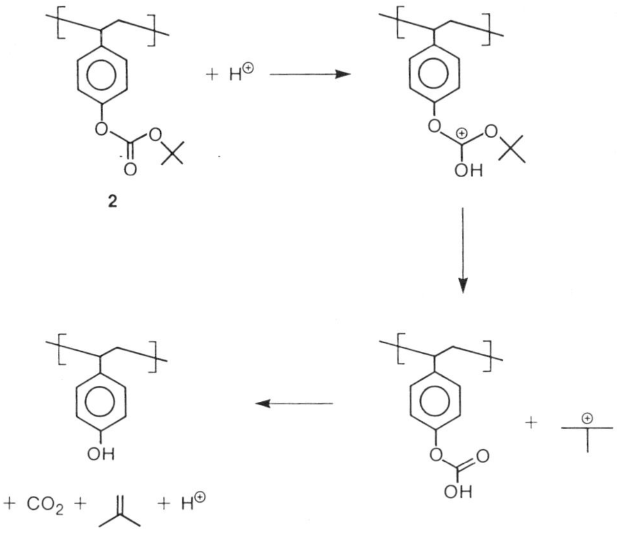

Scheme 2. Acid-catalyzed cleavage of t-BOC groups.

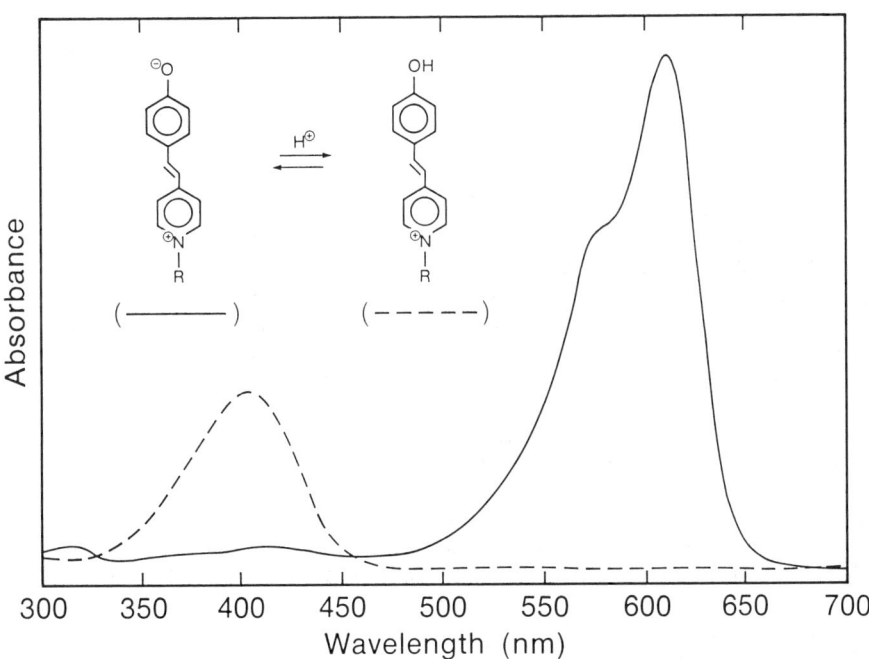

Figure 1. Absorption spectra of merocyanine dye and protonated dye form in dichloromethane.

The analysis of acid present *before* irradiation was determined to be 2×10^{-6} mmol for a one micrometer films on a 2 inch wafer. This is a significant fraction of total acid present after irradiation. For a 0.5 mJ/cm^2 dose, this is nearly 30% of the total acid content. However, the acid present before exposure is not significant for t-BOC thermolysis. No carbonyl infrared absorbance change was noted following softbake.

Analyses were performed on each of the resist components to determine the source of the background acid. Background acid was present in all three resist components – polymer, sulfonium salt, and casting solvent. However the sulfonium salt contained the highest concentration of acid by two orders of magnitude. Residual acid in sulfonium salt may be due to the method of preparation which involves an aqueous metathesis step to exchange the hexafluoroantimonate ion for bisulfate. This procedure is likely to produce hydrogen fluoride due to hydrolysis of hexafluoroantimonate salt (14).

After irradiation at 254 nm significant increases in Brönsted acid concentration were noted. The amount of acid generated by photolysis was determined by subtracting the unexposed acid content from the total acid content after irradiation. The results (Table I) show that the amount of photogenerated acid is very small – approximately 1×10^{-5} mmol for a 2 mJ/cm^2 dose on a two inch wafer. However it should be noted that the amount of sulfonium salt used for these experiments is quite small. Actual photoresist compositions might contain significantly more sensitizer. For example, Ito and Willson (2) performed imaging experiments on resin 2 containing 20% of diphenyliodonium salt. The generation of acid is fairly linear with dose (Figure 3) at conversions less than 25%.

Despite the small amount of acid generated in these experiments, the film dissolution behavior following postbake is dramatically affected. Acid content exceeding 5×10^{-6} mmol per 2 inch wafer is sufficient to change the solubility characteristics of the resist such that exposed resist film is no longer soluble in nonpolar developer solvent.

A quantum yield was determined for acid generation based on the total absorbance of the film and the reflectivity from single crystal silicon at 254 nm. Total film absorbance at 254 nm is 0.185 which corresponds to 35% absorption by the film. Reflectivity at 254 nm is 0.66 (17) and so an additional 15% of the incident light is absorbed after reflection. Total absorption was taken to be 50% of incident dose. It should be noted that the resin absorption without sulfonium salt is 0.155 and thus constitutes about 84% of the total film absorbance with sulfonium salt.

The quantum yields for acid generation vary from 0.26 to 0.40 (Table I). The highest quantum yield was observed for the lowest dose – a fact which cannot be rationalized on the basis of absorption change since no bleaching of film absorbance is observed during the irradiation. Quantum yield variation with dose may be due to inhomogeneous distribution of sulfonium salt in the resin causing some salt to generate acid at greater efficiency than the bulk distribution.

For comparison, the *solution* quantum yield was determined by the merocyanine dye technique. Acetonitrile solutions of triphenylsulfonium hexafluoroantimonate were irradiated with a 5 mJ/cm^2 dose. Dye solution was added and the acid content was determined by changes in dye absorption. The quantum yield for acid production was determined to be 0.8, which agrees reasonably well with the value (0.71) determined for the hexafluoroarsenate salt (8).

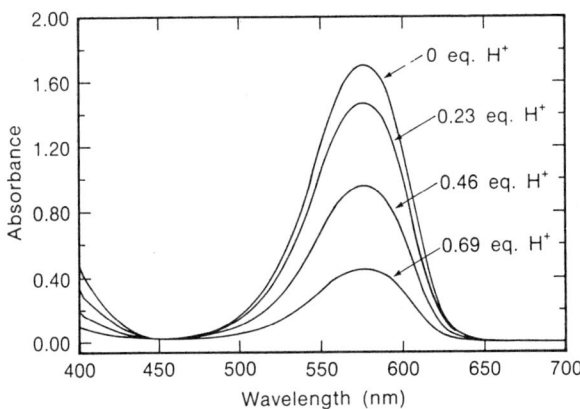

Figure 2. Bleaching of merocyanine dye absorption on addition of fractional equivalents of trifluoroacetic acid.

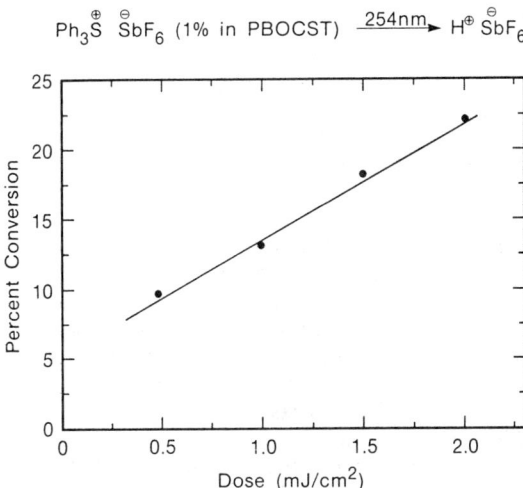

Figure 3. Acid photogeneration versus dose.

Table I. Acid Production and Quantum Yields from Irradiation at 254 nm of Polymer 3 Films Containing 1% Sulfonium Salt 4

Incident Dose (mJ/cm^2)[a]	Absorbed Dose (mJ/cm^2)[b]	Acid Production (mmol)[c]	Percent Conversion (%)	Quantum Yield[d]
0.5	0.25	4.8×10^{-6}	10	0.40
1.0	0.50	6.6×10^{-6}	13	0.30
1.5	0.75	9.2×10^{-6}	19	0.28
2.0	1.0	1.1×10^{-5}	23	0.26

[a] Irradiations and dose measurements were performed through 254 nm band pass filters.

[b] Taken as 50% of incident dose based on absorbance of 0.185 and reflectance of silicon of 0.66 at 254 nm.

[c] Amount of acid produced on irradiation of 1.1 micrometer films on 2 inch diameter silicon wafers. The acid present in the film before irradiation has been subtracted out. Results are averaged for four measurements with average error of less than 10%.

[d] 1 mJ/cm^2 of 254 nm light incident on a 2 inch diameter wafer corresponds to 2.6×10^{16} photons.

Thus the quantum yield for acid production from triphenylsulfonium salts is 0.8 in solution and about 0.3 in the polymer **2** matrix. The difference between acid generating efficiencies in solution and film may be due in part to the large component of resin absorption. Resin excited state energy may not be efficiently transferred to the sulfonium salt. Furthermore a reduction in quantum yield is generally expected for a radical process carried out in a polymer matrix due to cage effects which prevent the escape of initially formed radicals and result in recombination (18). However there are cases where little or no difference in quantum efficiency is noted for radical reactions in various media. Photodissociation of diacylperoxides is nearly as efficient in polystyrene below the glass transition point as in fluid solution (19). This case is similar to that of the present study since the dissociation involves a small molecule dispersed in a glassy polymer.

Catalytic Chain Length. After irradiation of an acid-catalyzed resist film a postbake is carried out to affect some chemical change. For the t-BOC containing polymer **2** postbake results in removal of the t-BOC functionality to give poly(4-vinylphenol). Under ordinary lithographic conditions the exposure dose and acid photogenerator concentration is adjusted to obtain complete removal of t-BOC functionality. For the current study we have chosen conditions to avoid complete loss of carbonate group in order to measure catalytic chain length for the process.

The resist films were irradiated with the same incident doses used for the determination of acid content. A postbake was performed at 100°C for one minute. The carbonyl absorption was measured before and after irradiation and postbake. The difference was used to determine the number of t-BOC groups removed for a given weight of resist film. By dividing the number of t-BOC groups by the amount of photogenerated acid, the catalytic chain length was determined (Table II).

For the incident doses studied, the catalytic chain length varies from about 800 to 1100. The lowest numbers were observed for the lowest and highest doses. At 2.0 mJ/cm^2 incident dose the catalytic chain begins to drop off as t-BOC group removal starts to approach completion.

Table II. Catalytic Chain Length for Acid Catalyzed t-Butylcarbonate Decomposition

Absorbed Dose (mJ/cm^2)[a]	Loss of Carbonyl (%)	Loss of Carbonyl (mmol)	Acid Produced[b] (mmol)	Catalytic Chain Length[c]
0.25	33	3.7×10^{-3}	4.8×10^{-6}	800
0.50	65	7.3×10^{-3}	6.6×10^{-6}	1110
0.75	82	9.2×10^{-3}	9.2×10^{-6}	1000
1.0	86	9.7×10^{-3}	1.1×10^{-5}	870

[a]Taken as 50% of incident dose based on absorbance of film at 254 nm (0.185) and reflectance of silicon at 254 nm (0.66).

[b]Average of four measurements with average error of less than 10%.

[c]Defined as mmol of carbonyl loss per mmol of photogenerated acid.

Acid diffusion. Acid catalyzed resist systems are particarly noteworthy for their high sensitivity toward radiation. However it has been suggested that the amplification effect observed with catalytic resist systems is achieved only at the expense of lost resolution. Some diffusion of catalyst is necessary to achieve sufficient loss of BOC groups in order to impart sufficient difference in polarity for discriminatory film dissolution. Yet unlimited acid diffusion would result in loss of resolution.

The present study permits the evaluation of acid diffusion in poly(4-t-butoxycarbonyloxystyrene) during the postbake step. Catalytic chain lengths of 1000 require acid diffusion over a volume of polymer film which encompasses 1000 repeat units. Since the film density is 1.1 g/cm^3, the volume of polymer containing 1000 repeat groups can be described by a sphere with a radius of 50Å. Thus concerns regarding resolution loss during postbake would appear unwarranted even at very small feature size in poly(4-t-butoxycarbonyloxystyrene). The resolution of features down to several hundred Å should not be adversely affected by diffusion during postbake. A recent study with scanning transmission electron beam exposure has achieved a resolution of 150 Å for this resist system (20).

Experimental

General. Dichloromethane was freshly distilled from calcium hydride prior to use. All glassware used for the analysis was first rinsed with dye solution and then with dichloromethane. The visible absorption spectra were recorded on a Hewlett-Packard Model 8450A UV/Visible Spectrometer. Measurements were conducted in yellow light due to the photosensitivity of the dye solutions. Exposures were performed on the OAI 30/5 Exposure Tool through 254 nm band pass filters (10 nm band width for transmission greater than 10%). The exposure doses were measured with OAI Exposure Monitor Model 355 equipped with a 254 nm sensor (monitor and sensor were calibrated with a National Bureau of Standards secondary standard). Infrared spectra were recorded on an IBM IR/32 (FTIR). Film thickness measurements were performed with an Tencor Alpha Step 200. Poly(4-t-butoxycarbonyloxystyrene) (21) and triphenylsulfonium hexafluoroantimonate (22) were prepared by known procedures.

Film Acid Analysis. A dye solution was prepared by dissolution of 4.4 mg of **3a** in dichloromethane (50 mL). The blank solution was prepared by transferring by pipette 2.0 mL of dye solution to a 25 mL volumetric flask. The solution was then diluted with dichloromethane to a total solution volume of 25 mL. Eight 2 inch wafers were spincoated with 1.1 micron of a polymer solution containing 1% triphenylsulfonium hexafluoroantimonate in poly(4-t-butoxycarbonyloxystyrene). Four of the films were used for background acid determination. The films were dissolved in dichloromethane and 2.0 mL of dye solution was added. The resulting solution was transferred to a 25 mL volumetric flask and the wafer was rinsed with two more portions of dichloromethane to ensure quantitative transfer of material. The combined dichloromethane solution was diluted with more dichloromethane until the final volume of the solution was 25 mL. The procedure was repeated for all four wafers and visible spectra were measured for the four wafer solutions and the blank solution. An average value was determined for the optical density of the wafer solutions and background acid content determined based on an extinction coefficient of 71,000 for the 610 nm absorption. The remaining four films were exposed through 254 nm band pass filters to predetermined amounts of radiation. After radiation, the acid content was determined as above. The acid production was computed by taking the difference between the acid content after irradiation and the background acid content.

Catalytic Chain Lengths. Resist films (1.1 microns) containing t-butoxycarbonate protected phenolic resin (**3**) and 1.0 weight percent of triphenylsulfonium hexafluoroantimonate (**4**) were spin coated on two inch silicon wafers (quantitative acid measurements, film thickness) and one inch sodium chloride plates (IR-analysis). Flood exposure with deep UV light was performed with a mercury-arc lamp through a 254 nm filter applying doses of 0.5, 1.0, 1.5 and 2.0 mJ/cm^2 to both coated silicon wafers and sodium chloride plates. Quantitative IR measurements (intensity change of carbonyl absorption at $1760 cm^{-1}$ were done before and after postbake at 100°C for 60 sec. The catalytic chain length for the acid catalyzed cleavage of t-butoxycarbonyl groups was calculated by dividing the number of removed carbonate groups by the number of photogenerated protons.

Solution Quantum Yield. A sulfonium salt solution was prepared by dissolving 20.4 mg of **1** in 50 mL of acetonitrile. The absorbance of the solution at 254 nm was measured to be greater than 2 and no bleaching was observed on irradiation. A dye solution was prepared by dissolving 5 mg of **3a** in 2 mL dichloromethane followed by dilution with acetonitrile to obtain a total solution volume of 50 mL. To 1.0 mL of sulfonium salt solution was added 2.0 mL of dye solution and the absorbance at 576 nm was measured. Sulfonium salt solution (1.0 mL) was then irradiated through 254 nm band pass filters with a 5.0 mJ/cm^2 dose. The sulfonium salt solution was irradiated in a cuvette cell held in a horizontal position using 254 nm filtered light from the OAI 30/5 exposure tool. The sides of the cell were protected from scattered light and only the top surface was exposed. Following irradiation, 2.0 mL of dye solution was added to the sulfonium salt solution. The absorbance at 576 nm was then determined. This procedure was repeated three more times for both exposed and unexposed solutions. From this data an average value for acid generation was calculated as above and the quantum yield determined.

Conclusions

The irradiation of films prepared from 1% triphenylsulfonium salts in poly(4-t-butoxycarbonyloxystyrene) with lithographically useful doses of 254 nm light generates acid which is less than 0.1% of the t-BOC groups. The efficiency of the photochemistry is several times less than the efficiency of acid generation from triphenylsulfonium salts in solution. The catalytic chain is about 1000 for the t-BOC deprotection step at 100°C. This implies that catalyst diffusion during postbake is on the order of 50Å

Acknowledgments

The authors wish to acknowledge George Gaines of General Electric Corporate Reasearch and Development and Robert Tweig of the IBM Research Research Division for providing samples of the merocyanine dyes.

Literature Cited

1. Vollenbroek, F. A.; Geomini, M. J. H. J. Proc. of the SPIE 1988, 920, 419-428.
2. Ito, H.; Willson, C. G. In Polymers in Electronics; Davidson, T., Ed.; American Chemical Society: Washington, DC, 1984; p 11.
3. Moreau, W. M. Semiconductor Lithography; Plenum: New York, 1988; pp. 68-73 and references cited therein.
4. Crivello, J. V. In Polymers in Electronics; Davidson, T., Ed.; American Chemical Society: Washington, DC, 1984; p 3.
5. Dectar, J. L.; Hacker, N. P. J. Chem. Soc., Chem. Commun. 1987, 1591.
6. Davidson, R. S.; Goodin, J. W. Eur. Polym. J. 1982, 18, 487.
7. Yagci, Y.; Schnabel, W. Makromol. Chem., Macromol. Symp. 1988, 13/14, 161.
8. Pappas, S. P.; Pappas, B. C.; Gatechair, L. R.; Schnabel, W. J. Polym. Sci., Polym. Chem. Ed. 1984, 22, 69.
9. Barton, J. W. In Protecting Groups in Organic Chemistry; McOmie, J. F. W., Ed,; Plenum: London, 1973, Chapter 2.
10. Ito, H.; Willson, C. G. Polym. Eng. Sci. 1983, 23, 1012.
11. Osuch, C. E.; Brahim, K.; Hopf, F. R.; McFarland, M. J.; Mooring, A.; Wu, C. J. Proc. of the SPIE 1986, 631, 68.
12. Houlihan, F. M.; Shugard, A.; Gooden, R.; Reichmanis, E. Proc. of the SPIE 1988, 920, 67.
13. McKean, D. R.; MacDonald, S. A.; Clecak, N. J.; Willson, C. G. ibid., 60.
14. Mazeika, W. A.; Neumann, H. M. Inorg. Chem. 1966, 5, 309.
15. Gaines, G. L. Jr. Anal. Chem. 1976, 48, 450.
16. March, J. Advanced Organic Chemistry: Reactions, Mechanisms and Structure; McGraw-Hill: New York, 1977; 2nd Ed.; pp. 225-231.
17. American Institute of Physics Handbook; Gray, D. E., Ed.; McGraw-Hill: New York, 1972; 6-147.
18. Guillet, J. Polymer Photophysics and Photochemistry; Cambridge University Press: Cambridge, 1985, pp 112-113.

19. Moore, J. W. Ph.D. Thesis, University of Toronto, Toronto, 1988.
20. Umbach, C. P.; Broers, A. N.; Koch, R. H.; Willson, C. G.; Laibowitz, R. B. IBM J. Res. Develop. 1988, 32, 454.
21. Frechet, J. M.; Eichler, E.; Ito, H.; Willson, C. G. Polymer 1983, 24, 995.
22. Crivello, J. V.; Lam, J. H. W. J. Org. Chem. 1978, 43, 3055.

RECEIVED June 14, 1989

Chapter 3

Chemically Amplified Resists

Effect of Polymer and Acid Generator Structure

Francis M. Houlihan, Elsa Reichmanis, Larry F. Thompson, and Regine G. Tarascon

AT&T Bell Laboratories, 600 Mountain Avenue, Murray Hill, NJ 07974

> *Deep-UV photolithography will have an important place in the semiconductor manufacturing arena by the mid 1990's. For this lithographic technology to achieve its ultimate capability, it will be necessary to have non-novolac based resists and new resist processes. Resist materials based on chemical amplification have been reported and shown to have most of the resist properties needed for deep-UV photolithography. The polymer and acid generator structures are critical issues in the design of an optimum chemically amplified deep-UV resist. This paper reports on the effect of polymer and acid generator structures on the lithographic performance of three polymers; poly(4-t-butoxycarbonyloxystyrene) (TBS), poly(4-t-butoxycarbonyloxy-α-methylstyrene) (TBMS), and poly(4-t-butoxycarbonyloxystyrene sulfone) (TBSS) and three photoactive acid generator materials; 2,6 dinitrobenxyl tosylate (Ts), triphenylsulfonium hexafluoroarsenate (Af), and triphenylsulfonium trifluoromethonesulfonate (Tf) The polymer TBSS is known to undergo radiation induced chain scission and provides an improvement in the sensitivity compared to resists formulated with polymers which do not undergo chain scission. The lithographic performance of a resist formulated from this polymer and 2,6-dinitrobenzyl tosylate acid generator is reported.*

"The more things change, the more they stay the same" is a well-known quote that is very applicable to photolithography. For nearly four decades, integrated circuits have been patterned with photolithography. The complexity and speed of IC's have increased by several orders of magnitude with a concomitant decrease in feature size and cost per bit. During these dramatic changes, photolithography has remained the dominant imaging technology. The cost and sophistication of exposure tools have been increasing, and their performance has kept up with device process demands for resolution and registration accuracy. These improvements have been accomplished through the availability of more precise tools and a decrease in the exposure wavelength. The most advanced i line (365 nm) stepper will produce features as small as 0.5 µm; but a shorter wavelength source will be required for the production of smaller features. To meet these future requirements, equipment manufactures have already made available prototype 5X reduction steppers with refractive optics based on 248 nm radiation produced by eximer lasers (1,2), in addition to 1X and 4X reflective optics system using a Hg arc source (3,4).

During these decades of change and evolution in photolithography, novolac-diazoquinone chemistry remained the basis of conventional positive photoresists (5). However, preliminary results indicate that the traditional positive resists are too optically opaque to be useful in the DUV wavelength regime where the novolac matrix resin has significant absorption (6). For the first time, new resists must be introduced into manufacturing in conjunction with a new photolithographic technology. Based on initial evaluations, we have established some performance criteria for an optimum deep-UV resist, and they are given in Table I. When one ex-

TABLE I: Performance Criteria for a Deep-UV Resist

Sensitivity	<50 mJ cm^{-2} *
Contrast	>4
Resolution	<0.35 µm
Optical density	<0.4 µm^{-1}
Etching resistance	$\cong$ novolac based positive photoresists
Shelf-life	> 1 year

* For tools using conventional Hg sources a sensitivity of < 10 mJ cm^{-2} may be required.

amines the available light sources and associated optical systems, it appears unlikely that a material with an overall quantum efficiency of one or less will provide the required sensitivity (7). A new class of resists that achieve differential solubility from an acid catalyzed chemical reaction were discovered by Ito, et al (8,9). The first systems reported were based on poly(4-t-butoxycarbonyloxystyrene)(TBS) and arylsulfonium or iodonium salts as the photoactive acid generators (AG). These systems meet many of the requirements listed in Table I; however, they may have certain practical and processing drawbacks that result from the ionic character of the AG. Ionic materials have limited solubility in organic polymers thus limiting the choice of polymers available for the matrix resin. In addition, many of these "onium" salts contain metal atoms such as arsenic and antimony that are known to be device contaminants.

A nonionic, non-volatile photoactive acid generator, 2,6-dinitrobenzyl tosylate has been recently reported and shown to be effective in chemically amplified resist systems (10). This ester is a nonionic compound that has a much wider range of solubility in matrix polymers and does not contain undesirable inorganic elements. While it is known to exhibit a lower sensitivity to irradiation than the onium salt materials, many structural variations can be produced to precisely vary the acid properties of the molecule and to control the diffusion of the AG in the polymer matrix (11).

Much less work has been focused on the effect of polymer structure on the resist performance in these systems. This paper will describe and evaluate the chemistry and resist performance of several systems based on three matrix polymers: poly(4-t-butoxycarbonyloxy-α-methylstyrene) (TBMS) (12), poly(4-t-butoxycarbonyloxystyrene-sulfone) (TBSS) (13) and TBS (14) when used in conjunction with the dinitrobenzyl tosylate (Ts), triphenylsulfonium hexafluoroarsenate (As) and triphenylsulfonium triflate (Tf) acid generators. Gas chromatography coupled with mass spectroscopy (GC/MS) has been used to study the detailed chemical reactions of these systems in both solution and the solid-state. These results are used to understand the lithographic performance of several systems.

EXPERIMENTAL

Materials

The dinitrobenzyl tosylate, (15) triphenylsulfonium hexafluoroarsenate (16), and triphenylsulfonium triflate (17) were prepared as described in the literature. The monomers, 4-t-butoxycarbonyloxy-α-methylstyene (t-BOC-α-methylstyrene), and 4-t-butoxycarbonyloxystyrene (t-BOC-styrene) and their respective homopolymers, TBS and TBMS were prepared as described in the literature (12,14). TBSS was prepared by conventional, free-radical methods (13,18). The composition of this polymer (ratio of SO_2 to t-BOC styrene) is controlled by changing the polymerization temperature and/or initiator concentration (Table II).

TABLE II: Composition and Reaction Conditions for the preparation of TBSS Polymers

Composition Ratio	t-BOC styrene (ml)	SO_2 (ml)	AIBN $x10^3$ gm	Temp °C	Yield %
2:1	10.0	4.0	57	55	41
3:1	15.5	6.0	85	60	74
5:1	20.0	8.0	115	62	43

The substituted styrene monomer and azo-*bis*-isobutyronitrile initiator were placed in a glass reactor which then was attached to a vacuum manifold. The reactor was evacuated and cooled in a dry-ice-acetone bath. The appropriate quantity of SO_2 monomer was distilled into the reactor, the reactor was then sealed under vacuum and placed in a constant temperature bath for three to eight hours. After the desired conversion was reached, the reaction was terminated by cooling to -20°C, opening the reactor and diluting its contents with acetone. The polymer was precipitated into methanol, the product filtered, dissolved and reprecipitated. After the final precipitation, the polymer was filtered, dried and weighed to determine the yield. The composition was determined by elemental and IR analysis. Table III contains polymer molecular weight and molecular weight dispersity data for the materials used in the lithographic evaluations.

TABLE III: TBS, TBMS and TBSS Polymer Molecular Properties

Polymer	$M_w \times 10^{-5}$	M_w/M_n
TBS	0.27	1.53
TBMS	3.45	2.32
TBSS[a]	1.96	2.1

[a]The ratio of t-BOC styrene to SO_2 was 3:1.

GC/MS Studies

GC/MS experiments were done using an HP5995c mass spectrometer equipped with a crosslinked methyl silicone column (0.2 mm diameter, 12 m length), and helium carrier-gas at a flow rate of 0.5 mL per min. A split interface was used with a split ratio of 1/100. The column temperature was held at -10°C for 2 mins followed by a rapid ramp at 30°C min^{-1} to 250°C. Samples of polymer were heated on a Chemical Data Systems platinum ribbon probe. Polymers were dissolved in cyclohexanone or xylene (1 wt%), and 2-nitrobenzyl tosylate, tosic acid hydrate, 2,6-dinitrobenzyl tosylate, or triphenylsulfonium hexafluoroarsenate (5 wt% relative to the polymer) were added to the polymer solutions. Samples for analysis were prepared by casting 10μL of solution onto the platinum ribbon and heating to either 80°C (2-nitrobenzyl tosylate samples), or 120°C (all other samples) to remove excess solvent. In the case of the samples containing 2-nitrobenzyl tosylate, GC/MS spectra were taken of the volatile products evolving from cast and baked films which were heated in the GC/MS at 130°C for 2 mins. In the case of the samples containing 2,6-dinitrobenzyl tosylate or triphenylsulfonium hexafluoroarsenate, GC/MS spectra were taken of the volatile products evolving from the samples that were irradiated at 248 nm with a dose of 195 mJ/cm^2 followed by rapid heating to 120°C in the GC/MS. In all cases, spectra of the volatile material given off during the pyrolysis of remaining material at 725°C for 2 min were taken. The percent depolymerization of the polymer during the initial heating period is defined as the ratio of the amount of the products given off during the post-irradiation bake at 120°C (or a simple bake in the case of the samples treated with 2-nitrobenzyl tosylate) and the total amount of products given off during both this bake and the final pyrolysis. Authentic samples were used to calibrate the GC results.

GPC Studies

GPC analyses were performed with a Waters Model 244 chromatograph using Microstyragel columns. Both differential refractive index and UV (254 nm) detectors were used. THF was the eluant with a flow rate of 2 ml min^{-1}. A benzene internal standard was employed to correct for flow variations and for normalization of the integrated peak areas. The column set was calibrated using nearly monodispersed polystyrene standards and all molecular data are reported as polystyrene-equivalent molecular weights.

Samples for analysis were prepared by spin-coating Si substrates from cyclohexanone solutions of 2,6-dinitrobenzyl tosylate and TBMS or TBS. The samples were prebaked at 90°C for 15 min in a convection oven, and analyzed after exposure and heating at 120°C for 30 min. The films on each substrate were dissolved with 5 ml of THF and a 1 ml aliquot of 0.1 vol % benzene in THF added. Finally, the solutions were diluted to 10 ml with THF, and 100µl injected onto the GPC for analysis.

Lithographic Evaluation

Resist solutions were prepared by dissolving the appropriate matrix polymer in cyclohexanone (10 g of polymer in 100 ml of solvent) followed by the addition of the required amount of acid generator. These solutions were filtered through teflon membrane filters with the last filtration being done with 0.2 µm pore size filters. The resists were spin-coated onto freshly cleaned, oxidized Si substrates and prebaked at 105°C for 30 minutes. The resist-coated substrates were exposed using a Süss model MA56M contact aligner equipped with a Lambda-Physik excimer laser operating at 248nm. Immediately after exposure, the wafers were post-exposure baked (PEB) on a CEE Model 2000 hot-plate equipped with a vacuum hold-down chuck at 105°C for one to three minutes. The PEB times were adjusted for optimum performance for each acid generator-matrix polymer resist combination. The exposed and baked wafers were developed in an aqueous-base media such as tetramethylammonium hydroxide. Sensitivity and contrast were determined by measuring film thickness with a Nanometrics Nanospec/AFT thickness gauge and plotting normalized film thickness remaining as a function of log dose. The sensitivity is the dose at 0% film remaining (with no thinning of the unexposed film), and the contast (γ) is the slope of the linear portion of the exposure response curve. Resolution and image quality were determined by examining developed resist patterns with a Hitachi Model S-2500 scanning electron microscope equipped with a LaB$_6$ gun and low voltage column.

RESULTS AND DISCUSSION

Three matrix polymers were chosen to evaluate the effect of polymer structure on the performance of chemically amplified resists. TBS was used as the reference material, while TBMS and TBSS were selected because of their ability to undergo photo-induced chain scission (19-21) in addition to acid catalyzed deprotection. Additionally, TBMS undergoes acid catalyzed chain scission. It was reasoned that even a small amount of chain scission would lead to an incremental increase in solubility affording improved sensitivity. Enhancements in sensitivity are desirable for the 2,6-dinitrobenzyl tosylate based systems since the ester is known to exhibit a lower sensitivity to irradiation than the onium salts due to a lower quantum yield for acid generation, reduced catalytic chain length and reduced acidity.

Thermally Initiated Depolymerization

Heating pure TBMS, TBS and TBSS films at 130°C gave no volatile products. Pyrolysis at 725°C gave rise to both deprotection (as determined by the evolution of isobutene and carbon dioxide), and depolymerization to afford the respective monomers, sulfur dioxide, 4-hydroxystyrene, or 4-hydroxy-α-methylstyrene. The compounds, 4-hydroxystyrene and 4-hydroxy-α-methylstyrene, were identified on the basis of their mass spectra, which were consistent with those reported in the literature for these materials (22,23). Additionally, TGA analysis confirmed that all three polymers undergo complete volatilization upon heating to >400°C.

When films of the substituted styrene and styrene-sulfone polymers containing 5 wt% 2-nitrobenzyl tosylate were heated to 130°C, only products arising from the decomposition of the protecting group (carbon dioxide, isobutene, oligomers of isobutene) were detected. Films cast from the α-methylstyrene analog gave products arising not only from the deprotection process, but also from a depolymerization reaction. Two products of depolymerization were observed, the expected 4-hydroxy-α-methylstyrene, and an unknown compound subsequently identified as 5-hydroxy-1,1,3,3-tetramethylindan. Similar results were observed when tosic acid hydrate was used as source of acid, indicating that the process is acid initiated. In the case of tosic acid hydrate, a third compound, 4-isopropylphenol, was also observed as a depolymerization product.

The GC/MS spectra were used to estimate the % depolymerization of TBMS under various conditions (see Table IV). The % depolymerization of polymer samples containing 2-nitrobenzyl tosylate is dependent upon the solvent used to cast films. When xylene is used, a higher % depolymerization is observed than with cyclohexanone. Moreover, the ratio of depolymerization products also varies with casting solvent. Specifically, the indan thermolysis product is more predominant when films were cast from xylene. The reaction with tosic acid hydrate resulted in a high degree of depolymerization; this is most likely a consequence of depolymerization occurring during sample preparation.

Photochemically Initiated Depolymerization

Having established the tendency of TBMS to undergo acid catalyzed depolymerization, the propensity for photochemically generated acid to effect the same phenomenon was evaluated. The resistance of TBS and TBSS to acid catalyzed depolymerization was also evaluated. Onium salts are well known photogenerators of acid, (16,17) while 2,6-dinitrobenzyl tosylate has been shown recently (15) to efficiently generate acid upon irradiation. Unirradiated samples of all the polymers containing either 2,6-dinitrobenzyl tosylate or onium salt showed no detectable sign of polymer depolymerization or deprotection when subjected to a PEB temperature of 120°C.

TABLE IV: Depolymerization of TBMS
Initiated by Thermal and Photochemical Generators of Acid at 120°C

Acid Generator (5wt%)	Mode	Casting Solvent[a]	Atmosphere[b]	%Depoly-merization	Composition[c] Mole%		
					I	II	III
Nitrobenzyl Tosylate	Thermal	C	AD	11	87	13	-
Nitrobenzyl Tosylate	Thermal	X	AD	16	64	36	-
Tosic Acid Hydrate	Thermal	C	AD	77	45	45	11
Dinitrobenzyl Tosylate	hv	C	AD	16	88	12	-
Dinitrobenzyl Tosylate	hv	C	AS	4	66	34	-
Dinitrobenzyl Tosylate	hv	C	AP	10	83	17	-
Triphenylsulfonium Hexafluoroarsenate	hv	C	AD	14	77	23	-
Triphenylsulfonium Hexafluoroarsenate	hv	C	AW[d]	8	31	37	32

a. C denotes cyclohexanone and X denotes xylene.
b. Experiments done on days of high or low atmospheric moisture are denoted as air(AW) or air(AD), respectively. Samples stored during transit on high humidity days in a P_2O_5 glove bag are denoted AP and samples stored during transit in a glove bag saturated with water are denoted AS.
c. The products are identified as follows: I is 4-hydrxy-α-methylstyrene, II is 5-hydroxy-1,1,3,3- tetramethylindan and III is 4-isopropyl phenol.
d. Substantial amounts of isobutene were detected on pyrolysis of these samples at 725°C.

Irradiation followed by PEB of TBMS, TBS and TBSS polymer films containing acid photogenerators, resulted in complete deprotection of the t-butoxycarbonyl group providing sufficient dose was used for the exposure. In agreement with the thermal experiments using 2-nitrobenzyl tosylate, depolymerization resulting from acid catalyzed chain scission is observed only in the case of TBMS. Both deprotection and depolymerization of the α-methylstyrene polymer occur upon irradiation and PEB in the presence of both the dinitro and onium salt compounds (Table IV). The deprotection reaction is essentially quantitative as evidenced by the fact that only trace amounts of isobutene are detected during the final pyrolysis of the samples at 725°C. It was found that the extent of depolymerization, and product ratio is dependent upon atmospheric humidity. The presence of the moisture tends to give a higher ratio of the indan compound, and for films containing the onium salt, reaction in a humid atmosphere results in the appearance of a third depolymerization product, 4-isopropylphenol. Additionally, samples prepared under high humidity resulted in substantial amounts of isobutene during the final pyrolysis. This may be due to either incomplete removal of the protecting group upon PEB, or, alternatively, to the generation of polyisobutylene during the PEB step. Since carbon dioxide was not detected, and substantial quantities of oligomers derived from isobutylene were observed, the latter hypothesis appears more likely. Additionally, IR analysis of films treated in a similar manner detected the absence of the t-BOC group.

The tendency of the protected styrene and α-methylstyrene homopolymers to depolymerize in the presence of photogenerated acid was also evaluated using GPC analysis. While quantitative data regarding deprotection and/or depolymerization could not be obtained using polystyrene equivalent molecular weights, qualitative information, confirming the tendency of TBMS to depolymerize was obtained. The analyses for TBMS show striking changes in molecular weight after exposure which are amplified upon PEB to effect a 2.5 fold reduction in molecular weight. This is accompanied by the appearance of low molecular weight peaks (Figure 1) in the region where the 2,6-dinitrobenzyl tosylate acid generator elutes. Two of these new peaks differ in retention time from the ester and they appear to have aromatic groups as judged by their response with the 254nm detector. Extensive depolymerization, with formation of monomeric products is consistent with these results. Low molecular weight products are also formed when tosic acid is used, but their number increased and molecular weights were higher than with the the photolyzed ester, presumably because of depolymerization occurring during sample preparation. Alternatively, only a slight decrease (~11%) in the polystyrene equivalent molecular weight was found for the styrene based system after both exposure and PEB and no low molecular weight products were observed.

Poly(α-methylstyrene) and poly(styrene-sulfone) have been reported to undergo both chain scission and depolymerization upon irradiation with[60] Co γ-rays (21,24). It was demonstrated that for poly(α-methylstyrene), chain scission and the formation of monomer and dimer (1,1,3-trimethyl-3-phenylindan) occur through a cationic mechanism. It follows, that addition of a strong acid to poly(α-methylstyrene) should promote chain cleavage. The propensity of poly(4-hydroxy-α-methylstyrene) to undergo acid catalyzed chain scission in solution has been reported (12) and the ability of the parent polymer to depolymerize in the presence of acid in the solid-state was confirmed here. Thermolysis of poly(α-methylstyrene) in the presence of 2-nitrobenzyl tosylate at 120°C effects 5% depolymerization and exclusive formation of the dimer, 1,1,3-trimethyl-3-phenylindane. Based on this evidence, it is clear that the solid-state depolymerization of TBMS and formation of 5-hydroxy-1,1,3,3-tetramethylindan observed in this study proceeds through cationic mechanisms as outlined in Schemes I and II respectively. While poly(styrene-sulfone) does not undergo an acid catalyzed chain

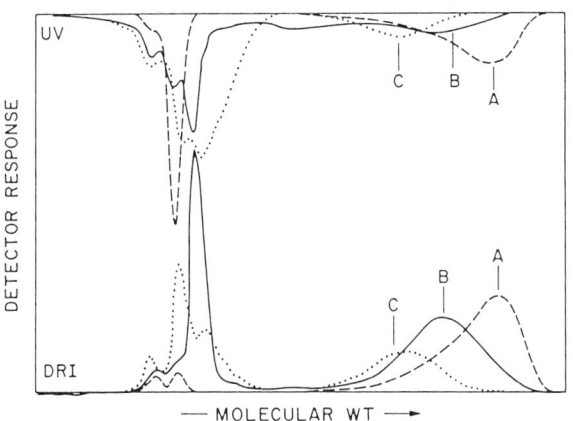

Figure 1: GPC chromatograms of poly (t-BOC-α-methylstyrene) formulated with 2,6-dinitrobenzyl tosylate (A, B) and p-toluenesulfonic acid (C). The processing conditions were (A) as coated, (B) after post-exposure-baked and (C) after post-exposure-bake. The top, inverted traces were obtained with the UV detector.

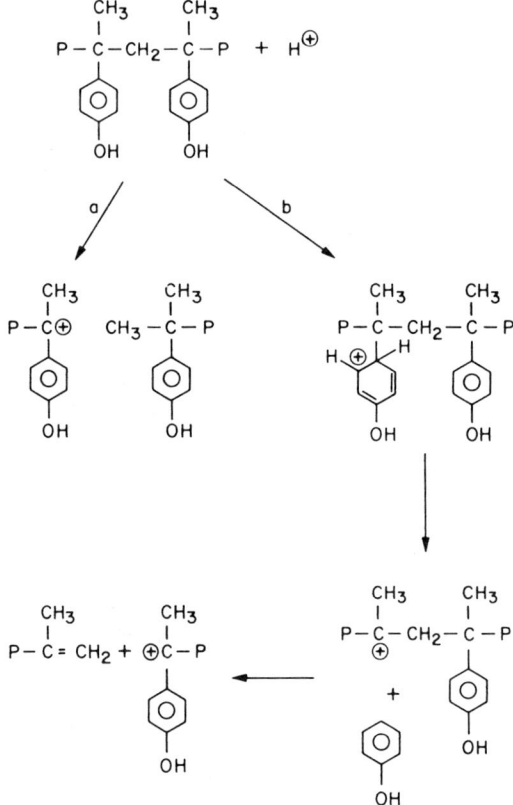

Scheme I

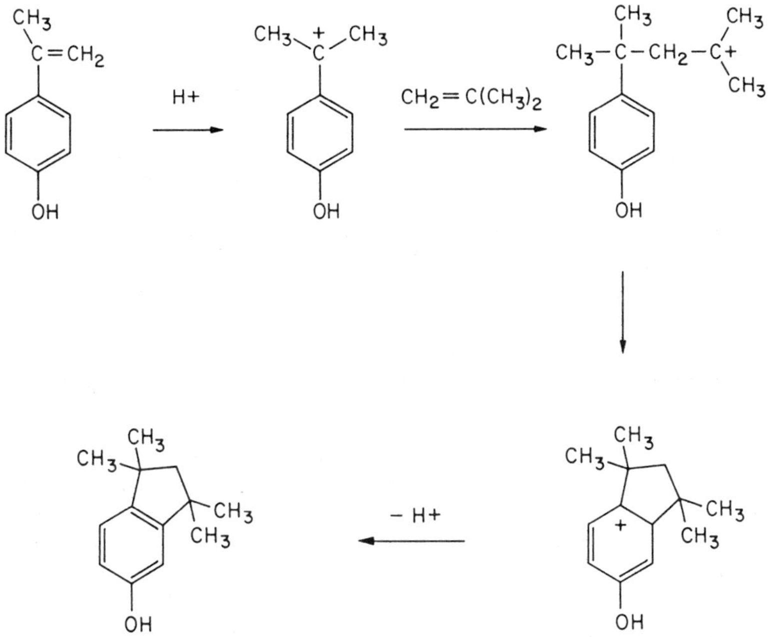
CH3
C=CH2
OH
H+
CH3 + CH3
C
OH
CH2=C(CH3)2
CH3 CH3
CH3-C-CH2-C+
CH3
OH
CH3
CH3
CH3
CH3
+
OH
- H+
CH3
CH3
CH3
CH3
OH

Scheme II

scission process, aromatic sulfones are known to undergo radiation induced chain scission and have been demonstrated to be effective e-beam and deep-UV resists (20).

Process Considerations

The chemically amplified resists reported here for deep-UV applications require a post-exposure thermal treatment process step to effect the deprotection reaction. This step has proven to be critical, and in order to understand the processing considerations it is instructive to discuss, qualitatively, the various primary and secondary reactions that occur with these systems during both exposure and PEB, ie:

$$AG + h\nu \rightarrow AH + G \xrightarrow{\Delta} A^- + H^+ + G \quad (1)$$

$$A^- + C^+ \rightarrow AC \quad (2)$$

$$H^+ + D^- \rightarrow HD \quad (3)$$

$$H^+ + \textit{p-poly} \rightarrow \textit{poly}\text{-OH} + H^+ \quad (4)$$

where AG is the radiation sensitive acid generator, C^+ is a contaminant cation, H^+ is the proton that is regenerated after each deprotection reaction (4), D^- is a contaminant anion, *p-poly* is the protected polymer and *poly*-OH is the alkali soluble product polymer. The "contaminants" addressed above (C^+ and D^-) are not present in every system, their concentrations have not been established and their presence is inferred from observed lithographic effects. Even with these uncertainties, it is the authors experience that the qualitative kinetic equations are useful in process optimization and in guiding fundamental studies to gain insight into individual process steps.

The AG molecule is converted to a strong acid (AH) upon absorption of a photon and the rate of this reaction is fast, with the extent of reaction being governed by the quantum effeciency of the particular acid generator and flux. The acid proton affects the desired deprotection reaction (4) with a finite rate constant. This rate is a function of the acid concentration, $[H^+]$, the temperature and most importantly, the diffusion rate of the acid in the polymer matrix. The diffusion rate in turn, depends on the temperature and the polarity of the polymer matirx. At room temperature, the rate of this reaction is typically slow and it is generally necessary to heat the film to well above room temperature to increase reaction rates and/or diffusion to acceptable levels. The acid (H^+) is regenerated (reaction 4) and continues to be available for subsequent reaction, hence the *amplification* nature of the system.

From these highly idealized reactions, one can gain an understanding of some potential diffculties and process related concerns. For this system to work satisfactorily, it would be necessary for the radiation generated acid concentration, $[H^+]$, to remain constant. However, in most chemically amplified systems, undesired side reactions occur that prematurely destroy the acid, i.e., reactions with contaminants such as water, oxygen, ions or reactive sites on the polymer (reactions **2 and 3**).

The rates of these reactions depend upon the contaminant concentration and the inherent rate constants of the reactions. While the exact nature of these reactions differ for each type of chemically amplified system and are not fully understood, this generalized discussion is sufficient to understand many of the process issues.

The process control of the post-exposure bake that is required for chemically amplified resist systems deserves special attention. Several considerations are apparent from the previous fundamental discussion. In addition for the need to understand the chemical reactions and kinetics of each step, it is important to account for the *diffusion* of the acid. Not only is the reaction rate of the acid-induced deprotection controlled by temperature but so is the diffusion distance and rate of diffusion of acid. An understanding of the chemistry and chemical kinetics leads one to predict that several process parameters associated with the PEB will need to be optimized if these materials are to be used in a submicron lithographic process. Specific important process parameters include:

1. *Time between exposure and post-exposure bake*

The rate and extent of the deprotection reaction (**4**) is critically dependent on the acid concentration, $[H^+]$. Side reactions (**2 and 3**) reduce the effective acid concentration and must be controlled. All of these reactions are thermally activated, however they do occur at a finite rate at room temperature. In order to assure a constant total extent of deprotection (reaction 4) it is necessary to control the elapsed time between exposure and PEB.

2. *Temperature of post-exposure bake*

The PEB temperature and temperature uniformity must be tightly controlled for the same reasons discussed above. It has been found that it is feasible to drive the deprotection reaction in t-butoxycarbonyl protected systems to completion, providing the side reactions are minimized or controlled. This is a necessary requirement for satisfactory lithographic performance.

3. *Time of post-exposure bake*

Since it is important to control the thermally activated side reactions in addition to the temperature, the time of the PEB must also be controlled. It should be noted that excessive time at an elevated temperature can result in undesirable additional chemical changes in the polymer such as crosslinking or oxidation that result in solubility changes of both the exposed and unexposed regions.

Lithographic Characterization

The lithographic performance, as measured by sensitivity and contrast for the TBS, TBMS and TBSS polymers containing two onium salt AG systems and the tosylate AG is given in Table V. The absorbance of these films at 248nm was ~0.3 μm^{-1} in each case.

TABLE V: Lithographic Performance of Resist Systems Formulated with TBS, TBSS and TBMS Matrix Polymers and 5 wt% Tf, Ar and TS Acid Generators

Polymer	Acid Generator	Sensitivity mJ cm^{-2}	Contrast γ
TBS	Tf	3	3
TBMS	Tf	5	3
TBSS	Tf	4	5
TBS	Ar	20	2
TBMS	Ar	18	5
TBSS	Ar	6	6
TBS	Ts	170	2
TBMS	Ts	90	3
TBSS	Ts	65	6

The triphenylsulfonium trifluoromethanesulfonate (Tf) photoactive acid generator affords the highest sensitivity (3-5 mJ cm^{-2}) for all polymer systems studied. The contrast for these systems ranged between 2 and 6 and sub-micron resolution was obtained with all the materials. Resist systems using the triphenylsulfonium hexafluoroarsenate (Ar) precursor exhibited slightly lower sensitivities (16-20 mJ cm^{-2}) while contrast values were similar, i.e., 2-6. Upon formulation with 5 wt% 2,6-dinitrobenzyl tosylate (Ts) the substituted styrenes exhibited still lower sensitivities (65-170 mJ cm^2) and contrast remained in the range of 2-6.

Differential solubility of chemically amplified resists is the result of two sequential reactions: photoinduced decomposition of the acid generator followed by a thermally driven catalytic reaction that decarboxylates the protected matrix polymer. When comparing different polymers and acid generator resist systems, it is important to ensure that the PEB results in equivalent degrees of deprotection (near complete). While some effect on the sensitivity and contrast of the styrene based polymers formulated with an onium salt acid generator is observed when an inherently photodegradable polymer is used, the small differences in sensitivity could easily be attributed to process related variables. Resist exposure dose is reduced (sensitivity increased) by as much as a factor of 2.5 when the tosylate ester is used in conjunction with TBSS or TBMS, polymer systems known to undergo radiation induced main chain cleavage. This increased sensitivity is likely the result of an increase in solubility that arises from a reduction in polymer molecular weight due to chain scission. This effect would work synergistically with the change in polymer solubility resulting from the catalytic deprotection of the t-butyloxycarbonyl groups to afford the hydroxystyrene derivative.

The sensitivity of both TBSS and TBMS can be further improved by the addition of higher weight fractions of Ts (Table VI). The absorbance of the TBMS/Ts resist compositions is also given in Table VI.

TABLE VI: Lithographic Performance of a 2:1 TBSS and TBMS Polymer with the Ts Acid Generator

Polymer	% Ts	Absorbance 248nm μm^{-1}	Sensitivity mJ cm^{-2}	Contrast γ
TBSS	5	.27	65	6
TBSS	10	.40	52	10
TBSS	15	.53	26	20
TBMS	5	0.24	90	2
TBMS	10	0.37	38	4
TBMS	15	0.50	18	5

Note that incorporation of 15 wt% Ts in either polymer affords resists that require an irradiation dose as low as 20 mJ cm^{-2}. Other lithographic properties such as adhesion and crack resistance are improved with TBSS vs. TBMS. The best overall lithographic performance was achieved with the poly(4-t-butoxycarbonyloxystyrene-sulfone) polymers. Although the sensitivity of resist systems formulated with 2,6-dinitrobenzyl tosylate was lower, the contrast, resolution and process latitude were superior in comparison to the onium salt formulations, and the alternate matrix resins. Figure 2 depicts typical contrast curves for TBSS containing 5, 10 and 15 wt% of the 2,6-dinitrobenzyl tosylate acid generator, and Table VI lists the sensitivity and contrast values taken from such curves. The remarkably high contrast for the 10 and 15 wt% formulations indicates a very non-linear relationship between the extent of deprotection and dissolution rate, a phenomenon not observed with the onium salt systems. Although this is not fully understood, it is perhaps due to an increase in dissolution inhibition imparted by the nonionic character of the tosylate acid generator.

Scanning electron microscopy confirmed sub-0.5μm resolution capabilities in TBSS -acid generator resist films (Figure 3). Note that the edge profiles are nearly vertical. Preliminary results indicate that the plasma etching resistance is satisfactory for semiconductor device processing.

SUMMARY

Deep-UV resists comprised of matrix polymers and a 2,6-dinitrobenzyl tosylate photoactive acid generator have been described and compared to previously reported onium salt systems. Although these resists exhibited lower sensitivity than onium salt-based materials, the contrast and processibility are superior. The use of a matrix polymer capable of radiation-induced chain scission improves the sensitivity and allows the 2,6-dinitrobenzyl tosylate acid generator to more nearly

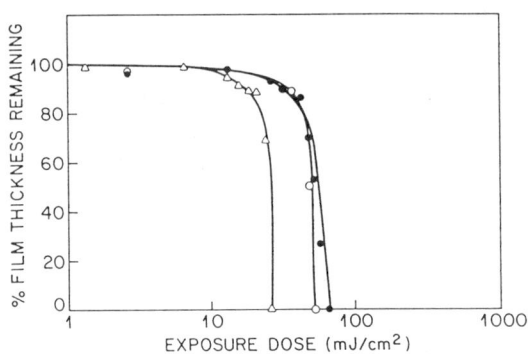

Figure 2: Exposure curve for PBSS containing 5(●), 10(o) and 15(Δ) wt % TS.

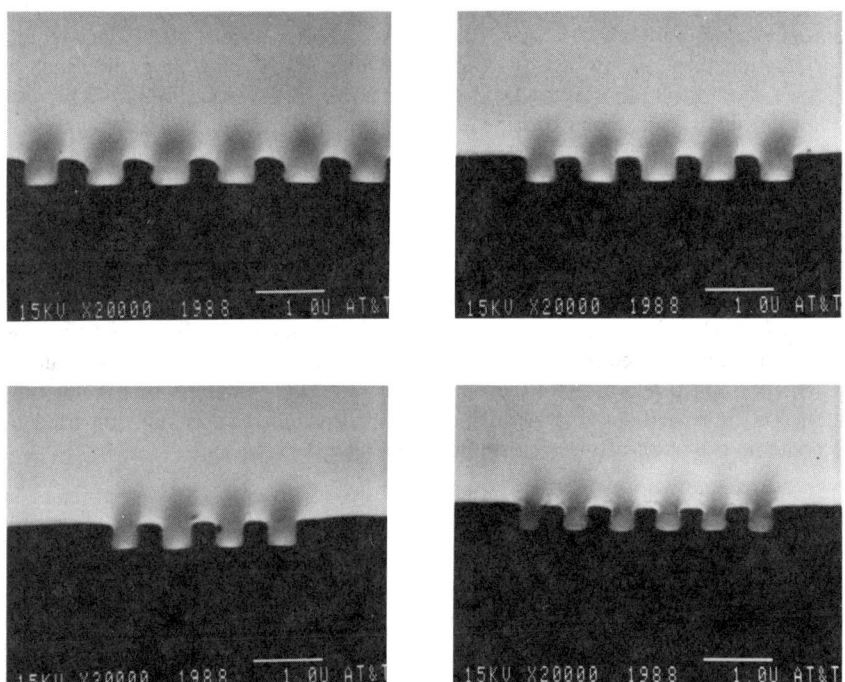

Figure 3: SEM micrographs depicting 0.5, 0.45, 0.4 and 0.35μm coded line/space images printed in PBSS-As resist.

meet sensitivity requirements of deep-UV steppers. Work is in progress to further optimize both the molecular properties and processing conditions for this system.

ACKNOWLEDGEMENTS

The authors would like to acknowledge E. A. Chandross, G. N. Taylor, L. E. Stillwagon and M. Y. Hellman for useful discussions concerning the chemical aspects of this work and to S. Vaydia, V. Pol and J. T. Clemens for input and consultation about the deep-UV resist requirements and process evaluation.

LITERATURE CITED

[1] Pol, V., Bennewitz, J. H., Escher, G. C., Feldman, M., Firtion, V. A., Jewell, T. E., Wilcomb, B. E. Clemens, J. T., *Proc. SPIE*, 1986, **633**, 6.
[2] Pol, V., *Solid State Technology*, 1987, **30** (1), 71.
[3] Buckley, J. D., Karatzas, C., *Proc. SPIE*, 1989, **1088**.
[4] Ruff, B., Tai, E., Brown, R. *Proc. SPIE*, 1989, **1088**.
[5] Willson, C. G. in *Introduction to Microlithography*. Thompson, L. F., Willson, C. G., Bowden, M. J., eds., ACS Symposium Series, **219**, American Chemical Society, Washington, DC, 1983, pp.111-117.
[6] Moreau, W. M., "Semiconductor Lithography, Principles, Practices and Materials", Plenum, NY, 1988, pg.372.
[7] Willson, C. G., Bowden, M. J. in *Electronic and Photonic Applications of Polymers*. Bowden, M. J., Turner, S. R., eds., ACS Advances in Chemistry Series, **218**, American Chemical Society, Washington, DC, 1988, p 87.
[8] Ito, H., Willson, C. G., in *Polymers in Electronics*. Davidson, T., ed., ACS Symposium Series, **242**, American Chemical Society, Washington, DC, 1984, p. 11.
[9] Willson, C. G., Ito, H., Frechet, J. M. J., Tessier, T. G., Houlihan, F. M. *J. Electrochem. Soc.*, 1986, **133(1)**, 181.
[10] Houlihan, F. M., Shugard, A., Gooden, R., Reichmanis, E., *Proc. SPIE*, 1988, **920**, 42.
[11] Neenan, T. X., Houlihan, F. M., Kometani, J. M., Tarascan, R. G., Reichmanis, E., Thompson, L. F., *Proc. SPIE*, 1989, **1086**, 2.
[12] Ito, H., Willson, C. G., Frechet, J. M. J., Farrall, M. J., Eichler, E., *Macromolecules*, 1983, **16**, 510.
[13] Tarascon, R. G., Reichmanis, E., Houlihan, F. M. Shugard, A., Thompson, L. F., *Polymer Engineering and Science*, 1989.
[14] Frechet, J. M. J., Eichler, E., Ito, H., Willson, C. G. *Polymer*, 1983, **24** 995.
[15] Houlihan, F. M., Shugard, A. Gooden, R., Reichmanis, E., *Macromolecules*, 1988, **21**, 2001.
[16] Crivello, J. V., Lam, J. H. W. *J. Poly. Sci., Sci., Poly, Chem. Ed.*, 1979, **17**, 977.
[17] Crivello, J. V. In "Advances in Polymer Science", Springer-Verlag, Berlin, 1984 pp.1-48.
[18] Matsuda, M., Iino, M., Hirayama, T., Miyashita, T., *Macromolecules*, 1972, **5(3)**, 240.

[19] Bowden, M. J., Thompson, L. F., *J. Electrochem. Soc.*, 1974, **121**, 1620.
[20] Bowden, M. J., Chandross, E. A., *J. Electrochem. Soc.*, 1975, **122**, 1370.
[21] Hayashi, K., Yamamoto, Y., Miki, M., *Macromolecules*, 1977, **10**(5), 1316.
[22] Aliev, S. M., and Pokindin, V. K., *Vopr. Neftekhim.* 1977, 137.
[23] Boon, J. L., Wetzel, R. J., and Godshalk, G. L., *Limnol. Oceanorg.*, 1982, **27** (5), 839.
[24] Brown, J. R., and O'Donnell, J. H., *Macromolecules*, 1971, **5**, 109.

RECEIVED July 13, 1989

Chapter 4

Copolymer Approach to Design of Sensitive Deep-UV Resist Systems with High Thermal Stability and Dry Etch Resistance

Hiroshi Ito[1], Mitsuru Ueda[2], and Mayumi Ebina[2]

[1]IBM Research Division, Almaden Research Center, 650 Harry Road, San Jose, CA 95120-6099
[2]Department of Polymer Chemistry, Yamagata University, Yonezawa, Yamagata 992, Japan

> A sensitive deep UV resist was designed by copolymerizing α,α-dimethylbenzyl methacrylate with α-methylstyrene by radical initiation. The electron-rich α-methylstyrene lacks self-propagation and tends to undergo alternating copolymerization with electron-poor monomers such as methacrylates, especially at high feed ratios. Intramolecular anhydride formation that occurs upon heating of certain polymethacrylates and poly(methacrylic acid) is suppressed in such alternating copolymers. Thus, a high glass transition temperature of 210°C is observed for the 1:1 copolymer after deesterification. When mixed with an "onium salt" photochemical acid generator, the dimethylbenzyl ester moiety provides a high resist sensitivity and acid-catalyzed polarity changes. The methacrylate units incorporated in the polymer chain give excellent UV transmission, while the α-methylstyrene units provide good dry etch resistance and high thermal stability.

As the trend toward the higher circuit density in microelectronic devices continues, there has been an increasing interest in lithographic technologies utilizing short wavelength radiations such as electron beam, X-ray, and deep UV (<300 nm). Deep UV lithography employing KrF excimer lasers (248 nm) appears to be emerging as a major technology for ULSI fabrication.

Poly(methyl methacrylate) (PMMA) has been known to provide a high resolution ever since it was first used as an e-beam and deep UV resist. However, its lack of sensitivity precludes its use in semiconductor manufacturing. In order to improve the sensitivity, various PMMA analogs and copolymers have been prepared. For example, incorporation of fluorine into the polymethacrylate structure has proved useful in enhancing the sensitivity (_1_). In addition to the radiation-induced main chain scission mechanism, acid-catalyzed deprotection of polymethacrylates containing ester functionalities sensitive to A_{AL}-1 acidolytic thermolysis was successfully utilized for the design of sensitive resist systems incorporating "chemical amplification" (_2_). In this scheme, the lipophilic polymethacrylates are converted, releasing olefin and a proton upon postbake, to hydrophilic poly(methacrylic acid) (PMAA) by reaction with strong acids generated by irradiation of "onium salt" cationic photoinitiators. The radiation-induced polarity change allows either positive or negative imaging, depending on the choice of the developer. However, although the sensitivity requirement has been met and their optical properties are excellent for deep UV lithography, polymethacrylates are not very useful for device fabrication because aliphatic polymers are not resistant to dry etching.

Diazonaphthoquinone/novolac resists are widely used in semiconductor manufacturing due to their high dry etch durability, and because aqueous base development provides high contrast and resolution. The resolution capability of such photoresists has significantly improved in conjunction with the development of high numerical aperture g-line (436 nm) and i-line (365 nm) step-and-repeat exposure tools. However, application of such two-component resists to deep UV lithography has been hampered by the non-bleachable absorption of diazonaphthoquinone and the poor deep UV transmission of novolac resins. Crosslinking negative resist systems based on chlorinated polystyrene derivatives (*3-6*) and an aqueous base developable, negative deep UV resist (MRS, RD2000N) based on commercial poly(p-vinylphenol) and diazide (*7*) both offer a high sensitivity and good dry etch durability, but the former suffers from swelling during development and the latter from its high deep UV opacity.

Viable deep UV resist materials must possess high sensitivity, high dry etch durability, high thermal stability, and high resolution without swelling, as well as good transmission in the 250 nm region. One successful combination of these properties can be seen in the tBOC resist (*8,9*), which is based on an onium salt acid generator and poly(p-t-butoxycarbonyloxystyrene) (PBOCST) which is transparent in the 250 nm region (optical density = 0.1-$0.13/\mu m$). The resist can be imaged in a positive or negative mode without swelling owing to the acid-catalyzed conversion of the lipophilic PBOCST to hydrophilic poly(p-hydroxystyrene) (PHOST) with T_g of ca. 180°C, is very sensitive owing to chemical amplification, and is resistant to dry etching because of the aromatic nature of the polymer.

In this paper, we report an alternative approach to the design of deep UV resist systems combining the desired properties, which involves copolymerization of methacrylic ester with styrenic comonomer and the use of the acid-catalyzed deprotection chemistry.

Experimental

Materials. α,α-Dimethylbenzyl methacrylate (DMBZMA) was synthesized according to the previously reported procedure (*2*). Styrene (ST) and α-methylstyrene (MST) were commercially obtained and purified by conventional methods. The methacrylate was copolymerized with ST or MST with α,α'-azobis(isobutyronitrile) (AIBN) in toluene at 60°C (Scheme I). Copolymer compositions were determined by elemental analysis. Poly(DMBZMA$_{0.48}$-ST$_{0.52}$) had $M_n = 55,600$ and $M_w = 193,000$. For MST copolymers, $M_n = 15,900$ and $M_w = 29,800$ (x = 0.57), $M_n = 11,200$ and $M_w = 24,100$ (x = 0.75), $M_n = 20,200$ and $M_w = 79,800$ (x = 0.77), $M_n = 38,200$ and $M_w = 115,000$ (x = 0.85). The onium salt used in this study was triphenylsulfonium hexafluoroantimonate (Ph$_3$S$^+$·SbF$_6^-$) (*10*). The polymers (ca. 17 wt%) and the sulfonium salt (4.7-4.9 wt% of total solid) were dissolved in Arcosolv PM Acetate (propylene glycol monomethyl ether acetate) or in cyclohexanone.

Measurements and Lithographic Evaluation. IR spectra were measured on an IBM IR/32 FT spectrometer. UV spectra were recorded on a Hewlett-Packard Model 8450A UV/VIS spectrometer using thin films cast on quartz plates. Molecular weight determinations were made by gel permeation chromatography (GPC) using a Waters Model 150 chromatograph equipped with 6 μStyragel columns at 30°C in tetrahydrofuran. Thus, the molecular weights reported in this paper are polystyrene-equivalent. Thermal analyses were performed on a Du Pont 1090 thermal analyzer at a heating rate of 5°C/min for TGA and 10°C/min for DSC under inert atmosphere. GC/MS analysis was carried out by using a Hewlett-Packard 5995A gas chromatograph/mass spectrometer. Film thickness was measured on a Tencor alpha-step 200. Spin-cast films were baked at 100 or 130°C for 10 min and exposed through a narrow bandpass filter (254 nm) to deep UV radiation from an Optical Associate Inc. exposure system. Lithographic imaging was carried out in a contact mode. A Plasma Therm parallel-plate etcher was used for CF_4 reactive ion etching (RIE) with a graphite cathode maintained at 25°C (10 sccm, 35 mtorr, 0.25 W/cm^2, -195 V).

Results and Discussion

Thermal Analysis. Ito et al. have reported the effect of ester structure on ease of thermolysis and A_{AL}-1 acidolysis of poly(p-vinylbenzoates) (*11*) and polymethacrylates (*2*). The polymethacrylates shown in Scheme II release olefins upon heating or by reaction with acid and are converted to PMAA except for poly(benzyl methacrylate). As TGA curves presented in Figure 1 indicate, the deprotection (deesterification) temperature depends on the structure of the ester group and is in general a good measure of ease of acidolysis. Poly(α,α-dimethylbenzyl methacrylate) (PDMBZMA) and the polymethacrylate of dimethylcyclopropyl carbinol deesterify at much lower temperatures than poly(t-butyl methacrylate) (PTBMA). Though the cyclopropyl carbinol ester is more susceptible to thermolysis than PDMBZMA, it undergoes rearrangement to form a thermally stable ester upon heating, especially in the presence of acid (*2*). Thus we have selected DMBZMA for its high susceptibility to acidolysis. PDMBZMA (x = 1.0 in Scheme I) is converted at ca. 200°C to PMAA releasing MST and then to poly(methacrylic anhydride) (PMAN) (*2,12*) with a glass transition temperature (T_g) of 160°C through intramolecular dehydration as TGA in Figure 1 and DSC in Figure 2 indicate. PDMBZMA has been shown to release a small amount of MST upon exposure to electron beams (*12*). Atactic PTBMA undergoes deesterification and dehydration almost simultaneously at temperatures above 200°C (*2,13-15*). Because of the dehydration of these polymethacrylates and PMAA, the resist systems based on acid-catalyzed deprotection of polymethacrylates cannot realize the potentially high T_g of PMAA (228°C).

As MST or ST is introduced into PDMBZMA as a comonomer, the intramolecular dehydration becomes less significant according to GC/MS analysis (Figure 3) and IR studies (Figure 4). The IR spectrum of a thin film (ca. 1 μm) of P(DMBZMA$_{0.57}$-co-MST$_{0.43}$) shown in Figure 4 indicates that anhydride formation is minor with a large amount of PMAA units remaining after heating the film at 180°C for 65 min. This should be contrasted with the complete dehydration observed with the DMBZMA homopolymer heated under the same conditions (*2*). The limited anhydride formation in the copolymers results in an increased 2nd run T_g after deprotection and some dehydration, as DSC in Figure 2 demonstrates. The almost alternating copolymer of DMBZMA with MST (x = 0.57) exhibits a high T_g of 210°C after deprotection and minor dehydration, which is in contrast to the 1:1 copolymer with ST (x = 0.48) showing T_g at 140°C after the thermal events. The 1:1 DMBZMA-ST copolymer produces more anhydride with a smaller concentration of PMAA units remaining after heating at 180°C for 65 min than the MST counterpart. In radical copolymerization, both DMBZMA and ST can self-propagate and therefore the 1:1 copolymer is not alternating but random, and contains a significant amount of DMBZMA diad sequences which facilitate the intramolecular dehydration. Since MST does not self-propagate in radical copolymerization, the 1:1 copolymer of DMBZMA with MST is almost alternating, with few DMBZMA diad sequences, which minimizes anhydride formation. Incorporation of MST into polymethacrylates tends to increase T_g due to the high

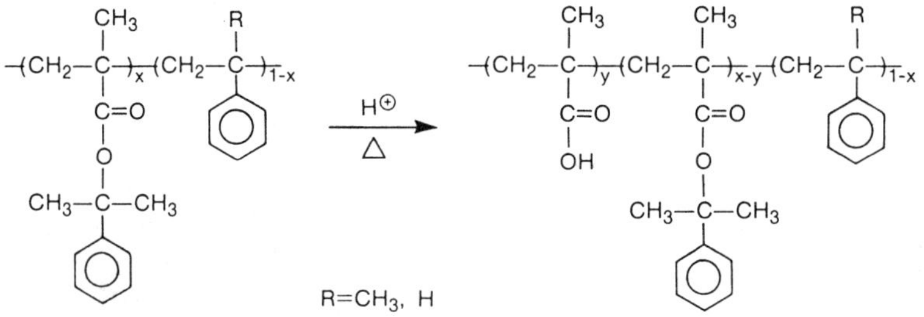

Scheme I Acid-catalyzed deesterification of DMBZMA copolymers.

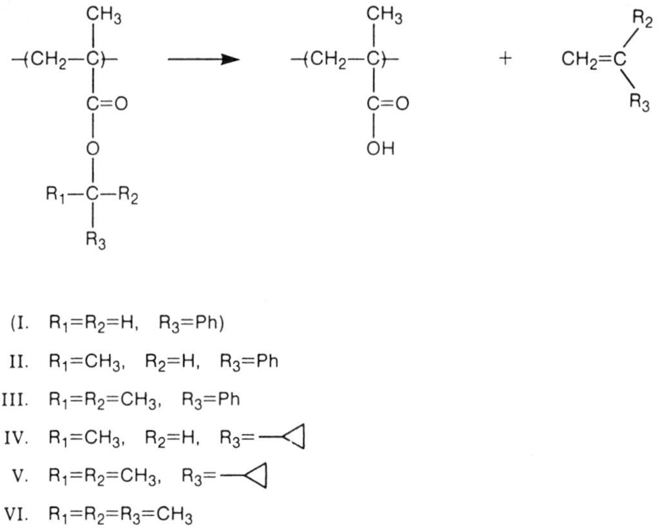

(I. $R_1=R_2=H$, $R_3=Ph$)
II. $R_1=CH_3$, $R_2=H$, $R_3=Ph$
III. $R_1=R_2=CH_3$, $R_3=Ph$
IV. $R_1=CH_3$, $R_2=H$, $R_3=$ ◁
V. $R_1=R_2=CH_3$, $R_3=$ ◁
VI. $R_1=R_2=R_3=CH_3$

Scheme II Thermal and acid-catalyzed deesterification of polymethacrylates.

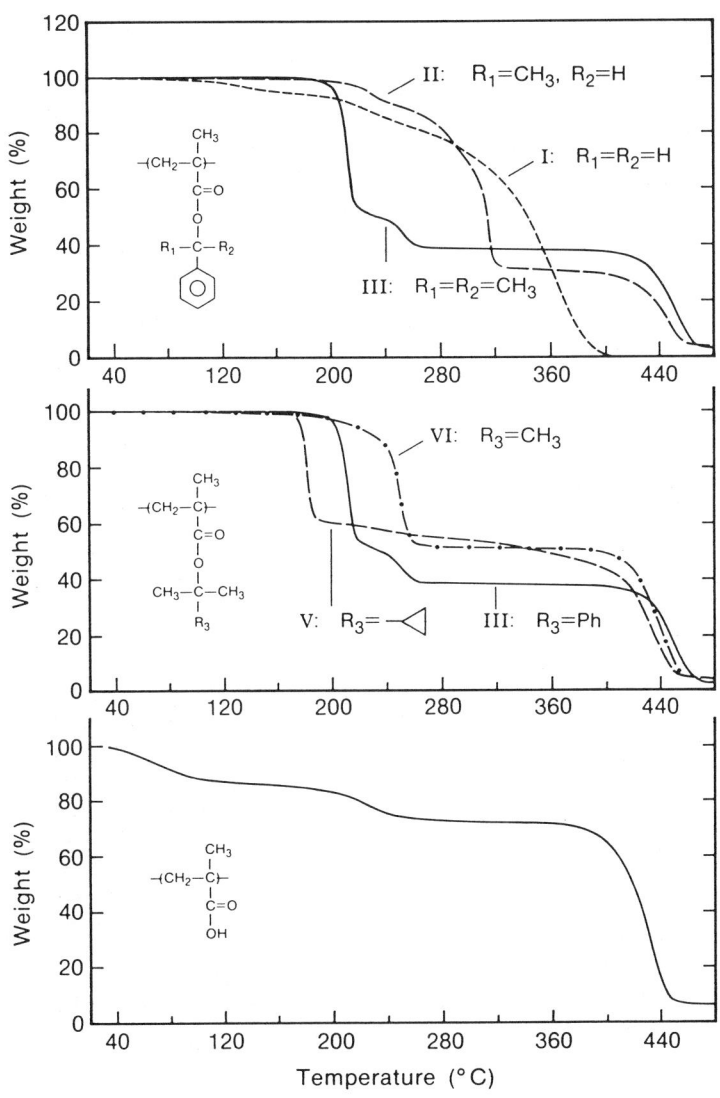

Figure 1. TGA of polymethacrylates and PMAA (heating rate: 5°C/min).

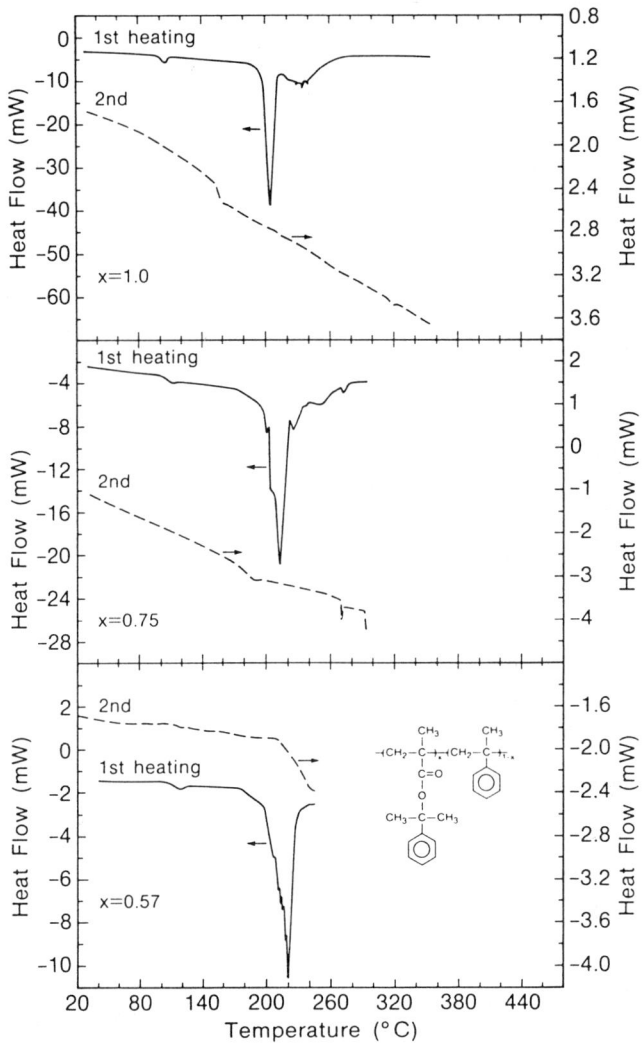

Figure 2. DSC of DMBZMA-MST copolymers (heating rate: 10°C/min).

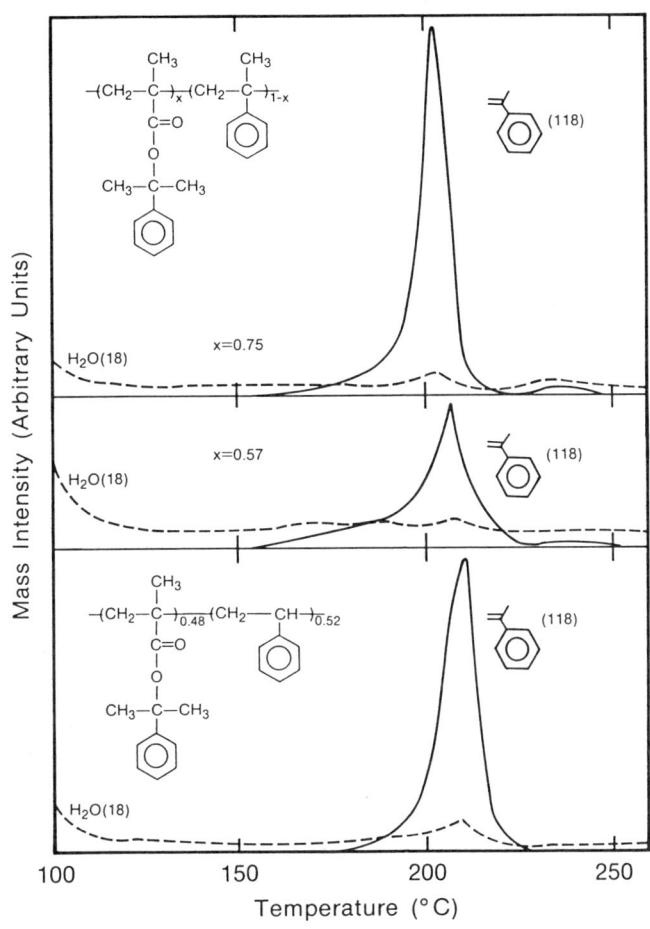

Figure 3. GC/MS of DMBZMA copolymers (heating rate: 5°C/min).

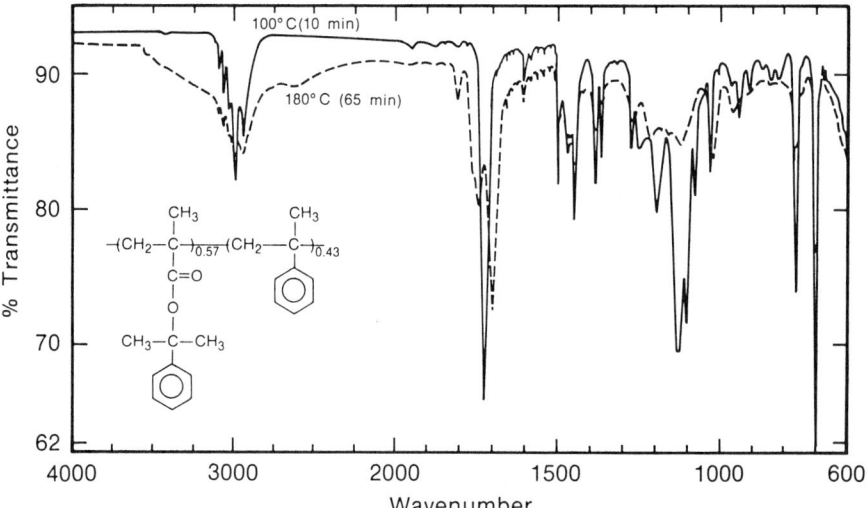

Figure 4. IR spectra of poly(DMBZMA-co-MST) before and after heating at 180°C for 65 min.

T_g (168°C) of poly(α-methylstyrene) (PMST) whereas ST incorporation has a smaller effect due to the lower T_g of polystyrene (ca. 100°C). In addition, maintaining a high concentration of methacrylic acid units in the copolymer by minimizing dehydration helps increase T_g because of the high T_g (228°C) of PMAA.

Thus, we have chosen the alternating copolymer of DMBZMA with MST as our resist material for its high thermal stability (T_g = 210°C). TGA curves of the DMBZMA-MST copolymers are compared with that of PDMBZMA in Figure 5. Although incorporation of MST does not affect the deprotection temperature, the copolymers exhibit lower main chain stability than PMAN (PDMBZMA becomes PMAN above 260°C) and behave like PMMA and poly(α-methylstyrene) in terms of their main chain stability.

Resist Imaging. The UV spectrum of a 1 μm thick film of poly(DMBZMA$_{0.57}$-MST$_{0.43}$) (Figure 6) indicates that the film is very transparent in the 250 nm region with an OD of 0.11 and 0.10/μm at 254 and 248 nm, respectively. Addition of 4.7 wt% of Ph_3S^+-SbF_6 to the copolymer results in an OD of 0.24 at 248 nm (58 % transmission). The resist was exposed to 4.8 mJ/cm² of 254 nm radiation and postbaked at 120°C for 2 min, providing ca. 70 % deprotection as demonstrated by IR spectra in Figure 7.

The copolymer resist can be imaged in either a positive or negative mode owing to the polarity change incorporated in the design. The resist film (1.3 μm thick) was exposed to 0.15-5 mJ/cm² of 254 nm radiation and postbaked at 100°C or 130°C for 2 min. Positive images were obtained by development with AZ2401/H_2O = 1/7 or 400K/H_2O = 1/4 while negative images were produced by using chlorobenzene or anisole as a developer. Contrast curves are presented in Figure 8. The film thickness was measured after postbake (●) and after development (□ for positive and ▲ for negative development). The exposed film shrinks upon postbake due to liberation of α-methylstyrene, and thus the thickness loss that occurs upon postbake is a good measure of conversion of the ester to the acid. The maximum shrinkage expected for this system is 41.7 %. The resist begins to become soluble in aqueous base at about 20 and 10 % shrinkage when postbaked at 130 and 100°C, respectively. Full positive development is accomplished at 2-3 mJ/cm² with a very high contrast (γ) ranging from 11 (130°C postbake) to 4.8 (100°C postbake).

The negative development is unusual since the exposed film becomes insoluble in an organic developer at much lower conversions (at a much lower dose of 0.4 mJ/cm²) than for positive development. Thus, the thickness remaining in the exposed area after negative development is greater at lower doses as long as the imaging dose is greater than the threshold value of 0.4 mJ/cm², simply because the thickness loss in the exposed regions occurs during postbake but not in the development step. PDMBZMA containing 4.9 wt% of Ph_3S^+-SbF_6 behaves similarly, losing upon postbake at 130°C 30 and 45 % of its thickness at 0.34 and 2.30 mJ/cm², respectively, with no additional thickness loss during development. When the PDMBZMA resist is postbaked at 100°C, the shrinkage in the exposed regions amounts to 17 and 40 % and the thickness remaining after development is 80 and 57 % at 0.36 and 2.3 mJ/cm², respectively. This negative development behavior is completely different from resist systems based on poly(p-t-butoxycarbonyloxystyrene) ([16]), poly(t-butyl p-vinylbenzoate) ([16]), and copolymers of TBMA with ST, which can be properly imaged in a negative mode only when near-maximum shrinkage is attained upon postbake. The P(DMBZMA-co-MST) resist exhibits 38 % conversion when postbaked at 130°C for 2 min after exposure to 0.48 mJ/cm² according to our IR study. This conversion should correspond to 15.6 % shrinkage, which agrees well with the contrast curve shown in Figure 8. The copolymers of DMBZMA and MST are expected to undergo main chain scission upon exposure to high energy radiation. However, since the thickness loss that occurs upon postbake corresponds to the

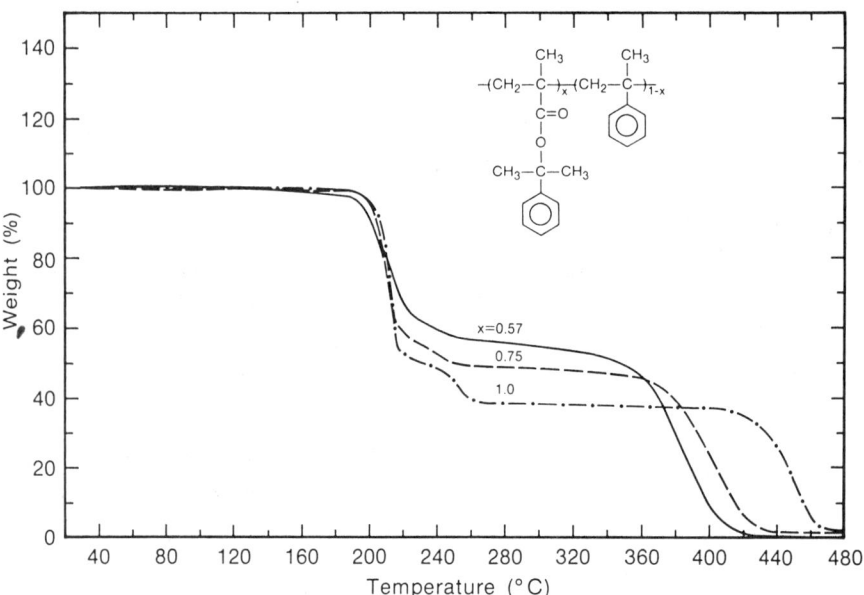

Figure 5. TGA of copolymers of DMBZMA with MST (heating rate: 5°C/min).

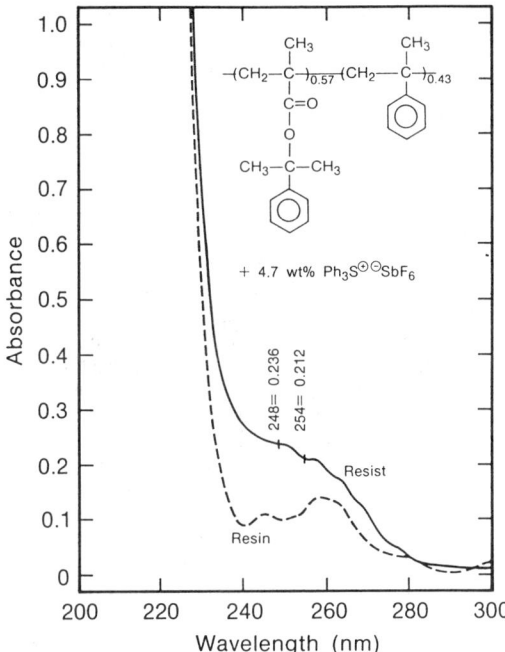

Figure 6. UV spectra of 1 μm thick films of the alternating copolymer and deep UV resist containing 4.7 wt% of $Ph_3S^+\cdot SbF_6^-$.

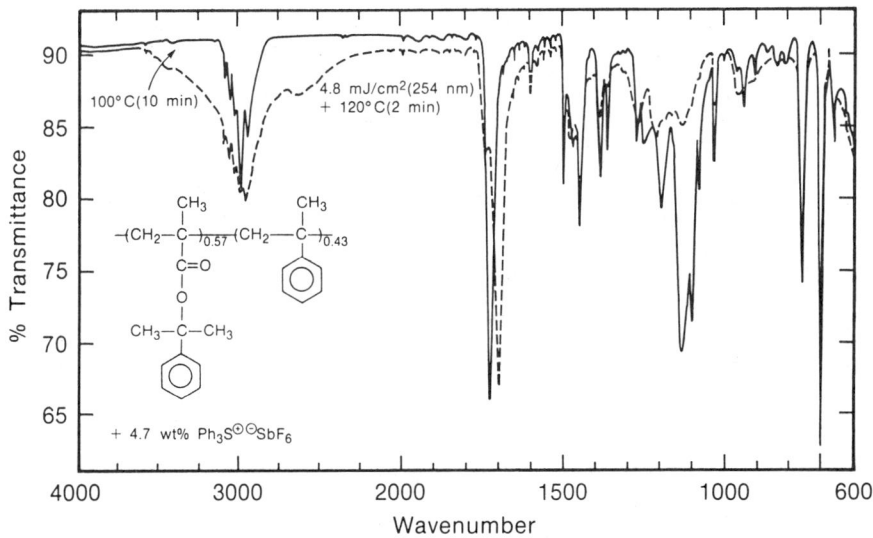

Figure 7. IR spectra of 1 μm thick film of the copolymer resist before and after deep UV exposure.

degree of deprotection in this case, it is unlikely that radiation-induced depolymerization contributes to the imaging mechanism within the range of exposure doses we studied. The cross-section of negative images obtained by exposure to 0.52 mJ/cm^2 followed by postbake at 130°C is presented in Figure 9, clearly indicating that the thickness loss is only about 10 % (1.32 →1.2 μm). It was difficult to determine the contrast for the negative imaging because the films exposed to <0.2 mJ/cm^2 were uneven in thickness or peeled off the substrate.

Thus, the sensitivity of the copolymer resist is extremely high and it is possible to image resist systems based on the polarity change even when the concentration of the polar units produced is less than 50 mole%. The concentration of the polar methacrylic acid units is only 19 mole% in the case of the P(DMBZMA-co-MST) resist exposed to ca. 0.5 mJ/cm^2 and postbaked at 130°C, which still allows negative imaging as demonstrated in Figure 9. As discussed with the PTBVB resist (_16_), in addition to ease of acidolysis, the structure of the protecting groups seems to affect the solubility differentiation, which in turn governs the resist sensitivity as well.

The as-developed negative images are thermally stable owing to the high T_g (210°C) of the alternating copolymer of methacrylic acid and MST as discussed earlier. The thermal stability of the positive image (ca. 110°C) can be readily improved to the range of the negative image simply by exposing the imaged film to a small UV dose, followed by brief postbake at ca. 100°C, which converts the ester copolymer to the more stable acid copolymer. Scanning electron micrographs of positive (3.5 mJ/cm^2) and negative images (3.0 mJ/cm^2) heated at 200°C for 30 min are presented as Figure 10. Cross-sections of the heat-treated positive (3.0 mJ/cm^2) and negative images (2.2 mJ/cm^2) in Figure 11 clearly reveal the high thermal stability of the resist.

Alkyl methacrylates such as t-butyl methacrylate (TBMA) could be incorporated in the place of DMBZMA. In fact, a copolymer of TBMA (0.49) with ST (0.51) containing 4.9 wt% of $Ph_3S^+\cdot SbF_6^-$ can be imaged at 5-8 mJ/cm^2 of 254 nm radiation as demonstrated by the scanning electron micrograph of negative images in Figure 12. In this case, the sensitivity is reduced as well as the thermal stability.

<u>Dry Etch Resistance.</u> Aromatic polymers are more stable than aliphatic polymers in dry etching environments (_17,18_). Alternating copolymers of ST with olefins tri- or tetra-substituted with electron-withdrawing groups behave like PMMA in electron beam and UV irradiation but are as resistant as polystyrene to CF_4 plasma (_19_). Similarly, increased etch resistance has been observed with copolymers of MMA and MST (_20_). CF_4 RIE rates of the copolymers with and without $Ph_3S^+\cdot SbF_6^-$ are compared with those of PMMA, PDMBZMA, and PMST in Figure 13. PDMBZMA is etched at a slower rate than PMMA because of the aromatic ester group but loses its thickness much faster than PMST. Incorporation of 25 % of MST into PDMBZMA does not improve the etch resistance much. However, the copolymer containing 43 % MST is as stable as PMST in CF_4 plasma and the presence of the sulfonium salt does not affect the etch rate. Conversion of the ester to the acid form by UV exposure and postbake does not alter the etch rate either, indicating that the positive and negative images are both as durable as PMST in the plasma.

Summary

We have prepared a sensitive deep UV resist by copolymerizing DMBZMA with MST. The resist is designed such that each component provides specific functions as shown in Scheme III. The methacrylate unit in the polymer chain provides good UV transmission, allowing the triphenylsulfonium chromophore to absorb the deep UV light. The α,α-dimethylbenzyl ester moiety provides facile acidolysis and therefore a high sensitivity as well as a polarity change for dual tone imaging. The

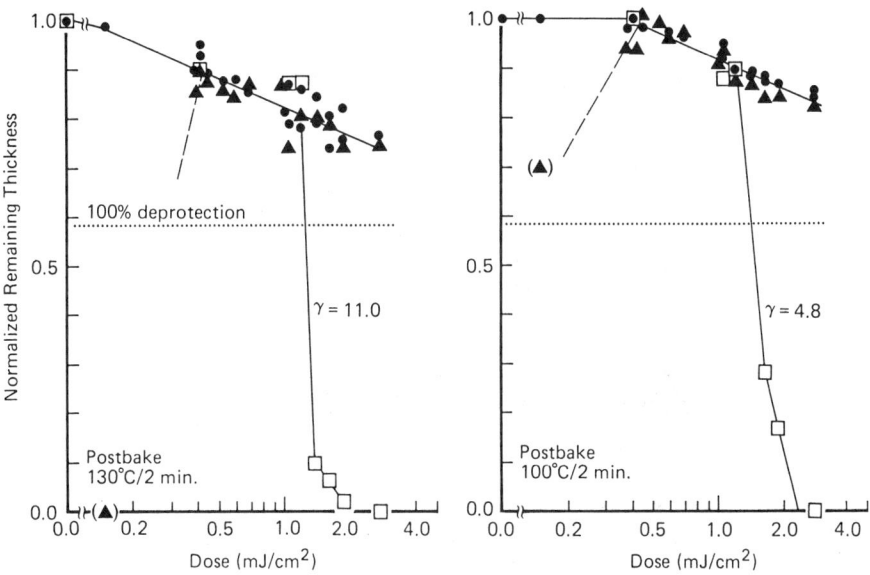

Figure 8. Deep UV contrast curves for poly(DMBZMA-co-MST) containing 4.78 wt% of $Ph_3S^+ \cdot SbF_6^-$: thickness measured (•) after postbake, (□) after development with $400K/H_2O = 1/4$, and (▲) after development with anisole.

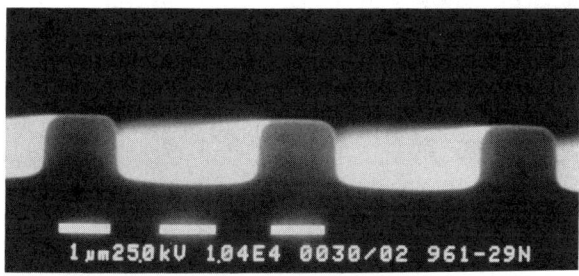

Figure 9. Cross-section of negative images obtained by exposure to 0.52 mJ/cm² of 254 nm radiation (130°C postbake).

Figure 10. Scanning electron micrographs of positive (top, 3.5 mJ/cm^2) and negative images (bottom, 3.0 mJ/cm^2) heated at 200°C for 30 min (the positive image was re-exposed to 2.8 mJ/cm^2 of 254 nm radiation and baked at 130°C for 2 min prior to the 200°C bake).

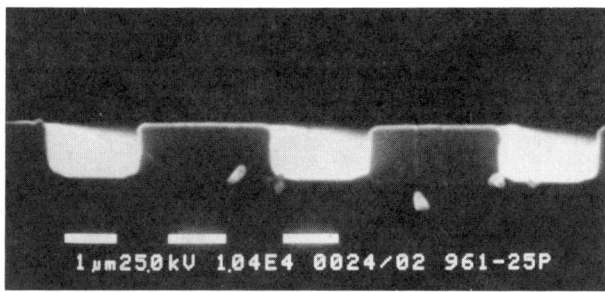
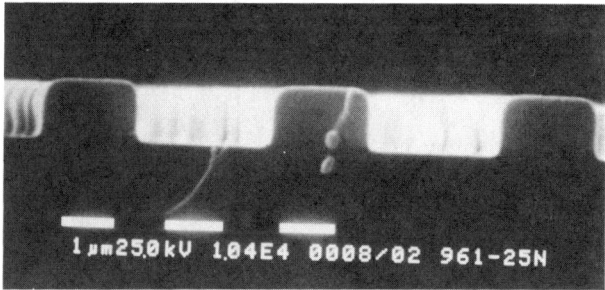

Figure 11. Cross-section of positive (top, 3.0 mJ/cm^2) and negative images (bottom, 2.2 mJ/cm^2) heated at 200°C for 30 min.

Figure 12. Scanning electron micrograph of negative images delineated in poly(TBMA-co-ST) resist at 7.6 mJ/cm^2 of 254 nm radiation.

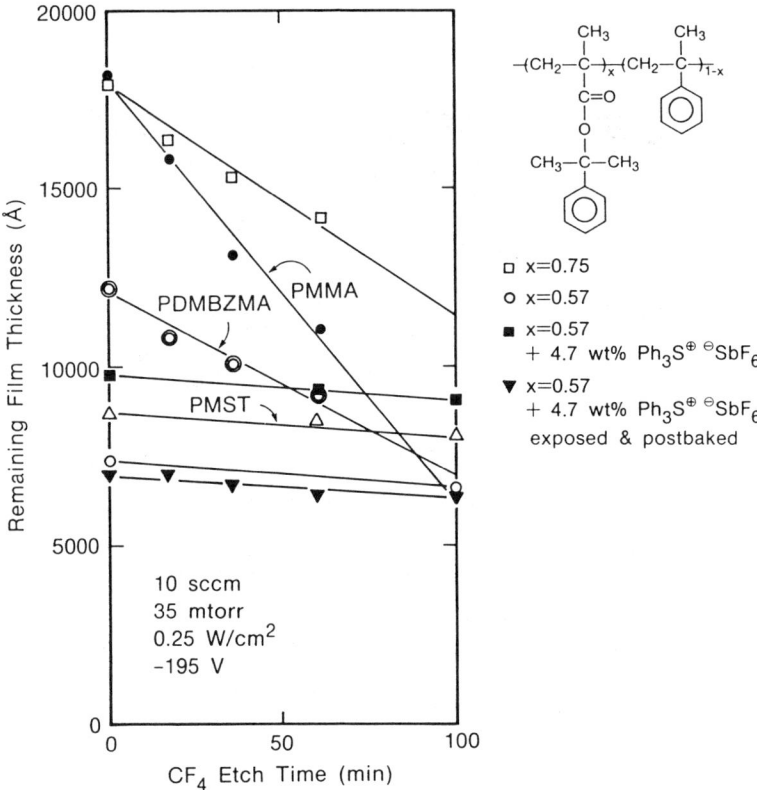

Figure 13. CF_4 etching of the copolymers and reference polymers.

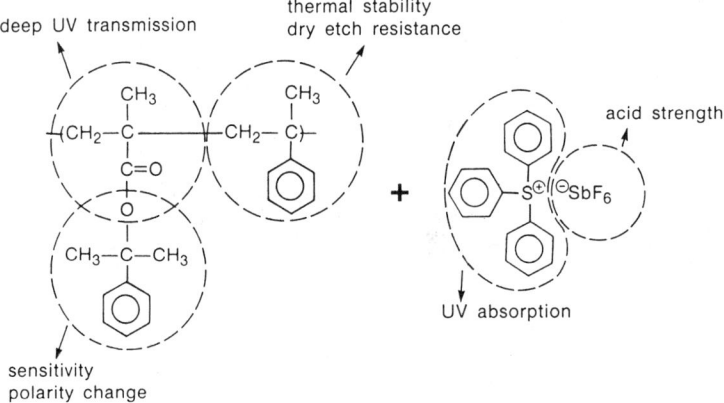

Scheme III Two-component copolymer resist.

MST unit in the polymer chain offers dry etch durability and high thermal stability in conjunction with the alternating nature. The sulfonium salt generates a strong Brönsted acid upon irradiation, with the sulfonium cation absorbing the deep UV light and with the gegen anion determining acid strength thereby contributing to the resist sensitivity.

Acknowledgment

We would like to thank C. Weidner, E. Hadziioannou, and R. Siemens for their analytical support.

Literature Cited

1. for example, FBM resist: Miura, H. et al. *Jpn. J. Appl. Phys.* 1978, **17**, 541; Kakuchi, M. et al. *J. Electrochem. Soc.* 1977, **124**, 1648.
2. Ito, H.; Ueda, M. *Macromolecules* 1988, **21**, 1475.
3. Imamura, S. *J. Electrochem. Soc.* 1979, **126**, 1628.
4. Harita, Y.; Kamoshida, Y.; Tsutsumi, K.; Koshiba, M.; Yoshimoto, H.; Harada, K. *SPSE 22nd Symp. Unconventional Imaging, Sci. & Technol.* 1982, 34.
5. Liutkus, J.; Hatzakis, M.; Shaw, J.; Paraszczak, J. *Polym. Eng. Sci.* 1983, **23**, 1047.
6. Novembre, A. E.; Masakowski, L. M.; Hartney, M. A. *Polym. Eng. Sci.* 1986, **26**, 1158.
7. Iwayanagi, T.; Kohashi, T.; Nonogaki, S.; Matsuzawa, T.; Douta, K.; Yanazawa, H. *IEEE Trans. Electron Devices* 1981, **ED-28**, 1306.
8. Ito, H.; Willson, C. G. *Polym. Eng. Sci.* 1983, **23**, 1012.
9. Ito, H.; Willson, C. G. In *Polymers in Electronics*; Davidson, T., Ed.; ACS Symposium Series No. 242; American Chemical Society: Washington, DC, 1984; p. 11.
10. Crivello, J. V.; Lam, J. H. W. *J. Org. Chem.* 1978, **43**, 3055.
11. Ito, H.; Willson, C. G.; Fréchet, J. M. *J. Proc. of SPIE* 1987, **771**, 24.
12. Hatada, K.; Kitayama, T.; Danjo, S.; Tsubokura, Y.; Yuki, H.; Morikawa, K.; Aritome, H.; Namba, S. *Polym. Bull.* 1983, **10**, 45.
13. Grant, D. H.; Grassie, N. *Polymer* 1960, **1**, 125.
14. Matsuzaki, K.; Okamoto, T.; Ishida, A.; Sobue, H. *J. Polym. Sci., Part A* 1964, **2**, 1105.
15. Lai, J. H. *Macromolecules* 1984, **17**, 1010.
16. Ito, H.; Pederson, L. A.; Chiong, K. N.; Sonchik, S.; Tsai, C. *Proc. of SPIE* 1989, submitted.
17. Pederson, L. A. *J. Electrochem. Soc.* 1982, **129**, 205.
18. Gokan, H.; Esho, S.; Ohnishi, Y. *J. Electrochem. Soc.* 1983, **130**, 143.
19. Ito, H.; Hrusa, C.; Hall, H. K. Jr.; Padias, A. B. *J. Polym. Sci., Part A, Polym. Chem.* 1986, **24**, 955.
20. Sugita, K.; Ueno, N.; Funabashi, M.; Yoshida, Y.; Doi, Y.; Nagata, S.; Sasaki, S. *Polymer Journal* 1985, **17**, 1091.

RECEIVED June 14, 1989

Chapter 5

Nonswelling Negative Resists Incorporating Chemical Amplification

The Electrophilic Aromatic Substitution Approach

Jean M. J. Fréchet[1], Stephen Matuszczak[1], Harald D. H. Stöver[1,3], C. Grant Willson[2], and Berndt Reck[2]

[1]Department of Chemistry, Baker Laboratory, Cornell University, Ithaca, NY 14853-1301
[2]IBM Research Division, Almaden Research Center, 650 Harry Road, San Jose, CA 95120-6099

> Photoinduced crosslinking of polymers through electrophilic aromatic substitution has been achieved with a family of styrenic polymers or copolymers containing both latent electrophiles and activated aromatic groups. These polymers can be used in combination with a photoacid generator to design a non-swelling negative multipurpose resist which can be used for deep-UV, X-ray or E-beam imaging and has a very high sensitivity. For example, in the deep-UV, sensitivities of less than 1 mJ/cm^2 are obtained with very high contrasts. Irradiation of the two-component resists results in the generation of strong acid which, upon baking, activates the latent electrophile to a carbocationic species that couples to neighboring activated aromatic moieties in a crosslinking process. Vinyl-phenol units are incorporated in the copolymer formulation to provide activated aromatic sites, solubility of the resist in aqueous base, and lack of swelling during image development. Alternate three-component formulations in which the latent electrophile is separate from the activated aromatic moiety are also suitable.

A number of new resist materials which provide very high sensitivities have been developed in recent years [1-3]. In general, these systems owe their high sensitivity to the achievement of **chemical amplification**, a process which ensures that each photoevent is used in a multiplicative fashion to generate a cascade of successive reactions. Examples of such systems include the electron-beam induced [4] ring-opening polymerization of oxacyclobutanes, the acid-catalyzed thermolysis of polymer side-chains [5-6] or the acid-catalyzed thermolytic fragmentation of polymer main-chains [7]. Other important examples of the chemical amplification process are found in resist systems based on the free-radical photocrosslinking of acrylated polyols [8].

The seminal work on deep-UV resist materials which incorporate chemical amplification was started at IBM San Jose's Research Laboratory in 1979 when Fréchet and Willson first prepared poly(4-t-butyloxycarbonyloxy styrene) and end-capped copolymers of o-phthalaldehyde and 3-nitro-1,2-phthalic dicarboxaldehyde.

[3]Current address: Department of Chemistry, McMaster University, Hamilton, Ontario L8S 4M1, Canada

The former material was designed for its ability to lose its t-BOC protecting groups under a variety of conditions thereby affording a free phenolic polymer and imaging by differential dissolution. The second, designed as a self-developing imaging system, owed its activity to the presence of photocleavable o-nitrobenzyl groups which allowed its partial depolymerization upon exposure to UV light due to a ceiling temperature phenomenon.

These interesting new approaches were later extended and improved by Ito et al. [5] with the introduction of appropriately chosen triarylsulfonium or diaryliodonium salts as the photoprecursors of catalytic amounts of strong acid for t-BOC group removal or cleavage of the poly(phthalaldehyde).

The use of phenolic polymers in photocrosslinkable systems usually involves multicomponent systems which incorporate polyfunctional low molecular weight crosslinkers. For example, Feely et al. [9] have used hydroxymethyl melamine in combination with a photoactive diazonaphthoquinone which produces an indene carboxylic acid upon irradiation to crosslink a novolac resin. Similarly, Iwayanagi et al. [10] have used photoactive bisazides in combination with poly(p-hydroxy-styrene) to afford a negative-tone resist material which does not swell upon development in aqueous base.

Preparation of the Resists and Lithographic Evaluation.

Design of the Resist Material. Our approach to resists that operate via electrophilic aromatic substitution is outlined in Scheme I.

The reaction sequence which is used can be summarized as follows: In the first step acid is generated by photolysis of a triarylsulfonium salt. Subsequent reaction with a latent electrophile, such as a substituted benzyl acetate, produces a carbocationic intermediate while acetic acid is liberated. The carbocationic intermediate then reacts with neighboring aromatic moieties in a coupling reaction which liberates a proton thus ensuring that the overall process is catalytic and that chemical amplification is achieved. Several approaches are possible as the latent electrophile and the activated aromatic compound may be part of the same or of different molecules. However, it is necessary that at least one of the components of the mixture be a polymer with good coating, solubility, and optical properties and that the different components of the mixture be compatible. The paragraphs below will describe first an approach in which both the latent electrophile and the electron-rich aromatic components are part of the same polymer, then a second approach in which non-polymeric difunctional latent electrophile is used with a phenolic polymer. In both cases the source of photogenerated acid is a triarylsulfonium salt, other sources of photogenerated acid are available [5].

Preparation of Copolymers Containing Both Electrophilic and Nucleophilic Groups. Our first implementation of this reaction scheme involved the preparation of a series of copolymers incorporating both a latent electrophile and an electron-rich aromatic moiety which, being phenolic, also provides access to swelling-free development in aqueous medium. The copolymers are prepared as shown in Figure 1 by copolymerization of 4-t-butyloxycarbonyloxy-styrene with 4-acetyloxymethyl-styrene. Although the reactivity ratios of these two monomers are different [11], our study of this system has confirmed that they copolymerize essentially in random fashion.

Removal of the t-BOC protecting groups from the copolymer is best done by refluxing in glacial acetic acid, a process which does not affect the acetoxymethyl pendant groups or the molecular weigh distribution of the final polymer. Figure 2 shows the GPC trace for a copolymer containing 80% free phenolic groups and 20% 4-acetoxymethyl groups (80/20 copolymer). Curve (a) shows the polymer before deprotection with M_w = 62,000 and M_n = 28,000 (polydispersity = 2.2), while curve (b) shows the same polymer after deprotection with M_w = 45,000 and M_n = 20,000

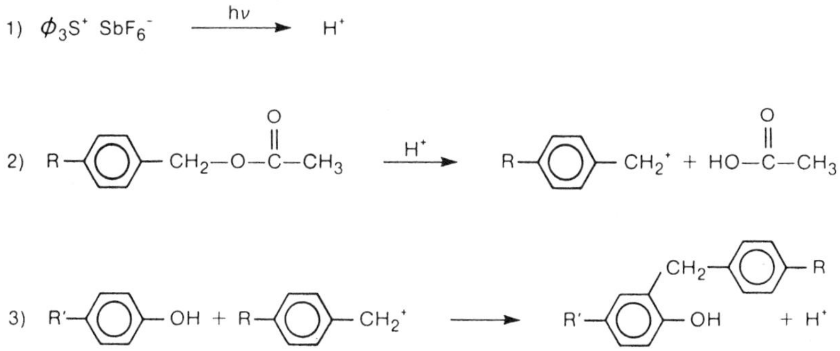

1) $\phi_3S^+ \, SbF_6^- \xrightarrow{h\nu} H^+$

2) $R-\underset{}{\bigcirc}-CH_2-O-\overset{O}{\underset{\|}{C}}-CH_3 \xrightarrow{H^+} R-\underset{}{\bigcirc}-CH_2^\bullet + HO-\overset{O}{\underset{\|}{C}}-CH_3$

3) $R'-\underset{}{\bigcirc}-OH + R-\underset{}{\bigcirc}-CH_2^\bullet \longrightarrow R'-\underset{}{\bigcirc}(CH_2-\underset{}{\bigcirc}-R)-OH + H^+$

Scheme I. Resist design based on electrophilic aromatic substitution.

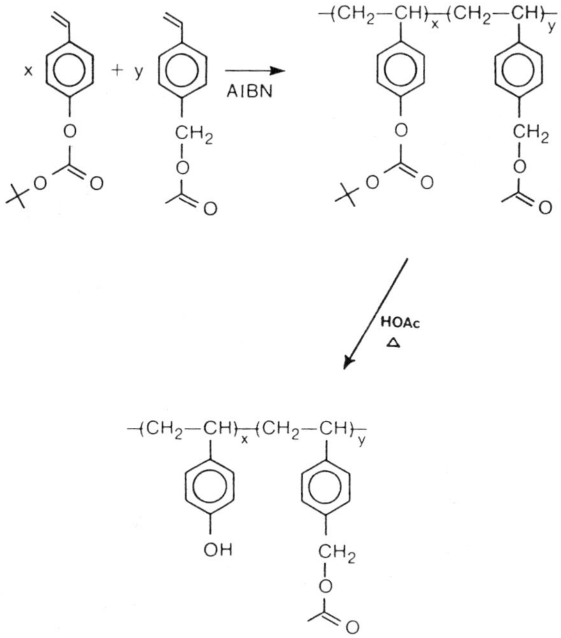

Figure 1. Preparation of the copolymers of 4-vinylbenzyl acetate and 4-vinylphenol.

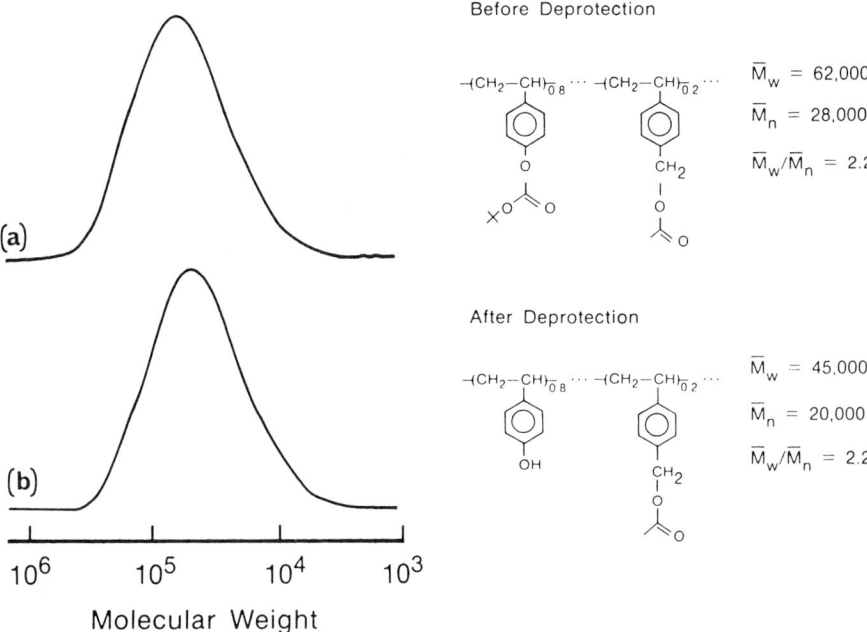

Figure 2. Gel Permeation Chromatogram of the copolymer (a) before and (b) after removal of the *t-BOC* protecting groups

(polydispersity = 2.2) confirming the clean nature of the deprotection process. Figure 3 shows the UV spectrum of a 1 μm thick film of a resist consisting of 90 wt% of the same 80/20 copolymer and 10 wt% of triphenylsulfonium hexafluoroantimonate; it is seen that the film is suitable for imaging at 254nm since the absorbance of the resist does not exceed 0.6 per micrometer of film thickness.

<u>Sensitivity and Contrast Measurements.</u> Imagewise exposure of films of the various copolymers containing from 5-10% triphenylsulfonium hexafluoroantimonate to UV light at 254 nm resulted in the crosslinking of the exposed areas as shown in Figure 4.

The characteristic curves [1] for the various copolymers were determined **at a constant loading of 10 wt% sulfonium salt** using 1um thick films and exposure through a narrow bandwidth Hg-line filter to varying doses of 254 nm radiation. The characteristic curve shows the thickness of the insolubilized regions of the film remaining after development as a function of log[exposure dose]. These measurements provide access to D_g^i or gel dose, the minimum dose required to observe the formation of an insoluble residue, as well as D_g^o the minimum dose required to produce an insolubilized film of thickness equal to that of the starting film (1μm in this instance). The characteristic curve of a 65/35 copolymer is shown in Figure 5. This Figure shows that the lithographic sensitivity of the resist material based on a 65/35 copolymer having M_n = 22,000 and M_w = 46,000 is approximately 0.6 mJ/cm^2 while its contrast (slope of the curve) is close to 4. Measurements of the lithographic characteristics of a series of copolymers having different compositions and essentially the same molecular weights and polydispersities are summarized in Table 1.

It can be seen in Table 1 that the lithographic sensitivity of the copolymers blended with 10% sulfonium salt increases as the percentage of latent electrophile (vinylbenzyl acetate) units is increased. For a 50/50 copolymer the lithographic sensitivity is approximately 0.5 mJ/cm^2 with a very high contrast of over 4. It should be noted however that aqueous development is no longer possible for the 50/50 copolymer for which some isopropanol must be added to the aqueous base developer.

Table 1: Lithographic sensitivity and contrast data for various copolymers

Copolymer[a]	M_w	M_n	M_w/M_n	D_g^i	D_g^o	γ
95/5	44,000	20,000	2.2	0.68	1.0	> 4
90/10	45,000	21,000	2.1	0.56	1.0	~ 4
80/20	45,000	20,000	2.2	0.56	1.0	~ 4
65/35	46,000	22,000	2.1	0.45	0.85	3.6
50/50	41,000	21,000	2.0	0.51	0.6	> 4
100/0	39,000	20,000	1.9	0.86	1.2	> 4

[a]*Copolymer composition x/y: the first number x indicates the mole % of 4-hydroxystyrene units and the second number y indicates the mole % of 4-acetoxymethylstyrene units in the copolymer. All measurements were made at a constant 10 wt% loading of triphenylsulfonium hexafluoroantimonate.*

<u>Use of a Difunctional Crosslinker.</u> An alternate approach to chemically amplified imaging through electrophilic aromatic substitution is shown in Figure 6 below. In this approach a polyfunctional low molecular weight latent electrophile is used in a three component system also including a photoactive triaryl sulfonium salt and a phenolic polymer. In this case again crosslinking of the polymer is observed upon

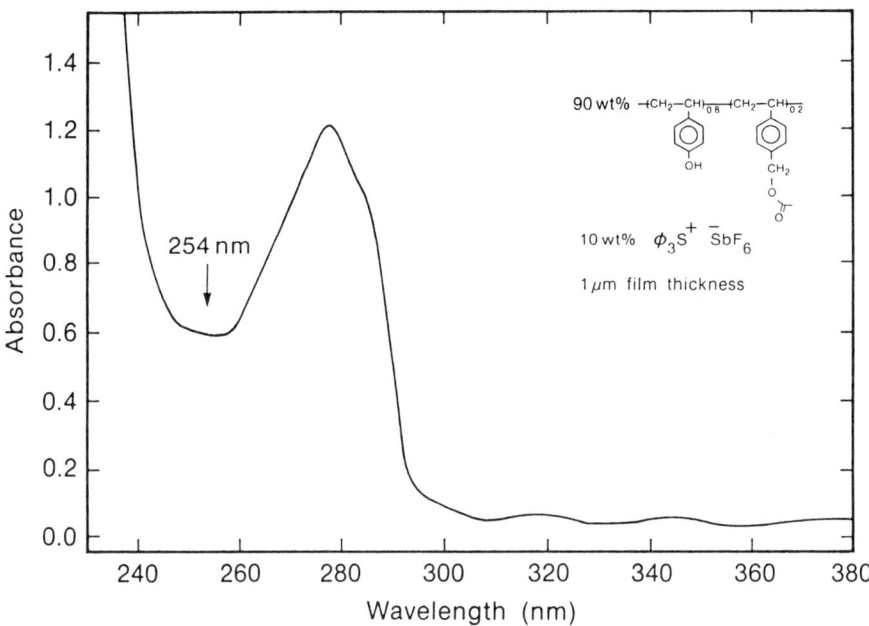

Figure 3. UV spectrum of a 1um thick film of 80/20 copolymer containing 10 wt% triphenylsulfonium hexafluoroantimonate.

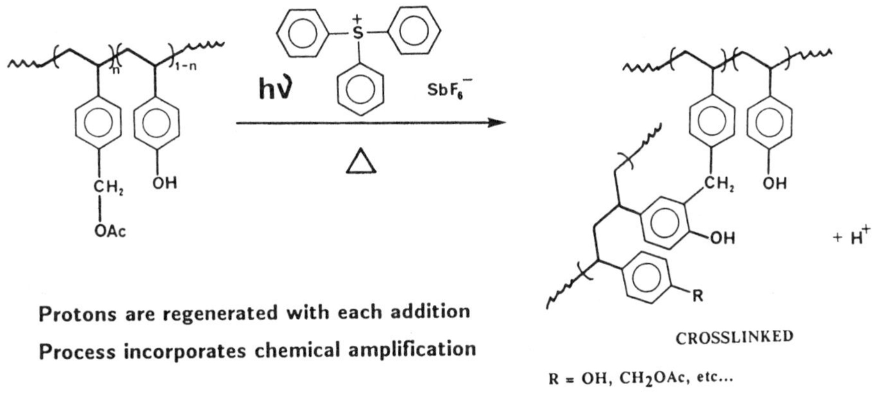

Protons are regenerated with each addition
Process incorporates chemical amplification

Figure 4. Crosslinking process via electrophilic aromatic substitution.

1.0
0.5
0.0
Normalized Film Thickness after Development
0.25
0.5
1.0
2.0
4.0
Log Dose [mJ/cm2]
OH
CH2OAc

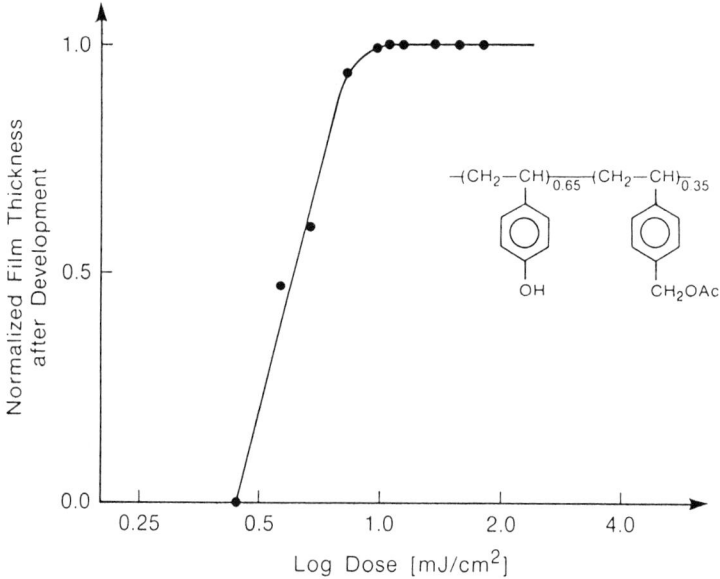

Figure 5. Characteristic curve for sensitivity measurement on the 65/35 copolymer.

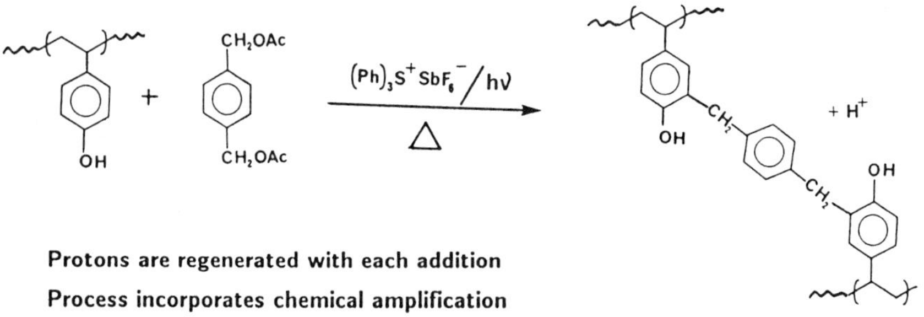

OH
CH2OAc
CH2OAc
(Ph)3S+ SbF6- / hν
Δ
OH
+ H+
OH
Protons are regenerated with each addition
Process incorporates chemical amplification

Figure 6. Crosslinking via a non-polymeric multifunctional latent electrophile.

irradiation. Sensitivity measurements for such systems are still in progress. Another example of this approach involves the use of a novolac as the phenolic component [12].

Imaging experiments

The resist material was prepared using 90 wt% of the 80/20 copolymer and 10 wt% of the triphenylsulfonium hexafluoroantimonate. After spin-coating onto silicon wafer to 1μm thickness and baking 5 min at 105°C the wafer was exposed to <2 mJ/cm^2 of filtered 254nm radiation through a mask in a contact printing process, then post-baked 2 min at 125°C and developed using aqueous base developer. With polymers containing 5-10% vinylbenzyl acetate units the developer was a 1:1 mixture of commercial developer MF312 and water. MF312 developer was used for formulations containing 20 and 35% of vinylbenzyl acetate units, while a 9:1 mixture of MF312 and isopropanol was used for the 1:1 copolymer. Finally, the vinylbenzyl acetate homopolymer was developed with a 3:4 mixture of anisole and isopropyl alcohol. Figure 7 shows a scanning electron micrograph of the contact printed image which was obtained using a resist containing 20 mole% vinylbenzyl acetate and 80% 4-hydroxystyrene, formulated with 10% triarylsulfonium hexafluoroantimonate and developed with MF312 (aqueous base). Similarly, exposure using 5 inch wafers and a Perkin-Elmer 500 projection printing tool operating with a narrow bandwidth filter at 254nm afforded images such as that shown in Figure 8 after development under the conditions outlined above for a polymer containing 20mole% vinylbenzyl acetate. In the latter case, the resist material was formulated with 5% of the onium salt.
We have also performed preliminary imaging experiments usin E-beam exposure, these experiments indicate that the 80:20 copolymer is a sensitive E-beam resist material which requires an exposure dose of < 1μC/cm^2. Further experiments involving both E-beam and X-ray exposure are in progress.

Mechanistic and model studies.

We have carried out extensive modeling studies of this imaging system using NMR spectroscopy and liquid chromatography to monitor the reaction. These studies (Figure 9) involving for example electrophilic reactions between 4-isopropyl phenol **1** and benzyl acetate or substituted benzyl acetate **2** have provided us with a good understanding of the overall reaction and may furnish valuable insights on ways to improve our new imaging system. The rate-determining step in the alkylation process is the formation of the benzylic carbocationic species **3**. This may then react either by C-alkylation or O-alkylation to afford **4** or **5** respectively. While the final products of the reaction are C-alkylated, it appears that O-alkylation may be an important pathway towards these final products. In fact, a small amount of **5** is always seen *during* the reaction, consequently, our model studies have been extended to a number of appropriately functionalized benzyl phenyl ethers. For example, the reaction of 4-isopropylbenzyl phenyl ether with a catalytic amount of triflic acid proceeds *at least* three orders of magnitude faster than the same reaction with 4-isopropylbenzyl acetate. This is somewhat surprising in view of the relative nucleophilicities of the leaving groups (acetic acid or phenol) and it suggests that the reaction pathways available to the ether and to the acetate are different. In particular, cleavage of the benzyl ether may well be promoted by the formation of an intramolecular sandwich pi-complex [13, 14], a pathway which is not available to the acetate. The detailed results of these modeling studies will be reported shortly [15].

Figure 7. Scanning electron micrograph of a contact printed image.

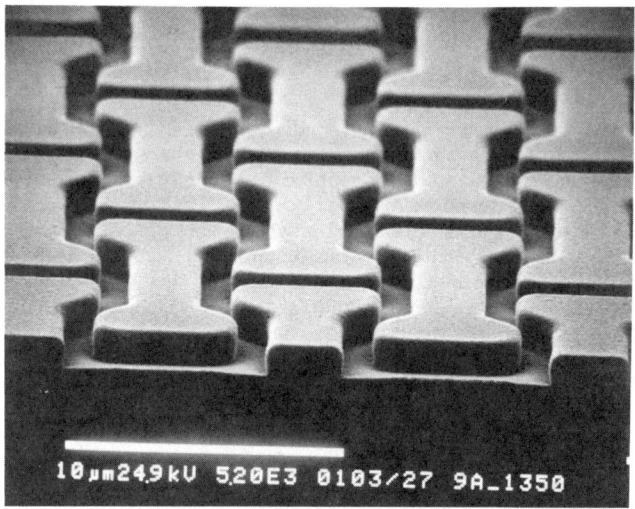

Figure 8. Scanning electron micrograph of an image obtained by projection printing using a Perkin-Elmer 500 tool.

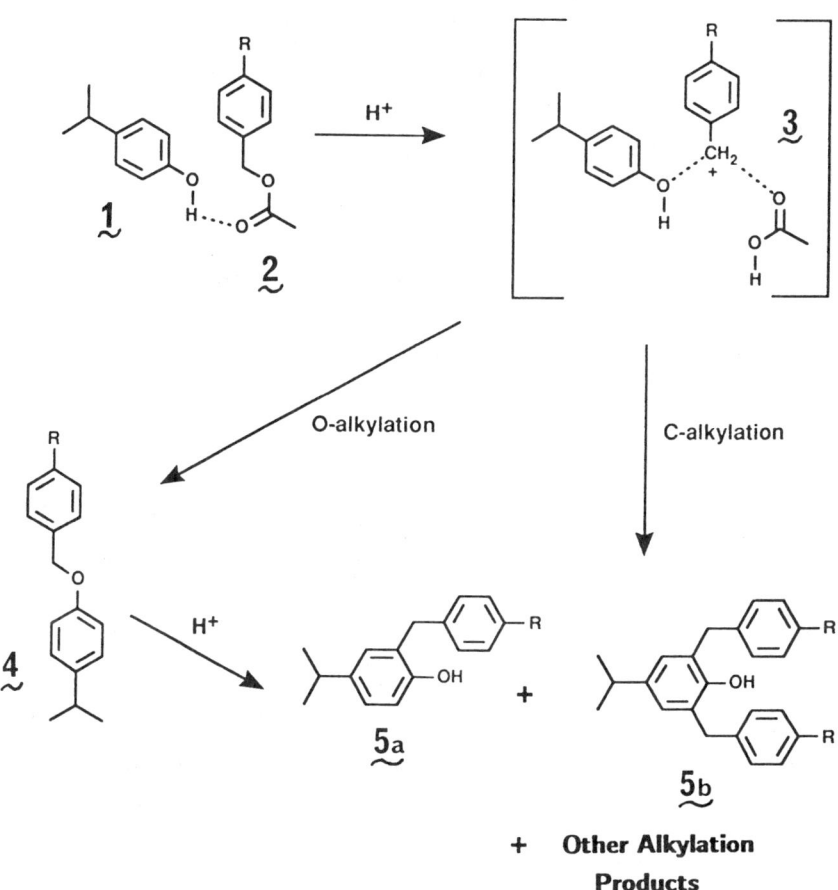

Figure 9. C-Alkylation and O-alkylation in the mechanism of crosslinking.

Experimental Procedure

Preparation of Copolymers of Vinylbenzylacetate and 4(t-BOC)styrene. 4-Vinylbenzyl acetate was obtained from Eastman Kodak or prepared from 4-chloromethylstyrene by displacement with acetate under phase transfer conditions. A solution of 5.0g of 4-t-butyloxycarbonyloxystyrene and 1.20g of 4-vinylbenzyl acetate in 6mL toluene containing 0.060g of azobisisobutyronitrile was heated overnight in an oil bath at 75°C. The polymer was diluted with toluene and precipitated into hexane. After drying the polymer (5.3g) was analyzed by NMR spectrometry and found to contain 20 mole% of vinylbenzyl acetate units.

Removal of the t-BOC Protecting Groups. The polymer prepared above was dissolved in glacial acetic acid and the solution was heated at reflux for 3 hours under nitrogen atmosphere. After cooling the deprotected polymer was recovered in near quantitative yield by precipitation into water. Spectroscopic analysis confirmed the complete removal of the t-BOC phenolic protecting groups while the acetate groups remained on the polymer.

Resist Modeling Experiments.
These experiments were carried out mainly by NMR spectrometry (^{1}H and ^{13}C) and high performance liquid chromatography. Stopped-flow and low temperature techniques were used in some of the NMR experiments. In all cases the reactions were aimed at monitoring the half-life of the starting materials as well as monitoring the products which were formed. Detailed procedures will appear shortly [15].

Conclusion

This study demonstrates clearly that excellent resist materials incorporating chemical amplification can be obtained using systems which operate on the basis of radiation-induced electrophilic aromatic substitution. The copolymers we have described show excellent sensitivities both under UV and E-beam exposure due to the catalytic nature of the reactions which are involved. The design we have used involving copolymers containing large amounts of free phenolic groups is particularly well suited to negative imaging since it provides high light transmission in the deep-UV and allows development in aqueous base without image distortion due to swelling. Resist modeling experiments have provided us with a thorough understanding of the key variables in the overall process and are useful in guiding the design of new and even more active materials based on the same principle. This study is continuing as several new polymers incorporating new design features have been created.

Acknowledgments

Financial support for this work was provided by IBM Corporation Materials and Processing Sciences Program. This support is gratefully acknowledged.

Literature Cited

[1]   C.G. Willson in "Introduction to Microlithography" L.F. Thompson, C.G. Willson, M.J. Bowden, Editors) ACS Symposium Series #219, Chapter 3, 87 (1983). See also "Materials for Microlithography", L.F. Thompson, C.G. Willson and J.M.J. Fréchet (Editors), ACS Symposium Series #266 (1984).

[2]   R.W. Blevins, R.C. Daly, S.R. Turner, "Lithographic Resists" Encyclopedia of Polymer Science & Engineering 2nd Ed., Vol. 9, John Wiley (1987)

[3]   C.G. Willson, H. Ito, J.M.J. Fréchet, T.G. Tessier, F.M. Houlihan; J. Electrochem. Soc. 133, 181-187 (1986).

[4] L.F. Thompson, E.D. Feit; R.D. Heidenreich, Polym. Eng. & Sci., 14 (7), 529 (1974).
[5] J.M.J. Fréchet, H. Ito, C.G. Willson; Proc. Microcircuit Engineering 1982, 260. C.G. Willson, H. Ito, J.M.J. Fréchet, F. Houlihan; Proc. IUPAC 28th Macromolecular Symposium 443, 1982. H. Ito, C.G. Willson, J.M.J. Fréchet; Proc. SPIE Int. Soc. Opt. Eng., 771, 24-31 (1987). H. Ito, C.G. Willson, J.M.J. Fréchet; U.S. Patent 4,491,628 (1985).
[6] J.M.J. Fréchet, N. Kallman, B. Kryczka, E. Eichler, F.M. Houlihan, C.G. Willson, Polymer Bulletin, 20, 427 (1988). J.M.J. Fréchet, E. Eichler, S. Gauthier, B. Kryczka, C.G. Willson; ACS Symposium Series, "The Effect of Radiation on High Technology Polymers" E. Reichmanis and J. O'Donnell, Editors, ACS Symposium Series #381 155 (1989).
[7] J.M.J. Fréchet, F. Bouchard, E. Eichler, F.M. Houlihan, T. Iizawa, B. Kryczka, C.G. Willson; Polymer Journal, 19, 31-49 (1987). J.M.J. Fréchet, F. Bouchard, F.M. Houlihan, B. Kryczka, E. Eichler, C.G. Willson, N. Clecak; J. Imaging Science, 30, 59-64 (1986). J.M.J. Fréchet, F. Bouchard, F.M. Houlihan, E. Eichler, B. Kryczka, C.G. Willson; Die Makromol. Chem. Rapid Commun., 7, 121-126 (1986).
[8] C.E. Hoyle, M. Keel, K.-J. Kim; Polymer 29, 18 (1988).
[9] W.E. Feely, J.C. Imhof, C.M. Stein, T.A. Fisher, M.W. Legenza; SPIE Regional Technical Conference on Photopolymers, Ellenville, N.Y., 49 (1985); and Polym. Eng. Sci., 16, 1101 (1986). W.E. Feely, Proc. SPIE 631, 48 (1986).
[10] T. Iwayanagi, T. Kohashi, S. Nonogaki, T. Matsuzawa, K. Douta, H. Yanazawa; IEEE Trans. Electron Devices, ED-28, 1306 (1981). S. Nonogaki, M. Hashimoto, T. Iwayanagi, H. Shiraishi; Proc. SPIE 539, 189 (1985).
[11] S. Matuszczak, B. Reck, H.D.H. Stover, C.G. Willson, J.M.J. Fréchet; Manuscript in preparation.
[12] B. Reck, R.D. Allen, R.J. Twieg, C.G. Willson, H.D.H. Stover, N.H. Li, J.M.J. Fréchet, Proceedings SPE Regional Technical Conference on Photopolymers. Ellenville, N.Y. 1988.
[13] M.J.S. Dewar, in "Molecular Rearrangements" (P. DeMayo, Editor), Interscience, NY, 295 (1963).
[14] L.S. Hart, C.R. Waddington, J. Chem. Soc., Perkin II, 1607 (1985).
[15] H.D.H. Stöver, S. Matuszczak, R. Chin, K. Shimizu, C.G. Willson, J.M.J. Fréchet, Polym Mat. Sci. Eng. Vol. 61 (1989) in press.

RECEIVED August 18, 1989

Chapter 6

Acid-Catalyzed Cross-Linking in Phenolic-Resin-Based Negative Resists

A. K. Berry[1], K. A. Graziano[1], L. E. Bogan, Jr.[1], and J. W. Thackeray[2]

[1]Rohm and Haas Company, 727 Norristown Road, Spring House, PA 19477
[2]Shipley Company, Inc., 2300 Washington Street, Newton, MA 02162

> Modelling studies of acid hardening resin (AHR) resists have shown that (a) 10^{-6} mole acid/g resist are required to induce sufficient crosslinking to effect a fifty-fold decrease in dissolution rate, and (b) the activation energies of crosslinking (E*) for 4 experimental crosslinkers are in the range of 21-57 kJ/mol. A crosslinker showing potential for high sensitivity, contrast and wide process latitude based on these studies also failed to provide high resolution. Diffusion of acid within the resist film is suggested to account for this observation. Studies involving different strong acid catalysts show a correlation between size of the acid molecule and its efficiency in crosslinking.

Chemical amplification has been employed successfully in the development of a new class of high sensitivity, high resolution resists. These systems are catalytic in nature in that a single photoevent, often the generation of a strong acid, gives rise to multiple chemical reactions (1). In a positive resist formulation, for example, photogenerated acid catalytically destroys a dissolution inhibitor (2). Alternatively, there exists a class of chemically amplified negative resists in which photogenerated acid catalyzes a series of chemical reactions leading to crosslinking (3,4). In this case, the photoacid is not consumed in the crosslinking reaction and is "amplified" in proportion to the number of crosslinks that one molecule of catalyst can produce.

Negative tone resists based on acid hardening resin (AHR) chemistry are three component systems (3,4) comprised of a novolak resin, a melamine crosslinking agent, and a radiation sensitive acid generator (RSAG). Classes of compounds which have been used to generate a strong acid upon exposure include the well studied onium salts (5), a family of nitrobenzyl esters (6), and a variety of halogenated organic compounds including 1,1-bis(p-chlorophenyl)-2,2,2-trichloroethane (DDT) (7) and substituted s-triazine derivatives (8). The sequence leading to crosslink formation begins with the generation of acid upon exposure. The protonated melamine liberates a molecule of alcohol upon heating to leave a nitrogen stabilized carbonium ion, as shown in Scheme 1. Alkylation of the novolak then occurs at either the phenolic oxygen (as shown) or at a carbon on the aromatic ring, and a proton is regenerated. There are several reactive sites on the melamine, allowing it to react more than once per molecule to give the crosslinked polymer.

The crosslinking chemistry of AHR resists is fundamentally similar to that found in thermoset coatings (9). Although reactions based on hydroxy-functional resins and melamine crosslinkers have been widely investigated, controversy still exists concerning the crosslinking mechanism. Model studies in both solution (10) and film (11-13) have demonstrated the effects of acid catalyst concentration and strength and the effects of cure temperature on the rate constants of the crosslinking reaction. These studies have led to the determination of the reaction order in acid and the activation energy for a variety of acid catalysts and hydroxy-functional polymers.

The effects of acid catalyst concentration and post exposure bake (PEB) temperature and time are of great concern in AHR resists as well. It has been shown that the sensitivity of these resists depends strongly on PEB conditions (14). Recently Seligson and Das (15) proposed a model for thermally activated resists which explains the observed variations of AHR resist sensitivity with PEB conditions. Their model also allows prediction of any dose-dependent parameter in terms of changes in PEB time and temperature. They have shown for example, that Microposit SAL601-ER7 exhibits low enough activation energy to be fairly insensitive to small changes in the PEB step.

In the present study we use the acid catalyzed crosslinking reaction directly (Scheme 1) as a means to estimate the amount of acid generated in a typical AHR resist. We then use the model proposed by Seligson and coworkers to determine the activation energy of crosslinking for a variety of crosslinking resins as they might be used in a resist formulation. Where

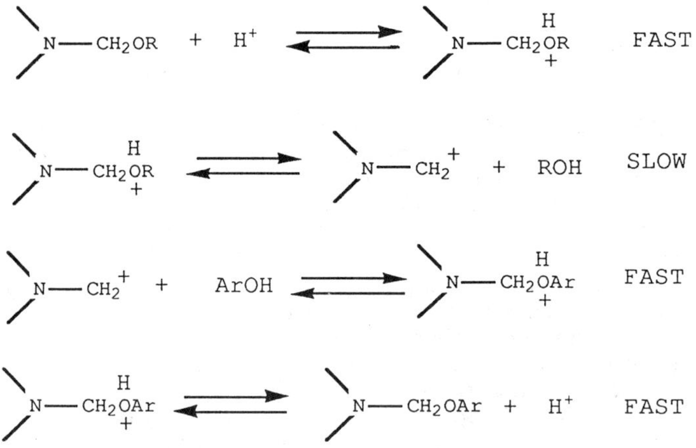

$$\ce{>N-CH2OR + H+ <=> >N(+)(H)-CH2OR} \quad \text{FAST}$$

$$\ce{>N(+)(H)-CH2OR <=> >N-CH2+ + ROH} \quad \text{SLOW}$$

$$\ce{>N-CH2+ + ArOH <=> >N(+)(H)-CH2OAr} \quad \text{FAST}$$

$$\ce{>N(+)(H)-CH2OAr <=> >N-CH2OAr + H+} \quad \text{FAST}$$

Scheme 1. Mechanism of crosslink formation.

possible, resist activation energies will be compared to lithographic performance.

Experimental

The polymer used in this study was a m,p-cresol novolak (M_w=14000, M_n =1300). Melamine resins were obtained from American Cyanamid Co. Polymer/crosslinker formulations were prepared using 23 wt % novolak in cellosolve acetate and 15 parts per hundred crosslinker based on novolak solids. p-Toluenesulfonic acid was added as a 1 wt % solution in cellosolve acetate. Films were spun to 1.0 micron thickness and baked on a Tempchuck System, TC100 vacuum hotplate. Wafers were developed in metal ion free tetramethylammonium hydroxide (TMAH) at concentrations giving 300-500 Å/sec uncrosslinked dissolution rates over a 95 - 120°C temperature range. Dissolution rates were measured by laser interferometry.

Results and Discussion

The crosslinkers examined in this study were aminoplast resins **1-4** selected from melamine-formaldehyde, urea-formaldehyde, benzoguanamine-formaldehyde, and glycoluril-formaldehyde resins, all of which undergo the crosslinking sequence shown in Scheme 1. The response of these crosslinkers to acid catalysis in thin films is compared on a relative basis to the well studied methylated melamine, **1** (9-11).

1

Acid Concentration Required for Crosslinking Resist Films. The strong acid catalyzed reaction of methylated melamine **1** with novolak in a film was studied first in order to estimate the amount of acid generated in experimental AHR resists upon exposure. Incremental amounts of a 1% solution of p-toluenesulfonic acid (pTSA) were added to a resist solution which contained no RSAG. The solutions were coated onto wafers and subjected to a typical bake cycle. The dissolution rate was then monitored by laser interferometry. By using several concentrations of acid, the lithographic potential curve (LP = log Ru - log Rc, where Ru = dissolution rate of uncrosslinked resist and Rc = dissolution rate of crosslinked resist) shown in Figure 1 was constructed. The acid concentration corresponding to 90% film retention, or LP = 1.69, was found to be 4.3 x 10^{-6} mole of acid per gram of resist. Figure 1 also shows a DRM generated LP curve for exposure of a DUV resist containing crosslinker **1**.

The response of crosslinkers **1-4** to pTSA catalysis as a function of acid concentration is shown in Figure 2 for a 75°C/1 minute softbake and 105°C/1 minute hardbake cycle. The concentration of acid required to crosslink these films to a given LP is an indication of the relative resist sensitivities, while the steepness of the LP curves reflects resist contrast. Crosslinkers **3** and **4** are about twice as sensitive to pTSA catalysis as **1**, while **2** requires a higher concentration of pTSA for crosslinking. Furthermore, the steepness of the curve for **3** suggests that it would show higher contrast in a resist formulation.

Determination of Activation Energy of Crosslinking. In any class of thermally activated AHR resists, it is important to determine how resist sensitivity is affected by variations in the post exposure bake step. If a resist is relatively insensitive to slight temperature variations, then photospeed and linewidth can be easily controlled from wafer to wafer. A kinetic model describing the crosslinking chemistry of acid hardening resist systems has been proposed by Seligson and coworkers (_15_). This model was shown to be a useful analytical tool in predicting resist response to changes in incident dose, PEB time and PEB temperature in AHR resists. In addition, two kinetically useful parameters, the apparent activation energy of crosslinking and the reaction order for the acid catalyst, can be determined. We use this model as a tool in the early stages of resist component evaluation as one indicator of potential resist performance.

The activation energy, E*, and the kinetic order, m, may be derived from the following equation:

$$D_{eff} = D \exp(-E^*/RT) \, (t/t_o)^{1/m} \qquad (1)$$

where D_{eff} is the effective dose (_15_), D is the incident dose, T is PEB or hardbake temperature (K), R is the Gas constant, t/t_o is the normalized PEB time, E* is the activation energy of crosslinking, and m is the reaction order in acid. Experimentally determined values of E* (42 kJ/mol) and m (2.5 - 3.0) using this model (_16_) are

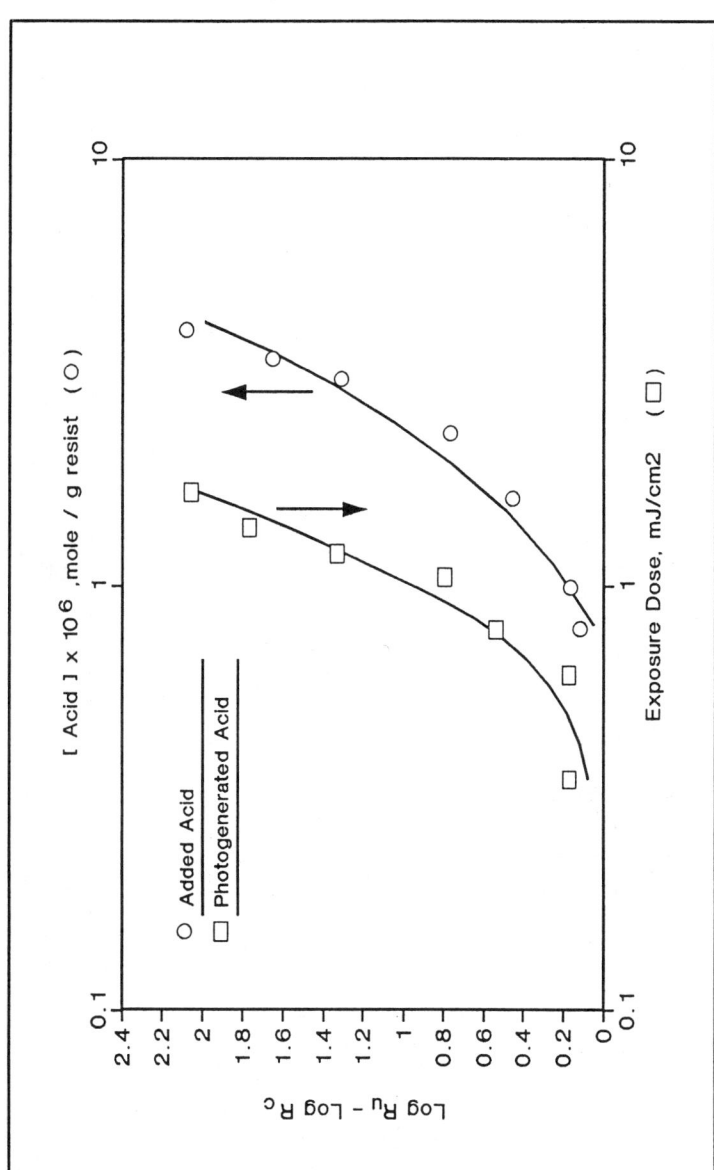

Figure 1. Comparison of crosslinking catalyzed by added acid and photogenerated acid for crosslinker **1**.

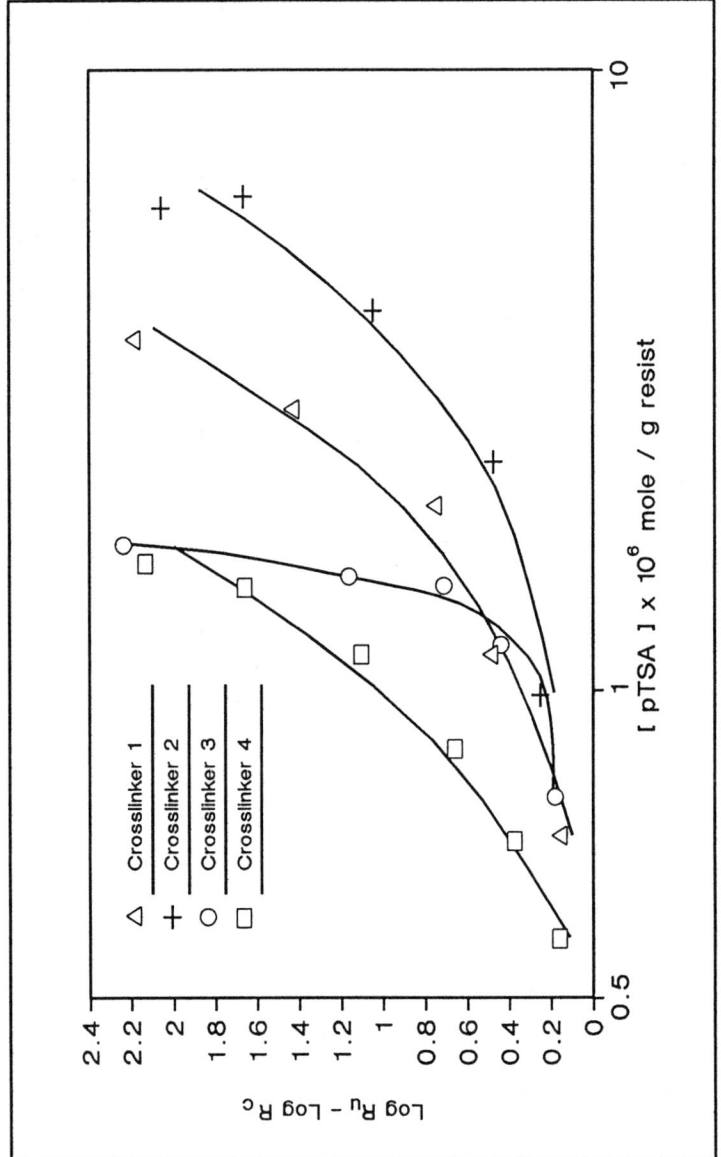

Figure 2. Lithographic potential for crosslinkers **1-4** with added acid.

within the range found in other model studies in films (11-13).

Evaluation of E* using Equation 1 required determining the concentration of acid, or "dose" of acid, necessary to crosslink the polymer over a range of processing conditions. The hardbake temperature was varied from 95 to 120°C and the bake time from 30 to 240 seconds. The relative degree of crosslinking in each sample was determined based on dissolution rate measurements. A 50-fold decrease in dissolution rate between the uncrosslinked and crosslinked films was chosen to correspond to lithographically useful crosslinking for the purpose of this study. E* was determined by generating an Arrhenius type plot of log $[H^+]$, corresponding to log dose in Equation 1, versus $1/T$ at a constant hardbake time.

Crosslinker **1** was examined first to test the validity of the Seligson model in conjunction with adding acid to polymer/crosslinker solutions. The concentration of pTSA (moles acid / gram resist solution) required for LP = 1.69 (a 50-fold decrease in Rc) was determined over the bake time and temperature range stated. From a log-log plot of [pTSA] versus hardbake time at constant temperature, the order of reaction in acid, m, is calculated to be 2.6, consistent with other studies (10). An Arrhenius treatment of the data plotting log [pTSA] versus $1/T$ is linear, supporting the activation energy description of the reaction. The activation energy is calculated from the slope of the line, and is found to be 54 kJ/mol for **1**, in agreement with other model studies in films (11).

Analogous experiments were performed using crosslinkers **2 - 4** and the results are shown in Figure 3 and Table I. The order in acid, m, is similar for all crosslinkers. The E* value found for the butylated crosslinker **3** is considerably lower than those found for the other crosslinkers. This result was surprising in view of the crosslinking mechanism shown in Scheme 1, in which the removal of ROH is the slow step. While **3** is in the aminoplast family and expected to undergo crosslinking as shown in Scheme 1, the particularly low E* could suggest another lower energy reaction path available to **3** which also leads to crosslinking (17).

The kinetics of melamine crosslinking have been found to depend strongly on the reaction medium (9-11). Model studies in solution for example reveal an activation energy of 96 kJ/mol for the reaction of **1** with primary and secondary alcohols, compared with 52 kJ/mol in a polymer network. Meijer (10) has suggested that the key to this difference lies in the removal of the condensation product ROH by evaporation, particularly in crosslinking reactions involving equilibria. Further

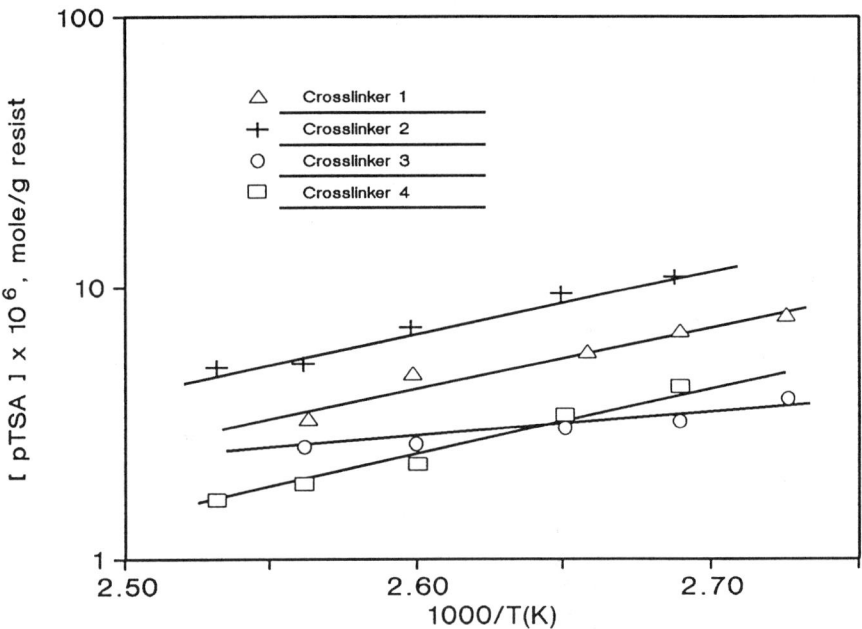

Figure 3. Arrhenius treatment of data for experimental crosslinkers.

TABLE I. Activation Energy of Crosslinking in Thin Films

$$\ce{>N-CH_2-OR + ArOH ->[H^+][\Delta] >N-CH_2-OAr + ROH}$$

Crosslinker	R	m	E*
1	Me	2.6	54 ± 5 kJ/mol
2	Me	2.7	57 ± 5 kJ/mol
3	Bu	3.1	21 ± 2 kJ/mol
4	Bu	2.8	43 ± 4 kJ/mol

evidence for the importance of the evaporation rate is seen in a comparison of methoxymethyl and butoxymethyl melamine crosslinking reactions in films (11). Although the latter might be expected to have a faster forward reaction, it is found to react more slowly than **1**. The slow diffusion of BuOH out of the film compared to MeOH has been invoked to account for this difference. In previous studies of aminoplast crosslinking reactions in films, the film thickness has been 25 to 50 microns, however for the thinner 1 micron films in this study, it is possible that the contribution of ROH evaporation to the overall rate becomes less significant.

From a lithographic standpoint, the implications of low E* are better control of critical dimension (CD), and a liberal process window. It is evident from Figure 3 that the slope of the Arrhenius plot gives a relative indication of the PEB latitude of a resist containing that crosslinker. Seligson and coworkers have used the kinetic model (Equation 1) along with linewidth measurements to quantify resist process requirements (15,16). They have found that a resist with E* = 42 kJ/mol has acceptable PEB latitude within a range of 2°C, while a resist with E* = 20 kJ/mol may be processed within a 4°C range. Of the crosslinkers evaluated here, **3** appears to offer the widest process window.

Lithographic Characteristics. Based on the potential of crosslinker **3** to show high sensitivity and contrast and wide process latitude, it was of interest to evaluate its lithographic capability, using crosslinker **1** as the standard for comparison. Crosslinkers **1** and **3** (equal weight) were each incorporated into otherwise identical experimental AHR resist formulations. E-beam exposures were performed so that differences in DUV absorbance characteristics of the crosslinkers could be ignored. The e-beam sensitivities of the resists containing crosslinkers **1** and **3** were 6.2 and 4.2 μC/cm^2, respectively, and the results of optimal e-beam exposure are shown in Figure 4. Submicron features with steep sidewalls are readily obtained with the resist containing crosslinker **1**, while the 0.5 micron lines are rounded and resolution is generally poor with crosslinker **3**. In fact, the lithographic results are opposite to predictions based on these studies.

The diffusion of acid within a resist film is a critical issue with respect to resolution capability, and this is something which the model studies did not directly address. In a chemically amplified system a single photogenerated acid molecule initiates many chemical events. Thus there has to be some mobility associated with the acid. The low E* value and high sensitivity found for crosslinker **3** suggest that crosslinking occurs rapidly compared to crosslinker **1**. It seems that under identical process conditions, the same acid is more mobile in the film containing **3**. The poor resolution associated with crosslinker **3** then could be due to the higher rate of acid diffusion within that film.

Effect of Acid on Lithographically Useful Crosslinking.
The crosslinking reaction shown in Scheme 1 is known to respond best to strong acid catalysis. Since the nature of the acid generated upon exposure varies from resist to resist, we have examined the following strong acid catalysts, many of which are utilized in various chemically amplified resists: pTSA, CH_3SO_3H, HCl, HBr, and HPF_6 ([5-8]). The sulfonic and halogen acids are strong acids with pK_a values less than 0, ([18,19]) while HPF_6 is a much stronger super acid ([20]).

These acids were compared on the basis of their efficiency to crosslink a novolak / crosslinker **1** film. The amount of each acid required to give LP = 1.69 with a 105°C/1 minute postbake are summarized in Table II.

TABLE II. Acid Concentration for Crosslinking

Acid	pK_a ([18,19])	[H+], mole/g resist
pTSA	−1	4.3×10^{-6}
CH_3SO_3H	−1	3.6×10^{-6}
HPF_6	([20])	2.5×10^{-6}
HBr	−9	1.7×10^{-6}
HCl	−7	1.5×10^{-6}

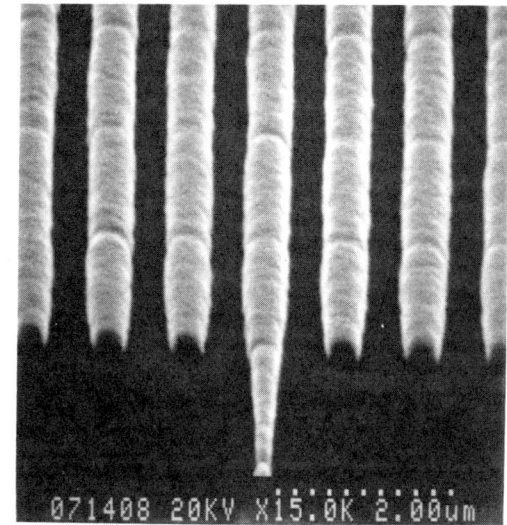

Crosslinker 1

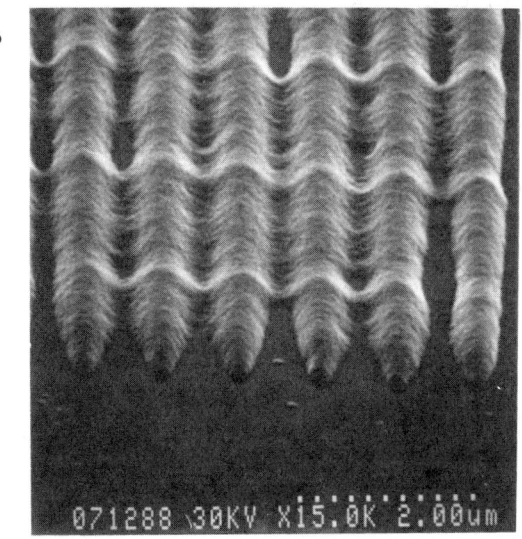

Crosslinker 3

Figure 4. SEM micrographs depicting 0.5 μm l/s with e-beam exposure of (a) resist formulated with crosslinker **1** (6.2 μC//cm^2), (b) resist formulated with crosslinker **3** (4.2 μC/cm^2).

Based on these data, HCl seems to be the most effective at catalyzing crosslinking. This is somewhat surprising, since HPF_6 is considered to be a much stronger acid. It is possible that the size of the acid influences the efficiency of crosslinking, as HCl is a much smaller molecule and could be more mobile in a film and thus capable of initiating more crosslinks. In addition, it is likely that counterion effects in the novolak /crosslinker medium also play an important role in determining the overall crosslinking efficiency. A similar set of experiments with varying the identity of the acid was performed using crosslinker **3**. Surprisingly, HCl did not seem to initiate crosslinking, while the amounts of HBr and pTSA required for crosslinking were comparable, 2.3×10^{-6} mole acid/g resist. When 5×10^{-6} mole HCl/g resist was added to the crosslinker **3** system and baked in the presence of HCl vapor, the resulting film was crosslinked and showed negligible dissolution. This suggests that either the amount of HCl required for crosslinking with **3** is greater than 10^{-6} mole/g resist, or that under normal bake conditions HCl readily diffuses out of a film containing **3** before it has a chance to react. If other acids show similar high mobility in films containing crosslinker **3**, this could explain both the high sensitivity and poor resolution associated with resists containing crosslinker **3**.

Conclusions

The acid catalyzed crosslinking reaction of aminoplasts with hydroxy functional polymers has been used to determine the amount of acid generated in a typical AHR photoresist and to evaluate the activation energy and PEB latitude associated with several crosslinkers. A crosslinker which demonstrated the greatest potential for high sensitivity and contrast and a wide process window, also failed to provide high resolution. It is suggested that the mobility of acid within the film contributes to both of these findings. We are continuing to investigate the effects of both acid and crosslinker on resist performance.

Literature Cited

1. Ito, H.; Willson, C.G.; Frechet, J.M.J. *SPE Reg. Tech. Conf. on Photopolymers*, Nov. 1982.
2. Willson, C.G.; Ito, H.; Frechet, J.M.J.; Tessier, T.G.; Houlihan, F. *J. Electrochem. Soc.* 1986, *133*, 181.
3. Feely, W.E.; Imhof, J.C.; Stein, C.M. *Polym. Eng. Sci.* 1986, *26*, 1101.

4. Burns, A.; Luethje, H.; Vollenbroek, F.A.; Spiertz, E.J. *Microelectronic Eng.* 1987, *6*, 467.
5. Crivello, J.V. *SPE Reg. Tech. Conf. on Photopolymers*, Nov. 1982.
6. Houlihan, F.M.; Shugard, A.; Gooden, R.; Reichmanis, E. *Proc. of SPIE*, 1988, *920*, 67.
7. Feely, W.E. Eur. Patent. Appl. 232,972, 1987.
8. Buhr, G. U.S. Patent 4,189,323, 1980.
9. Blank, W.J. *J. Coat. Technol.* 1979, *51*(656), 61.
10. Meijer, E.W. *J. Polym. Sci.* 1986, *A24*, 2199, and references therein.
11. Bauer, D.R.; Budde, G.F. *J. Appl. Polym. Sci.* 1983, *28*, 253.
12. Bauer, D.R.; Dickie, R.A. *J. Polym. Sci.: Polym. Phys. Ed.* 1980, *18*, 1997.
13. Bauer, D.R.; Dickie, R.A. . *Polym. Sci.: Polym. Phys. Ed.* 1980, *18*, 2015.
14. DeGrandpre, M.; Graziano, K.; Thompson, S.D.; Liu, H.; Blum, L. *Proc. of SPIE* 1988, *923*, 19.
15. Seligson, D.; Das, S.; Gaw, H.; Pianetta, P.; *J. Vac. Sci. Tech.* 1988, *6*(6), 2303.
16. Seligson, S.; Das, S. *SPE Reg. Tech. Conf. on Photopolymers* Nov. 1988.
17. Petersen, H. *Text. Res. J.* 1971, *41*, 239.
18. Albert, A.; Serjeant, E.P. *The Determination of Ionization Constants*; Chapman and Hall: New York, 1984, 144-64.
19. Perrin, D.D. *Ionisation Constants of Inorganic Acids and Bases in Aqueous Solution*; Pergamon Press: New York, 1982.
20. Olah, G.A.; Prakash, G.K.S.; Sommer, J. *Superacids*; John Wiley and Sons, Inc.: New York, 1985, 24-7.

RECEIVED August 10, 1989

Chapter 7

New Design for Self-Developing Imaging Systems Based on Thermally Labile Polyformals

Jean M. J. Fréchet[1], C. Grant Willson[2], T. Iizawa[3,4], T. Nishikubo[3], K. Igarashi[3], and J. Fahey[1]

[1]Department of Chemistry, Baker Laboratory, Cornell University, Ithaca, NY 14853–1301
[2]IBM Research Division, Almaden Research Center, 650 Harry Road, San Jose, CA 95120–6099
[3]Department of Applied Chemistry, Kanagawa University, Yokohama 221, Japan

>The polycondensation of bis-allylic or bis-benzylic diols with dibromomethane under phase transfer catalysis may be used to generate some polyformal polymers and copolymers which depolymerize readily when heated or when subjected to acidolysis. The polymers are best prepared by condensation of diols such as 1,4-dihydroxy-2-cyclohexene or 1,4-dihydroxy-1,2,3,4-tetrahydronaphthalene and CH_2Br_2 under phase transfer conditions. The polymers which are obtained by this process generally have fairly broad polydispersities and are stable to temperatures higher than 150°C. Under acid catalysis, they decompose below 80°C to afford mainly the aromatic elimination product (benzene or naphthalene), and formaldehyde hydrate. As the polymers absorb very weakly even in the deep-UV ($E_{max} < 300$), imaging is possible through formulations incorporating both the polyformals and a small amount of a triarylsulfonium salt or similar photoacid generator. The polyformals can be used as self-developing and chemically amplified imaging systems which operate in the deep-UV and possess good sensitivities due to the catalytic nature of their photoinitiated thermal decomposition.

The initial work on a new generation of deep-UV and E-beam resist materials which incorporate chemical amplification [1] was carried out in 1979 when Fréchet and Willson first prepared poly(4-t-butyloxycarbonyloxystyrene), poly(4-allyloxystyrene, end-capped poly(o-phthalaldehyde), and copolymers of o-phthalaldehyde and its 3-nitro derivative. The first of these materials was designed for its ability to lose its t-BOC phenolic protecting groups under thermolytic conditions, the second for its potential to undergo a catalyzed Claisen rearrangement, while the two cyclic polyacetals would undergo chain depolymerization under a variety of conditions. In particular the copolymers containing 3-nitro-phthalaldehyde units have o-nitrobenzyl acetal groups which are readily cleaved by UV irradiation [2] thereby rendering the copolymers photodepolymerizable (Figure 1).

This basic approach to chemical amplification was subsequently extended by Ito et al. [3] through resist formulations incorporating appropriately chosen triaryl sulfonium or diaryliodonium salt. For example, end-capped poly(phthalaldehyde) used in combination with an onium salt photoacid generator is an excellent self-

[4]Current address: Department of Chemical Engineering, Hiroshima University, Japan

developing imaging material. While this material is interesting, its usefulness is restricted as depolymerization of the polymer after exposure to radiation does not require any thermal activation; therefore, the phthalaldehyde monomer may be evolved within the confines of the exposure tool where it might well interfere with optical and mechanical components.

This study will show an approach to materials which have many of the desirable features of the poly(phthaladehyde)-onium salt imaging system but do not suffer from the problem of spontaneous gaseous material evolution upon irradiation. The new polyformal-based imaging systems all have a "built-in" thermal activation requirement which allows for image self-development *outside* of the exposure tool.

Imaging via Thermolytic Main-Chain Cleavage.

A significant part of our recent work with imaging systems which incorporate chemical amplification has involved the design of polymers which can undergo *thermally activated* multiple main-chain cleavages as the result of a phototriggered process.

Imaging via Main-chain Cleavage of Polycarbonates, Polyesters and Polyethers.

We have designed, prepared, and tested dozens of new imaging materials based on polycarbonates [4-10], polyethers [11] and polyesters [12] which are all susceptible to *acid-catalyzed* thermolytic cleavage. With all of these systems, imaging is accomplished through irradiation of a film of the polymer containing a small amount of a photoactive compound which can generate acid upon irradiation. Imagewise exposure to a source of the appropriate wavelength causes formation of a *latent image* consisting of acid dispersed only in those areas of the polymer film which were exposed to radiation. The latent image can *subsequently* be developed by a baking step in which the latent image is provided with the activation energy which is required for the catalyzed thermolysis of the polymer to occur. The unexposed areas of the film do not undergo thermolysis as the uncatalyzed process requires much higher baking temperatures in order to take place. If the fragments resulting from multiple main-chain cleavages are somewhat volatile, self-development may be achieved by carrying out the thermolytic development step in an evacuated environment.

Mechanism of Cleavage of Polyesters and Polycarbonates.

It is useful at this point to review the mechanism of cleavage of esters and carbonates [13]. For the sake of simplicity, this will be considered only in the context of a polycarbonate, and only a single cleavage step will be considered. It must be remembered that a similar mechanism would also apply to appropriately designed polyesters, and that, in the case of our polymers, the cleavage step would eventually involve all the carbonate functionalities which constitute the polymer chain; thus a thermolytic cleavage would result in complete depolymerization of a chain. The *uncatalyzed* thermolytic cleavage of a carbonate proceeds through a cis-elimination which requires that the carbonate possess a β-hydrogen. Early studies of this reaction have shown that the transition state has significant polarity, (suggesting much E_1-like character for the elimination) especially in the case of carbonates. While it is well-known that the ease of elimination increases from primary to tertiary carbonates [13], we have based our novel designs of polymers containing bis-allylic or bis-benzylic carbonates on the premise that *strong allylic or benzylic stabilization* of the polar transition state would occur, thereby greatly facilitating the reaction. This assumption proved to be correct as the elimination reaction of allylic (Scheme I) and benzylic polycarbonates is qualitatively as facile as that of polycarbonates of tertiary alcohols [7, 9]. If a catalytic amount of acid is present, the thermolysis reaction is notably facilitated and therefore occurs at a much reduced temperature. This is key to our resist

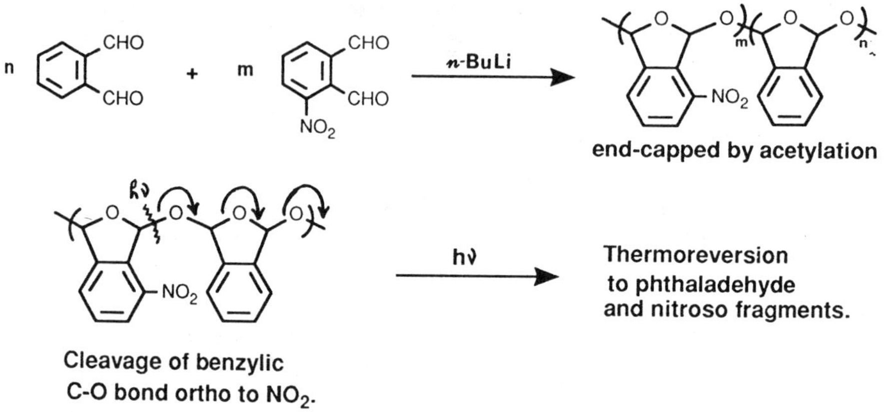

Figure 1. Preparation and photocleavage of a poly(phthalaldehyde)

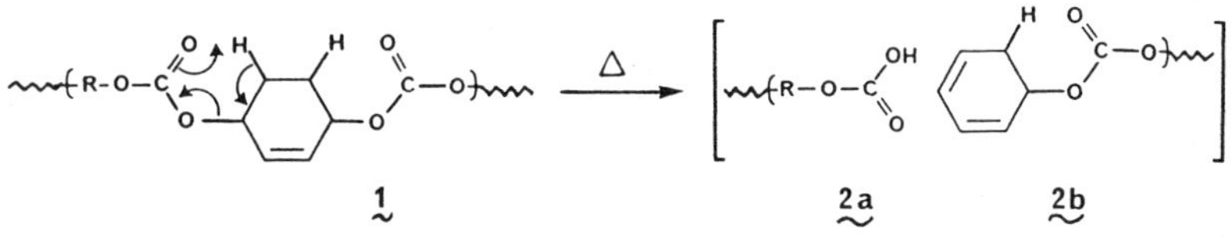

Scheme I

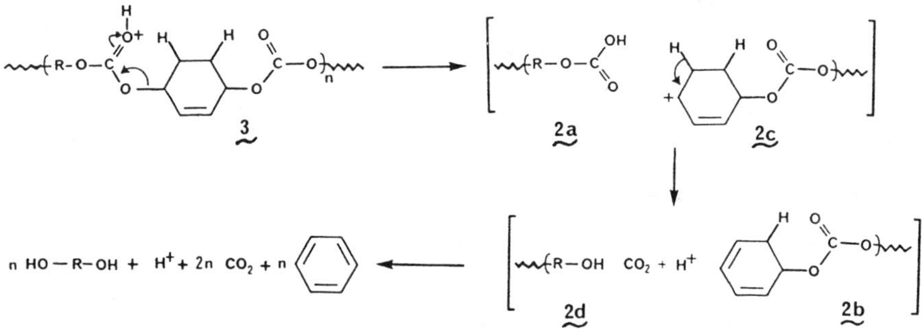

Scheme II

design as acid is generated within the coating by irradiation rendering those areas which have been irradiated more susceptible to low temperature thermolytic decomposition than areas which have remained unexposed.

Scheme II shows the E1-like elimination process which may prevail in the case of the acid-catalyzed thermolytic cleavage of polycarbonate **1**. The reaction starts with protonation of a carbonate carbonyl group to afford **3**, this is followed by breaking of the adjacent allylic carbon-oxygen bond to produce two fragments: a fragment containing a monoester of carbonic acid **2a**, and another containing an allylic carbocationic moiety **2c**. Elimination of a proton from this carbocationic moiety results in regeneration of the acid catalyst (the "eliminated" proton) and formation of a terminal diene-containing fragment **2b**. The unstable monoester of carbonic acid **2a** decarboxylates releasing a terminal alcohol fragment **2d**. The process continues at other carbonate sites with complete breakdown of the polymer chain resulting in the eventual release of benzene (two successive eliminations on the bis-allylic moiety), additional carbon dioxide, and a diol HO-R-OH. The protons initially generated by irradiation are not consumed in the process (except by possible side-reactions or impurities) which explains the high quantum efficiency or chemical amplification of this resist design.

In contrast, the *clean* thermolysis of appropriately chosen ethers into alkene and alcohol components was little known until our work in this area demonstrated its applicability to transformations involving both side-chain ether groups [14] or the main chain of polyethers [11].

<u>Thermolytic Cleavage of Allylic and Benzylic Ethers and Polyethers.</u> The thermal lability of the types of ethers which are of interest in the context of this study was discovered during a thorough study of the application of certain benzylic carbonates as labile protecting groups for alcohols and phenols [15]. It was observed that, under acid catalysis, the bis-methylcarbonate derivative of 1-(4-hydroxyphenyl)ethanol (**4**, in Equation 1) was transformed into the benzylic ether (**5**) which then underwent clean acid-catalyzed thermolysis to the corresponding styrene (**6**) in very high yield [14].

In this reaction, only the benzylic carbonate can react readily since cleavage of the benzylic carbon-oxygen bond leads to a stabilized carbocation while no such stabilization would exist for the hypothetical products which would be obtained by cleavage of the phenyl carbonate. The benzylic carbocation intermediate which is formed can either eliminate to the corresponding styrene **6** or recombine with the nucleophilic methanol which is formed by decarboxylation. It is the latter reaction which appears to prevail at room temperature as the ether **5** can be isolated in excellent yield. Subsequent heating in the presence of acid catalyst drives the reaction to the elimination product **6** and free methanol.

The study of several other model ethers derived from benzylic and allylic alcohols for which α-elimination of a proton is possible confirmed that this reaction is general and occurs at low temperatures in the presence of strong acid.

The suitability of ethers derived from 1,4-dihydroxy-1,2,3,4-tetrahydronaphthalene (**DHTN**) in the design of polymers susceptible to catalyzed thermolytic cleavage is demonstrated by the behavior of its bis-p-nitrophenyl ether derivative upon treatment by a trace of acid. Figure 2, curve A, shows the ^{1}H-NMR spectrum of the starting compound, while curve B shows the product which is obtained upon addition of triflic acid. It is readily seen from these spectra that quantitative cleavage into naphthalene and p-nitrophenol is obtained as elimination occurs easily to afford the aromatic product. The driving force in this reaction is the facile aromatization which produces naphthalene.

Recent work [16] with a variety of polyethers derived from DHTN as well as from 1,4-dihydroxy-2-cyclohexene (**DHCH**) by phase transfer catalyzed polycondensations has shown that these ethers were readily subjected to acid-catalyzed thermolysis and were therefore suitable for use in dry-developing imaging systems. During the study of these polyethers (e.g. structure **7**) it was found that their preparation under phase transfer catalysis in dichloromethane solution frequently resulted in the incorporation of some dioxymethylene units as in structure **8**. Such incorporation of a few polyformal moieties [17,18] in polyether **7** does not detract from its ability to undergo acidolytic or thermolytic cleavage as polymer chains containing units such as **8** can depolymerize readily to produce benzene, bisphenol A and formaldehyde hydrate. This finding suggests that polymers obtained from the reaction of allylic or benzylic diols with dihalomethane should also meet our basic requirements for facile thermolytic elimination.

<u>Preparation of the Active Polyformals under Phase-transfer Catalysis.</u> The polycondensation of DHTN or DHCH with dibromo- or dichloro-methane was carried out under phase transfer catalysis as shown in equations 2 and 3. Best results were obtained with dibromomethane, 60% aqueous KOH and tetrabutyl ammonium bromide.

Table 1 contains details of experimental conditions used in the preparation of the polyethers. It can be seen that the reduced viscosity of the polymers which were obtained increased with the concentration of the aqueous KOH solution and was inversely related to the amount of dibromomethane used in the reaction.

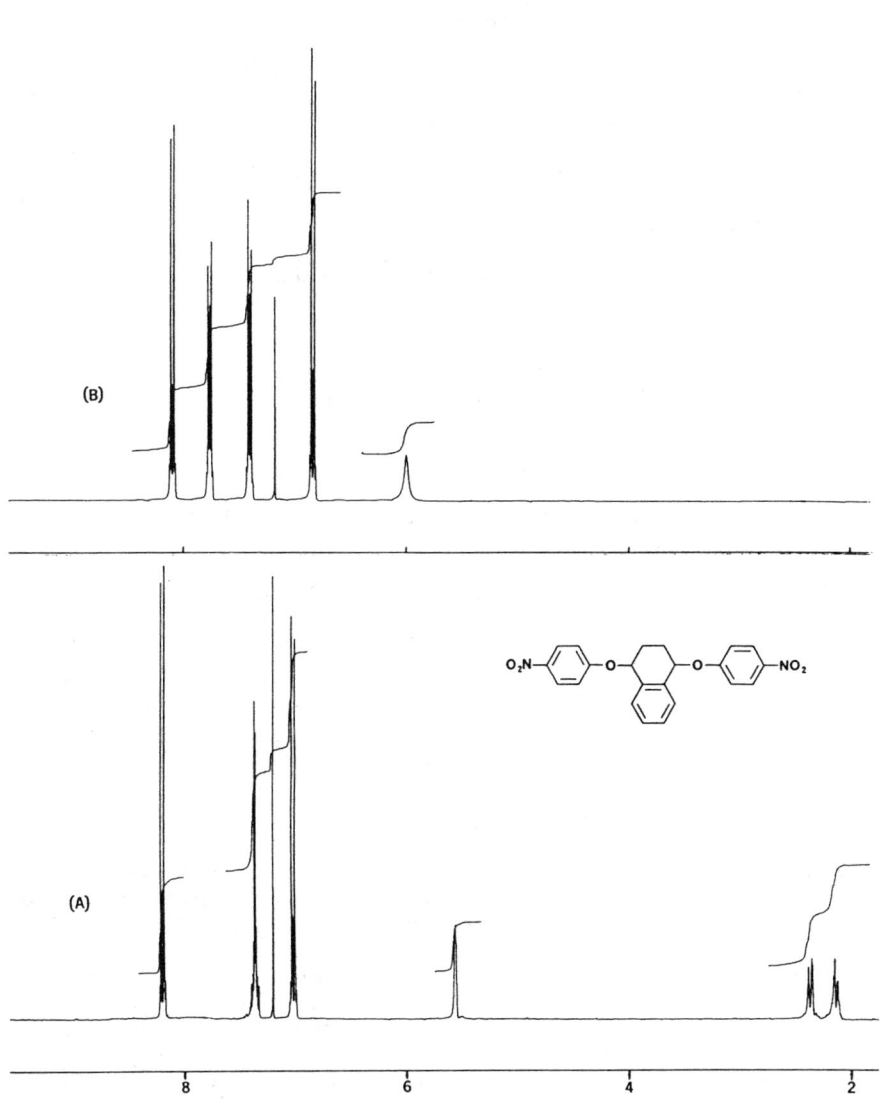

Figure 2. Acidolysis of a bis-benzylic ether model compound

Thermogravimetric analysis of polymer **9** (Figure 3) showed 5% loss of weight at 250°C, 50% loss at 273°C and 3.9% weight remaining at the end of the heating cycle (350°C). Polymer **10** had a higher decomposition temperature with 5% loss at 260°C and 50% loss at 305°C while 6.3% of the starting mass remained at 400°C. DSC studies of polymers **9** and **10** show that they have very low glass transition temperatures near 62 and 30°C respectively. These low Tg values do not detract from the intrinsic value of the polymers but preclude their use in imaging processes where these temperatures might be exceeded significantly. GPC measurements (polystyrene standards) showed that the polyformals had relatively high molecular weights with a broad dispersity. Typically, polymer **9** having a reduced viscosity of 0.27 showed a peak molecular weight of 30,000 with M_n = 6,000 and M_w = 54,000. Similarly, polymer **10** (reduced viscosity = 0.25) had M_n = 7,500 and M_w = 37,000 (by GPC with polystyrene standards).

Table 1. Polycondensation of DHTN and DHCH with dihalomethane

Diol[a]	Dihalomethane	Base[b]	T (°C)	TBAB (mole%)	Yield (%)	n_{sp}/C[c]
DHTN	CH_2Br_2 (20mL)	50% KOH	80	100	trace	-
DHTN	CH_2Br_2 (20mL)	50% KOH	60	100	14	0.06
DHTN	CH_2Br_2 (20mL)	50% KOH	25	100	46	0.08
DHTN	CH_2Br_2 (20mL)	60% KOH	25	100	91	0.13
DHTN	CH_2Cl_2 (20mL)	60% KOH	25	100	48	0.06
DHTN	CH_2Br_2 (20mL)	60% KOH	25	10	94	0.13
DHTN	CH_2Br_2 (2mL)	60% KOH	25	10	95	0.27
DHCH	CH_2Br_2 (20mL)	60% KOH	25	10	35	0.08
DHCH	CH_2Br_2 (2mL)	60% KOH	25	10	62	0.25

[a] *Experiment on 10 mmole scale.*
[b] *Aqueous solution, 20mL.*
[c] *Reduced viscosity measured at 0.5 g/dL in DMF at 30°C.*

While polymer **9** itself only absorbs weakly in the deep-UV, the UV spectrum of a dichloromethane solution of **9** treated with a catalytic amount of triflic acid changes rapidly with time as naphthalene is being produced by cleavage of the polymer chains. Figure 4 shows the change in UV absorbance observed when 3.5 mL of a 3.0 x 10^{-4} mole/L solution of polymer **9** in CH_2Cl_2 is treated with 10μL of triflic acid at 27.6°C. Figure 5 shows the first order dependence of the same acidolysis reaction. The acidolysis reaction was also monitored by other techniques including NMR Spectroscopy and Thermogravimetry-GC-Mass Spectrometry; all of these techniques combined to indicate that the acid-catalyzed decomposition was complete and afforded only naphthalene, formaldehyde and water as shown in equations 4 and 5. Similar studies are currently underway with polymer **10** which decomposes to benzene, formaldehyde and water, as well as other polyethers and polyformals based on the same general design.

Testing of polymer 9 as a dry-developing imaging system.
The imaging process involving polymers **9** or **10** used in combination with a photoacid generating compound is outlined in Figure 6. Spin-coated films of polymer **9**

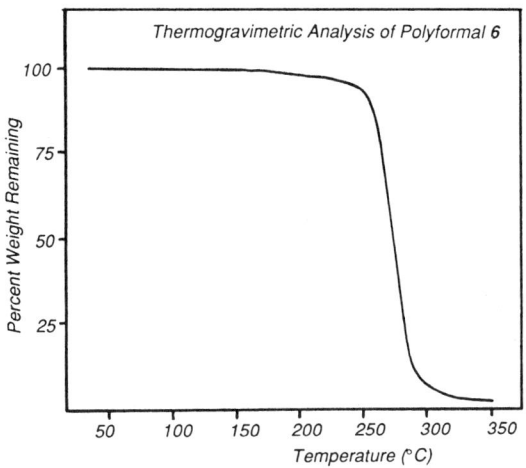

Figure 3. Thermogravimetric analysis of polymer 9.

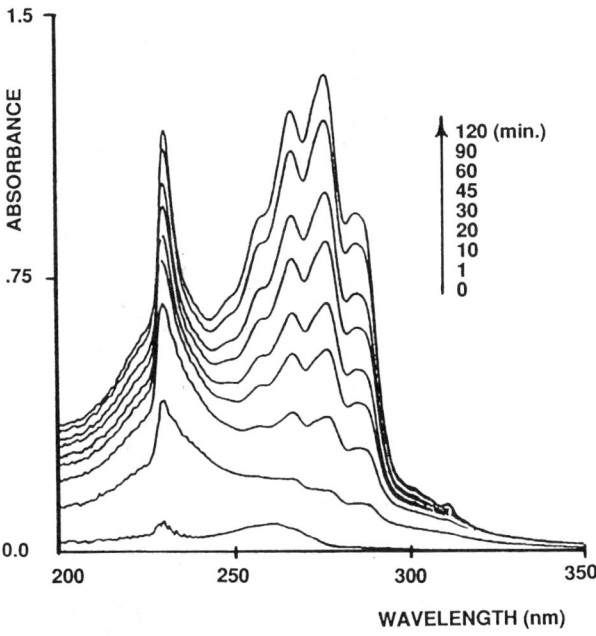

Figure 4. Change in UV absorbance as a function of time in the acidolysis of 9

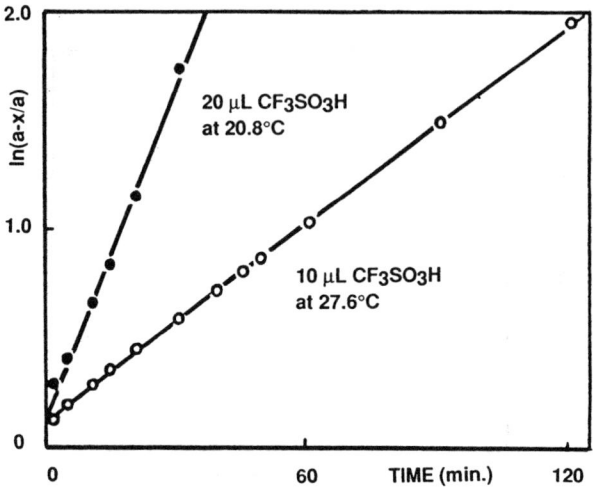

Figure 5. First order dependence of the acidolysis of 9 in solution.

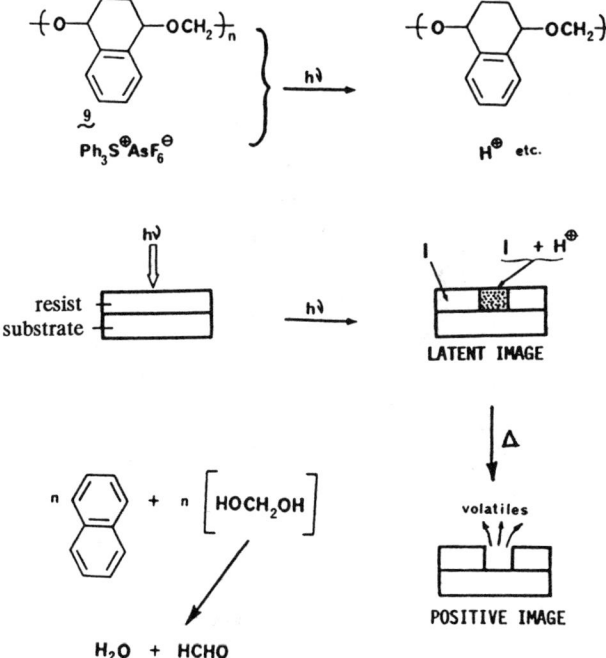

Figure 6. Schematics of the imaging process of polymer 9 with 10% onium salt

$$\text{\textbf{9}} \quad \xrightarrow[(2)\ \Delta]{(1)\ H^+} \quad n\ \text{(naphthalene)} + n\ [HOCH_2OH] \quad \text{(eq. 4)}$$

$$\searrow H_2O + HCHO$$

$$\text{\textbf{10}} \quad \xrightarrow[(2)\ \Delta]{(1)\ H^+} \quad n\ \text{(benzene)} + n\ [HOCH_2OH] \quad \text{(eq. 5)}$$

$$\searrow H_2O + HCHO$$

containing 10% triphenyl sulfonium hexafluoroantimonate were exposed to UV radiation using a special wedge mask to determine the sensitivity of the material. At 254nm an exposure dose of approximately 10-12 mJ/cm^2 was required to produce a fully self-developed image after baking at 65°C; a much lower dose (ca. 2mJ/cm^2) being sufficient to image the polymer in conventional (wet development) fashion. Polymer **9** has a very low glass transition temperature and therefore baking cannot be carried out at higher temperatures to effect self-development without diminishing seriously the quality of the image which is obtained. Figure 7 shows an electron micrograph of a fully self-developed image obtained from polymer **9**. Some loss of image quality is due both to overexposure and some softening of the edges of the image features.

Experimental Section

<u>Polycondensation of DHTN with dibromomethane under phase transfer conditions.</u>
The polycondensation of DHTN [13.13g, 0.080 mole] with 32mL dibromomethane was carried out in the presence of tetrabutyl ammonium bromide [2.48g or 0.008 mole] and 160mL 60 wt% aqueous KOH solution with vigorous stirring at room temperature for 24h. The polymer was precipitated in methanol and purified by reprecipitation from THF solution into distilled water to afford 13.51g (94.8%) of the desired polyether **9** which exhibits a reduced viscosity of 0.27 dL/g [measured at 30°C in DMF at a concentration of 0.5g/dL]. The polymer had an IR spectrum with a prominent C-O-C peak at 1210 cm^{-1}. The ^{1}H NMR spectrum of the polyether [CDCl$_3$] showed peaks at δ = 1.7-2.5 (C-CH$_2$-CH$_2$-C), 4.6-5.3 (O-CH$_2$-O), and 7.3-8.0 (aromatic).
Similar procedures were used in the preparation of the other polymers; the use of mixtures of DHTN and DHCH resulted in the formation of copolymers with T$_g$ values intermediate between those of **9** and **10**.

<u>UV monitoring of the acidolysis of polyether 6</u>
A standard UV cell was filled with 3.5mL of a 3 X 10^{-4} M dichloromethane solution of polyether **9**. The solution was then treated with 10 uL of trifluoromethanesulfonic acid and the changes in UV absorption of the mixture were monitored. Once the reaction was complete the molar extinction coefficient of the product at 276nm was identical to that of naphthalene, therefore conversions during acidolysis were calculated directly from absorption measurements (A_t/A_ϕ).

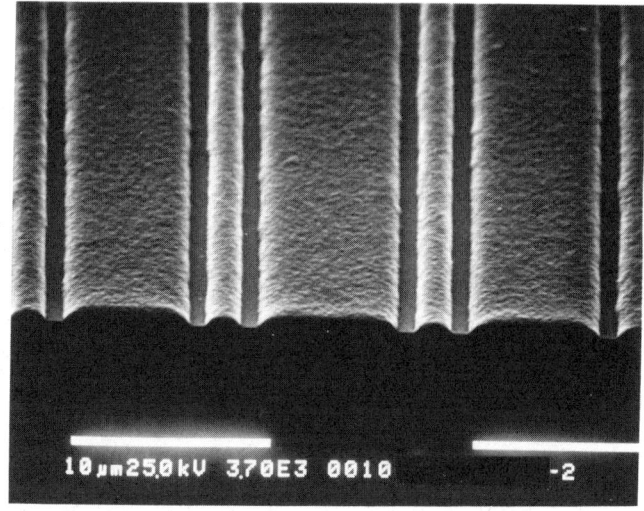

Figure 7. Scanning Electron Micrograph of a self-developed image of polymer 9

Imaging experiments.

A stock solution of the polymer was prepared by dissolving 2.0g of polymer 9 and 0.2g triphenyl sulfonium hexafluoroantimonate in 20g dry cyclohexanone. The solution was filtered through a 0.45um membrane filter and spin-coated onto 5" silicon wafers and baked at 100°C for 2 min. Exposure was achieved using a Perkin-Elmer 500 projection scanner at a scan speed of 2400 and aperture setting of 3 with a UV-2 (254nm) filter. After exposure and post-bake at 65°C, the exposed areas of the film had lost over 80% of their thickness by self-development. The surface residue containing the inorganic salt by-products could be removed by spraying with isopropanol to afford a clean positive image of the mask.

Acknowledgments

Financial support for part of this research was provided by IBM Corporation's Materials and Processing Science Program; this support is gratefully acknowledged. In addition the authors acknowledge the Natural Sciences and Engineering Research Council of Canada for partial support of Dr. Iizawa while in Prof. Fréchet's laboratory at the University of Ottawa.

Conclusion

It is possible to design self-developing imaging systems based on the concept of catalyzed thermolysis of polymer main-chain. One requirement for such a system is the availability of an efficient cleavage reaction which produces volatile fragments. This study has demonstrated that polyformals derived from 1,4-dihydroxy-1,2,3,4-tetrahydronaphthalene are particularly convenient as the cleavage reaction is driven in part by the gain of aromatization energy with formation of free naphthalene, with water and formaldehyde as the other by-products. This system has greater potential than our original poly(phthalaldehyde)-onium salts based materials [3] as its development requires thermal activation of the latent image, thereby eliminating all risks of chemical contamination of the exposure tool. It may be possible to extend this approach to other systems which incorporate structures with more resistance to typical semiconductor fabrication conditions.

Literature cited

[1] C.G. Willson, H. Ito, J.M.J. Fréchet, T.G. Tessier, F.M. Houlihan; J. Electrochem. Soc. 133, 181-187 (1986).
[2] J. Hébert, D. Gravel; Can. J. Chem., 52, 187 (1974).
[3] J.M.J. Fréchet, H. Ito, C.G. Willson; Proc. Microcircuit Eng., 260 (1982). *Ibid.*, 261 (1982). C.G. Willson, H. Ito, J.M.J. Fréchet, F. Houlihan; Proc. IUPAC 28th Macromol. Symp., Amherst, Mass., 443 (1982). H. Ito, C.G. Willson, J.M.J. Fréchet; Proc. Tokyo Conf. on VLSI, (1982); and U.S. Patent 4,491,628 (1985).
[4] J.M.J. Fréchet, F.M. Houlihan, F. Bouchard, B. Kryczka, C.G. Willson, J. Chem. Soc. Chem. Commun. 1514 (1985). Polym. Mat. Sci. Eng. 53, 263 (1985).
[5] F.M. Houlihan, F. Bouchard, J.M.J. Fréchet, C.G. Willson, Macromolecules 19, 13 (1986).
[6] J.M.J. Fréchet, F.M. Houlihan, F. Bouchard, B. Kryczka, E. Eichler, C.G. Willson, N. Clecak, J. Imaging Sci., 30, 59 (1986).
[7] J.M.J. Fréchet, F.M. Houlihan, F. Bouchard, E. Eichler, B. Kryczka, C.G. Willson, Makromol. Chem. Rapid Commun., 7, 121 (1986)

[8] J.M.J. Fréchet, T. Iizawa, F. Bouchard, M. Stanciulescu, C.G. Willson, N. Clecak, Polym. Mat. Sci. Eng. 55, 299 (1986).
[9] J.M.J. Fréchet, F. Bouchard, E. Eichler, F.M. Houlihan, T. Iizawa, B. Kryczka, C.G. Willson, Polymer Journal, 19, 31 (1987).
[10] J.M.J. Fréchet, E. Eichler, M. Stanciulescu, T. Iizawa, F. Bouchard, F.M. Houlihan, C.G. Willson, ACS Symposium Series #346, 138 (1987).
[11] J.M.J. Fréchet, 2nd SPSJ International Conference, Tokyo (1986), paper #2CIL4.
[12] M. Stanciulescu, PhD. Dissertation, University of Ottawa (1988).
[13] R. Taylor in "Chemistry of Carboxylic Acids and Derivatives" Suppl. #2, (Patai, Editor) Chapter 15, 860, Wiley, N.Y. (1979).
[14] J.M.J. Fréchet, N. Kallmann, B. Kryczka, E. Eichler, F.M. Houlihan, C.G. Willson, Polymer Bulletin, 20, 427 (1988). J.M.J. Fréchet, E. Eichler, S. Gauthier, B. Kryczka, C.G. Willson, ACS Symposium Series #381 "Radiation Chemistry of High-Technology Polymers" (E. Reichmanis and J. O'Donnell, Eds.), 155 (1989).
[15] F.M. Houlihan, B. Kryczka, J.M.J. Fréchet, unpublished data. See also ref. 14.
[16] J.M.J. Fréchet, M. Stanciulescu, T. Iizawa, C.G. Willson; Polym. Mat. Sci. Eng., 60, 170 (1989). J.M.J. Fréchet, J. Fahey, C.G. Willson, T. Iizawa, K. Igarashi, T. Nishikubo; Polym. Mat. Sci. Eng., 60, 174 (1989).
[17] V. Percec, B.C. Auman, Polymer Bulletin, 10, 385 (1983)
[18] A.S. Hay, F.J. Williams, H.M. Relles, B.M. Boulette, P.E. Donahue, D.S. Johnson, J. Polym. Sci., Polym. Lett. Ed., 21, 449 (1983).

RECEIVED August 18, 1989

MULTILEVEL RESIST CHEMISTRY AND PROCESSING

MULTILEVEL RESIST CHEMISTRY AND PROCESSING

The current trend towards increased integrated circuit density and decreased feature size has created a need for lithographic processes which are no longer single level in nature. Device design rules are fast approaching the 0.5μm level, and delineation of this lateral dimension over existing topography approaches the limits of what can be done using conventional single level processes.

To circumvent this problem, various schemes making use of multi-level processes are capable of defining submicron features over the most difficult of pattern topographies. Concurrently equipment which enables the patterns defined in the top resist imaging layer to be transferred through the remaining structure by dry-etching means has been developed. Depending upon the composition of each layer in the multi-level structure a variety of plasma types have been investigated and formulated to minimize the change in lateral dimension of each feature as it is transferred through the structure. While standard novolac type resists used in single-layer processes are compatible with trilevel processing, they lack the necessary properties to be used in simplified multi-level schemes. This has stimulated researchers to propose materials which impart resistance to the harsh dry-etching environments existing in the pattern transfer processes.

The use of organometallic polymers in which the metal atom forms a refractory oxide in an oxygen plasma has provided a beginning to the development of materials which are etching resistant. To date, numerous metals, i.e. tin, titanium, germanium and silicon have been investigated, and the latter appears to be the metal of choice. The extensive literature available on silicon based chemistry, and the relative high stability of these organometallic polymers in IC manufacturing environments have led to the synthesis of innumerable materials which may have application as multi-level resist materials. These so called bi-level resists have been shown to respond favorably to the forms of radiation used in the circuit pattern delineation process, and exhibit resolution beyond present circuit pattern design rules. Further simplification in the bi-layer process has recently been made possible by a gas-phase surface functionalization of thick (~ 1.0 μm) single layer resist. In this scheme, irradiation effects a change in reactivity of the thick resist to a metal containing reagent allowing for the selective incorporation of a high concentration of the metal atom into the exposed areas. These areas then form a dry-etch mask for the remaining part of this layer. The pressing uncertainty in the aforementioned schemes lies in the ability to hold the total linewidth loss during pattern transfer to within device design specifications. For ≤ 0.5μm images, typical tolerances must be ≤ 10 percent of the given dimension, and this constraint puts enormous pressure on the process engineers to devise truly anisotropic etching steps such that the dimensional stability of the resist features is maintained. Future development in the areas of resist chemistry, and plasma etching equipment and processes will be necessary to incorporate multi-level resist dry process technology as a standard practice in integrated circuit fabrication.

Anthony E. Novembre
AT&T Bell Laboratories
600 Mountain Avenue
Murray Hill, NJ 07974

Chapter 8

Polysilanes: Solution Photochemistry and Deep-UV Lithography

R. D. Miller[1], G. Wallraff[1], N. Clecak[1], R. Sooriyakumaran[1], J. Michl[2], T. Karatsu[2], A. J. McKinley[2], K. A. Klingensmith[2], and J. Downing[2]

[1]IBM Research Division, Almaden Research Center, 650 Harry Road, San Jose, CA 95120-6099
[2]Center for Structure and Reactivity, Department of Chemistry, University of Texas at Austin, Austin, TX 78712-1167

> The mechanism of the photochemical decomposition of a variety of alkyl polysilanes in solution has been examined. At short wavelengths, silylene extrusion is a major pathway although silyl radicals are also generated. At long wavelengths, the silylene mode is shut down and chain degradation proceeds primarily by bond homolysis. The persistent silyl radicals which are produced at room temperature are not those generated by initial chain cleavage, but result instead from the loss of one alkyl group from the chain. Evidence is presented against a simple photochemical silicon-carbon bond cleavage and in support of a more complex route initiated by a 1,1-reductive chain cleavage to generate initially chain silylene derivatives followed by rearrangement. The results of a DUV lithographic study in a bilayer configuration using poly(cyclohexylmethylsilane) as an imaging and barrier layer are reported. 0.75 μm features were resolved with a commercial 1:1 projection printer utilizing a deep-UV source and compatible optics. Sub 0.5 μm images could be resolved and transferred by O_2-RIE using an excimer laser projection printer operating at 248 nm. The resolution using poly(cyclohexylmethylsilane) and excimer laser exposure far exceeds the 0.8 μm limit previously reported for some polysilane copolymers.

Soluble polysilane derivatives (1) represent a new class of radiation sensitive materials for which a number of new applications have recently appeared. In this regard, they have been utilized as: (i) thermal precursors to ceramic materials (2,3); (ii) a new class of oxygen insensitive photoinitiators for vinyl polymerizations (4); (iii) single component media for charge conduction (5,6); (iv) radiation sensitive materials for microlithographic applications (7-9); and recently as (v) materials with interesting nonlinear optical properties (10,11).

$$(RR'Si)_n$$

Historically, the polysilanes are old materials and the first diaryl derivatives were probably prepared in 1924 by Kipping (12). The simplest dialkyl representative poly(dimethylsilane) (PDMS) was described in 1949 (13). These materials were, however, highly crystalline, insoluble and intractable and elicited little scientific interest until Yajima and co-workers demonstrated that PDMS could be utilized as a thermal precursor to β-SiC by prior conversion to a tractable carbosilane derivative (2). The modern era in polysilane chemistry began about 10 years ago with the synthesis of the poly(methylphenylsilane) (14) and some soluble copolymers (15,16). Since this time, interest in these materials has heightened as evidenced by the explosion of patents and publications in the area.

Although new organometallic procedures (17,18) have been reported recently, the method of choice for the production of linear, high molecular weight polymers is still the modified Wurtz coupling of substituted dichlorosilanes using alkali metals and alloys (1). The mechanism of this heterogeneous polymerization is quite complex and the intermediacy of both radical (19) and anionic chain carrying intermediates have been postulated (20). In addition, substantial solvent effects have been reported which vary with the pendant substituents (8,19,21). Ultrasonic agitation has also been utilized to facilitate polymerization at lower temperatures (22,23).

$$RR'SiCl_2 + Na \xrightarrow{\Delta} (R'RSi)_n + 2 NaCl$$

It was the curious electronic spectra of oligosilanes derivatives which originally attracted the interest of theoreticians and experimentalists. Unlike saturated carbon catenates, oligosilanes absorb in the UV spectral region (24). This characteristic was originally quite unexpected for a sigma bonded system. In a similar fashion high molecular weight silicon catenates absorb strongly in the UV, suggesting extensive sigma electron delocalization in the polymer backbone (24). Atactic, noncrystalline alkyl polysilane derivatives absorb from $300-325$ nm. Aryl substituents attached directly to the backbone cause spectral red shifts of $20-30$ nm attributable to the interaction of the substituent π orbitals with the silicon backbone orbitals. The electronic spectra of polysilanes also depend on molecular weight, with both the absorption maxima and the extinction per SiSi bond approach limiting values around a degree of polymerization of $40-50$. Interestingly, the spectral characteristics also depend strongly on the backbone conformation and the planar zigzag form is the most red shifted of all (22,25-30). The conformation of the polymer backbone and, hence, the polymer absorption spectrum in the solid state depends strongly on the nature of the substituents. Recently reported soluble poly(diarylsilanes) are the most red shifted of all the polysilanes, apparently due to a combination of electronic and conformational effects (31).

Polysilanes are photolabile and sensitive to ionizing radiation. Figure 1 shows the absorption spectrum of a typical polysilane and its response to irradiation. The strong spectral bleaching suggested that considerable chain degradation was occurring upon irradiation. This was confirmed by GPC examination of the irradiated samples which showed the progressive formation of lower molecular weight fragments. Although the rate of photochemical bleaching depends on many features (e.g., the nature of the substituents, polymer T_g, ambient atmosphere, etc.), chain scission is always a major pathway. Polymer crosslinking becomes more competitive in the presence of pendant unsaturation (vide infra).

Polysilane Photophysics
--

The strong dependence of the spectral properties on of the backbone conformation was initially unexpected for a sigma bonded system. This phenomenon was probed computationally for acyclic systems containing up to 20 silicon atoms using the INDO/S procedure and the results shown in Figure 2 (32,33). A number of features are particularly obvious from the figure. First, it is clear that an "all-trans" arrangement would be expected to absorb at lower energies than the comparable

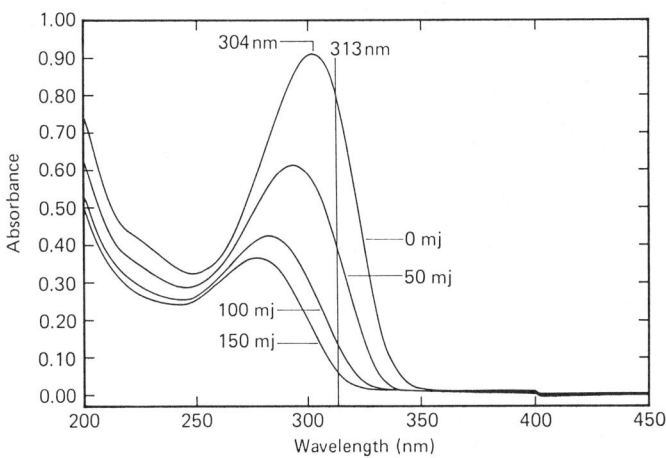

Figure 1. Photobleaching of a film of poly(n-hexylmethylsilane) at 313 nm.

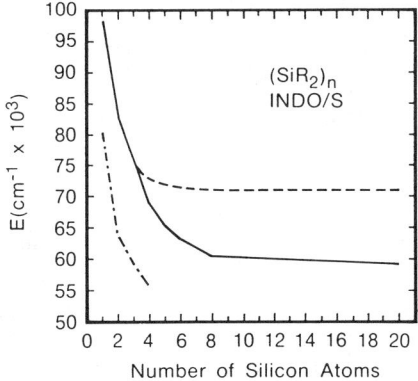

Figure 2. Calculated (INDO/S) first singlet excitation energy: all-*trans* $[(CH_3)_2Si]_n$ (dot-dash); all-*trans* $[H_2Si]_n$ (full); all-gauche $[H_2Si]_n$ (dashed).

"all-gauche" conformation. This observation is readily rationalized in terms of long range orbital interactions (specifically between the 1,4 sp^3 hybrid orbitals of the sigma bonded silane chain) which in turn depend on the dihedral angles of the backbone. Second, we see that while the difference in the excitation energies between the trans and gauche forms increase initially with the degree of catenation, eventually this effect is expected to saturate with the difference remaining roughly constant for chains longer than approximately twelve silicon atoms. Finally, it is also apparent from the figure that the HOMO-LUMO excitation energies are also predicted to saturate rapidly with increased catenation. The computational results are, therefore, consistent with the observed spectral and structural properties of polysilane derivatives.

The nature of the photoexcited state of poly(di-n-alkylsilane) derivatives in solution has been probed by fluorescence polarization techniques using poly(di-n-hexylsilane) (PDHS) as a model system (32,33). While the absorption spectra of the poly(di-n-alkylsilanes) in solution are broad and featureless, the emission spectra are much narrower. The emission is short-lived (75-200 ps) (32,33,38) and the shape of the emission band suggests extensive delocalization of the excitation. This has led to the suggestion that the photochemistry occurs from an electronic state other than the singlet. In this regard, the triplet state has been considered (34,35), and the shape of the weak, broad phosphorescence emissions observed for some dialkyl polysilane derivatives is suggestive of a more localized excited state. Although quenching studies have confirmed a photoreactive triplet state for poly(methyl-n-propylsilane), studies on other derivates have been less conclusive (34).

Returning again to the polarized fluorescence studies, it has been reported that the degree of polarization (P) is remarkably wavelength dependent in solution (32,36). For PDHS, it ranges from a value of near the theoretical limit of 0.5 for long wavelength excitation to near zero (no correlation between the absorbing and emitting transition moments) at short excitation wavelengths. Such an observation would be consistent with the concept of a polysilane chain in solution consisting of a collection of weakly interacting chromophores communicating by rapid energy transfer. A possible model explaining these results would be a chain composed of a series of trans (or nearly trans) links of different lengths and hence varying excitation energies, which are insulated to some extent from one another by conformational twists or kinks (32-38). Irradiation into the red edge of the absorption band would address only the longest of the trans chromophores, thus leading to prompt emission and a strong correlation between the absorbing and emitting transition moments. On the other hand, irradiation at the blue edge of the absorption band would excite preferentially the short, high energy chromophores leading to rapid transfer of the excitation energy to the longer chromophores, resulting in the lower values for the polarization degree observed for short wavelength excitation. It is also possible that the short segments are more photochemically reactive which would rationalize the substantial decreases observed in the measured fluorescence quantum yields for short wavelength excitation (33-35) . INDO/S calculations support this model for poly(di-n-hexylsilane) in solution and suggest that the primary chromophores are trans or near trans segments separated by strongly non-planar kinks (32-34). Furthermore, these studies indicate that considerable localization of the excitation occurs even on very short trans segments (33,34).

Photochemistry

Photochemical transformations of cyclic and short chain polysilane oligomers have been intensively investigated (39). Irradiation of these materials in the presence of trapping reagents, such as silanes or alcohols, has suggested that substituted silylenes and silyl radicals are primary reactive intermediates. The former have been implicated by the isolation of appropriate trapping adducts and the latter by the formation of linear silanes containing fewer silicon atoms than present in the starting materials. Silylene intermediates from the photolysis of cyclic oligosilanes also have been detected spectroscopically (40-42). Although related studies on polymeric systems are rare, recent results on polymeric main and pendant side chain disilane derivatives indicate that bond homolysis to produce silyl radicals is an important photochemical pathway (43,44). Mechanistic studies on low molecular weight silyl radicals indicate that radical disproportionation is competitive with recombination when α-hydrogens are available for abstraction (39,45,46).

The photodecomposition of a number of soluble substituted silane high polymers in the presence of trapping reagents, such as triethylsilane (TES), has been described (47) and some results are shown in Table 1. In every case, the major products from exhaustive irradiation at 254 nm were the corresponding silylene trapping adducts, and these adducts are primary photoproducts produced directly from the high polymer. The latter was demonstrated by the observation that 1,1,1-triethyl-2,2-dibutyldisilane is produced from poly(di-n-butylsilane) with no induction period (48). Similar results were obtained using alcohols as trapping reagents. Consistent with the suggestion that dialkyl silylene intermediates are produced, the dialkyl polysilanes $(RR'Si)_n$ yielded deuterated adducts ($Et_3SiSiRR'D$) when Et_3SiD was employed as a trapping reagent.

The formation of silyl radicals in the exhaustive photolysis of the silane polymers was indicated by the isolation of disilanes of general structure (HSiRR'SiRR'H) as shown in Table 1. These materials accumulate in the photolysate and are photostable as they absorb only weakly at the irradiation wavelength (254 nm). Longer chain silanes are presumably continuously degraded under the conditions of the exhaustive irradiation.

In summary, the production of substituted silylenes and silyl radicals upon exhaustive irradiation at 254 nm of polysilane high polymers suggested that the polymer photochemistry resembled that previously reported for short chain acyclic and cyclic oligomers (39). More recent experiments, however, have suggested that the photochemical mechanism for the degradation of the high polymers is more complex than first envisioned (vide infra) (48).

$$- SiRR' - (SiRR')_2 - SiRR' - \xrightarrow{h\nu} - SiRR' - SiRR' - SiRR' - + RR'Si: \quad [1]$$

$$- SiRR' - (SiRR')_2 - SiRR' - \xrightarrow{h\nu} - SiRR' - SiRR' \cdot + \cdot SiRR' - SiRR' - \quad [2]$$

The results of the exhaustive irradiation studies, which suggest that both silylenes and silyl radicals are produced, are consistent with the processes [1] and [2], each of which has precedent in silane oligomer photochemistry (39). Thermodynamic considerations suggest that the silylenes and radicals are not produced in a single step from the absorption of a single photon of 254 nm light, but are most probably produced competitively. Conformation of this comes from recent studies which indicate an unusual wavelength dependence for the photolyses of two typical symmetrical dialkyl silanes (48). In this regard, the irradiation of either poly(di-n-hexylsilane) (PDHS) or poly(di-n-butylsilane) (PDBS) at 248 nm (pulsed) or 254 nm (cw) in the presence of triethylsilane-cyclohexane (1:1) produces the trapping adducts $Et_3SiSiR_2H(R = Bu$ or $Hx)$ and the homolytic cleavage products $H(SiR_2)_nH$ ($n = 2,3, R = Bu$ or Hx), respectively. The formation of the adduct Et_3SiSiR_2H is, however, quite wavelength dependent and the quantum yield drops rapidly to zero for irradiation wavelengths greater than 300 nm (see Figure 3). Polymer degradation still proceeds rapidly at the longer wavelengths as evidenced by the formation of persistent radicals (ESR, vide infra) and the bleaching of the original polymer absorption at 316 nm. GPC examination of the solutions irradiated at $\lambda > 300$ nm also shows considerable reduction in the original polymer molecular weight. It is important to note that the short chain disilanes

Table 1. Trapping products from the exhaustive irradiation of some polysilane derivatives [(SiR'R^2)$_n$] at 254 nm in the presence of triethylsilane; (a) yields were less than 2%

Product	R^1 = n-C$_4$H$_9$ R^2 = n-C$_4$H$_9$	R^1 = n-C$_6$H$_{13}$ R^2 = Me	R^1 = c-C$_6$H$_{11}$ R^2 = Me
Et$_3$SiR^{1}R^{2}SiH	59%	70%	71%
HSiR^{1}R^{2}SiR^{1}R^{2}H	11%	11%	14%
Et$_3$SiSiR^{1}R^{2}SiEt$_3$	a	3%	a
HSiR^{1}R^{2}OSiR^{1}R^{2}H	a	a	2
Et$_3$SiOSiR^{1}R^{2}SiR^{1}R^{2}H	a	a	3

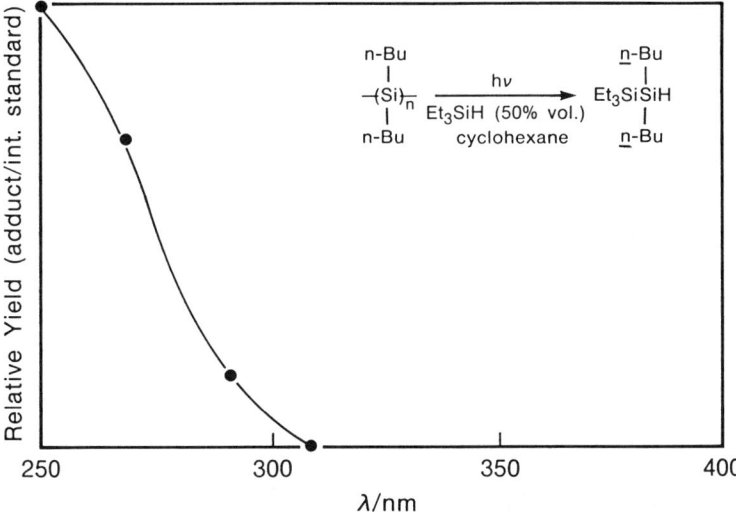

Figure 3. Yield of adduct Et$_3$SiSiBu$_2$H from irradiation of poly(di-n-butylsilane) in the presence of triethylsilane as a function of the irradiation wavelength.

$H(SiR_2)_nH(n = 2,3)$ are also not formed at $\lambda > 300$ nm, presumably because oligomers of $H(SiR_2)_nH$ are transparent for $n \leq 7$ (49). When $(Et)_3SiD$ is employed, irradiation at 254 nm produced Et_3SiSiR_2D, but the silanes of structure $H(SiR_2)_nH(n = 2,3)$ remain undeuterated. Similarly, irradiation in the presence of Et_3SiD at $\lambda > 300$ nm results in no deuterated fragments (IR analysis). This result suggests that silyl radicals, presumably produced by process [2], apparently abstract hydrogen from the polymer itself rather than from the solvent. A reasonable source of hydrogen from the polymer is from the side chain carbon α to the radical site in a classical radical disproportionation reaction.

The source of the hydrogen atoms in the abstraction reaction was confirmed by irradiation of PDHS (α-d_2) at 308 nm. This material in pentane, pentane-d_{12}, cyclohexane, cyclohexane-d_{12} or Et_3SiD produces products containing $-SiD$ but no SiH bonds. It appears, therefore, that reactions [1] and [2] are decoupled and that radical disproportionation is occurring. The long wavelength results which show no detectable silylene while polymer degradation and radical formation are still operative, also indicate that the further dissociation of silyl radicals into silylenes is not occurring to any significant extent in fluid solution at room temperature. This result is also consistent with recent studies reported for pentamethyldisilyl radicals (46).

While photochemical wavelength effects in solution are unusual, they are not unprecedented (50). In fact, the results described here fit nicely with the photophysical picture of dialkyl polysilanes in solution presented earlier. In this model, the polysilane chain was described as a weakly interacting, statistical collection of chromophores composed of trans or near trans units partially electronically decoupled by conformational kinks. Adopting this model, the wavelength dependent photochemistry could be explained as follows. Irradiation at long wavelengths excites selectively the longest chromophores which undergo a series of radical cleavage processes. Excitation at short wavelengths excites the shorter or high energy chromophores which can either rapidly transfer energy to longer chromophores and proceed via radical cleavage or decompose directly by the extrusion of a silylene fragment. Homolytic cleavage could conceivably be occurring in both long and short chromophores or exclusively from the former. In the latter case, rapid energy transfer from the shorter to the longer chromophores would be required.

Equation [2] suggests that silicon centered radicals are produced by the chain cleavage of the polymer backbone. The trapping experiments and the results using deuterated substrates indicate that disproportionation and presumably recombination are also occurring. Evidence for the formation of silyl radicals in the solid state upon irradiation in the presence of oxygen comes from the observation of fragments containing $-SiOSi-$, $-SiOH$ and $-SiH$ linkages (9,51,64). In solution, the isolation of short silane fragments upon photolysis (47) and the facile photoinitiation of the polymerization of vinyl monomers (4) constitute evidence that silyl radicals are produced. Recently, Todesco and Kumat (52) have reported a complex ESR signal upon irradiation of poly(methyl-α-naphthylsilane-co-dimethylsilane) at 77 K and have tentatively suggested on the basis of the apparent g value that the signal is due to silicon radical(s) produced by chain scission.

We have reported that ESR and ENDOR (53) examination of degassed, irradiated samples of a number of polymers (PDHS, PDBS, poly(di-n-tetra-decylsilane) (PDTDS), poly(di-n-octylsilane) (PDOS), poly(di-n-decylsilane) (PDDS) and poly(di-4-methylpentylsilane) (PDMPS) shows the formation of radical species which are persistent for hours and even days. The radical spectra (see Figure 4) are, however, clearly not consistent with those expected from the simple cleavage of the polymer backbone as described by Equation [2]. The structure of these radicals has now been assigned as $-SiR_2\overset{\bullet}{Si}RSiR_2-$. As shown in Figure 4, the ESR spectra produced from different symmetrical poly(di-n-alkylsilanes) are all very similar. In addition, these spectra are temperature dependent as also shown in the figure for the radical derived from (PDMPS). Temperature studies were conducted using this polymer because of the improved solubility at low temperatures relative to the other n-alkyl polymers. At low temperatures, the spectrum is composed primarily of two doublets due to splitting between the silicon radical center which is presumably nearly planar and two non-equivalent hydrogens on the α-carbon of the substituent. At higher temperatures, the spectra begin to average due to the rotational rocking of the alkyl side chain attached to the radical center around the SiC bond, although complete averaging is never reached even at the highest convenient temperature. The spectra of the early radicals produced from all of the symmetrical di-n-alkyl derivatives were virtually identical except for differences caused by motional averaging, the rate of which at a given temperature is influenced by the size of the alkyl substituent. The following values for the g value and the hyperfine couplings were obtained for the radical derived from PDHS, $(C_2^\beta Si^\alpha)_2 \dot{Si}^i - C^\alpha H^{\alpha'} H^{\alpha''} C^\beta H_2^\beta CH_2^\gamma -$: (a) $g = 2.00472$ (consistent with a branched polysilylated silyl radical), (b) ^{29}Si satellites: $a_\alpha = 5.8G(2Si)$, $a_i = 75G$, assigned to $-(Si^\alpha)_2 - Si^i$, (c) proton coupling: $a_{\alpha'} = 11G$, $a_{\alpha''} \simeq 3G(1H)$ near the low temperature limit and $a_\alpha \approx 6.99G(2H)$ near the high temperature limit, $\alpha_\beta = 0.34G$, $\alpha_\gamma = 0.13G$, with sgn a_α = sgn a_γ, = − sgn α_β (d) carbon coupling, $\alpha_\beta = 4.12G$.

The origin of the persistent radicals which are produced in low yields is of some interest. Simple photochemical silicon-carbon bond cleavage lacks precedent and is not consistent with the failure to observe any alkane or 1-alkene in the photolysis mixture. While the latter could not be expected to survive in the irradiated solution, the former would. A possible route involving photochemical 1,1-reductive elimination is described below.

$$- RR'Si - SiRR' - SiRR' - RR'Si - \xrightarrow{h\nu} \begin{cases} - SiRR'_2 - RR'_2 Si + R\ddot{S}iRR'SiRR'Si - \quad [3A] \\ - SiR_2R' - R_2R'Si + R'\ddot{S}iRR'SiRR'Si - \quad [3B] \end{cases}$$

$$R\ddot{S}iRR'SiRR'Si - \rightarrow - RR'SiSiR = SiRR' \quad [4]$$

$$- RR'SiSiR = SiRR' + \cdot SiRR' - \rightarrow - RR'SiSiRSiRR'SiRR' - \quad [5]$$

Regarding this proposal, it should be noted that while 1,1-eliminations on Si-Si-C units to generate silylenes are well known thermal processes (54) the photochemical variant seems not to have been described. The rearrangement of silylsilylenes (4) to disilenes is known to be rapid (55), and silyl radical addition at the least hindered site would produce the observed persistent radical. Preliminary evidence for the operation of 1,1-photoelimination processes in the polysilane high polymers has been obtained, in that the exhaustive irradiation at 248 nm of poly(cyclohexylmethylsilane) (PCHMS) produces ∼10-15% volatile products which contain trialkylsilyl terminal groups. For example, the following products were produced and identified by GC − MS: (R = cyclohexyl, R' = methyl) $H(RR'Si)_2H$ (49%), $H(RR'Si)_3H$ (19%), $R_2R'SiH$ (2%), $R'_2RSiRR'SiH$ (5%) and $R_2R'SiRR'SiH$ (7%).

On the basis of the results described, there seem to be at least three processes which are responsible for the molecular weight reduction in substituted silane high polymers upon irradiation in solution: (i) Chain abridgement by silylene extrusion which occurs only at short wavelengths; (ii) chain scission by silicon-silicon bond homolysis; and (iii) chain scission by 1,1-photochemical reductive elimination.

The photochemical studies on polysilane high polymers indicate that they constitute a new class of radiation sensitive materials which undergo primarily scissioning upon irradiation, although crosslinking reactions become progressively more important with pendant unsaturation. For polymers which simultaneously undergo scission and crosslinking, the quantum yields may be determined by an analysis of both the number average and weight average molecular weights as a function of the irradiation dose (56). Application of this technique to the photolyses of a variety of substituted silane high polymers yields the data described in Table 2 (24,57). Examination of Table 2 shows that in each case the quantum yields for scission (Φ_s) are quite high in solution ($\Phi_s \sim 0.5 - 1.0$) while crosslinking becomes more important with pendant unsaturation, particularly when the unsaturated substituents are directly attached to the polymer backbone (entries 6, 7). In each case examined, the quantum yields for both processes decrease markedly in going from solution to the solid state. This would be expected where reactive intermediates such as silyl radicals are involved due to the solid state cage effects which decrease the mobility of reactive sites. The numbers listed in Table 2 are approximate due to uncertainties such as variations in the polymer molecular weight distributions, the use of polystyrene calibration standards for the molecular weight determinations, uncertainties in the actinometry, etc.. While it is recognized that the relative order of the numbers is probably correct, the absolute values will depend on the accuracy of the molecular weight determinations. In this regard, it has been noted that the values of $\overline{M}_w$ obtained by GPC and by light scattering often differ by as much as a factor of 2 – 3 (58) .

Polysilane Lithography

The curious electronic properties of high molecular weight polysilanes have suggested many potential applications, some of which have been described earlier in this paper. In this section, we focus on the lithographic potential and a number of unique properties which make certain polysilanes ideally suited for such purposes. In this regard, the polysilanes are: (i) thermally and oxidatively stable yet photochemically labile; (ii) strongly absorbing over a broad spectral range yet bleachable; (iii) good film formers and the polymers are incompatible with most common polymers; (iv) high in silicon content and hence very etch-resistant in oxygen plasma environments (7,59). Lithographically the polysilanes have been exercised as: (i) soluble, castable O_2-RIE etch barriers in trilevel schemes; (ii) as the combined imaging and etch barrier layer in a bilayer configuration for both wet (7,60) and dry (7,9,61,62) developing processes; (iii) short-wavelength contrast enhancement layers (63); and (iv) as resists for ionizing radiation (22,64,66).

The drive toward higher density microcircuitry requires the generation of smaller high resolution patterns often with large feature aspect ratios. One approach for improving resolution is to utilize shorter wavelength exposure sources. This has led to the development of projection printing exposure tools and resists operating in the mid-UV (300 – 350 nm) (65) and more recently in the deep-UV (220 – 280 nm) range.

We have previously reported the use of poly(cyclohexylmethylsilane) (PCHMS) in a bilayer configuration to generate high resolution images upon exposure to mid-UV (MUV) radiation (7,8,60) .

In principle, there is no reason why selected polysilanes could not be used in the DUV in a bilayer configuration in the same manner as previously described for the MUV. Although the dialkyl polysilanes have strong absorption maxima around 300 – 325 nm, there is enough residual absorption in the DUV to allow efficient light absorption in thin films. In addition, many of the dialkyl polysilanes bleach significantly upon exposure in the DUV, although the bleaching rate is significantly slower than observed at the absorption maximum. For the example shown in Figure 5, although the first 50 mJ/cm^2 of exposure results in little change in the optical density around 254 nm due to the blue shift in the original absorption maximum with chain degradation, subsequent exposures produce significant

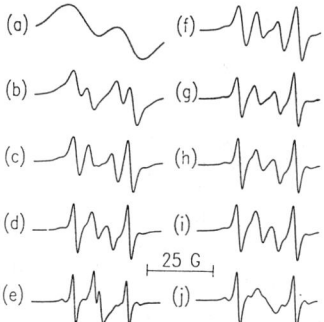

Figure 4. The ESR spectra of irradiated polysilanes $(R_2Si)_n$ in n-octane [(a)-(c) in n-pentane]; a-e: R = 4-methylpentyl (a: 200 K, b: 220 K, c: 260 K, d: 300 K, e: 350 K); f: R = n-$C_{14}H_{29}$; g: R = n-$C_{10}H_{21}$; h: R = n-C_8H_{17}; i: R = n-C_6H_{13}; j: R = n-C_4H_9; 300 K.

Table 2. Quantum yields for polymer scission and crosslinking for some representative polysilanes $(SiR^1R^2)_n$; (a) Molecular weights were measured by GPC using polystyrene calibration standards; (b) Ban, H; Sukegawa, K. J. Polym. Sci., Polym. Chem. Ed., 1988, 26, 521

				$\overline{M}_w^o{}^a$		Solution		Solid Film	
	R^1	R^2	Solvent	$\times 10^{-3}$	λ_{ex}	Φ_s	Φ_x	Φ_s	Φ_x
1.	(4-methylpentyl)	—CH_3	Toluene	41.4	355	1.2	0.06	0.022	0
2.	C_6H_{13}	C_6H_{13}	Toluene	44.6	353	0.6	0	—	—
3.	Ph-CH_2CH_2—	—CH_3	Toluene	208.9	347	0.97	0.08	—	—
4.	$C_{12}H_{25}$—	—CH_3	Toluene	185.1	347	0.54	0	—	—
5.	C_3H_7—	—CH_3	—	3900		—	—	0.011	0.0013[b]
				190		—	—	0.015	0.0013[b]
6.	Ph-	—CH_3	THF	245	313	0.97	0.12	0.015	0.002
7.	t-Bu-Ph-	—CH_3	Toluene	633.6	367	1.00	0.18		

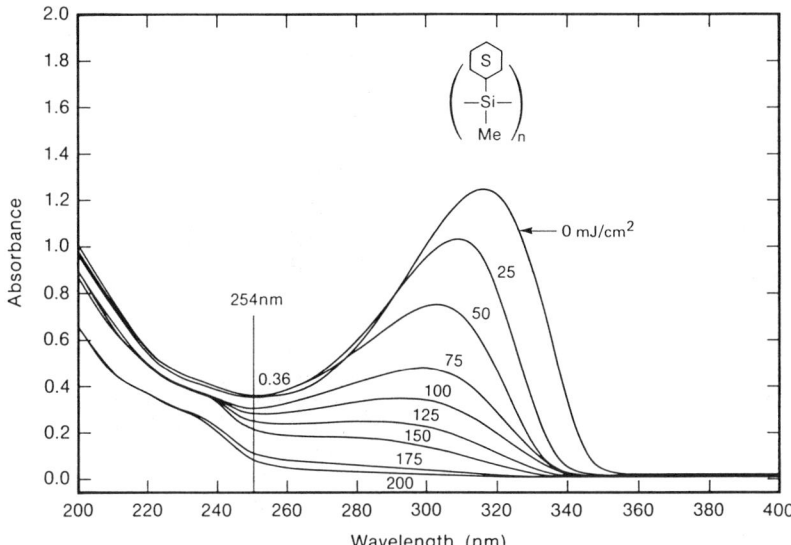

Figure 5. Spectral bleaching of a film of PCHMS upon irradiation at 254 nm.

changes. However, for DUV bilayer applications where the lower optical density ($1.2/\mu m$, 254 nm versus $4.23/\mu m$, 318 nm) assures the homogeneous irradiation of thin films ($0.2 - 0.3$ μm), bleaching is less critical than in the MUV. In spite of this, an ititial study of a number of polysilane copolymers for DUV lithography produced disappointing results with resolution limited to $0.8 - 1.0$ μm for the most promising candidate and the observation of residues and image webbing upon development (66). We now suggest that the conclusions from this initial study were unduly pessimistic and report the results of our study of poly(cyclohexylmethylsilane) (PCHMS) as a material for bilayer DUV lithography. For this study, we surveyed samples in the molecular weight range ($\overline{M}_w$ 25 – 45K) which had been end-capped with chlorotriethylsilane during the polymerization. We now report the results of our examination of the DUV lithographic characteristics of a typical dialkylpolysilane poly(cyclohexylmethylsilane) used in a bilayer configuration.

In any bilayer application employing O_2-RIE for image transfer, the imaging layer must also be resistant to the plasma etch conditions. Figure 6 shows a comparison of the etch stability of PCHMS with that of a sample of a hardbaked AZ-type photoresist under typical conditions employed for anisotropic oxygen etching. Although the PCHMS initally etches slightly, subsequent thickness changes are neglibible. Under similar conditions, a hardbaked AZ-type photoresist etches continuously at a almost constant rate over entire the etch period. After 20m where only 8% of the original thickness of the hardbaked photoresist remains, approximately 93% of the PCHMS layer remains. The etch rate ratio (PCHMS/photoresist) ranges from ~35:1 after 5m to a value of >100:1 after 20m. In any case, the etch-resistance of PCHMS is more than adequate for a successful bilayer process.

Contact exposures were performed using a 254 nm band pass filter and a stepwedge to determine the resist contrast, and the contrast curves for PCHMS in a number of developers are shown in Figure 7. Even though cyclohexanol proved to be a high contrast developer, its physical characteristics (high viscosity and melting point around room temperature) were inconvenient. In this regard, either cyclohexanol containing approximately 5% isopropanol or a 80/20 mixture of butylacetate/isopropanol was more convenient and performed adequately.

Two types of imaging experiments were conducted. The first, using a PE-500 Microalign 1X projection printer operating in the UV-2 mode, resulted in the delineation of 0.75 μm line space arrays (see Figure 8), which could be transferred by anisotropic oxygen reactive ion etching (O_2-RIE). These patterns were the smallest features available on the mask. As expected, the pattern resolution could be improved using a KrF excimer laser 10X step and repeat tool operating at 248 nm. Figure 9 shows sub 0.5 patterns generated with this tool and transfered by O_2-RIE. The dose required to generate these patterns was between 125-150 mJ/cm^2.

In summary, poly(cyclohexylmethylsilane) homopolymer used in a bilayer configuration has been demonstrated to be useful for high resolution DUV pattern generation and O_2-RIE image transfer using either a commercial 1:1 projection printing tool operating in the DUV or a 10:1 excimer laser stepper at 248 nm. With the latter, images smaller than 0.5 μm have been printed representing a considerable improvement over the $0.8 - 1.0$ μm limit recently proposed for the best of a series of polysilane copolymers (66).

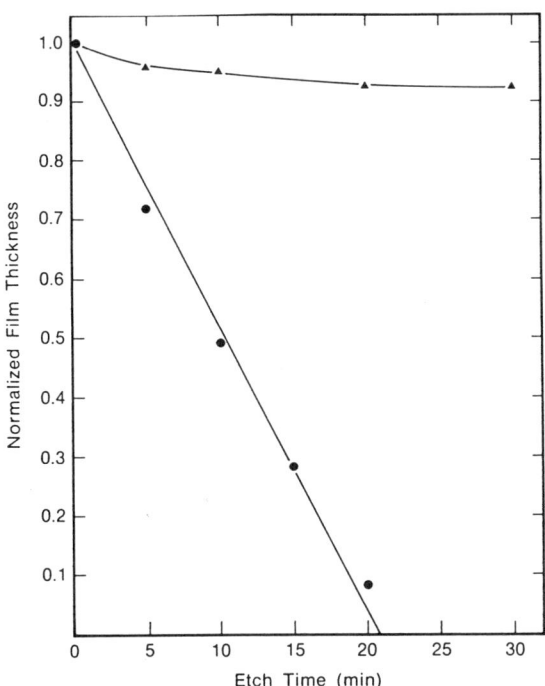

Figure 6. O$_2$-RIE etching of poly(cyclohexylmethylsilane) (PCHMS) — ▲ —, and a hardbaked AZ-type photoresist (— • —); etch conditions 10 mTorr, 40 SCCM, −232V, 110W.

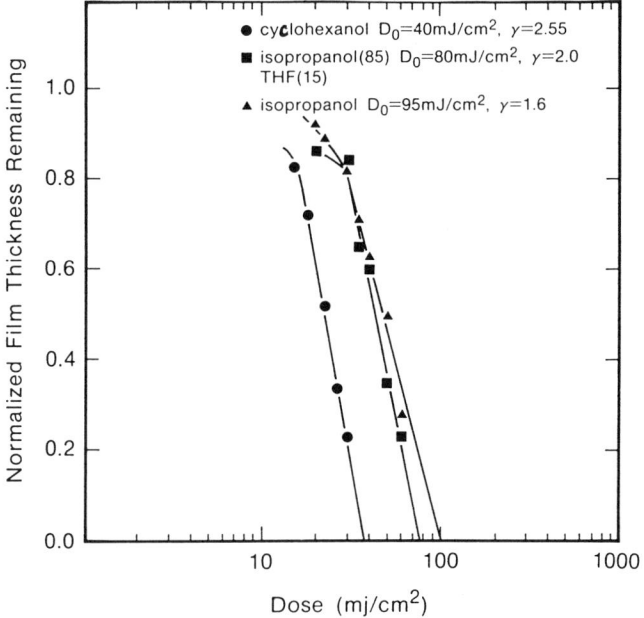

Figure 7. Stepwedge lithographic contrast curves for PCHMS irradiated at 254 nm.

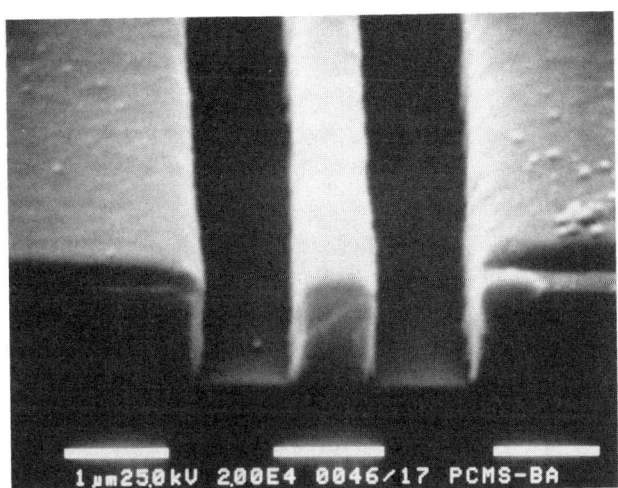

Figure 8. Bilayer imaging of PCHMS using a PE-500 Microalign 1:1 projection printer (UV-2 mode). 0.16 μm PCHMS over 1.0 μm of a hardbaked AZ-type photoresist. O_2-RIE image transfer; 0.75-μm images.

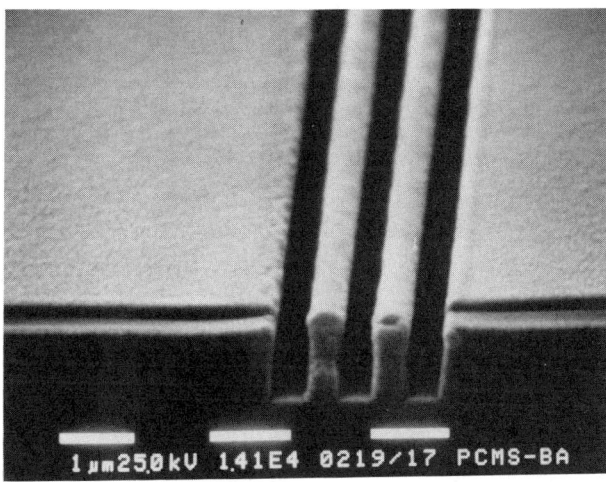

Figure 9. Bilayer imaging of PCHMS using a 10:1 excimer laser step-and-repeat tool at 248 nm. 170 nm PCHMS over 1.0 μm of a hardbaked AZ-type photoresist; 125 mJ/cm^2. O_2-RIE image transfer; 0.4-μm images.

Acknowledgments

R. D. Miller gratefully acknowledges partial financial support for this work from the Office of Naval Research. Similarly, J. Michl acknowledges partial financial support from the NSF (87-96257) and from IBM Corporation.

References

1. For recent reviews of polysilane high polymers see West, R., J. Organometallic Chem. 1986, 300, 327; West, R. In The Chemistry of Organic Silicon Compounds; Patai, S.; Rappoport, Z. Eds.; Wiley-Interscience: New York, 1989; Vol. 2, Chapter 19, 1207.
2. Yajima, S.; Hayashi, J.; Omori, M. Chem. Lett. 1975, 931.
3. West, R. In Ultrastructure Processing of Ceramics, Glasses, and Composites; Hench, L.; Ulrich, D.C., Eds.; John Wiley and Sons Inc.: New York, 1984.
4. West, R.; Wolff, A. R.; Peterson, D. J. J. Rad. Curing 1986, 13, 35.
5. Kepler, R. G.; Zeigler, J. M.; Harrah, L. A.; Kurtz, S. R. Phys. Rev. B 1987, 35, 2818.
6. Stolka, M.; Yuh, H. J.; McGrane, K.; Poi, D. M. J. Polym. Sci. Part A: Polym. Chem. 1987, 25, 823.
7. Miller, R. D.; Hofer, D.; Fickes, G. N.; Willson, C. G.; Marinero, E.; Trefonas, III, P.; West, R. Polym. Eng. Sci. 1986, 26, 1129.
8. Miller, R. D.; Hofer, D.; Rabolt, J. F.; Sooriyakumaran, R.; Willson, C. G.; Fickes, G. N.; Guillet, J. E.; Moore, J. M. Polymers for High Technology, ACS Symposium Series, No. 346, American Chemical Society, Washington, D.C., 1987, 170, and references cited therein.
9. Zeigler, J. M.; Harrah, L. A.; Johnson, A. W. Proc. of SPIE 1985, 539, 166.
10. Kajzar, K.; Messier, J.; Rosilio, C. J. Appl. Phys. 1986, 60, 3040.
11. Baumert, J. C.; Bjorklund, G. C.; Lundt, D. H.; Jurich, M. C.; Looser, H.; Miller, R. D.; Rabolt, J. F.; Swalen, J. D.; Twieg, R. J. Appl. Phys. Lett. 1988, 53, 1147.
12. Kipping, F. S. J. Chem. Soc. 1924, 125, 2291.
13. Burkhard, C. A. J. Am. Chem. Soc. 1949, 71, 963.
14. Trujillo, R. T. J. Organomet. Chem. 1980, 198, C27.
15. Wesson, J. P.; Williams, T. C. J. Polym. Sci., Polym. Chem. Ed. 1980, 18, 959.
16. West, R.; David, L. D.; Djurovich, P. I.; Stearly, K. L.; Srinivasan, K. S. V.; Yu, H. G. J. Am. Chem. Soc. 1981, 103, 7352.
17. Aitken, C. T.; Harrod, J. F.; Samuel, E. J. Organomet. Chem. 1985, 279, C11.
18. Aitken, C. T.; Harrod, J. F.; Gill, V. S. Can. J. Chem. 1987, 65, 1804, and references cited therein.
19. Zeigler, J. M. Polym. Preprints 1986, 27, 109.
20. Worsfold, D. J. In Inorganic and Organometallic Polymers, ACS Symposium, Series No. 360; Zeldin, M.; Wynne, K. J.; Allcock, H. R., Eds., American Chemical Society, Washington, D.C., 1988, 101.
21. Miller, R. D.; Hofer, D.; McKean, D. R.; Willson, C. G.; West, R.; Trefonas, III, P. T. In Materials for Microlithography, ACS Symposium, Series No. 266, Thompson, L. F.; Willson, C. G.; Fréchet, J. M. J., Eds., 1984, 293.
22. Miller, R. D.; Rabolt, J. F.; Sooriyakumaran, Fleming, W.; Fickes, G. N.; Farmer, B. C.; Kuzmany, H. In Inorganic and Organometallic Polymers, ACS Symposium, Series No. 360, Zeldin, M.; Wynne, K. J.; Allcock, H. R., Eds., American Chemical Society, Washington, D.C., 1988, 43.
23. Kim, H. K.; Matyjaszewski, K. J. Am. Chem. Soc. 1988, 110, 3321.
24. Trefonas, III, P.; West, R.; Miller, R. D.; Hofer, D. J. Polym. Sci., Polym. Lett. Ed. 1983, 21, 823, and references cited therein.
25. Miller, R. D.; Hofer, D.; Rabolt, J. F.; Fickes, G. N. J. Am. Chem. Soc. 1985, 107, 2172.

26. Rabolt, J. F.; Hofer, D.; Miller, R. D.; Fickes, G. N. Macromolecules 1986, 19, 6111.
27. Kuzmany, H.; Rabolt, J. F.; Farmer, B. L.; Miller, R. D. J. Chem. Phys. 1986, 85, 7413.
28. Lovinger, A. J.; Schilling, F. C.; Bovey, F. A.; Zeigler, J. M. Macromolecules 1986, 19, 2657.
29. Miller, R. D.; Farmer, B. L.; Fleming, W.; Sooriyakumaran, R.; Rabolt, J. F. J. Am. Chem. Soc. 1987, 109, 2509.
30. Harrah, L. A.; Zeigler, J. M. Macromolecules 1987, 20, 601.
31. Miller, R. D., Sooriyakumaran, R. J. Polym. Sci., Polym. Lett. Ed. 1987, 25, 321.
32. Klingensmith, K. A.; Downing, J. W.; Miller, R. D.; Michl, J. J. Am. Chem. Soc. 1986, 108, 7438.
33. Michl, J.; Downing, J. W.; Karatsu, T.; Klingensmith, K. A.; Wallraff, G. M.; Miller, R. D. In Inorganic and Organometallic Polymers, ACS Symposium, Series No. 360, Zeldin, M.; Wynne, K. J.; Allcock, H. R., Eds., American Chemical Society, Washington, D.C., 1988, 61.
34. Michl, J.; Downing, J. W.; Karatsu, T.; McKinley, A. J.; Poggi, G.; Wallraff, G. M.; Sooriyakumaran, R.; Miller, R. D. Pure Applied Chem. 1988, 60, 959.
35. Harrah, L. A.; Zeigler, J. M. In Photophysics of Polymers, Hoyle, C. E.; Torkelson, J. M., Eds., ACS Symposium Series, No. 358, American Chemical Society, Washington, D.C., 1988, Chapt. 35.
36. Johnson, G. E.; McGrane, K. M., ibid., Chapt. 36.
37. A somewhat different photophysical model has been proposed recently for poly(methylphenylsilane) whose emission behavior is also different with the implication that it might apply generally (38).
38. Kim, Y. R.; Lee, M; Thorne, J. R. G.; Hochstrasser, R. H.; Zeigler, J. M. Chem. Phys. Lett. 1988, 145, 75.
39. Ishikawa, M.; Kumada, M. Adv. Organomet. Chem. 1981, 19, 51, and references cited therein.
40. Gaspar, P. P.; Holter, D.; Konieczny, C.; Corey, J. Y. Acc. Chem. Res. 1987, 20, 329, and references cited therein.
41. Drahnak, T. J.; Michl, J.; West, R. J. J. Am. Chem. Soc. 1979, 101, 5427.
42. Shizuka, H.; Lanaka, H.; Tonokura, M. Chem. Phys. Lett. 1988, 143, 225.
43. Nate, K.; Ishikawa, M.; Imamura, N.; Murakami, V. J. Polym. Sci. Part A: Polym. Chem. 1986, 24, 1551.
44. Ishikawa, M.; Hongzhi, N.; Matsusaka, K.; Nate, K.; Inoue, T.; Yokono, H. J. Polym. Sci., Polym. Lett. Ed. 1984, 22, 669.
45. Boudjouk, P.; Roberts, J. R.; Golino, C. M.; Sommer, L. H. J. Am. Chem. Soc. 1972, 94, 7926.
46. Hawari, J. A.; Griller, D.; Weber, W. P.; Gaspar, P. P. J. Organomet. Chem. 1987, 326, 335.
47. Trefonas, III, P.; West, R.; Miller, R. D.; J. Am. Chem. Soc. 1985, 107, 2737.
48. Karatsu, T.; Miller, R. D.; Sooriyakumaran, R.; Michl, J. J. Am. Chem. Soc. 1989, 111, 1140.
49. Boberski, W. G.; Allred, A. C. J. Organomet. Chem. 1974, 71, C27.
50. Turro, N. J.; Ramamurthy, V.; Cherry, W.; Farneth, W. Chem. Rev. 1978, 78, 125.
51. Ban, H.; Sukegawa, K. J. Appl. Polym. Sci. 1987, 33, 2787.
52. Todesco, R. V.; Kamat, P. V. Macromolecules 1986, 19, 196.
53. McKinley, A. J.; Karatsu, T.; Wallraff, G. M.; Miller, R. D.; Sooriyakumaran, R.; Michl, J. Organometallics 1988, 7, 2567.
54. Walsh, R. Organometallics 1988, 7, 75, and references cited therein.
55. Raabe, G.; Michl, J. J. Chem. Rev. 1985, 85, 419, and references cited therein.

56. Willson, C. G. In Introduction to Microlithography, ACS Symposium, Series No. 219, Willson, C. G.; Bowden, M. J., Eds., American Chemical Society, Washington, D.C., Chapt. 6.
57. Miller, R. D.; Guillet, J. E.; Moore, J. Polym. Preprints 1988, 29, 552.
58. Cotts, P. M.; Miller, R. D.; Trefonas, III, P. T.; West, R.; Fickes, G. N. Macromolecules 1987, 20, 1047.
59. Reichmanis, E.; Smolinsky, G.; Wilkins, Jr., C. Solid State Technol. 1985, 28(8), 130.
60. Hofer, D. C.; Miller, R. D.; Willson, C. G. Proc. of SPIE 1984, 469, 16.
61. Marinero, E. E.; Miller, R. D. Appl. Phys. Lett. 1987, 50, 1041.
62. Hansen, S. G.; Robitalli, T. E. J. Appl. Phys. 1987, 62, 1394.
63. Hofer, D. C.; Miller, R. D.; Willson, C. G.; Neureuther, A. R. Proc. of SPIE 1984, 469, 108.
64. Miller, R. D. In Advances in Chemistry Zeigler, J. M.; Fearon, F. G., Eds., American Chemical Society, Washington, D.C., 1988 (in press).
65. Miller, R. D.; McKean, D. R.; Tompkins, T. C.; Clecak, N.; Willson, C. G. In Polymers in Electronics, ACS Symposium, Series No. 242, Davidson, T., Ed., American Chemical Society, Washington, D.C., 1984, 25.
66. Taylor, G. N; Hellman, M. Y.; Wolf, T.; Zeigler, J. M. Proc. of SPIE 1988, 920, 274.

RECEIVED July 21, 1989

Chapter 9

Syntheses of Base-Soluble Si Polymers and Their Application to Resists

Shuzi Hayase, Rumiko Horiguchi, Yasunobu Onishi, and Toru Ushirogouchi

Chemical Laboratory, Toshiba Corporation, Komukai-Toshiba-cho, Saiwai-ku, Kawasaki 210, Japan

Three kinds of Si-polymers, which have Si in the main chain and are soluble in basic aqueous solutions, were synthesized. The first one is polysiloxane with a phenol group directly bonded to Si(Polymer(I)). The second one is polysilane substituted with a phenol group(polysilane(II)). The third one, polysilanes substituted with carboxylic acids-(polysilane(III)s) were synthesized from polysilane(II) and carboxylic acid anhydrides Polysilane(II) and polysilane(III)s were soluble in polar solvents such as alcohols, acetonitrile and basic aqueous solutions, however, they were insoluble in non-polar solvents such as toluene and hexane. On photolysis, the polysilane(II) and the polysilane(III)s behaved almost the same as alkylpolysilanes, except for polysilane(III)s with a chloride group which accelerated the photolysis. The polysilane(II) showed blue shifts in solvents making the stronger hydrogen bonding with phenol groups. The blue shift was also seen on cooling. Thermal stability of the polysilane(II) was the same as that of methylphenylpolysilane, however, the polysilane(III)s were somewhat unstable, compared to the polysilane(II), and the stability depended on the substituents. The solubility of polysiloxane(I) in basic solutions depended on the polymerization temperatures and the substituents on Si. The solubility decreased with an increase in the softening point. Polysilane(III)s and polysiloxane(I) were each exposed with UV light and developed with TMAH aqueous solution.

Organosilicon polymers are especially important as imaging layers in bilayer resist processes(1). A wide range of organosilicon resist materials, in which the organosilicon compound has been incorporated either into the polymer main chain or into pendant groups, has appeared in the literature(2-8). However, almost all polymers have been developed with organic solvents. Considering current VLSI production lines, the process with basic aqueous development will be preferred.

Some novolac resins to which alkylsilyl groups are bonded have been reported(9-13). However, these novolac resins did not have adequate oxygen reactive ion etching stability because the solubility in the basic solution decreased with an increase in the Si content, which limited the amount of Si groups which could be bonded to phenol groups. Since the oxygen RIE durability of the polymers containing Si in the main chain turned out to be superior when compared with that of the polymer containing Si in the side chain(14), polysiloxanes soluble in basic solutions have attracted interest(15-19).

Other polymers containing Si in the main chain are the polysilanes. The discovery that polysilane derivatives were radiation sensitive(20), coupled with their stability in an oxygen plasma, has led to a number of microlithographic applications(14, 21-24). For the polysilanes, the development was carried out with organic solvents such as THF or methyl ethyl ketone or alcohols. In some cases, it has been reported(14) that the distortion of the pattern occurs and residue is left. So far, many polysilanes have been synthesized and reported(20), however, almost all polysilanes had hydrophobic substituents such as aromatic or/and aliphatic groups. Our first interest was whether or not polysilane with a hydrophilic groups would be able to be synthesized and be stable. The physical properties as well as the resist properties were unknown. We tried to synthesize polysilanes substituted with a phenolic group, because the polysilanes fill our requirement and moreover polysilanes with various functional groups on the side chain should be accessible, starting from the polysilane under mild conditions.(Figure 1)

We now wish to report the syntheses and physical properties of polysilanes substituted with a phenolic group and carboxylic groups, and polysiloxanes with a phenolic group directly bonded to Si(Figure 2). We also describe basic aqueous development of these materials.

Experimental Section

Polysilane(II) was prepared by the method described in the previous literature(25). Polysilane(III)s were prepared by reacting polysilane(II) with carboxylic acid anhydrides in THF catalyzed by triethylamine at 20-50°C(Table 2)(26). Polysiloxane(I) was synthesized as follows. To 3.3g of NaOH in 40mL of water, 10g of M-7(Table 3) in 30ml of toluene was added. The mixture was stirred at 120°C for 3h, removing toluene. The reaction mixture was neutralized with dilute HCl aqueous solution. The polymer which was extracted with ether, was heated at 180°C under reduced pressure(run 2 in Table 4). PM-5(run6

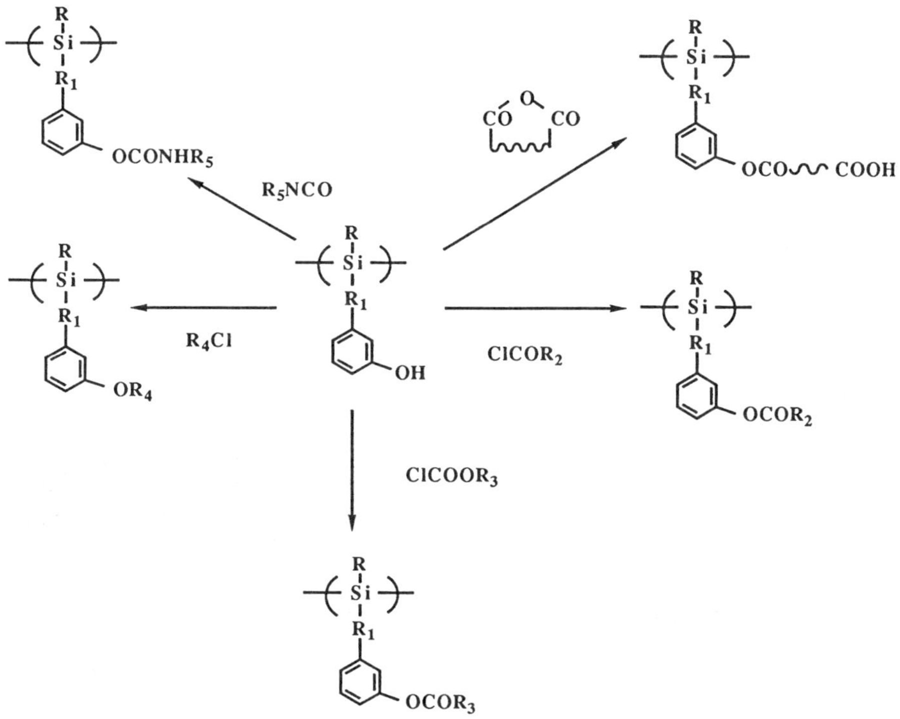

Figure 1. Potential reactions of polysilane(II).

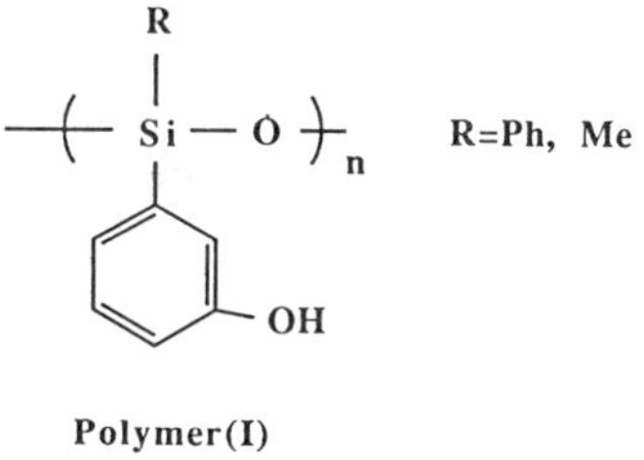

Polymer(I)

Figure 2. Polymer structure of I.

R=Ph, Me

in Table 4) was obtained by precipitating the crude polymer in toluene after hydrolysis of M-5 for 8h at 120 °C, followed by drying at 100°C.

Solubility of polymers in tetramethyl ammonium hydroxide aqueous solution was measured by dipping the wafer on which the polymer solution was spin-coated, for 1 min. at 25°C. The prebake was carried out at 90°C for 5min. Sensitivity of resists was measured after the exposure with CA 800(Cobilt) or KrF excimer laser($0.9 mJ/cm^2$/1 pulse). The polymer structure was determined by ^{1}H-NMR, ^{13}C-NMR(FX90Q apparatus,JEOL) and ^{29}Si-NMR. The molecular weight distribution was determined with a Toyo Soda Model 801 gel permeation chromatograph at 40°C. The four columns were connected in series, each packed with G-2000H8x3 and G-400H8(Toyo Soda polystylene gel), respectively.

Results and Discussion
Polysilanes substituted with a phenol group.
Synthesis
We tried to synthesize various polysilanes with a phenol group(25), however, the only one we were able to obtain was Polysilane(II). The synthetic route is shown in Figure 4. We chose a trimethylsilyl group as the protecting group of the phenol moiety, because it is easy to remove after the polymerization without damaging the Si-Si main chain, however, it has been reported that in some reactions, the Si-O-C bond cleavage takes place with Na dispersion(27).

The crude polysilane(II) was bimodal with a high molecular weight portion and a low molecular weight portion. The higher molecular weight portion(about 100,000) was fractionated easily as it is insoluble in toluene. The lower moleculer weight polymer was contaminated by a large amount of compounds containing Si-O-Si or Si-O-C linkages, which suggested that some cleavage of trimethylsilyloxy linkage took place during the polymerization. Conventional polysilanes are often bi or trimodal in their molecular weight distribution(20). Some improvements were employed in order to get mono-modal polymers(31-33). The mechanism of polymer formation is not known with certainty, but, some evidence has been presented to support the involvement of radicals or/and anions in the polymerization process(28-31). Our experimental result might imply that the polymerization process proceeds by more than one mechanism, one of which yields the polysilane(II) successfully.

We have concluded that in order to obtain the stable polysilane containing a phenol group our synthesis requires that a small alkyl group such as methyl must be a companion substituent on silicon, the phenolic group must be separated from Si by two carbons or more, and the position of the OH group on the aromatic ring must be preferably meta.(Figure 5).

^{1}H-NMR,^{13}C-NMR,^{29}Si-NMR and IR showed that the

polysilane(II) has the structure shown in Figure 3 and the main Si-Si chain was not disrupted by oxygen atom.
Table 1 shows the copolymerization results. The copolymerization with alkyl silanes gave good results, however, low molecular weight polymers were obtained upon copolymerization with phenylmethyl dichlorosilane. One possible explanation is that the mechanism of the propagation is changed when phenylmethylsilyl group is incorporated. Kim and Matyjaszewski have reported that sonochemical homopolymerization of dichlorosilanes is successful only for monomers with aryl substituents in nonpolar aromatic solvents. Dialkyldichlorosilanes do not react, but are copolymerized with phenylmethyl dichlorosilane. They explained this phenomena by electron transfer, changing radical species to anionic one(29). Another possible explanation is that the introduction of phenyl substituents decreases the access of the monomer to the active chain end, which increases the chance of side reactions involving in the trimethylsilyloxy group. On the other hand, the reaction of aryldichlorosilanes with Na is much more exothermic than that of dialkyldichlorosilanes. The more bulky the aryldichlorosilane, the more reaction giving by-products will come to compete with that giving polysilanes.
Solubility of the polysilane(II) was completely unlike conventional polysilanes with hydrophobic substituents. The polysilane(II) was soluble in acetone, alcohols and acetonitrile, and insoluble in toluene and hexane. Table 1 also shows the ratio(quantum yield of scission(Q(S))/quantum yield of crosslinking(Q(X)))(34). The value was similar to those of conventional polysilanes already reported(14,34).

Spectral properties and photochemistry.
Some solvent effects on the UV absorption have already been reported(35,36). Harrah et.al. reported that the UV absorption in hexamethyldisiloxane solution is not very different from that in THF(35). They also reported that toluene solution gives a somewhat different UV absorption than hexane. Recently, polysilane copolymers were reported to become more expanded in chloroform than in THF(52). There is no discription of the UV absorption properties of polysilanes in polar solvents.
Because the polymer has a hydrophilic substituent in one side and a hydrophobic substituent in the other side, the polymer may change the shape, depending on the characteristics of the solvents. In order to anticipate the polymer shape, intramolecular hydrogen bonding, intramolecular Van der Waals Force, hydrogen bonding between solvents and the polymer, and Van der Waals Force between them should be taken into account. In hydrophilic solvents, the OH in the polymer would face the solvent to make hydrogen bonding and the methyl group would interact with each other through Van der Waals Force

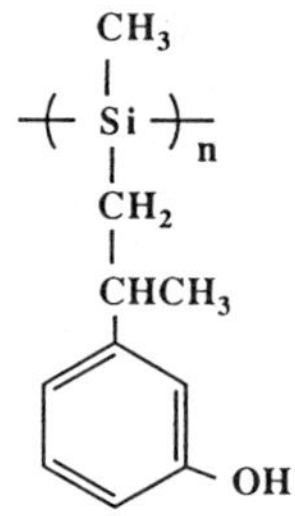

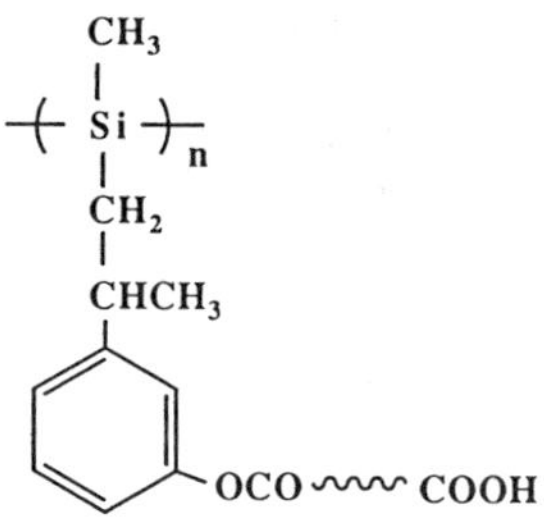

Polysilane(II) Polysilane(III)s

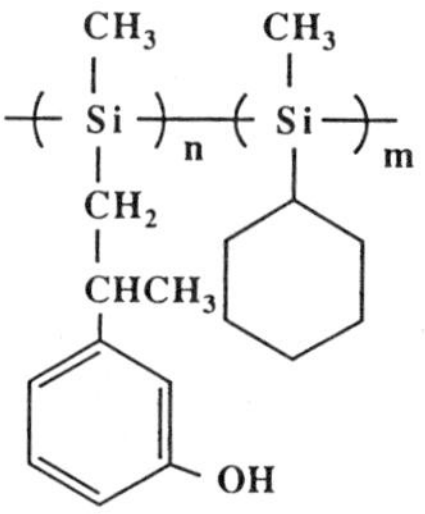

Polysilane(II-C)

Figure 3. Polymer structure of II and III.

Table 1 Molecular Weight and Quantum Yield of Copolymers

R	n	$Mw \times 10^{-5}$	$Q(s)/Q(x)$[1]
homopolymer		5.6	4.3
methyl	0.3	10.0	4.9
cyclohexyl	0.6	8.3	5.0
n-hexyl	0.3	1.5	4.0
phenyl	2.0	0.09	1.2

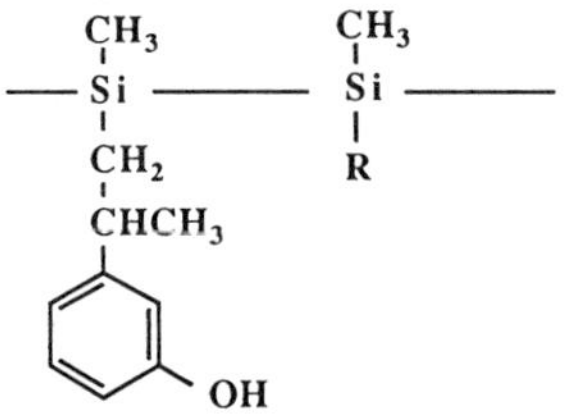

1 : n

1) Quantum efficiencies for scission / quantum efficiencies for crosslinking

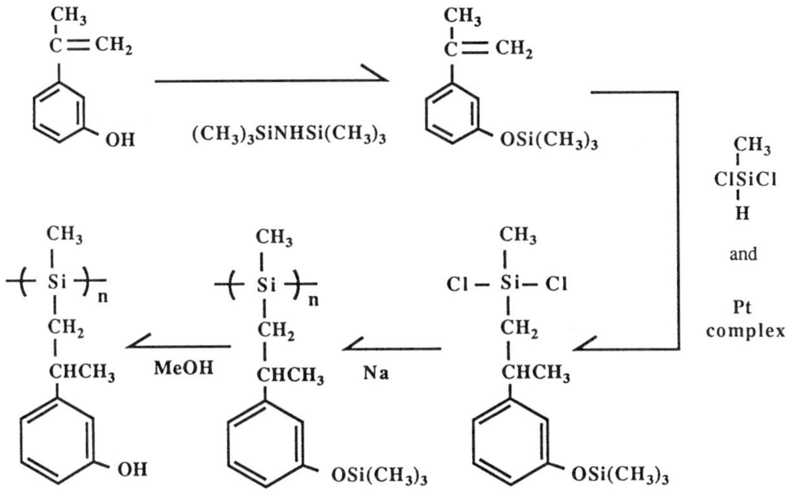

Figure 4. Synthetic route of II.

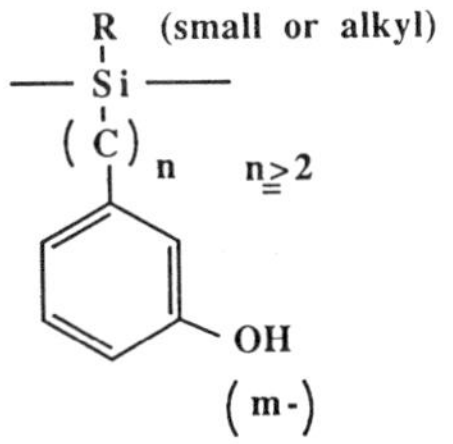

Figure 5. Structure of stable polysilane with a phenol group.

Table 2 Synthesis of Polysilanes with Carboxylic Acids

Polymer	P-1	P-2	P-3	P-4	P-5	P-6	P-7	P-8
Yield(%)	90	94	93	5	90	gel	gel	95
Functionality	0.30	0.44	0.50	0.30	0.16			0.30

P-1 -CO-C$_6$H$_{10}$-COOH

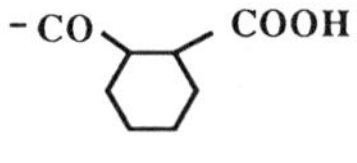

P-2 -CO-C$_6$H$_{10}$-COOH

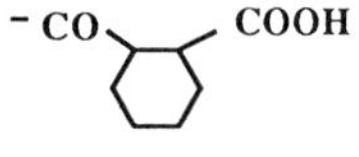

P-3 - COCH$_2$CH$_2$COOH

P-4 - COCH=CHCOOH

P-5 -CO-C$_6$H$_9$(CH$_3$)-COOH

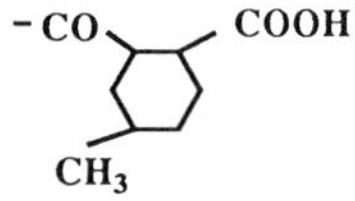

P-6 - COC(Cl)=CHCOOH

P-7 - CH$_2$CH=CH$_2$

P-8 - COCH$_2$Cl

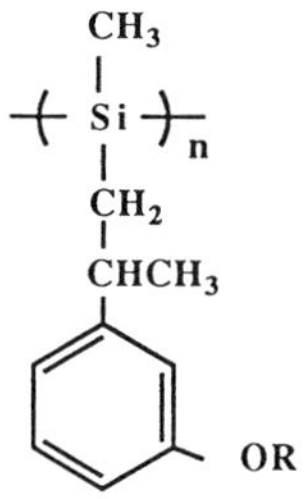

$$+\!\!\left[\begin{array}{c} \text{CH}_3 \\ | \\ \text{Si} \\ | \\ \text{CH}_2 \\ | \\ \text{CHCH}_3 \\ | \\ \text{C}_6\text{H}_4\text{-OR} \end{array}\right]_n$$

Table 3 Syntheses of monomers

abbreviation	monomer structure	monomer yield
M-6	EtO–Si(CH₃)(OEt)–[C₆H₄-OSiMe₃]	19%
M-5	EtO–Si(Ph)(OEt)–[C₆H₄-OSiMe₃]	45%
M-7	MeO–Si(Ph)(OMe)–[C₆H₄-OSiMe₃]	8%

$$R_1O-\underset{\underset{Cl}{|}}{\overset{\overset{R_2}{|}}{Si}}-OR_1 \; + \; \underset{Cl}{[C_6H_4\text{-}OSiMe_3]} \; \xrightarrow{Na} \; R_1O-\underset{\underset{[C_6H_4\text{-}OSiMe_3]}{|}}{\overset{\overset{R_2}{|}}{Si}}-OR_1$$

M-4

	R₁	R₂
M-1	Et	Ph
M-2	Et	Me
M-3	Me	Ph

Table 4 Polymerization of Alkoxysilanes

run	polymer	monomer	polymn. first[1] (C°)	(h)	polymn. second[2] (C°)	(h)	polymer yield(%)
1	PM-6	M-6	120	3	180	1	98
2	PM-7	M-7	120	3	180	1	86
3	PM-5	M-5	120	6	150	1/4	84
4	PM-5	M-5	120	3	180	1	81
5	PM-5	M-5	120	8	100	1	56
6[3]	PM-5	M-5	120	8	100	1[3]	33

1) Mixture of NaOH solution and monomer in toluene(experimental section)
2) After neutralization, 2mmHg
3) The polymer insoluble in toluene was dried(experimental section)

intramoleculerly in dilute solutions. In hydrophobic solvent, the shape would be unlike the above. Between them, a intermadiate shape may be present(Figure 6). The intermadiate shape may be rather expanded, compared to the other ones, because, the intramolecular interaction is weakened by the interaction with solvents. Assuming that the population of trans-conformation is larger in the expanded molecular shape, the UV absorption will be affected by the polymer shapes.

Figure 7 shows the λ_{max} of copolymer(II-C)(Table 1, R;cyclohexyl) in THF/methanol or THF/3-methyl pentane mixture. λ_{max} of copolymer(II-C) in THF was 301nm at 25°C. As methanol, more hydrophilic solvent, was added in the polymer solution in THF, the polymer solution showed a blue shift up to about 290nm. On the other hand, as 3-methylpentane, more hydrophobic solvent, was added in the polymer solution in THF, the polymer solution also showed a blue shift. This is explained very well by our model. Assuming that the polymers have the intermadiate shape (expanded shape) in THF, we would be able to see blue shift as more hydrophobic solvent or more hydrophilic solvents are added, if the population of trans conformation is larger in more expanded polymer chain.

λ_{max} of copolymer(II-C) in dioxane, diethylether, THF, ethanol, methanol and acetonitrile was 298, 297, 301, 296, 290, 290nm respectively. These solvent effect may be also explained by the model in Figure 7.

Much has been reported about the UV absorption changes on cooling. Trefonas and coworkers reported three kinds of bathochromic shifts, depending on the substituents on the Si backbone(37). Harrah and Zeigler found that the absorption of symmetrically and asymmetrically substituted polysilanes shift abruptly over a certain temperature range, which is attributed to rod to coil transition(36,38-40). It has also been reported that the transition temperature is not related to solvent polarity but correlates well with a solvent coupling constant(41-44). They also reported that polysilanes with branched or very large substituents bonded to the silicon backbone show blue-shifts on cooling(36,40). They explained this phenomena as a decrease in the average persistence length due to the progressive break-up of all-trans sequences in the polymer backbone(36,40). Johnson and McGrane also reported almost the same results(45,50). The suggestion that the polysilane chain in solution could be modeled by a statistical collection of weakly interacting chromophores of different length interrupted by conformational kinks was first proposed by Klingersmith and coworkers(62). Trefonas and his coworkers explained the gradual bathochromic shift on cooling as increasing relative populations of trans-conformations in the backbone(37). Bathochromic shifts were also seen for solid films(46-49).

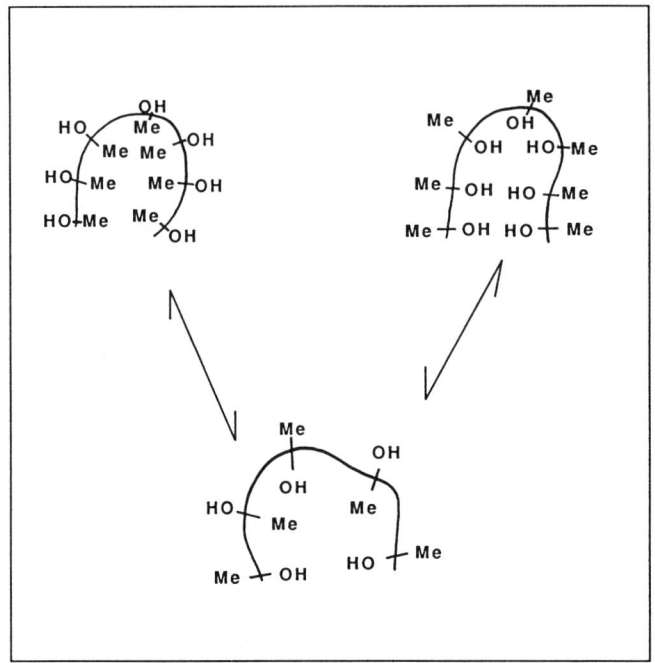

Figure 6. Relative molecular structure in polar solvents.

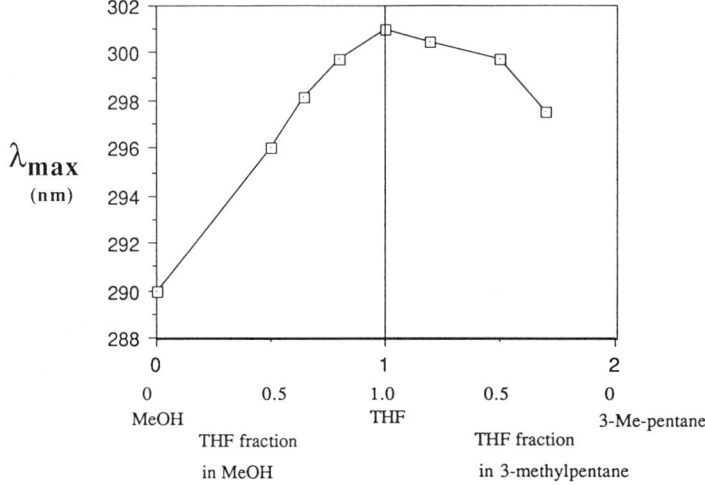

Figure 7. λ_{max} of polysilane(II-C) in methanol/THF or 3-methylpentane/THF. Polysilane(II-C): copolymer of polysilane(II) unit and cyclohexylmethylsilane unit.

UV shift of polysilane(II-C) on cooling were compared with that of poly(methyl cyclohexyl)polysilane in THF(Figure 8). Polysilanes with blanched substituents have been repoorted to show a slight blue shift on cooling(36). On our experiment, poly(methyl cyclohexyl)polysilane in THF showed 2 or 3 nm blue shift from 25 ºC to -196ºC. Polysilane(II-C) showed very large blue shift from 301nm at 25ºC to 290nm at -120 ºC. This suggests that stronger hydrogen bonding between solvent and polymer or increasing intramolecular hydrogen bonding makes the polymer conformationally distorted from the trans conformation.

^{29}Si-NMR spectrum of the polymer(II) shows a single peak at 31.20ppm assigned to Si backbone. It has been reported that polysilanes substituted asymmetrically, such as poly(hexylmethylsilane), showed multiplets containing from five to seven main peaks, which is caused by nonstereogenetic silicon center(54). This paper also reported that poly(phenethylmethylsilane) gave only broad single resonance at 32.16ppm, which is attributed to unresolved splitting. Peak width at half maximum in the polysilane(II) was 80Hz which was similar to that of poly(phenethylmethylsilane)(100Hz)(54). Therefore, it seems likely that the polysilane(II) is atactic.

This paper also reported that ^{29}Si chemical shifts for alkylpolysilanes decreased with increasing steric bulk of the substituents, varying inversely with the electronic excitation energy(54). It is also well known that electronic shielding of gauche is larger than that of trans. Figure 9 shows the result of ^{29}Si-NMR chemical shift measured in various solvents at various temperatures. As shown in Figure 9, there is a slight tendency toward a spectral blue-shift with increases in the ^{29}Si chemical shifts. The high field shift on cooling suggests that the population of gauche increased on cooling. This phenomena are consistent with blue shift on cooling.

Considering the results about solvent effects on the UV absorption and temperature dependence on the UV absorption, it seems likely that in our polysilanes, the popuration change of the trans conformation, which is caused by the relative strength bweteen intramolecular and intermolecular interaction involving hydrogen bonding, is responsible for the shift of the UV absorption.

Polysilanes substituted with carboxylic acids(polysilanes(III)).

Carboxylic acids were introduced by the reaction of the polysilane(II) with carboxylic acid anhydrides in the presence of an amine. Table 2 shows the results. In order to examine the effect of unsaturated bonds and halides in the side chain on photolysis, double bonds or chloride groups were introduced with the carboxylic acid. Polysilanes(III) were also soluble in polar solvents and basic aqueous solution. The solubility in tetramethyl ammonium hydroxide aqueous solution(TMAH) depended strongly

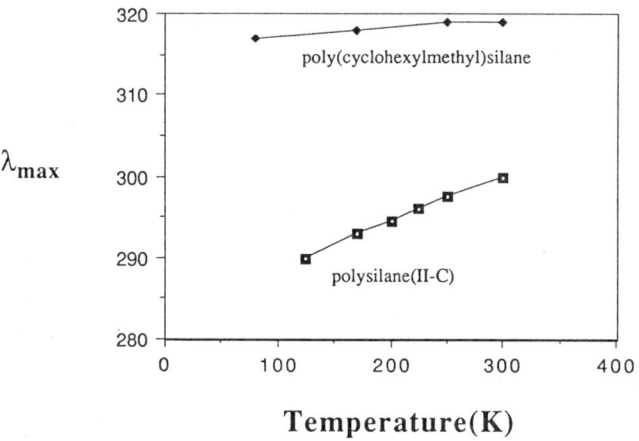

Figure 8. λ_{max} shift of polysilane(II-C) on cooling. For the structure of polysilane(II-C), see Figure 7.

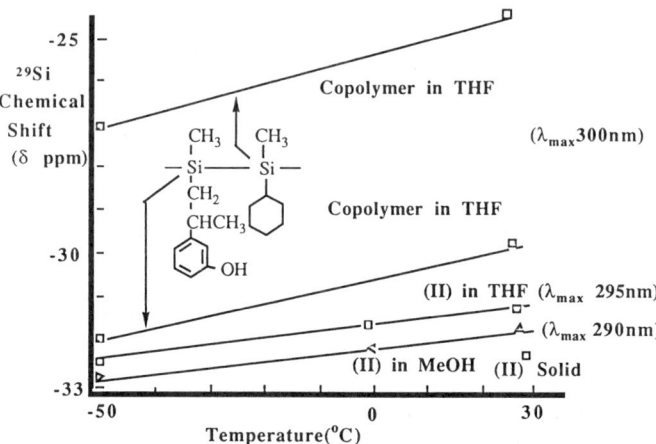

Figure 9. ^{29}Si NMR chemical shifts at various temperatures in polar solvents.

on the molecular weight as shown in Figure 10. Upon irradiation, the molecular weight decreased and the solubility increased rapidly when the molecular weight reached about 30000 in case of polymer(II) and 50000 for polysilane(P-3), which made the development using TMAH possible. The introduction of carboxylic acid groups increased the solubility of the polysilanes in base. The solubility of P-3 in 2.38% TMAH aqueous solution increased from 0nm/sec to 109nm/sec when the moiety content increased from 0 to 50mol%.

During photolysis, the double bond content of the polysilane(P-1)(15mol% in this experiment) decreased to 10mol%, as measured by ^{1}H-NMR spectroscopy. However, the ratio, quantum yield of scission($Q(S)$)/quantum yield of crosslinking($Q(X)$), was not affected by the reaction of the double bond. West and his coworkers have reported that poly((2-(3-cyclohexenyl)-ethyl)methylsilane-co-methylphenylsilane) crosslinked upon irradiation(55). The difference between our results and West's may lie in the amount of the double bond and inhibition of the radical closslinking by the phenol moiety. Polysilane with a halogen moiety, P-8, photodecomposed rapidly, compared with P-1 or P-3. The introduction of a chloride moiety was effective for the sensitization of the photodegradation. Similar results has already been reported(55).

Figure 11 shows the UV absorption change of P-3 during the photolysis. Alkyl polysilanes substituted with n-propyl, i-butyl or hexyl groups have been reported to show no large UV absorption change at 248nm during photolysis(14). The UV change of polysilane(P-3) was large at 248nm.

Thermal stability of polysilane(II) and (III)s.
Figure 12 shows DSC curves for poly(phenylmethylsilane-co-dimethylsilane), polysilane(II) and polysilane(III)s. The copolymer began to decompose thermally at around 250 °C. The introduction of the phenol moiety did not decrease the decomposition temperature. The behavior of P-1 and P-5, which have unsaturated bonds, were somewhat different. Both polymers had exothermic reactions at 110-120 °C, as well as around 250°C. The exothermic reactions at 110-120°C were not detected in case of the polysilane(II). The reaction at lower temperatures was attributed to cross-linking of the double bond in the side chain, which gave THF insoluble matter. P-3, without the double bond showed no change in solubility and the GPC curves were unaffected, however, P-3 turned out to form an acid anhydride at around 200°C by ^{1}H-NMR(26). Similar thermal cross-linking has been reported by West, et. al. in case of poly((2-(3-cyclohexenyl)ethyl)-methylsilane-co-methylphenylsilane(55). The presence of the double bond played a significant role in thermal reaction compared to photo-reaction.

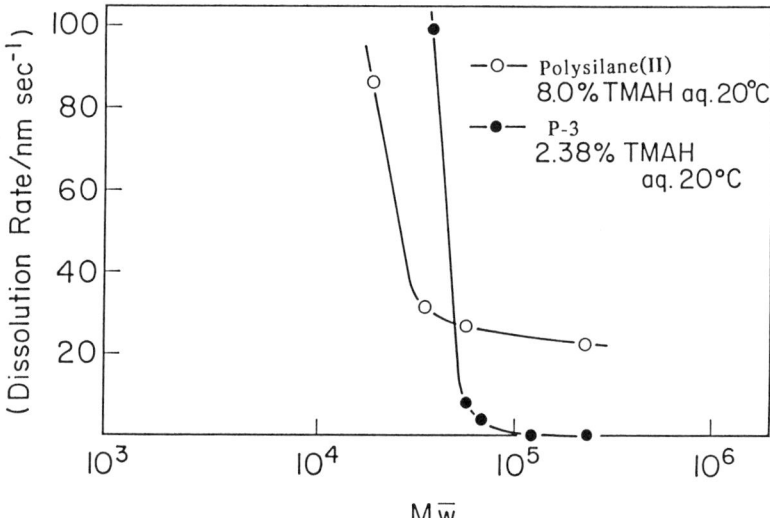

Figure 10. Solubility of some functionalized polysilanes in TMAH aqueous solutions during photolysis.

9. HAYASE ET AL. *Syntheses of Base-Soluble Si Polymers* 149

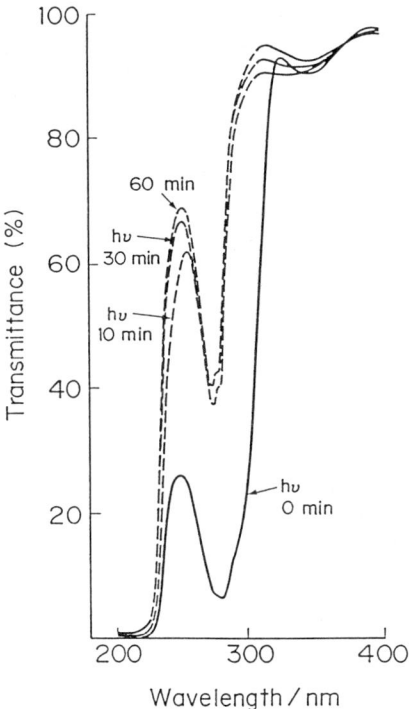

Figure 11. UV absorption change (P-3) during photolysis. Exposure, high-pressure mercury lamp (400 W). For the structure of P-3, *see* Table 2.

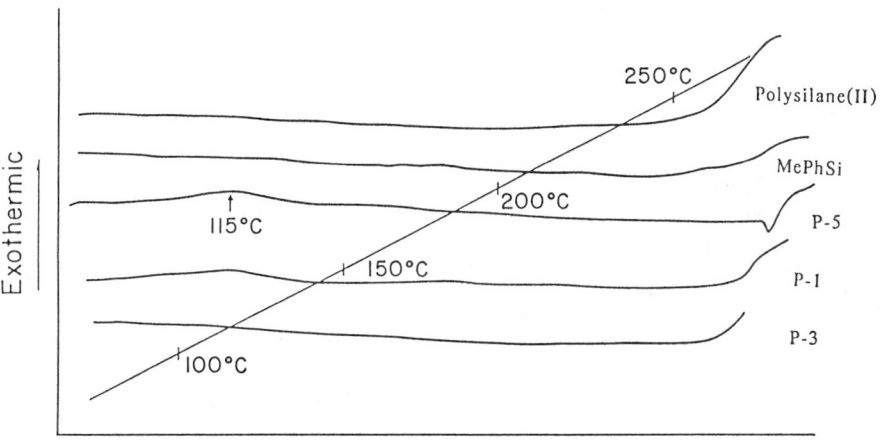

Figure 12. DSC curves of various polysilanes. MePhSi means methylphenylsilane copolymer.

Resist properties of polysilane(P-3).

20 weight % of P-3 in 2-acetoxyethyl ethyl ether solution was spin-coated to 0.7 micron thickness on a Si wafer, and baked at 90ºC for 5 min. Then, the wafer was exposed through a mask with a Kr/F excimer laser reduction step and repeat system (reduction ratio,10:1, NA, 0.37). Figure 13 shows the relation between the film thickness and the exposure dose when developed with a TMAH aqueous solution. When the resist thickness was normalized, the sensitivity was 1.5-2.5J/cm^2 and γ value was about 0.72.

The Figure 14 shows the cross-sectional pattern of the resist developed with a 2.38% TMAH aqueous solution. The mask had an equal line-space pattern. All the resist patterns shown in the photograph were obtained by 1.2 J/cm^2 exposure and 50 second, 25ºC development. The resist patterns, from 0.600micron line and space to 0.325 micron line and space, were the same as that of the mask size. However, the 0.300 micron line and space pattern was under-developed or under-exposed. The 0.275 micron line and space pattern was not opened. Figure 15 shows the relation between the mask size and the resist pattern size fabricated.

The resist pattern was triangular, which was probably due to phenyl silane unbleachable absorbance.

Polysiloxanes with phenol moiety directly bonded to Si.

So far, many syntheses of p, or m-silylphenol derivatives were reported(56-60). p-Trimethylsilyloxy trimethylsilyl benzene was synthesized from trimethylsilyl chloride and p-trimethylsilyloxychloro(or bromo)benzene in the presence of Mg,or Na(56,57,59,60). A similar compound was also synthesized from triphenylchlorosilane and p-lithio phenoxy lithium(58). m-Trimethylsilyloxytrimethylsilylbenzene was prepared from m-trimethylsilyloxy bromobenzene and trimethylsilylchloride(59).

We chose trimethylsilyloxy group as a protecting moiety for the phenol groups because it is easy to release after or during polymerization. Speier has reported that both ethyldimethyloxy-p-trimethylsilylbenzene and trimethylsilyloxy-p-dimethylethylsilylbenzene formed when p-trimethylsilyloxyphenylchloride was reacted with dimethylethylsilyl chloride(58). In o, or p-trimethylsilyl phenol, the thermal rearrangement of the trimethylsilyl group and the scission of Si-Ph in acid solutions were reported to take place(56). For the above reasons, the reaction of m-trimethylsilyloxy chlorobenzene with methyl or phenyl chlorosilane in the presence of Na were tried.

In order to avoid the formation of polysilanes, two of the three chlorides in the phenyl(or methyl)trichlorosilane were alkoxylated and the resultant phenyl(or methyl)dialkoxychlorosilane was reacted with the chloride using Na dispersion.

Figure 16 shows the synthetic route. Table 3 shows the

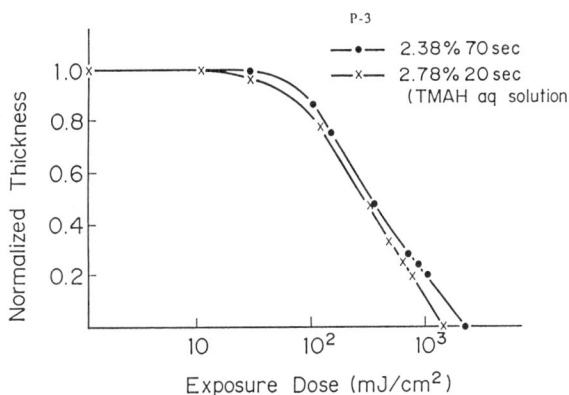

Figure 13. Sensitivity curve of P-3. Exposure, KrF excimer laser; development, TMAH aqueous solution.

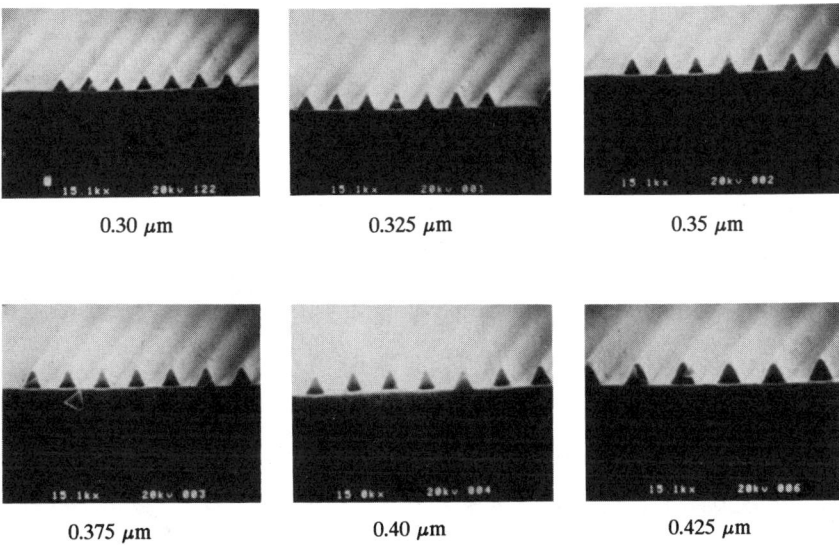

Figure 14. Cross section of polysilane (P-3) resist pattern exposed to KrF excimer laser. Development, 2.34% TMAH aqueous solution.

152   POLYMERS IN MICROLITHOGRAPHY

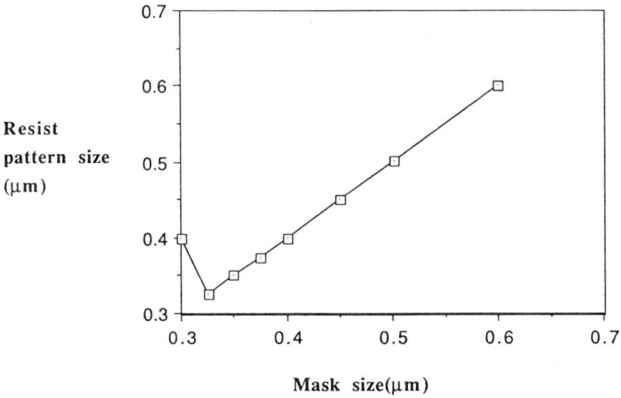

Figure 15. Relation between mask size and resist pattern size (*see* Figure 14).

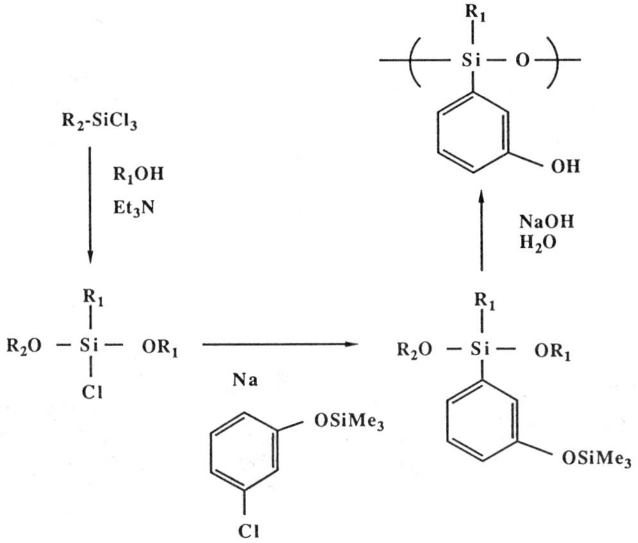

Figure 16. Synthesis of polysiloxane with phenol group directly bonded to Si.

yield of each monomer. The yield of M-7 from M-3 was only 8%, compared to that of 45% in case of M-5 from M-1. The yield of M-6 from M-3 was 19% which was lower than that of M-5. The large steric hindrance around Si-Cl of M-1 would prevent the M-1 from self-condensation which forms the disilane.

Polymerization was carried out in NaOH aqueous solution. Table 4 summarizes the reaction conditions and polymer yields. The first step of the polymerization was hydrolysis and oligomerization of the silylethers at low temperatures from 90°C to 150°C. In order to increase the polymer yield and the softening point of the polymers, the second step was carried out at higher temperatures under reduced pressure, removing alcohol and water. ^{1}H-NMR, ^{13}CNMR and IR spectroscopy of each polymer show that these polymers have the polysiloxane structure substituted with a phenol group.

The softening point and the solubility in TMAH aqueous solution varied with the polymerization temperaures and monomer structures, as shown in Table 5. When PM-5(run 5) was treated at 100°C, no condensation occurred. Further condensation started at 150°C(run3) and the polymer became insoluble in the TMAH aqueous solution at 180 °C treatment(run4). PM-7 and PM-6 did not become gel even at 180°C(run 1,2). Since the gellation is supposed to be caused by the both reaction of SiOH-SiOH(or SiOR) condensation and SiOH(or SiOR)-phenol condensation, in M-7 and M-6, which are less hindered around Si, the Si-O-Si formation would surpass the Si-O-C formation. On the contrary, in M-5, both reactions would take place competitively, which would cause the gellation. The increase in the molecular weight during the second process was seen by GPC. Fish and his coworkers have been reported that the hydrolysis of methyl(3-(1-oxypyridinyl)-dichlorosilane in aqueous ammonia gave polysiloxane with molecular weights ranging from 1000 to 10000, depending on the time of heat drying under vacuum(61).

As shown in Table 5, the softening point increased with an increase in the condensation temperature from 59°C to 179°C, however, the solubility in basic aqueous solution decreased. Methylsiloxane was more soluble in the base solution, however, the softening point was not as high as that of the phenylsiloxanes(PM-6).

Resist properties
Figure 17 shows the sensitivity curves of the resist containing PM-5(run-6)(75 weight%) and 5-naphthoquinone diazide sulfonate of 2,3,4-trihydroxybenzophenone(25 weight%). The average esterification ratio of the photoactive compound was 2.5 units per three OHs in the benzophenone. The γ value was 1.6 and the sensitivity was 90mJ/cm^2 when developed with 0.34% TMAH aqueous solution for 120 sec at 20°C. The resist containing M-5(run 3)

Table 5 Solubility in Base and Softening Point of Polymer(I)

run	solubility(nm/sec)[1]			softening point °C
	2.38%	1.09%	0.95%	
1	500	179	93	55-59
2	366	79	37	65-75
3	238	105	60	65-75
4	insoluble			150-170
5	-	120	80	59-69
6	-	115	-	59-79

1) TMAH(tetramethylammonium hydroxide aqueous solution)

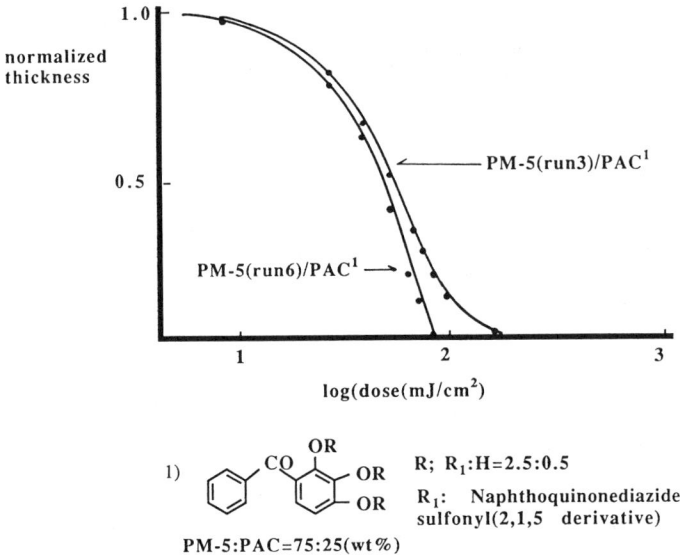

Figure 17. Sensitivity curve of resist containing polymer(I) and sensitizer. Exposure, CA800 (xenon lamp); development, 0.34% TMAH aqueous solution, 20 °C, 120 s.

which was treated at 150 °C, left some residue after development as seen in Figure 17. On the contrary, in the resist containing PM-6 or PM-7, no residue was produced. The introduction of much cross-linking to improve the softening point causes poorer development characteristics.

Conclusion

Polysilanes and polysiloxanes substituted with a phenol or carboxylic acids were soluble in TMAH aqueous solution. The resist containing the polysilanes or the polysiloxanes gave good patterns without leaving residue or distortion of the pattern. This investigation also showed that polysilanes with various functional groups can be synthesized.

Acknowledgment

We thank Professor Robert West in Wisconsin University for many helpful discussions. We also thank Dr. M. Nakase and T. Sato in Toshiba for excimer laser exposure.

References

1) Taylor,G.N.; Wolf,T.M. Polym. Eng. Sci. 1980, 20, 1087
2) Reichmanis,E,; Smolinsky,G.;Wilkins Jr,C.W. Solid State Technol. 1985, 28, 130.
3) Gozdz,A.S.;Carnazza,C.; Bowden,M.J. SPIE, 1986, 631, 28.
4) Bowden,M.J.; Gozdz,A.S. American Chemical Society, Polymer Preprint, 1987, 28(1), 317
5) MacDonald,S.A.; Ito,H.; Willson, C.G. Microelectronic Eng. 1983, 1, 269.
6) Babich,E.; Paraszczak,J.; Hatzakis,M.; Shaw,J. Microelectronic Eng. 1985, 3, 279.
7) Saigo,K.; Ohnishi, Y.; Suzuki,M.; Gokan,H. J. Vac. Sci., Technol. B, 1985, 3(1), 331.
8) Reichmans,E.; Smolinsky,G. SPIE, 1984, 469, 38.
9) Wilkins,C.W.; Reichmais,E.; Wolf,T.M.; Smith,B.C. J. Vac. Sci., Technol. B, 1985, 3(1), 306
10) Tarascon,R.G.; Shugard,A.; Reichmanis,E. SPIE, 1986, 631, 40.
11) Smith,B.C.; Hellman,M.Y.; E. Reichmanis, Paper presented at the "18th Middle Atlantic Regional Meeting, ACS" Newark, NJ. May 21-23, 1984.
12) Saotome,Y.; Gokan, H.; Saigo, K.; Suzuki, M.; Ohnishi, Y. J. Electrochem. Soc., 1985, 132, 909.
13) Miura,K.; Ochiai,T.; Kameyama,Y. SPIE, 1986, 631, 206.
14) For example, Taylor, G.N.; Hellman, M.Y.; Wolf,T.M. SPIE, 1988, 920, 274.
15) Onishi,Y.; Horiguchi,R.; Hayase,S. Proceedings of IUPAC chemrawn, Poster Presentation, IB01, May 1987, Tokyo Japan.
16) Sugiyama,H.; Inoue,T.; Mizushima,A.; Nate,K. SPIE, 1988, 920, 268.
17) Toriumi,M.; Shiraishi,H.; Ueno,T.; Hayasi,N.; Nonogaki,S.; Sato,F.; Kadota,K. J. Electrochem. Soc. 1987, 1345, 936.

18) Imamura,S.; Tanaka,A.; Onise, K. SPIE, 1988, 920, 291.
19) Noguchi,T.; Nito,K.; Seto,J. SPIE, 1988, 920, 168.
20) For example, West,R.; Maxka,J. ACS Sym. 1988, Series 360, 6
21) Hofer,D.; Miller,R.D.; Willson,C.G. SPIE, 1984, 469, 16.
22) Hofer,D.; Miller,R.D.; Willson,C.G.; Neureuther,A.R. SPIE, 1984, 469, 108.
23) Zeigler,J.M.; Harrah,L.A.; Johnson,A.W. SPIE, 1985, 539, 166.
24) Miller,R.D.; Hofer,D.; Rabolt,J.; Sooriyakumaran,R.; Wilson,C.G.; Fickes,G.N.; Guillet,J.E.; Moore,J. ACS Symp., 1987, Series 346, 170.
25) Horiguchi,R.; Onidhi,Y.; Hayase,S. Macromolecules, 1988, 21, 304.
26) Hayase,S.; Horiguchi,R.; Onishi,Y. Macromolecules, in press.
27) Speier,J.L. J. Am.,Chem. Soc. 1952, 74, 1003.
28) Worsfold,D.J.; American Chemical Society, Polymer Preprints, 1987, 28, 415.
29) Kim,H.K.; Matyjaszewski,K. J. Am. Chem. Soc. 1988,110, 3321.
30) Matyjaszewski,K.; Chem,Y.L.; Kim,H.K. ACS Symp. Ser. 1988, 360, 78.
31) Miller,R.D. et.al., in Materials for Microlithography, Thompson,L.F.; Willson,C.G.; Frechet,J.M.J, Eds., American Chemical Society, Washington, DC, 1984, 294.
32) Schilling,Jr.C.L.; Williams,T.C. American Chemical Society, Polymer Preprints, 1984, 25(1), 1.
33) Zeigler,J.M. American Chemical Society, Polymer Preprints, 1986, 27, 109
34) West,R.; Miller,R.D.; Hofer,D.C. J. Polymer Sci., Polym. Lett. Ed. 1983, 21, 823.
35) Harrah,L.A.; Zeigler,J.M. J. Polymer. Sci., Polym. Lett. Ed., 1985, 23, 209.
36) Harrah,L.A.; Zeigler,J.M. ACS Sym. Ser. 1986, 358, 482.
37) Trefonas,P.III; Damewood,J.R.; West,R.; Miller,R.D. Organometallics, 1985, 4, 1318.
38) Harrah, L.A.; Ziegler,J.M. J. Polymer Sci., Poly. Lett. Ed., 1985, 23, 209.
39) Harrah,L.A.; Zeigler,J.M. Bull. Am. Phys. Soc. 1985, 30, 540.
40) Zeigler,J.M.; Adolf,D.; Harrah.L.A. Proc. 20th Organosilicon Sym. 1986, Abs. p-2.25.
41) Schweizer,K.S. Chem. Phy. Lett. 1986,,125, 118.
42) Schweizer,K.S. J. Chem. Phy., 1986, 85, 1176.
43) Schweizer,K.S. J. Chem. Phys., 1986, 85, 1156,
44) Schweizer,K.S. American Chemical Society, Polymer Preprints, 1986, 27, 254.
45) Johnson,G.E.; MacGrane,K.M. ACS Symposium Ser. 1986, 358, 499.
46) Miller, R.D.; Hofer,D.; Rabolt,J.; Fickes,G.N. J. Am. Chem. Soc., 1985, 107, 2172.
47) Schilling,F.C.; Bovey,F.A. Americal Chemical Society, Polymer Preprint, 1988, 29, 72.

48) Zeigler,J.M. Macromolecules, 1986, 19, 2657.
49) Zeigler,J.M. Macromolecules, 1986, 19, 2660.
50) Johnson,G.M.; McGrane,K.M. Americam Chemical Society, Polymer Preprints, 1986, 27, 352.
51) Cotts,P.M.; Miller,R.D.; Trefonas III,P.T.; West,R., Fickes,G.N. Macromolecules, 1987, 20, 1046
52) Sawan,S.P.; Tsai,Y.; Huang,H; Muni,K.P. American Chemical Society, Polymer Preprint, 1988, 29(1), 252.
53) Cotts,P.M. Am. Chem. Soc. Polym. Mater. Sci. Eng. 1985, 53, 336.
54) Wolff,A.R.; Maxka,J.; West,R.; J. Polymer Sci., Part A. 1988, 26, 713.
55) Stuger,H.; West,R. Macromolecules, 1985, 18, 2349.
56) Fisch,K.C.; Shroff,P.D. J. Am. Chem. Soc. 1953, 75, 1249.
57) Speier,J.L. J. Am. Chem. Soc. 1952, 74, 1003.
58) Benkeser, R.A.; DeBoer,C.E.; Robinson,R.E.; Sauve,D.M. J. Am. Chem. Soc. 1956, 78, 6892.
59) Benkeser,R.A.; Krysiak, H.R. J. Am. Chem. Soc. 1953, 75, 2421.
60) Larson,G.L.; Nemeth,V.; Velente, H. Syn. Reac. Inorg. Metal-Org. Chem., 1976 6, 21
61) Fish,D.; Khan,I.M.; Smid,J. American Chemical Society, Polymer Preprint, 1988, 29, 8.
62) Klingensmith. K.A.; Downing, J.W.; Miller, R.D.; Michel, J. J. Am. Che. Soc., 1986, 108, 7438

RECEIVED August 10, 1989

Chapter 10

Lithographic Evaluation of Phenolic Resin–Dimethyl Siloxane Block Copolymers

M. J. Jurek and Elsa Reichmanis

AT&T Bell Laboratories, 600 Mountain Avenue, Murray Hill, NJ 07974

Reactive, functionally terminated poly(dimethyl siloxane) oligomers (PDMSX) have shown utility as the silicon-containing component in bilevel resists based on conventional novolac-diazonaphthoquinone chemistry. The ability to photoimage the novolac-siloxane block copolymer depends on the initial siloxane block length, the chemical structure of the novolac resin and the microphase separation of the system. Glass transition temperature measurements, transmission electron microscopy and Auger depth profiling experiments indicate that extensive microphase separation occurs if high molecular weight siloxane oligomers are used. When phase mixing of the two components was promoted in an o-cresol novolac-siloxane copolymer (PDMSX = 510 g/mole), 0.5 µm L/S patterns could be resolved using deep-UV exposures (248 nm) at a dose of 156 mJ/cm^2. Good solubility in aqueous tetramethylammonium hydroxide was observed with the copolymer at 10 wt % silicon and an O_2 reactive ion etching resistance of 1:18 compared to hard-baked HPR-204 was obtained. The 0.5 µm L/S patterns were transferred through 1.4 µm thick planarizing layer yielding features with high aspect ratios.

There is a current drive in microlithography to define submicron features in bilevel resist structures. The introduction of organometallic components, most notably organosilicon substituents, into conventional resists is one promising approach. To this end, organosilicon moieties have been primarily utilized in starting monomers (1-4) or in post-polymerization functionalization reactions on the polymer (5,6). Little work has been done on the reaction of preformed reactive oligomers to synthesize block copolymer systems.

Historically, block copolymers have found utility as a means to achieve a balance of properties between chemically different homopolymers (7-8).

Additionally, some properties unique to both systems may result. The majority of homopolymer blends are immiscible with one another and often experience poor interfacial adhesion between the separate phases. Since block copolymers are covalently linked together, macroscopic incompatibility at the interface is minimized. The macroscopic incompatibility of a two-polymer blend may be eliminated by the addition of a block copolymer derived from the two systems. Hence, copolymers can be used to strengthen blends of immiscible polymers by serving as emulsifiers (7-9).

The synthesis of block copolymers can be performed via several well-known synthetic procedures. The approach we have employed combines preformed, functionalized poly(dimethyl siloxane) oligomers with various novolac resins to form the block copolymer species. These copolymer systems can be envisioned more realistically as block copolymer-modified novolac resins since no fractionation steps designed to remove unreacted novolac homopolymer were used. Several molecular parameters were investigated, among them the block length of poly(dimethyl siloxane) and the molecular weight and chemical structure of the novolac resin. The choice of oligomer block length determines the final polymer's morphology, silicon content and aqueous base solubility. Earlier work with a chlorinated polymethylstyrene-polydimethyl siloxane system postulated that the microphase separated morphology of the copolymer could play a role in determining the resolution limits when used as a photoresist (10).

The work described in this paper reports the lithographic response of a number of novolac-siloxane copolymer systems. The chemical structure of the siloxane oligomer was held constant throughout the investigation. The number average molecular weight ($<M_n>$) was varied from 4400 (high molecular weight) to 510 g/mole (low molecular weight). The effect of novolac structure on the lithographic response of novolac-siloxane copolymers was studied using an o-cresol novolac, a 2-methyl resorcinol novolac and poly(hydroxystyrene). The o-cresol based phenolic resin could be reproducibly synthesized and had good aqueous base solubility; 2-methyl resorcinol exhibited improved solubility and higher T_g than the o-cresol system; and poly(hydroxystyrene) possessed a higher transmittance at 248 nm than either novolac. To prepare potential bilevel resist materials, the silicon incorporation was kept at >10 wt %.

EXPERIMENTAL

Tetramethylammonium hydroxide, TMAH, (Fluka Chemicals) was diluted with distilled water from a 25 wt % aqueous solution. In all cases the diazonaphthoquinone dissolution inhibitor used was Fairmont Positive Sensitizer #1009 (Fairmont Chemical Company). The syntheses of the PDMSX oligomers and novolac-PDMSX block copolymers have already been reported (11). The dimethylamine terminated poly(dimethyl siloxane), $<M_n>$ = 510 g/mole (Petrarch), was used as the PDMSX component or to prepare higher molecular weight analogs.

Resist solutions of o-cresol novolac-siloxane copolymers were prepared as 15 w/v % solutions of the polymer in 2-methoxyethyl acetate using 20 wt % (based on polymer) of the positive sensitizer. Poly(hydroxystyrene) and 2-methyl resorcinol copolymers were spun into films from 2-methyl tetrahydrofuran. Solutions were filtered through successive 1.0, 0.5 and 0.2 μm filters and stored in amber bottles. Resist films (0.5 µm thick) were spun onto four inch silicon substrates that had been coated with either 1.4 µm hard-baked Hunt Photoresist (HPR-204) or 0.4 µm thermally grown SiO_2 pretreated with hexamethyldisilazane. All films were prebaked at 105°C for 1 hour prior to exposure. Auger depth profiling and transmission electron microscopy samples were prepared without sensitizer as 0.1µm thick films on silicon substrates coated with 0.4µm SiO_2.

Deep-UV exposures were performed on a Karl Suss MA 56M contact aligner fitted with a Lambda Physik KrF excimer laser ($\lambda = 248.7$ nm) operating at an output of 13 mJ/cm^2/sec at 100 Hz. Resist films were dip developed in aqueous TMAH and rinsed in distilled water.

Pattern transfer through the HPR-204 planarizing layer via O_2 reactive ion etching (RIE) was achieved using a Plasma Technologies parallel plate etching unit at −340 V self-bias, 90 W power, 20 sccm O_2, at a pressure of 30 mtorr using a 10% overetch. All thickness measurements were performed on a Dektak IIA profilometer. Scanning electron microscope (SEM) photographs were obtained on a Hitachi S-2500 SEM.

Materials Characterization

The molecular weight ($<M_n>$) of the poly(dimethyl siloxane), PDMSX, was determined by both proton NMR and non-aqueous potentiometric titration (5). Proton NMR was used routinely to determine $<M_n>$ immediately after synthesis and prior to use in any copolymerization. Experimental confirmation of percent silicon in the copolymers was determined by elemental analysis (Galbraith Laboratories).

Novolac molecular weights were measured in THF at 35°C by high pressure size exclusion chromatography using a Waters Model 510 pump (flow rate = 1.0 ml/min), 401 differential viscometer detector and a set of Dupont PSM 60 silanized columns. A universal calibration curve was obtained with a kit of 10 narrow molecular weight distribution, linear polystyrene standards from Toya Soda Company. Data acquisition and analysis were performed on an AT&T 6312 computer using ASYST Unical 3.02 software supplied with the Viscotek instrument.

A Perkin-Elmer DSC-7 was used for thermal characterization of the starting phenolic resins and copolymers prepared. Scans were run at 10°C/min using sample weights of 15-25 mg. In all cases the amorphous, powdered polymers were used for the evaluation.

RESULTS AND DISCUSSION

The structures of the dimethylsiloxane block copolymers and respective parent homopolymers prepared for use as positive, bilevel resist materials are shown in Figure 1. Most copolymers were synthesized with >10 wt % silicon. The selection of PDMSX block length and novolac chemical composition proved to be the two most critical variables in achieving adequate resolution.

The quantitative nature of the silylamine-phenol reaction has been demonstrated for several different polymer systems (7). In our case, the charged PDMSX content was low to ensure that <1 phenolic group per novolac molecule reacted. This was done primarily to prevent extensive branching or crosslinking, and problems of insolubility and reproducibility associated with network formation

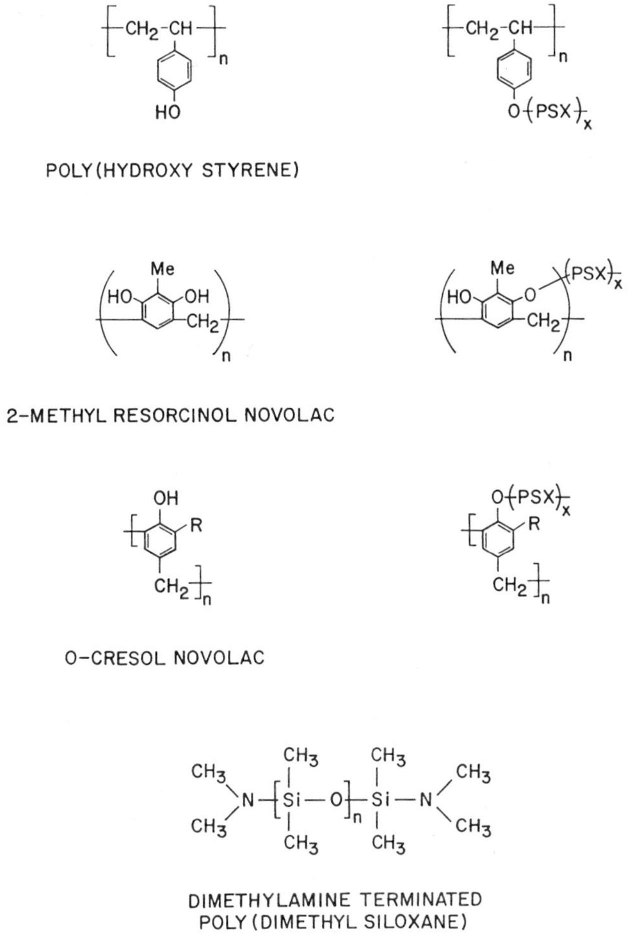

Figure 1   Structure of phenolic resin-poly(dimethyl siloxane) copolymers.

(11). No free PDMSX should be found in these systems; however, unreacted starting novolac will be present. Hence, each copolymer system consisted of a blend of novolac resin and novolac-PDMSX copolymer. The structure of these copolymer systems is complex. Since all phenolic groups possess equal reactivity towards silylamine groups, the difunctional PDMSX may react anywhere between two novolac chains (an A-B-A triblock species) or twice with the same novolac molecule. The second reaction type would yield a macrocyclic compound whose size would depend upon the proximity of the two phenolic groups to one another.

The molecular weight values for the starting oligomers and final copolymers were determined by gel permeation chromatography (GPC) and are shown in Table I. The initial molecular weight distributions (MWD = $<M_w>/<M_n>$) were less than two for both novolac resins. A portion of the low molecular weight components were removed during the three precipitations used during work-up, thus narrowing the observed MWD. The MWD of the PDMSX oligomers could not be determined by GPC due to reaction of the silylamine end groups with the column packing.

TABLE I

MOLECULAR WEIGHT DETERMINATION OF POLY(DIMETHYL SILOXANE) COPOLYMERS VIA GEL PERMEATION CHROMATOGRAPHY

PHENOLIC COMPONENT	SILOXANE $<M_n>$	wt % Si	POLYMER $<M_n>$	$<M_w>$	MWD
o-cresol	–	–	1450	2890	1.99
	4400	12.1	2200	3530	1.61
	4400	3.2	1870	3430	1.83
	510	10.2	1920	3480	1.82
	510	12.9	1960	4510	2.29
2-methyl resorcinol	–	–	1610	3050	1.89
	1770	10.1	2670	4800	1.80
	510	10.4	3190	6180	1.94
poly(hydroxy styrene)	–	–	11700	35900	3.07
	4400	12.3	14000	51800	3.71

Upon copolymerization of each novolac with the PDMSX oligomers, both $<M_n>$ and $<M_w>$ increased as expected, although not in a strictly additive manner since we have not formed a single block copolymer structure (7) but are measuring the molecular weights of a copolymer/novolac blend. The values obtained are complicated by the loss of additional low molecular weight material during work-up. The initial reaction mixture in tetrahydrofuran (THF) was coagulated in water giving 97% yield. A proton NMR of an ether extract of this water indicated the presence of novolac but no siloxane was observed. Only low molecular weight novolac would be expected to remain soluble in this aqueous THF solution.

The poly(hydroxystyrene) homopolymer (PHS) and PHS-PDMSX copolymer exhibited more predictable behavior due to the higher initial molecular weight of PHS. Both $<M_n>$ and $<M_w>$ increased and the MWD broadened after copolymerization. Incorporation of PDMSX increased the molecular weight averages. No low molecular weight material was found in the aqueous precipitating medium, indicating that all unreacted low molecular weight poly(hydroxystyrene) was precipitated.

The effect of the phenolic component's chemical structure (o-cresol vs. poly(hydroxystyrene) vs. 2-methyl resorcinol) on lithographic properties was studied. A high molecular weight (4400 g/mole) PDMSX precursor was used to maximize silicon content while minimizing the number of phenolic hydroxyl groups lost to copolymer formation. Each of the three phenolic resin-siloxane copolymers showed favorable resistance to O_2 plasma etching (at >10 wt % silicon) with selectivities of 1:12 or greater compared to a hard-baked planarizing layer (11). Additionally, each system exhibited good solubility in aqueous TMAH solutions. The lithographic response of each copolymer system was however, quite different. While the novolac materials exhibited positive resist behavior, the poly(hydroxystyrene-dimethyl siloxane) material behaved as a negative resist upon exposure to 248 nm radiation when used in conjunction with 10 or 20 wt % of the diazonaphthoquinone inhibitor. The sensitivity, expressed as the 50% gel dose, was 290 mJ/cm^2 with a contrast of 1.2. A 25:1 ethanol-water developer was employed but very poor resolution was observed.

A series of o-cresol novolac-PDMSX materials with a range of silicon contents (3.2-16.1 wt %) were prepared to examine the O_2 RIE response and to determine whether a solubility limit existed as a function of wt % silicon. All copolymer preparations exhibited good solubility in dilute TMAH and an asymptotic relationship (12) between wt % silicon and film thickness loss (3,12) during an O_2 plasma RIE process was found.

The thermal properties of the o-cresol novolac-PDMSX copolymers and the starting oligomers are shown in Table II. The T_g for the PDMSX oligomer (−123°C) did not change as the $<M_n>$ varied from 510 to 4400 g/mole. After copolymerization, a T_g for both the PDMSX component and the o-cresol novolac was obtained. Slight decreases in the high temperature T_g and increases in the low temperature T_g may be attributed to partial phase mixing between the two components (14, 15, 18). This becomes more evident as the PDMSX block length becomes smaller, and the thermodynamic solubility parameters become increasingly favorable. Additionally, the microphase separated domain sizes are expected to become smaller as PDMSX block lengths decrease (15).

The thermal properties of the 2-methyl resorcinol, poly(hydroxystyrene) and the PDMSX copolymers prepared with them are shown in Table III. For both copolymer systems using 4400 g/mole PDMSX blocks there was no significant

TABLE II

THERMAL CHARACTERIZATION OF O-CRESOL NOVOLAC-DIMETHYL SILOXANE COPOLYMERS

SAMPLE NUMBER	SILOXANE BLOCK		POLYMER	
	$<M_n>^1$	T_g (°C)^2	wt % Si^3	T_g (°C)^2
1	4400	−123	16.0	−122, 79
2	4400	−123	12.1	−121, 77
3	4400	−123	5.4	−122, 77
4	4400	−123	3.2	−122, 74
5	1770	−121	5.4	−95, 67
6	510	−121	10.2	−64^4, 51
7	510	−121	12.9	−21^4, 36
8	−	−	0	76

[1] $<M_n>$ determined from proton NMR.
[2] Values reported from differential scanning calorimetry.
[3] Determined by elemental analysis.
[4] Very weak transitions.

change in either the low or high temperature T_g values. As observed with the o-cresol novolac-PDMSX systems, the large PDMSX blocks were cleanly microphase separated. As the siloxane $<M_n>$ was decreased to 510 g/mole, no appreciable phase mixing occured in the resorcinol-based system as reflected by the lack of change in either the low or high T_g's of these copolymers. A second consideration is the solubility parameter (δ), which also contributes to this behavior. The δ for PDMSX (16) is 7.3 (cal/cm^3)^{½} while the calculated values for the o-cresol novolac, 2-methyl resorcinol and poly(hydroxystyrene) are 10.8, 11.9 and 9.4 (cal/cm^3)^{½} respectively (17). Differences in δ of greater than one between polymer blocks generally result in the observation of microphase separation (7,8). Although this phenomenon should occur in both the novolac and PHS systems, a greater degree of phase mixing should be possible in the o-cresol and hydroxystyrene materials. Based on the thermal data, one may conclude that the solubility parameter difference among the novolac resins results in less phase mixing for the 2-methyl resorcinol copolymer system in comparison to the o-cresol novolac-PDMSX copolymer. The results obtained for the PHS based material are less conclusive.

Transmission electron microscopy (TEM) was used to further explore the microstructure in each of the three copolymer systems. The 2-methyl resorcinol-PDMSX copolymer was expected to show the greatest degree of microphase

TABLE III

THERMAL CHARACTERIZATION OF POLY(DIMETHYL SILOXANE) COPOLYMERS WITH POLY(HYDROXY STYRENE) AND 2-METHYL RESORCINOL

PHENOLIC COMPONENT	SILOXANE BLOCK $<M_n>^1$	T_g (°C)^2	POLYMER wt % Si^3	T_g (°C)^2
2-methyl resorcinol	–	–	0	186
	4400	–123	11	–121, 184
	1770	–121	10	–119, 175
	510	–121	10	–120, 186
poly(hydroxy styrene)	–	–	0	172
	4400	–123	12	–119, 174

[1] $<M_n>$ determined by proton NMR.
[2] Values reported from differential scanning calorimetry.
[3] Determined from elemental analysis.

separation based on solubility parameter differences and thermal data. TEM photographs (207,000 x) reveal spherical domains of 400-600Å for copolymers prepared with 4400 g/mole PDMSX (Figure 2a) and 100-300Å domains with 510 g/mole oligomers (Figure 2b). The effect of decreasing PDMSX molecular weight on domain size was previously observed in a series of poly(methyl methacrylate)-PDMSX copolymers (15,18). The o-cresol novolac-PDMSX copolymers (Figure 3) exhibited a similar trend at high molecular weight PDMSX (Figure 3a); however, shorter siloxane block lengths resulted in a featureless surface (Figure 3b). This corroborates the thermal data which indicates that a large degree of phase mixing occurs in this copolymer. A TEM of the PHS-PDMSX copolymer (4400 g/mole PDMSX) is shown in Figure 4. The domain size in this copolymer is much smaller than that observed for the other copolymers at similar siloxane block lengths. This may be attributed to the small solubility parameter difference between the two components.

The microphase separation and preferential surface migration of PDMSX in various copolymer systems has been investigated (7,8,13-15,18). Auger depth profiling experiments were run to determine the surface-to-bulk film concentration of PDMSX in each copolymer system. Carbon, oxygen and silicon contents were monitored as a function of sputter depth, and the normalized percents for each element (±5%) are shown in Figure 5. The 2-methyl resorcinol-PDMSX copolymer (4400 g/mole PDMSX) had >50% silicon on the surface which rapidly decreased to a bulk film concentration of ~15%, a value that is comparable to the results obtained from elemental analysis. Figure 5b shows Auger depth profiling

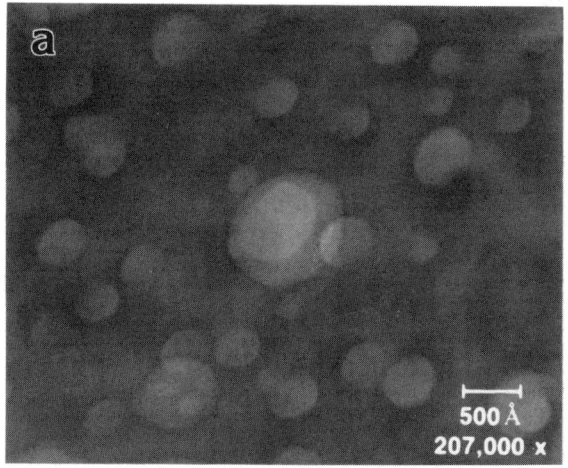

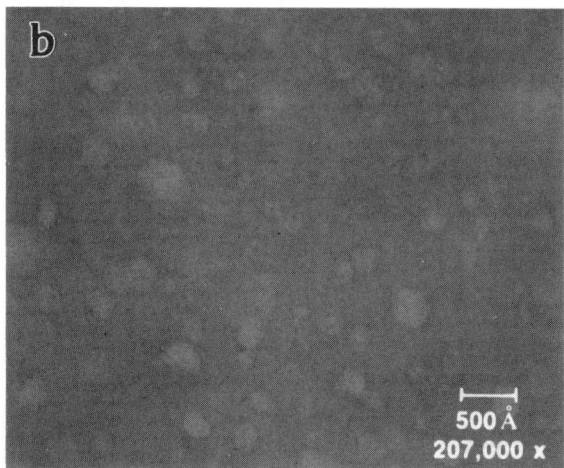

Figure 2   Transmission electron microscope photograph of 2-methyl resorcinol-PDMSX copolymers using (a) 4400 g/mole PDMSX and (b) 510 g/mole PDMSX.

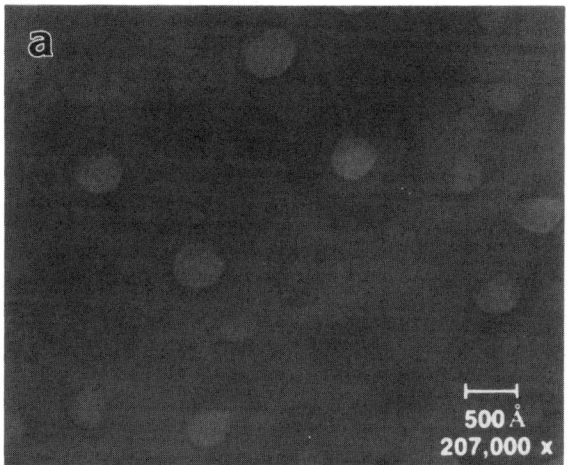

Figure 3. Transmission electron microscope photograph of o-cresol novolac-PDMSX copolymers using (a) 4400 g/mole PDMSX and (b) 510 g/mole PDMSX.

Figure 4 Transmission electron microscope photograph of poly(hydroxystyrene)-PDMSX copolymer.

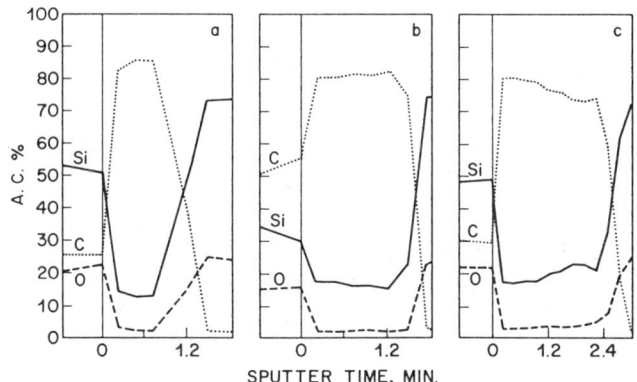

Figure 5 Auger depth profiling results of (a) 2-methyl resorcinol-PDMSX, (b) o-cresol novolac-PDMSX, and (c) poly(hydroxystyrene)-PDMSX.

results for the o-cresol novolac-PDMSX (510 g/mole PDMSX) copolymer. The surface enrichment of silicon by the PDMSX is less than that observed in the 2-methyl resorcinol copolymer, resulting in a more equal surface/bulk silicon distribution. This may result from the greater phase mixing between the two components as evidenced by TEM and DSC. The PHS-PDMSX copolymer (4400 g/mole PDMSX) shows a behavior similar to the 2-methyl resorcinol copolymer.

The O_2 reactive-ion etching (RIE) behavior of each copolymer was determined and compared to a hard-baked HPR-204 planarizing layer (Table IV). Each copolymer tested had ≥ 10 wt % silicon content and the etching rate selectivity vs. HPR-204 in all cases was 1:13 or greater. The O_2 RIE behavior may have been affected by the distribution of silicon through the thickness of the film as well as the quantity of silicon incorporated. The 2-methyl resorcinol-PDMSX copolymers exhibited the greatest O_2 RIE resistance, presumably due to microphase separation and surface migration of the PDMSX. This preferential incorporation of PDMSX

TABLE IV

OXYGEN REACTIVE ION ETCHING (RIE) OF NOVOLAC-SILOXANE BLOCK COPOLYMERS

COPOLYMER STRUCTURE	SILOXANE $<M_n>$	wt % Si	O_2 RIE rate[1] (Å/min)	O_2 ETCHING SELECTIVITY[2]
o-cresol novolac siloxane	4400	12	130	1:15
	1770	11	130	1:15
	510	10	110	1:18
	510	13	50	1:36
2-methyl resorcinol siloxane	4400	11	70	1:28
	1770	10	80	1:24
	510	10	80	1:24
poly(hydroxy styrene) siloxane	4400	12	150	1:13

[1] After 10 minutes plasma etching.
[2] Versus hard baked HPR-204.

at the air/resist interface would be expected to contribute to a more rapid build-up of an inorganic oxide layer. Auger analysis indicated >50% silicon at the surface which supports the O_2 plasma results. The PHS-PDMSX exhibited the poorest O_2 RIE resistance of the three copolymer systems which may result from the aliphatic backbone structure of PHS.

Optimization of the deep-UV exposure and aqueous TMAH development steps for all three parent phenolic resins formulate with the diazonaphthoquinone dissolution inhibitor resulted in the resolution of positive tone 0.75 µm L/S patterns at a dose of 156, 195 and 118 mJ/cm^2 for the o-cresol, 2-methyl resorcinol and PHS materials, respectively (Table V). The copolymers prepared with a 4400 g/mole PDMSX resulted in TMAH soluble films at >11 wt % silicon; however, the feature quality was extremely poor in each case. Figure 6 shows an SEM photomicrograph of a 2-methyl resorcinol-PDMSX copolymer using (a) 20 and (b)

TABLE V

EFFECT OF POLY(DIMETHYL SILOXANE) BLOCK LENGTHS ON RESOLUTION CAPABILITIES

POLYMER STRUCTURE	SILOXANE $<M_n>$	wt % Si	RESOLUTION of 0.75 µm L/S[1]
o-cresol novolac	–	0	YES
o-cresol novolac-siloxane	4400	12.1	NO
	1770	5.4	YES
	510	10.2	YES
	510	12.9	YES
2-methyl resorcinol	–	0	YES
2-methyl resorcinol-siloxane	4400	11.4	NO
	1770	10.1	NO
	510	10.4	NO
poly(hydroxy styrene)	–	0	YES
poly(hydroxy styrene) siloxane	4400	12.3	NO[2]

[1] Contact exposure at 248 nm using 20 wt % dissolution inhibitor.
[2] Behaves as a negative resist.

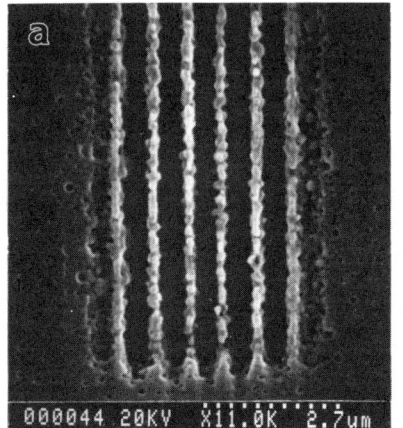 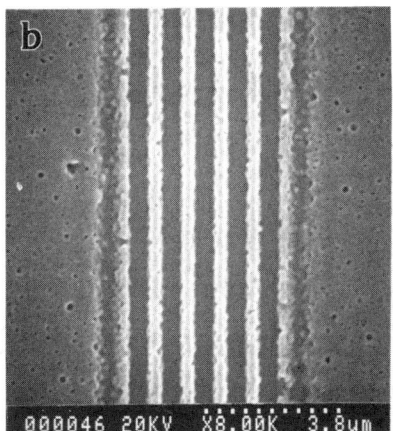

Figure 6 Scanning electron microscope photograph of coded 0.75 μm line-space images obtained with the 2-methyl resorcinol-PDMSX copolymer (<M_n> = 4400 g/mole) containing (a) 20 wt % and (b) 30 wt % diazonaphthoquinone dissolution inhibitor.

30 wt % dissolution inhibitor. At 20 wt %, coded 0.5 μm line/0.25 μm space patterns were resolved but only 75% of the initial film thickness (IFT) in the unexposed regions remained. At 30 wt % sensitizer loadings, coded 0.5 μm L/S patterns were resolved and only 6% of the IFT was lost in the unexposed areas. Both samples exhibited 'holes' in the unexposed region where preferential dissolution occurred. This is most likely the result of phase incompatibility within the novolac/novolac-siloxane blend. Since the dissolution rate of the parent novolac in aqueous TMAH is higher than that of the block copolymer it would not be unreasonable to expect that preferential dissolution of the novolac domains would occur. Similar solubility problems were observed with the o-cresol novolac-PDMSX copolymer using this high molecular weight poly(dimethyl siloxane) precursor.

Copolymers prepared from lower molecular weight PDMSX blocks were examined next. Regardless of the molecular weight of the PDMSX precursor, copolymers with 2-methyl resorcinol could not be imaged without substantial problems such as sample thinning and 'holes'. This may stem from the extensive microphase separated structures obtained regardless of PDMSX molecular weight (Figure 2b). In contrast, the o-cresol novolac copolymers exhibited improved imagibility when smaller PDMSX block lengths were used. This may result from a larger degree of phase mixing between the two components and/or smaller PDMSX domain sizes. The copolymer prepared from o-cresol novolac and a 510 g/mole PDMSX at 10 wt % Si resolved coded 0.5 μm L/S patterns at a dose of 156 mJ/cm^2 in 0.5 μm resist via contact exposure at 248 nm and subsequent etching through 1.4 μm HPR-204 (Figure 7). There was no evidence of preferential dissolution or 'holes' in the field. A coded 0.75 μm line/0.5 μm space pattern achieved with the same material before and after etching through 1.4 μm HPR-204 is shown in Figure 8. Steep sidewall profiles were obtained using a 10% overetch. A small amount of undercutting was observed, however, the etching conditions for pattern transfer were not optimized. Sloping sidewall profiles are evident on resist images after wet development. This may be due to the highly absorbant nature of the novolacs. The optical density of an unsensitized film at 248 nm is 0.59 A.U. for a 0.63 μm thick film. This is prohibitively high for deep-UV use; however, its utility at longer wavelengths is being explored.

CONCLUSIONS

The incorporation of PDMSX into conventional novolac resins has produced potential bilevel resist materials. Adequate silicon contents necessary for O_2 RIE resistance can be achieved without sacrificing aqueous TMAH solubility. Positive resist formulations using an o-cresol novolac-PDMSX (510 g/mole) copolymer with a diazonaphthoquinone dissolution inhibitor have demonstrated a resolution of coded 0.5 μm L/S patterns at a dose of 156 mJ/cm^2 upon deep-UV irradiation. A 1:18 O_2 etching selectivity versus hard-baked photoresist allows dry pattern transfer into the bilevel structure.

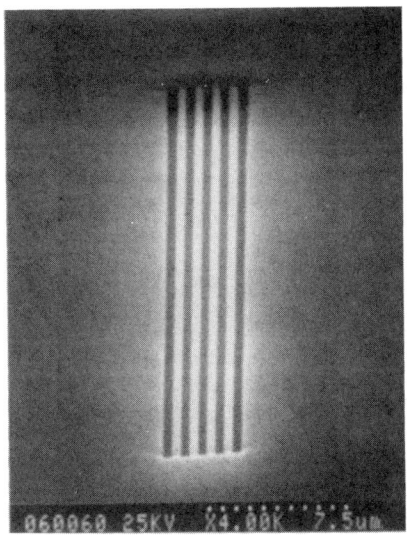 

Figure 7   Scanning electron microscope photographs of coded 0.5 μm line-space patterns obtained in the o-cresol novolac-PDMSX ($<M_n>$ = 510 g/mole) based resist followed by O_2 RIE pattern transfer.

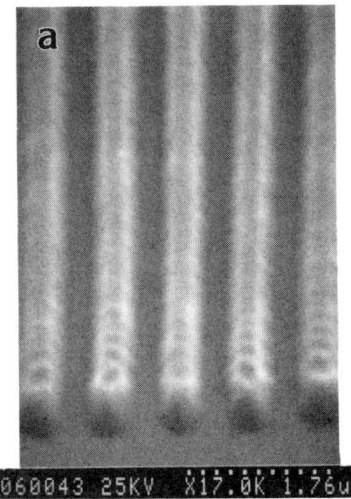 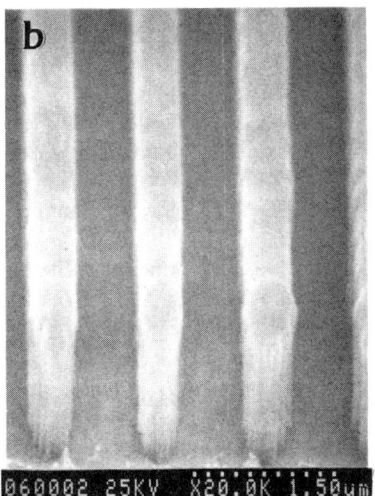

Figure 8   Scanning electron microscope photographs of coded 0.75 μm line /0.5 μm space images obtained in the o-cresol novolac-PDMSX ($<M_n>$ = 510 g/mole) (a) before and (b) after O_2 RIE pattern transfer.

ACKNOWLEDGMENTS

We thank A. E. Novembre and D. A. Mixon for help with GPC data, R. G. Tarascon for initial etching studies, S. A. Heffner for assistance in obtaining low temperature DSC and S. Nakahara for transmission electron microscopy.

LITERATURE CITED

1. Ohnishi, Y.; Suzuki, M.; Saigo, K.; Saotome, Y.; Gokan, H. *Proc. SPIE Advances in Resist Technology II*, 1985, *539*, 62.
2. Reichmanis, E.; Novembre, A. E.; Tarascon, R. G., Shugard, A. In *Polymers for High Technology: Electronics and Photonics*; Bowden, M. J.; Turner, S. R. Eds.; ACS Symposium Series No. 346; American Chemical Society: Washington, D.C., 1987; pp 110-121.
3. Taylor, G. N.; Wolf, T. M. *Polym. Eng. Sci.*, 1980, *20*, 1087.
4. Reichmanis, E.; Smolinsky, G.; Wilkins, C. W., Jr. *Solid State Technol.* Aug. 1985, 130.
5. Smith, B. C.; Hellman, M. Y.; Reichmanis, E. "ACS 18th Middle Atlantic Regional Meeting," Newark, N.J. 1984.
6. McColgin, W. C.; Daly, R. C.; Jech, J., Jr.; Brust, T. B. In *Proc. SPIE Advances in Resist Technology*, 1988, *920*, 260.
7. Noshay, A.; McGrath, J. E. *Block Copolymers: Overview and Critical Survey*, Academic Press: New York, 1977.
8. Olabisi, O.; Robeson, L. M.; Shaw, M. T. *Polymer-Polymer Miscibility*, Academic Press: New York, 1979.
9. Teyssie, Ph.; Broze, G.; Fayt, R.; Heuschen, R.; Jerome, R.; Petit, D. In *Initiation of Polymerization*; Bailey, F. E., Ed., ACS Symposium Series No. 212; American Chemical Society: Washington, D.C., 1984, pp 323-335.
10. Hartney, M. A.; Novembre, A. E.; Bates, F. S. *J. Vac. Sci. Technol.*, 1985, *B3*, 1346.
11. Jurek, M. J.; Tarascon, R. G.; Reichmanis, E. *Chem. Mater.*, 1989 *1*, 319.
12. Reichmanis, E.; Smolinsky, G. *Proc. SPIE Advances in Resist Technology*, 1984, *469*, 38.
13. Riffle, J. S.; Yilgor, I.; Banthia, A. K.; Tran, C.; Wilkes, G. L.; McGrath, J. E. In *Epoxy Resin Chemistry II*, Bauer, R. S., Ed.; ACS Symposium Series No. 221; American Chemical Society: Washington, D.C., 1983, p. 83.
14. Andolino-Brandt, P. J.; Webster, D. C.; McGrath, J. E. *ACS Polymer Preprints*, 1984, *25(2)*, 91.
15. Bowden, M. J.; Gozdz, A. S.; Klausner, C.; McGrath, J. E.; Smith, S. In *Polymers for High Technology: Electronics and Photonics*; Bowden, M. J.; Turner, S. R. Eds.; ACS Symposium Series No. 346; American Chemical Society: Washington, D.C., 1987; pp. 122-137.
16. Collins, E. A.; Bares, J.; Billmeyer, F. W., Jr. *Experiments in Polymer Science*; Wiley-Interscience: New York, 1973; p. 108.
17. Rudin, A. *The Elements of Polymer Science and Engineering*; Academic Press: New York, 1982; Chapter 12.
18. Smith, S. D.; McGrath, J. E. *ACS Polymer Preprints*, 1986, *27(2)*, 31.

RECEIVED July 17, 1989

Chapter 11

Preparation of a Novel Silicone-Based Positive Photoresist and Its Application to an Image Reversal Process

Akinobu Tanaka[1], Hiroshi Ban[1], and Saburo Imamura[2]

[1]NTT LSI Laboratories, Morinosato, Atsugi-shi, Kanagawa 243–01, Japan
[2]NTT Basic Research Laboratories, Midoricho, Musashino-shi, Tokyo 180, Japan

> We have developed a novel silicone-based positive photoresist (SPP) for two-layer resist systems. SPP is composed of an acetylated poly(phenylsilsesquioxane) (APSQ) and diazonaphthoquinone sensitizer. SPP can be developed with alkaline aqueous solutions, because the matrix resin, APSQ, is alkali-soluble due to the presence of silanol groups formed during synthesis of APSQ. SPP is useful not only for near UV lithography (positive mode), but also in negative mode using high energy sources for exposure. Negative process (image reversal) is capable of sub-halfmicron resolution using electron beam (EB), X-ray, and deep UV exposures. Resist sensitivities of SPP to EB, X-rays and deep UV are 5 $\mu C/cm^2$, 80-160 mJ/cm^2 and 10 mJ/cm^2, respectively. We suggest that a coupling of APSQ and the sensitizer occurs during EB and X-ray exposures, but it is absent during near UV exposures. This coupling reaction and the generation of indenecarboxylic acid are competing processes during the deep UV exposures.

Silicon-containing resists have been proposed as top imaging layers in two-layer resist (2LR) systems for high resolution lithography.[1] As they have high resistance to oxygen reactive ion etching (O_2 RIE), fine patterns formed in a very thin top resist can be transferred into a thick bottom organic polymer layer by O_2 RIE. Recently, alkali-developable silicon-containing positive photoresists have attracted much attention[2-6] due to their compatibility with practical VLSI fabrication processes using novolac-diazonaphthoquinone positive photoresists (AZ-type resists). We have synthesized an acetylated poly(phenylsilsesquioxane) (APSQ), and prepared an alkali-developable silicone-based positive photoresist (SPP)[7] composed of APSQ and a diazonaphthoquinone compound as a photosensitizer for near UV lithography.

0097–6156/89/0412–0175$06.00/0
© 1989 American Chemical Society

However, the continuing drive towards minimization in pattern sizes has created a demand for lithographic resolution higher than can be achieved with near UV lithography. Since one useful method for improving resolution is to use higher energy sources, we have applied SPP to EB, X-ray, deep UV lithography. Although SPP also exhibits a positive action when exposed to EB, X-rays and deep UV, the positive pattern requires a high dosage that is not acceptable for a practical use except for deep UV lithography. An image reversal process of AZ-type resists has been reported[8-10] to have several advantages such as improvement in resolution, sensitivity, and thermal stability. We have also found that an image reversal of SPP dramatically increases the sensitivity. Therefore, we have applied SPP to an image reversal process using high energy sources such as EB, X-rays and deep UV.

This paper describes the preparation of SPP and its application to an image reversal process, as well as the chemistry of the SPP image reversal.

Synthesis and characterization of APSQ

APSQ was synthesized by acetylation of PSQ in the presence of a Friedel-Crafts catalyst. A solution of poly(phenylsilsesquioxane) (PSQ) in acetyl chloride (AcCl) was reacted with a solution of anhydrous $AlCl_3$ in AcCl below 20°C. After stirring for 90 min, the solution was poured into ice water to obtain APSQ. The details of this process are described elsewhere.[11]

An attempt at conventional elemental analysis for APSQ failed because silicon carbide was irregularly produced during the measurement. Therefore, the molecular structure of APSQ was determined by NMR and IR. The IR data indicate that APSQ molecular structure is fundamentally similar to PSQ except the addition of acetyl (1650 cm^{-1}) and hydroxy (3400 cm^{-1}) groups. ^{29}Si NMR spectra are shown in Figure 1 together with assignments of ^{29}Si chemical shifts.[12] PSQ purchased from Petrarch Systems Inc. and Owens-Illinois Co. were used. Although ^{29}Si NMR spectra of these PSQ's differ, the APSQ's obtained from them have almost the same spectra. The total reactions are described in equation 1.

$$\left\{ \begin{array}{c} R \\ -Si-O- \\ O \\ -Si-O- \\ R \end{array} \right\}_m \xrightarrow[\text{ii) } H_2O]{\text{i) AcCl/AlCl}_3} HO \left\{ \begin{array}{c} R' \\ -Si-O- \\ O \\ -Si-O- \\ R' \end{array} \right\}_n H \quad (1)$$

$R = C_6H_5$ $R' = C_6H_5, C_6H_4COCH_3, OH$

An interesting issue is the simultaneous introduction of silanol groups during the acetylation of phenyl groups. As indicated by ^{29}Si NMR, some of the Si-phenyl bonds and framework siloxane bonds in PSQ are cleaved and chlorinated in the presence of Lewis acids. Si-C bonds have relatively low resistance to electrophilic attack and can be substituted by Si-Cl bonds in the presence of Lewis acids.[13] Although framework siloxane bonds are relatively strong and stable, they also undergo chlorination. The

Si-Cl groups produced are hydrolyzed by water to Si-OH groups during work-up.

APSQ is soluble in a dilute aqueous solution of tetramethylammonium hydroxide (TMAH). When APSQ was treated with trimethylsilyl chloride (TMSCl), the solubility of APSQ in the TMAH solution decreased, because silanol groups were terminated with TMS groups. This indicates that APSQ is alkali-soluble due to the presence of silanol groups. The solubility depended on the silanol content, which can be controlled by synthesis conditions or appropriate termination of silanol groups. APSQ obtained from Owens-Illinois PSQ was more soluble in TMAH solutions than that from Petrarch Systems, probably due to lower molecular weight (Mw). We used the former (Mw = 1,500) in this study.

Preparation of SPP and Application to near UV lithography

SPP was prepared by dissolving a novolac resin diazonaphthoquinone sulfonyl ester (DNQ) and APSQ in methyl isobutyl ketone. A concentration of 20 wt% DNQ relative to APSQ is sufficient for it to act as a dissolution inhibitor.

Figure 2 shows an SEM photograph of a 0.4 μm line and space pattern on a substrate with topographic features using the SPP 2LR system. A 0.2 μm thick SPP layer was spun onto a 1.5 μm thick bottom planarizing layer. The resist was exposed with a g-line (436nm) stepper equipped with a high numerical-aperture reduction lens (NA=0.6) and then dip-developed in a 1.6 wt% TMAH aqueous solution for 60 s at 25 °C. The pattern formed in the SPP layer was transferred to the bottom layer by O_2 RIE. The O_2 RIE etching rate of SPP was less than 3.5 nm/min, whereas that of the bottom layer was more than 90 nm/min. The selectivity of SPP to the bottom layer was more than 26.

Application to image reversal process
a) Electron beam lithography

A 0.3 μm thick SPP layer was exposed to a 20 kV electron beam followed by a flood exposure using near UV radiation with an integrated dose of over 500 mJ/cm^2. Such a dose was sufficient to convert the remaining DNQ to indenecarboxylic acid. The resist was then dip-developed in an aqueous TMAH solution for 60 s at 25°C.

Figure 3 shows the sensitivity curves for SPP (solid lines) compared with that of a novolac-based resist (dashed line). From these curves, we obtained the maximum clearing dose (D_0), the dose for 50% thickness remaining (D_{50}), and lithographic contrast (γ-value). These resist characteristics are summarized in Table I.

Table I Characteristics of SPP and novolac-based resist

Resist	TMAH (wt%)	D_0 ($\mu C/cm^2$)	D_{50} ($\mu C/cm^2$)	γ-value*
SPP	0.65	0.9	1.7	1.8
SPP	0.70	5.0	6.6	4.1
SPP	0.80	13.5	16.5	5.7
Novolac-based	1.20	6.2	14.5	1.4

* γ-value = $1/2(\log(D_{50}/D_0))^{-1}$

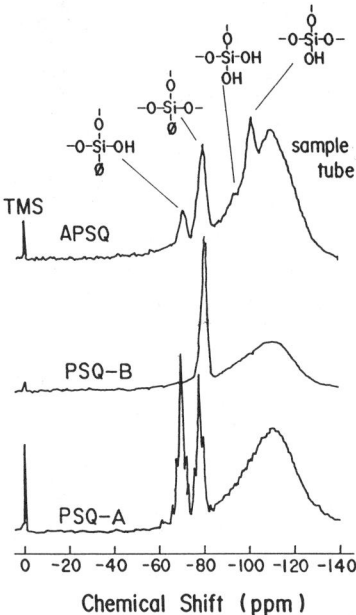

Figure 1. 39.7 MHz ^{29}Si NMR spectra of PSQ and APSQ obtained from PSQ-B in acetone-d_6. Chromium acetylacetonate was used as a relaxation reagent, and transients were 5000. PSQ-A (M_w = 900) and PSQ-B (M_w = 9500) were purchased from Owens-Illinois and Petrarch Systems, respectively.

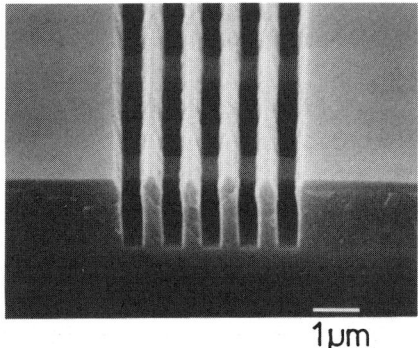

Figure 2. SEM photograph of 0.4-μm line-and-space pattern on a substrate with topographic features using SPP two-layer resist system. A 0.2-μm-thick SPP layer was exposed with a g-line stepper (NA = 0.6) at 350 mJ/cm^2 and then dip-developed in a 1.6 wt% TMAH solution for 60 s at 25 °C. The pattern formed in the SPP top layer was transferred to 1.5-μm-thick bottom layer by O$_2$ RIE.

A higher sensitivity of SPP can be obtained using a more dilute TMAH solution, but at the expense of lower contrast. A solution more dilute than 0.6 wt% cannot completely dissolve the resist. The SPP exhibited a higher sensitivity and contrast than the novolac-based resist.

Figure 4 shows an SEM photograph of 0.3 μm line and 0.5 μm space pattern delineated in an SPP 2LR system with a dose of 5 μ C/cm^2. The combination of this SPP image reversal process and EB direct wafer-writing technology represents a promising approach for achieving sub-halfmicron resolution.

b) X-ray lithography

A 0.4 μm thick SPP layer was exposed to X-rays followed by a flood exposure using near UV radiation. The resist was then dip-developed in a 0.8 wt% TMAH solution for 60 s at 25 °C. We used two x-ray exposure systems to evaluate the characteristics of the SPP resist. One is SR-1[14] which has a source composed of a molybdenum rotating anode with a 0.54 nm Mo-Lα characteristic line. The exposure was carried out in air. The other has a synchrotron radiation source with a central wavelength of 0.7 nm (KEK Photon Factory Beam Line, BL-1B). The exposure was carried out in vacuum (<10^{-4} Pa). A positive resist, FBM-G,[15] was used as a standard, because its sensitivity only weakly depends on the ambient.

Figure 5 shows X-ray sensitivity curves of SPP. The sensitivity (D_{50}) of the SPP was about 160 mJ/cm^2 when exposed in air. The sensitivity of 80 mJ/cm^2 in vacuum is relatively high. Surprisingly, SPP exhibited a negative action using image reversal process when exposed with the X-rays in air, indicating that X-ray exposure induces different chemical reactions from the well-known photochemical process of DNQ, which produces the indenecarboxylic acid. Figure 6 shows a 0.2 μm wide pattern with an aspect ratio of 5 obtained by X-ray exposure in vacuum. The fine pattern with vertical side walls can be replicated in a single-layer SPP resist.

c) Excimer laser (Deep UV) lithography

Excimer laser lithography is one of the most promising technologies for achieving sub-halfmicron lithography suitable for mass production. At almost the same dose of deep UV (248nm), SPP can be imaged in either a positive mode using a normal process, or a negative mode through an image reversal process (Figure 7). In a normal process (positive mode), a 0.4 μm thick SPP layer was exposed to deep UV and then dip-developed in a 1.6 wt% TMAH solution for 60 s at 25 °C. In an image reversal process (negative mode), SPP was exposed to deep UV followed by a flood exposure using near UV radiation and then dip-developed in a 0.7 wt% TMAH solution for 60 s at 25 °C. In the latter process, the sensitivity curves exhibit conversion from the initial negative mode to the subsequent positive mode with an increasing dose (Figure 7). This is very different from the situation in EB and X-ray exposures. This suggests that acid-producing reactions are involved during deep UV exposures. SPP image reversal has the advantage of higher pattern accuracy over the normal positive mode, because the effect of strong absorption at 248 nm on patterning is smaller in the negative mode than in the positive mode.[16]

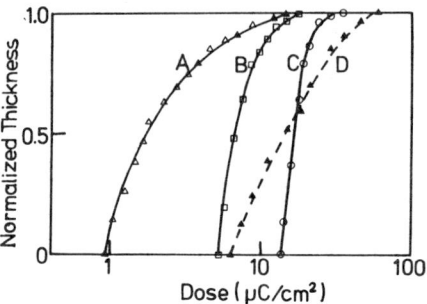

Figure 3. Sensitivity curves of SPP image reversal (solid line) after 20kV EB exposure compared with a novolac-based resist (dashed line). A 0.3 μm thick resist layer was exposed to EB followed by a flood exposure using near UV radiation and then dip-developed in an aqueous TMAH solution for 60 s at 25°C. TMAH concentration; A: 0.65 wt%, B: 0.70 wt%, C: 0.80 wt%, D: 1.2 wt%.

Figure 4. SEM photograph of 0.3 μm line and 0.5 μm space pattern using an SPP two-layer resist system. A 0.3 μm thick SPP layer was exposed to a 20 kV EB at 5 μC/cm^2 followed by a flood exposure and then dip-developed in a 0.65wt% TMAH solution for 60 s at 25 °C. The pattern formed in the top SPP layer was transferred to a 1.0 μm thick bottom layer by O_2 RIE.

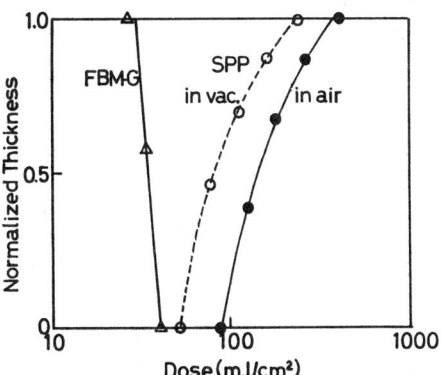

Figure 5. Sensitivity curve of SPP image reversal to X-rays compared with that of FBM-G. A 0.4 μm thick SPP layer was exposed to X-rays both in air and in vacuum followed by a flood exposure and then dip-developed in a 0.8 wt% TMAH solution for 60 s at 25 °C.

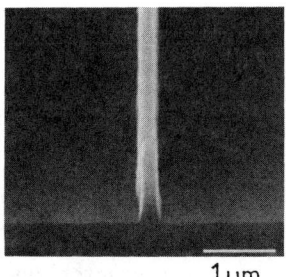

Figure 6. SEM photograph of a 0.2 μm wide pattern using an SPP single-layer resist. A 1.0 μm thick SPP layer was exposed to X-rays in vacuum at 160 mJ/cm^2 followed by a flood exposure and then dip-developed in a 0.8 wt% TMAH solution for 60 s at 25°C.

Chemistry of SPP image reversal

SPP chemistry is closely related to DNQ chemistry. The sensitivity became extremely low when SPP was flood-exposed to near UV radiation before EB exposure. This means that the conventional photoproduct from DNQ, indenecarboxylic acid, is very insensitive to EB exposure and DNQ plays a key role in this process. However, the actual chemical reactions have not precisely elucidated for high energy exposures. Esterification of DNQ and of phenol OH groups has been proposed as the probable high energy induced reaction in AZ-type resists through a Wolff rearrangement.[10] However, Pacansky has recently reported[17] that an EB exposure induced a reaction between DNQ and C-H or C-C bonds in aromatic rings. The reaction did not proceed through the Wolff rearrangement, although he confirmed that ketene intermediates were produced by EB at low temperatures.

We investigated X-ray induced reactions of SPP because these reactions are thought to be similar to those induced by EB in that both reactions are mainly initiated by the secondary electrons generated by X-ray and EB exposures. Therefore, X-ray results can be compared to EB-induced reaction mechanisms. When SPP was exposed to X-rays in vacuum, diazo groups of DNQ decomposed, as shown in Figure 8. The characteristic absorption at 2100-2200 cm^{-1} rapidly decreased as the X-ray dose increased. On the other hand, absorption bands at 1650-1730 cm^{-1} assigned to carbonyl groups barely changed. The X-ray dose needed for pattern fabrication was 160 mJ/cm^2. At this exposure dose, the decomposition of diazo groups was the most easily observed reaction.

According to gel permeation chromatography (GPC) data, the molecular weight of SPP increased after X-ray exposures both in air and in vacuum. The elution profiles in both cases were very similar, indicating that the coupling of APSQ and DNQ or the condensation of APSQ occurs irrespective of the ambient.

Along with the decomposition of diazo groups, the intensity of the characteristic absorption bands at 402 and 335 nm in the UV spectrum of DNQ also decreased (Figure 9). However, the absorption peak intensity ratio after X-ray exposure was slightly different from that after UV exposure, since the peak intensity at 335 nm was somewhat stronger than that at 402 nm. The intensities at both absorption maxima are plotted in Figure 10. Data are shown for near UV, deep UV, and X-ray exposures in air and in vacuum. For each radiation source, the plots are on the same line irrespective of the ambient, but the intercept characterizes exposure sources in a unique way, namely, ca. 0.03 for near UV, 0.06 for deep UV, and 0.1 for X-ray exposures for a 1 μm thick film. An increase in the intercept was attributed to the absorption increase in the tailing part of the adjacent strong absorption bands. The results after near UV exposures indicate that the ester of indenecarboxylic acid and phenol OH groups has no effect on the intercept change, because near UV exposure in vacuum produces the ester.[18] Therefore, this change in the intercept indicates the occurrence of coupling reactions of DNQ with aromatic rings as suggested by Pacansky et al.[17] The coupling reactions are induced not only by X-rays, but also by deep UV exposure. The reaction products of DNQ with aromatic rings

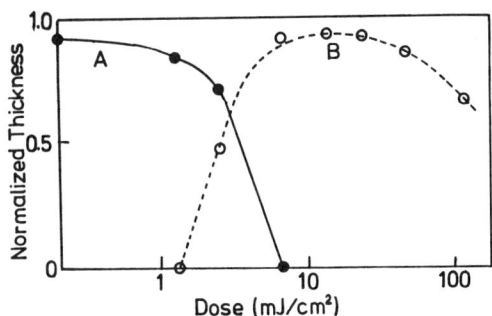

Figure 7. Deep UV sensitivity curves of SPP positive mode (A) compared with that of SPP negative mode (B). In positive mode, a 0.4 μm thick SPP layer was exposed to deep UV and then dip-developed in a 1.6 wt% TMAH solution for 60 s at 25 °C. In negative mode, SPP was exposed to deep UV followed by a flood exposure using near UV radiation and then dip-developed in a 0.7 wt% TMAH solution for 60 s at 25 °C.

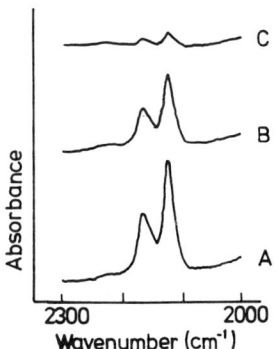

Figure 8. IR absorption changes of diazo groups in SPP with DNQ after X-ray exposure in vacuum. Exposure dose: A; 0 mJ/cm^2 (initial), B; 160 mJ/cm^2, C; 640 mJ/cm^2.

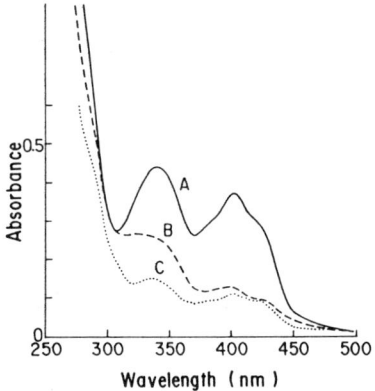

Figure 9. UV absorption changes in SPP with DNQ (20 wt% vs. APSQ) caused by exposures. Solid line: initial; dashed line: X-ray exposure in air (dose = 640 mJ/cm^2), dotted line: near UV exposure (dose = 100 mJ/cm^2).

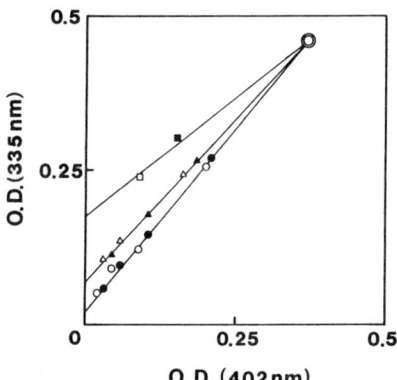

Figure 10. Relationship between absorption intensities at 402 and 335 nm in UV absorption spectra after exposures. ◎; initial spectrum of 1 μm thick unexposed SPP. Exposures: ○; UV in air, ●; UV in vacuum, △; deep UV in air, ▲; deep UV in vacuum, □; X-ray in air, ■; X-ray in vacuum.

are inert toward any further photochemical processing. This agrees with the observed negative action of SPP upon X-ray and deep UV exposures in air.

The deep UV induced reactions appear to be slightly different from X-ray and EB induced reactions. Deep UV exposure in air can induce an increase in solubility of SPP, indicating that indenecarboxylic acid is produced. IR spectra of SPP exposed to deep UV are shown in Figure 11. In this case, we used a mono-functional dissolution inhibitor, tert-amylphenol diazonaphthoquinone sulfonyl ester, instead of a multi-functional sensitizer, DNQ, because the IR spectrum of a mono-functional ester is easier to interpret than that of DNQ. The SPP containing this mono-functional ester also exhibits an image reversal reaction with almost the same characteristics as the SPP with DNQ.

Deep UV exposure in air produces photoproducts similar to those produced by near UV exposure (Figure 11). One slight difference is that absorption at 1680 cm^{-1} is a little stronger after UV exposure in air. After deep UV exposure in vacuum, the resist exhibited a new absorption shoulder at 1730 cm^{-1}, which appears to be a characteristic ester absorption. These peaks can probably be assigned to the products of a Wolff rearrangement.

Consequently, deep UV exposure in air induces several competing reactions leading to both positive and negative resist action; namely, the normal photoreaction to form an indenecarboxylic acid through a Wolff rearrangement, and the coupling reactions of DNQ and APSQ which may be reflected in the difference in the intercept in Figure 10. Therefore, the negative resist action of SPP after deep UV exposure at low doses in air changes to a positive mode action at higher doses (Figure 7).

Conclusion

We have developed an alkali-developable silicone-based positive photoresist (SPP) for a two-layer resist system. NMR and IR studies indicated that APSQ has a number of silanol groups which are introduced during the acetylation of PSQ. Due to the presence of silanol groups, APSQ is soluble in an aqueous alkaline solution. SPP is useful not only for near UV lithography (positive mode), but also in the negative mode using high energy sources such as electron beam, X-rays, and deep UV radiation. Resist sensitivities of SPP to EB, X-rays and deep UV are 5 $\mu C/cm^2$, 80-160 mJ/cm^2 and 10 mJ/cm^2, respectively. Sub-halfmicron resolution was demonstrated by EB and X-ray lithography. In X-ray lithography, a resolution of 0.2 μm with vertical side walls was obtained in a 1.0 μm thick SPP single layer resist. We think that the SPP image reversal mechanism involves the coupling of APSQ and DNQ that occurs during EB and X-ray exposures, rather than the Wolff rearrangement, and that this coupling reaction competes with the generation of indenecarboxylic acid during the deep UV exposures.

Experimental

Materials DNQ was synthesized by esterification of o-cresol formaldehyde novolac resin (Mw=900) and of 1,2-diazonaphthoquinone-5-sulfonyl chloride. The esterification rate was ca. 0.4. The novolac-based resist used in EB lithography to compare with SPP was

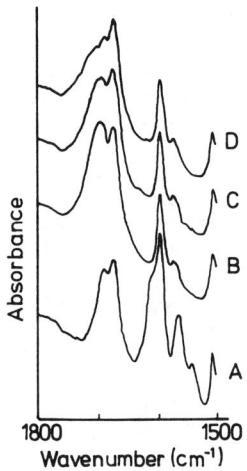

Figure 11. IR absorption changes in SPP with a mono-functional sensitizer caused by exposures. A; initial, B; near UV exposure in air (dose = 500 mJ/cm^2), C; deep UV exposure in air (dose = 200 mJ/cm^2), D; deep UV exposure in vacuum (dose = 200 mJ/cm^2).

composed of o-cresol formaldehyde novolac resin (Mw=3200) and DNQ (20wt% vs. novolac resin).
Measurements IR spectra were recorded on a Perkin Elmer Model 1800 spectrometer and UV spectra on a Hitachi Model 330 UV/VIS spectrometer. Samples were spun onto KRS or quartz plates with a film thickness of ca. 1.0 μm. NMR spectra were recorded on a Varian XL-200. Measurements were done at 23°C. GPC measurements were made on a Toyo Soda Model HLC802UR equipped with four GMHXL-type columns at 40°C in THF. The molecular weight (Mw) was determined by polystyrene standards.
Lithography A hard baked (at 200 °C for 60 min.) AZ-type resist, Microposit 1400 (shipley), was used as the bottom resist in 2LR application. SPP was spun onto the bottom resist layer and exposed to EB, X-rays and deep UV followed by a flood exposure using near UV radiation. Electron beam exposures were carried out with an Elionix ELS-5000 at 20kV. Deep UV exposures were carried out with our KrF excimer laser (248 nm) stepper.[19] Near UV flood exposures were carried out with Ushio ML-251A through a 350 nm filter. X-ray doses was normalized by the FBM-G sensitivity (D_0) of 40 mJ/cm^2 estimated from EB experiments. FBM-G was dip-developed in MIBK/IPA (1/150) mixture for 90 s at 25 °C. The power of KrF excimer laser was measured by a Model 210 power meter (COHERENT Co.). The O_2 RIE was carried out with a parallel plate-type reactive ion etcher (ANELVA DEM-451). The O_2 RIE conditions were O_2 gas flow of 50 sccm, gas pressure of 1.3 Pa, rf power of 0.1 W/cm^2, and dc bias of 500V.

Acknowledgment

The authors would like to thank Akira Yoshikawa and Hiroaki Hiratsuka for their valuable suggestions and encouragement. They are also grateful to Yoshio Kawai, Tadahito Matsuda, Toshiaki Tamamura, and Kimiyoshi Deguchi for their helpful information, and Katsuhide Onose and Keiko Iimura for their experimental help.

References

1) Ohnishi, Y.; Suzuki, M.; Saigo, K.; Saotome, Y.; Gokan, H. Proc. of SPIE, Advances in Resist Technology and Processing 1985, 539, 62.
2) Wilkins Jr, G.W.; Reichmanis, E.; Wolf, T.M.; Smith, B.C. J. Vac. Sci. Technol. 1985, B3(1), 306.
3) Saotome, Y; Gokan, H.; Saigo, K.; Suzuki, M; Ohnishi, Y.; J. Electrochem. Soc. 1985, 132(4), 909.
4) Toriumi, M.; Shiraishi, H.; Ueno, T.; Hayashi, N.; Nonogaki, S.; Sato, F.; Kadota, K. J. Electrochem. Soc. 1987, 134(4), 936.
5) Sugiyama, H.; Inoue, T.; Mizushima, A.; Nate, K. Proc. of SPIE, Advances in Resist Technology and Processing 1988, 920, 268.
6) Onishi, Y.; Horiguchi, R.; Hayase, S. IUPAC CHEMRAWN in Tokyo 1987, IB01.
7) Imamura, S.; Tanaka, A.; Onose,K. Proc. of SPIE, Advances in Resist Technology and Processing 1988, 920, 291.
8) Oldham, W.G.; Heike, E. IEEE Electron Device Lett. 1980, EDL-1 (10), 217.
9) Nagarajan, R.M.; Rask, S.D.; Lee, B.R. Proc. of SPIE, EB, X-ray and Ion-Beam Lithography 1987, 773, 83.

10) Mochiji, K.; Kimura, T. Microelectronic Eng. 1986, 4, 251.
11) Ban, H.; Tanaka, A.; Imamura, S. Polymer, in press
12) Williams, E.A.; Cargioli, J.D. Ann. Rep. NMR Spectroscopy 1979, 9, 221
13) Wilkinson, S.G.; Stone, F.G.A.; Abel E.W. Comprehensive Organometallic Chemistry (The Synthesis, Reactions and Structures of Organometallic Compounds) vol. 2; Pergamon Press: Oxford, England, 1982, p28
14) Hayasaka, T.; Ishihara, S.; Kinoshita, H.; Takeuchi, N. J. Vac. Sci. Technol. 1985, B3, 1581
15) Asakawa, H; Kogure, O. Digest of the 1982 Symp. VLSI Technol. (Oiso, Japan) 1982, 88
16) Kawai, Y.; Tanaka, A.; Ozaki, Y.; Takamoto, K.; Yoshikawa, A. Proc. of SPIE, Advances in Resist Technology and Processing in press
17) Pacansky, J.; Waltman, R.J. J. Phys. Chem. 1988 92, 4558
18) Pacansky, J,; Lyerla, J.R. IBM J. Res. Develop. 1979, 23(1), 42.
19) Ozaki, Y.; Takamoto, K.; Yoshikawa, A. Proc. of SPIE, Optical/Laser Microlithography 1988, 922, 444.

RECEIVED August 2, 1989

Chapter 12

Photooxidation of Polymers

Application to Dry-Developed Single-Layer Deep-UV Resists

Omkaram Nalamasu, Frank A. Baiocchi, and Gary N. Taylor

AT&T Bell Laboratories, 600 Mountain Avenue, Murray Hill, NJ 07974

Surface-functionalization resist schemes are an increasingly popular means for achieving submicrometer resolution by optical lithography. Several workers have been successful in utilizing such schemes to obtain high resolution, negative-tone, submicrometer patterns by selective introduction of Si in the exposed regions. However, because of silicon's moderate etch selectivity and the substantial diffusion depth and large concentrations needed to provide enough etch resistance, this method may not be able to resolve features smaller than 0.5 μm with reasonable sensitivity.

We have found that hydrophilic organic polymers treated with $TiCl_4$ have much higher etching selectivities than organosilicon polymers in an O_2 plasma. This paper examines some of the parameters that influence the reaction of $TiCl_4$ with a variety of polymers. We find that $TiCl_4$, readily functionalizes hydrophilic as well as moderately hydrophobic polymers, but fails to functionalize very hydrophobic films. Rutherford backscattering analysis reveals that $TiCl_4$ is hydrolyzed at hydrophilic polymer surfaces that have sorbed water. Lack of surface water on hydrophobic polymers explains the absence of a TiO_2 layer on these polymer surfaces.

From these observations, a photooxidative scheme has been developed in which a hydrophobic resist becomes hydrophilic upon oxidation induced by deep UV (248 and 193 nm) radiation. Subsequent treatment with $TiCl_4$ followed by oxygen reactive ion etching then affords high-resolution, negative-tone patterns. Studies are currently underway to minimize the line edge roughness and background residue present in such patterns.

Several years ago the concept of near-surface imaging was introduced[1,2]. The critical step that imparts the etch selectivity by introduction of refractive elements is termed gas-phase functionalization and involves the selective reaction of gaseous inorganic or organometallic compounds with exposed or unexposed photoresist films. The concept was tested with e-beam and photosensitive materials(2) using O_2 reactive ion etching (O_2 RIE) development (Figure 1). Since these initial studies, other workers have refined this technique to more selectively silylate the phenolic hydroxyl groups in the light exposed regions of positive photoresists(3-5). Development by O_2 RIE afforded high-resolution negative tone submicron patterns. While this approach may have sufficient resolution to permit 0.5 µm design rule circuits in a single layer resist with I-line exposure, it may not be optimal for printing features smaller than 0.5 µm because the moderate etching selectivity of Si requires a large diffusion depth for hexamethyldisilazane (HMDS) during gas-phase functionalization in order to provide the higher concentrations of Si needed for greater etch resistance during development.

Higher etching selectivities can be attained by substituting other inorganic elements for Si in the functionalization step. Taylor and coworkers(6-8) have found that the reaction of $TiCl_4$ with hydrophilic polymers made them highly etch resistant to an O_2 plasma with selectivities as high as 3000, whereas hydrophobic polymers etched at normal rates showing no selectivity. Further examination of the functionalization process revealed that it is dependent on process variables such as humidity, treatment times and temperature, etc. and that it resulted from the reaction of $TiCl_4$ with water sorbed on the hydrophilic polymer surfaces to give continuous layers of TiO_2(7). The 30-100 Å thick layers deposited in such a manner were efficient etch barriers in trilayer resist schemes since the O_2 RIE rate of TiO_2 is about 5% that of the SiO_2 under the normally used high-bias, trilayer etching conditions.

The original use envisioned for the $TiCl_4$ was in an organic-on-organic bilayer resist scheme (Figure 2) in which the thin topmost imaging layer is first patterned (exposed and developed). Then, selective reaction of $TiCl_4$ with the imaging or planarizing layers followed by O_2 RIE development would afford tone-retained or tone-inversed patterns, respectively. Although tone-retained versions of such a scheme were realized(6), practical aspects such as poor adhesion during wet development and non-selective deposition of TiO_2 when good adhesion was achieved made practical use difficult.

We have reexamined the TiO_2 deposition technique to see if it could be extended to single-layer imaging schemes (Figure 1). We found that more information was needed about the molecular properties required for selective sorption of water at the polymer surfaces. Consequently, we studied the reaction of $TiCl_4$ with polymer films which varied in their hydrophilicity and the acid strength of their hydroxyl groups. The analytical results for these materials were obtained by x-ray fluorescence spectroscopy (XFS), Rutherford backscattering spectroscopy and O_2 RIE and were compared to those for hard-baked (HB) HPR-206 hydrophilic films over a variety of treatment conditions. We summarize our results in this paper and utilize them to pattern sub-half micron images in single layer resists by doing exposures with 193 and 248 nm wavelengths from ArF and KrF excimer laser exposure tools.

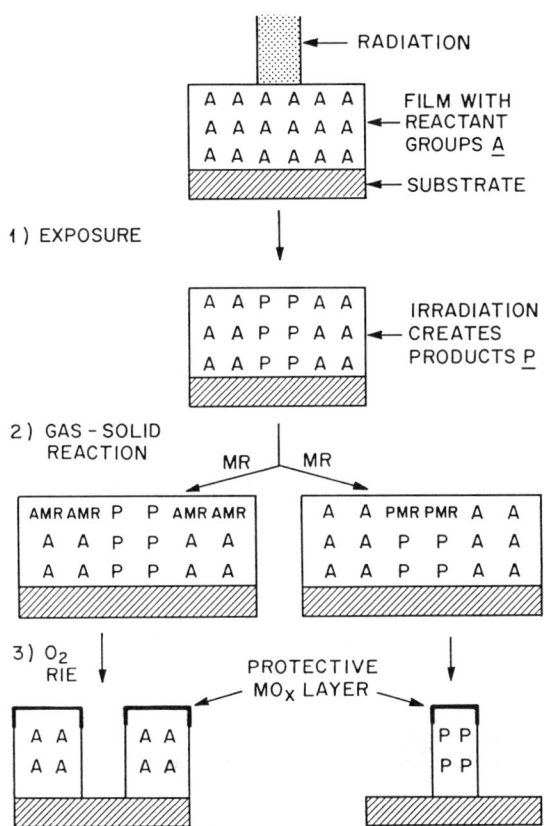

Figure 1. Gas phase functionalization scheme for single-layer resists.

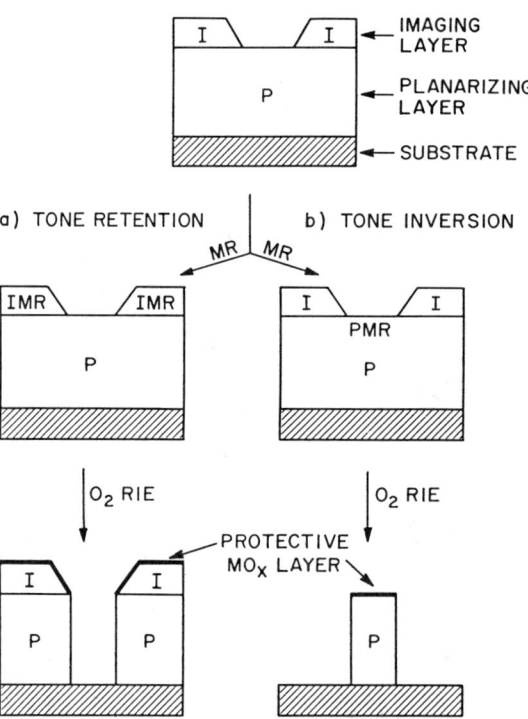

Figure 2. Bilevel resist schemes utilizing the gas-solid reactions of metal compounds (MR, M = metal, R = reactive group) with polymer films.

Experimental

Materials

Poly(methyl methacrylate)(PMMA), poly(vinyl acetate)(PVAc), poly(vinyl acetate-co-vinyl alcohol) and polystyrene (PS) were commercial samples from Aldrich Chemical Company. Poly(methyl methacrylate-co-methacrylic acid) polymers were prepared by the hydrolysis of PMMA with base. The following procedure is typical for preparing P(MMA-co-MAA) polymers. PMMA (10g, 0.1 mole) was dissolved in 100 ml. of THF and to this was added potassium hydroxide (10g, 0.18 mole) in 50 ml. of methanol and 10 ml. of water. The reaction mixture was refluxed with stirring for 2 days and the polymer was precipitated by acidification of the reaction mixture with dilute HCl. The polymer was purified by reprecipitation from THF by the addition of methanol. The acid content of the copolymers was determined by ^{1}H NMR using the peak areas of the ester methyl group and those of aliphatic methyl and methylene protons. The acid content was found to be ~10% from proton NMR. Polymers with 20-30% methacrylic acid content were prepared from PMMA by changing the reaction time, adding more base and THF to the reaction mixture after refluxing for 1 day or by conducting the hydrolysis in pyridine. ^{13}C NMR spectra also were recorded. The results from ^{13}C NMR agreed with those from ^{1}H NMR.

Acetylated *m*-cresol novolac copolymers were prepared by acetylation of *m*-cresol novolac with acetic anhydride in the presence of sodium hydroxide. The following acetylation procedure is typical. 3.2g of sodium hydroxide (50 mmol) was added to 4.8g of cresol novolac (40mmol) in 10 ml. of water. The reaction mixture was stirred for 10 mins. until all polymer went into solution. The required amount of acetic anhydride was then added, the reaction mixture was stirred for 10 more mins. and poured in 150 ml. of iced water. The polymer was filtered and purified by reprecipitation from a chloroform/benzene (5:2 v/v) solution by the addition of hexane. The acetylation content was determined by ^{1}H and ^{13}C NMR.

Chlorinated poly(styrene) samples were prepared by chlorination of PS with Cl_2 in trifluoroacetic acid(9) or by free radical chlorination using *t*-butyl hypochlorite(10) or by chloromethylation using chloromethyl actyl ether and $SnCl_4$(11).

Processing

Exposures

The 193 nm light from a Questek ArF excimer laser was used to obtain sensitivity data by exposing 1 cm^2 areas. Fine line patterning with 193 nm light was done on a Leitz IMS exposure apparatus with either 15X or 36X reflective objectives. The fluence was varied from 0.1 to 100 mJ/cm^2/pulse. All exposures at 248 nm were conducted on a deep-UV stepper(12) with a NA = 0.38 lens, 5X reduction, a minimum feature size of 0.4 μm and a fluence of ~0.3 mJ/cm^2/pulse.

TiCl$_4$ Treatment Procedure

Films were spin coated onto silicon wafers from solutions of the polymers and were baked either at 120°C for 1 hr. in a forced-air oven or on a hot plate inside a humidity controlled glove box containing the gas functionalization cell (GFC) (Figure 3). Wafers that were baked in a forced air oven were immediately transferred to the humidity controlled glove box and were allowed to equilibrate for at least 12 hrs. prior to treatment with TiCl$_4$. The relative humidity (RH) inside the glove box was maintained at 28-30% by circulating the vapor above a saturated aqueous potassium acetate solution through the glove box. The relative humidity outside the glove box varied from 40-65% but had no effect on the RH inside the box. The polymer films were treated for 30-120 sec. with TiCl$_4$ in the GFC after evacuating the cell to a low pressure (100-250 mtorr).

In the patterning experiments, chlorinated poly(styrene) films were baked at 120°C for 15 mins. in a forced air oven, equilibrated for ~12 hrs. inside the humidity-controlled glove box, exposed to deep-UV radiation and treated with TiCl$_4$ under usual conditions. No significant variation in the lithographic parameters was observed by varying the relative humidity in the 30-60% range in the glove box.

Functionalizations were usually conducted at room temperatures. In certain experiments, the bottom part of the cell was maintained at elevated temperatures in order to examine the effect of temperature on the functionalization process.

O$_2$ RIE Conditions

The O$_2$ RIE was conducted with a RF Plasma Products Inc. reactive ion etcher. Typical etching conditions were: bias voltage of -375 to -400V, 10-13 sccm of O$_2$, 10-15 mtorr pressure and 25-35 W power.

Analytical Methods

Ti content in the polymer films was measured with a Princeton Gamma Tech System 4 x-ray Fluorescence Spectrometer. The conditions employed were: Cr target, 50 keV source operating at 3 mA, 0.75 mm aperture, 4.8 mm beam stop, helium atmosphere and 100 sec. counting time. A calibration curve was constructed by plotting the fluorescence counts versus the amount of Ti in HB-HPR 206 films determined by Rutherford Backscattering Spectroscopic (RBS) analysis.

RBS spectra were obtained using a 2.120 MeV He^{+2} ion beam at a backscattering angle of 162°. The spectra were accumulated for a total ion dose of 40 uC using a 10 nA beam current. The number of Ti atoms/cm^2 in the sample was calculated by comparison to spectra for a standard Si wafer implanted with a known dose of Sb.

SEM photographs were taken on a Cambridge Instruments Stereoscan 60 Machine. UV spectra were recorded using a HP 8452A spectrometer, IR spectra were recorded on a Nicolet FT-IR spectrometer and NMR spectra were taken on Bruker AM-360. Chlorine and carbon elemental analyses were used to determine the chlorine content.

Results and Discussion

Etch selectivity is crucial to the gas-phase functionalized resist schemes. Since the thickness of the etch resistant TiO_2 layer that forms on the polymer film should depend on the amount of water sorbed on the polymer surface, we studied the influence of various processing parameters on the surface water content as measured by the amounts of Ti deposited.

More Ti was measured for longer treatment times, lower treatment temperatures and with higher background pressure in the gas functionalization cell. The increase in the thickness of TiO_2 layer with longer reaction times was more pronounced in hydrophilic HB-HPR 206 and PMMA films compared to PS films. More Ti was detected in PS films at shorter reaction times, but the Ti incorporation saturated after about 1.5 min. of reaction (Figure 4).

Ti incorporation was also found to be a function of the background (residual) pressure in the GFC. Table I lists Ti incorporations for 3 different m-cresol novolac polymers that were treated with $TiCl_4$ for 1 min. after evacuation of the reaction cell to 110 and 220 mtorr. The polymers that were treated at 220 mtorr had 1.5 times more Ti than those treated at 110 mtorr. Similar experiments conducted with HB-HPR 206 films show a linear relationship between the residual pressure in the GFC and thickness of the resulting TiO_2 layer.

Table I. Ti Incorporated at the Surface of Novolac Copolymers as a Function of Background Pressure in the Gas Functionalization Cell

Polymer	Mole % Acetyl	Ti Concentration Measured by XFS (atoms/cm^2)	
		110 mtorr	220 mtorr
m-Cresol Novolac	0	6.64×10^{15}	1.07×10^{16}
Acetylated m-Cresol Novolac	70	7.62×10^{15}	1.07×10^{16}
	80	7.63×10^{15}	1.10×10^{16}

To determine the number of equivalent Ti monolayers (1 equivalent Ti monolayer = 2×10^{15} Ti atoms/cm^2) needed on the polymer surface to protect the underlying organic film during O_2 RIE, several PMMA films were treated with $TiCl_4$ under different processing conditions. After treatment with $TiCl_4$ the films were etched in an O_2 plasma for different lengths of time and the etching rates were determined. The Ti concentration in the samples was measured both before and after etching (Table II).

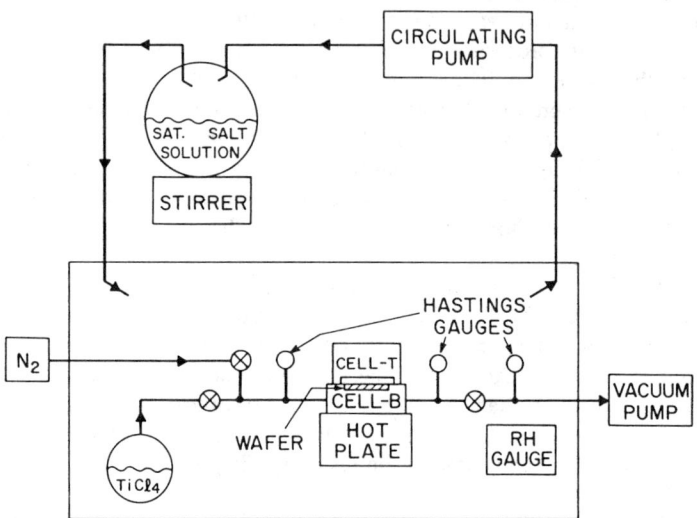

Figure 3. Schematic of a gas-solid reaction cell maintained in a humidity-controlled glove box.

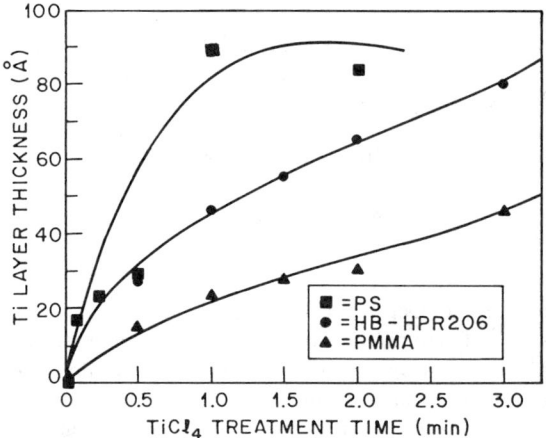

Figure 4. Ti layer thickness versus TiCl$_4$ treatment time for three polymer films.

Table II. O_2 RIE Behavior of $TiCl_4$-Treated PMMA Films as a Function of Ti on the Film Surface

Amount of Ti[a] (atoms/cm²)		Etching Time (min.)	PMMA Film Thickness (μm)		Etching Rate (Å/min)
Before Etching	After Etching		Initial	Final	
1.17×10^{16} (5.9)	6.28×10^{15} (3.1)	30	1.22	1.22	0
1.08×10^{16} (5.4)	5.77×10^{15} (2.9)	40	1.22	1.22	0
1.21×10^{16} (6.1)	3.86×10^{15} (1.9)	60	1.22	1.16	10
1.26×10^{16} (6.3)	1.30×10^{15} (0.7)	150	1.22	0.10	75
4.79×10^{15} (2.4)	2.83×10^{15} (1.4)	23	1.22	0.66	240
Untreated PMMA Film		10	1.22	0.52	1500

a. The values in parentheses are equivalent monolayers of TiO_2.
1 Ti Monolayer = 2×10^{15} Ti atoms/cm².
1Å Thickness = 3.12×10^{14} Ti atoms/cm².

PMMA films having 5-6 monolayers of TiO_2 lost no organic film upon 30-40 mins. etching, but lost 2.5-3 monolayers of Ti (etching rate of 0.5 Å/min.) resulting in an etch selectivity of 3000 as the untreated PMMA film etched at a rate of 1500 Å/min. under the identical etching conditions. Another sample of PMMA that was protected at the surface by ~6 monolayers of Ti lost ~4 monolayers during a 60 mins. etch. In this case, the polymer also etched slightly as indicated by a small decrease (0.06 μm) in the film thickness. Apparently, 2 monolayers of Ti is not quite capable of forming a continuous and tenacious etch resistant mask over the underlying organic film. This conclusion was further confirmed by etching a PMMA film protected by ~2.5 monolayers of Ti on its surface. Etching for 23 mins. not only reduced the Ti content by 40%, but also reduced the thickness of organic film by ~50%. The etching rate of PMMA protected by 2.5 monolayers of Ti is only 1/6 that of an untreated film. In contrast, PMMA films having 3-6 monolayers of Ti etched ~3000 times slower than the untreated film. It appears that at least 3 monolayers of Ti are required to form a tenacious and continuous etch resistant mask.

To determine the influence of polymer structure and the effect of hydrophilic and acidic functionalities on the reaction of $TiCl_4$ with organic polymers, several polymers with varying hydrophilic group content and acid strength were treated with $TiCl_4$ for 1 (Table III) and 2 mins. (Table IV) followed by O_2 RIE for 30 mins. Ti amounts and depth profiles were determined both before and after O_2 RIE. Examination of the results indicates that the etching rates of polymers are primarily a function of the Ti present on the polymer surface and are largely independent of polymer structure, molar content of hydrophilic groups and acid strength of the OH groups. For example, the Ti content in poly(vinyl alcohol), *m*-

cresol novolac and P(MMA-co-MAA) containing 25 mole % acid were very similar even though their acid strengths are very different. Also, Ti incorporation did not vary with the mole % of the hydrophilic groups in a given set of polymers. Ti amounts measured in poly(vinyl alcohol) and poly(vinyl acetate) as well as in acetylated m-cresol novolac copolymer series (Table V) were the same within experimental error.

Of the 8 polymers in Tables III and IV, only the PS and HB-HPR 206 films seemed to behave very differently. HB-HPR 206 films showed zero film thickness loss upon etching, while PS films exhibited the lowest etching selectivity in spite of having more Ti than any other film. For example, HB-HPR 206 films with

Table III. Influence of Polymer Structure on TiCl$_4$ Incorporation and O$_2$ RIE Rates of Various Polymer Films Treated for 1 Minute

Polymer	Ti Concentration Measured by XFS (atoms/cm$^2 \times 10^{-15}$)		Rate[a,c] (Å/min)	ΔTi Upon Etching (%)
	Before Etching	After Etching		
1. HB-HPR-206	14.5	11.8	0 (580)	18
2. m-Cresol Novolac	9.4	5.1	125 (660)	46
3. Acetylated-m-Cresol Novolac (100 mole %)	9.8	5.5	57	41
4. Poly(vinyl alcohol)	7.9	5.6	37	29
5. Poly(vinyl acetate)	6.1	4.0	298[b]	34
6. P(MMA-co-MAA) (25 mole % MAA)	7.3	-	<50 (1500)	-
7. PMMA	8.2	5.3	25 (1560)	35
8. Poly(styrene)	28.2	22.1	325[b] (610)	22

a. Etched for 30 min by O$_2$ RIE. b. Completely removed. c. Parenthesis values are for untreated films.

Table IV. Influence of Polymer Structure on $TiCl_4$ Incorporation[a] and O_2 RIE Rates of Various Polymer Films

Polymer	Ti Concentration Measured by XFS (atoms/cm^2 × 10^{-15})		Polymer Removal Rate[b] (Å/min.)	ΔTi Upon Etching (%)
Before Etching	After Etching			
1. HB-HPR 206	20.3	16.4	0	19
2. m-Cresol Novolac	13.0	6.2	15	52
3. Acetylated m-Cresol Novolac (100 mole %)	12.4	9.8	10	21
4. Poly(vinyl alcohol)	12.8	11.3	27	17
5. Poly(vinyl acetate)	9.9	7.4	6	26
6. P(MMA-co-MAA) (25 mole % MAA)	12.9	8.1	10	37
7. PMMA	14.0	7.4	20	47
8. Poly(styrene)	26.2	24.0	300	8

a. Treated for 2 mins.
b. Etched for 30 mins. by O_2 RIE.

1.45×10^{16} Ti atoms/cm^2 registered no film thickness loss after 30 min. O_2 RIE, but lost about 18% of Ti. Under the identical conditions PS films were completely etched away despite having 2.82×10^{16} Ti atoms/cm^2, but registered only 22% Ti loss (Table III). The etching rates of poly(vinyl acetate) and m-cresol novolac films in Table III were significantly higher than those treated for 2 min. (Table IV). Poly(vinyl acetate) and m-cresol novolac films treated for 1 min. had only 3 and 4.5 monolayers of TiO_2, respectively. A 30 mins. O_2 plasma treatment depleted 1-2 monolayers of TiO_2 leaving behind just ~2 monolayers of TiO_2, not enough to form a good, continuous etch mask. However, similar films treated for 2 mins. had 5 and 6.5 monolayers of TiO_2 which was reduced to about 3 monolayers after etching. This is still enough to form a reasonable etch mask.

Table V. Ti Concentrations on Acetylated *m*-Cresol Novolac Copolymers as a Function of mole % Acetyl Groups[1]

Polymer	Mole % Acetyl	Ti Concentration Measured by XFS Before Etching (atoms/cm^2)
1. *m*-Cresol Novolac	0	9.42×10^{15}
2. Acetylated Novolac Copolymer	20	9.46×10^{15}
3. "	60	1.10×10^{16}
4. "	70	9.69×10^{15}
5. "	80	1.05×10^{16}
6. "	90	9.39×10^{15}
7. "	100	9.83×10^{15}

1. TiCl$_4$ treatments were conducted for 1 minute at a background pressure of 220 mtorr

In order to understand the very poor etching selectivity of PS films, in spite of their very high Ti content, as well as to examine the effect of Ti amounts and depth profiles on the etching rates, RBS analysis was conducted on a representative group of hydrophilic and hydrophobic polymer films that were treated for 1 and 2 mins. with TiCl$_4$. The treated samples were broken into two pieces and one of each was subjected to O$_2$ RIE. Then both sets of samples were analyzed by RBS. Two generalizations can be made from these analyses. First, the Ti content is lower in etched samples as compared to unetched samples, as expected (Figures 5,6). Second, the Ti distribution is confined to a relatively thin layer at the surface of all polymers except PS, where it was found to be diffused through the entire thickness of the film (Figure 7) as indicated by the very broad Ti peak and more importantly, by the presence of a broad peak corresponding to Cl extending from channel numbers 315 to 354. The presence of chlorine through the entire thickness of the PS film, whether from diffused TiCl$_4$ or from HCl (hydrolysis product of TiCl$_4$), attests to the diffusion process in PS. In the case of hydrophilic and moderately hydrophobic polymer films, the Ti peak shapes and widths for the etched and unetched specimens are nearly identical leading to the conclusion that there is no significant diffusion of Ti into the bulk of the film and that Ti is present within the top 300 Å, the resolution limit of the RBS measurement. In

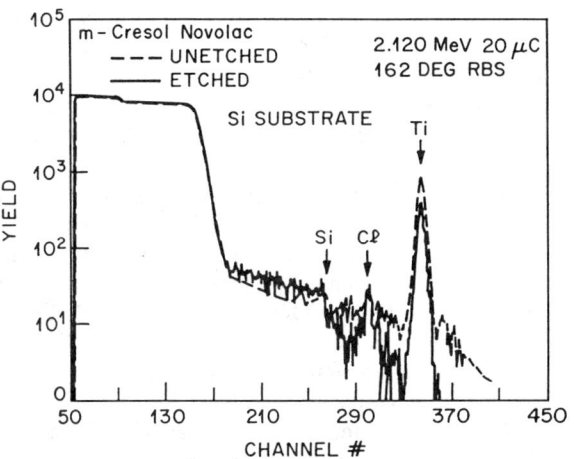

Figure 5. RBS spectra of a *m*-cresol novolac film treated with TiCl$_4$ for 2 mins. (dotted line) and then etched for 30 mins. by O$_2$ RIE (solid line).

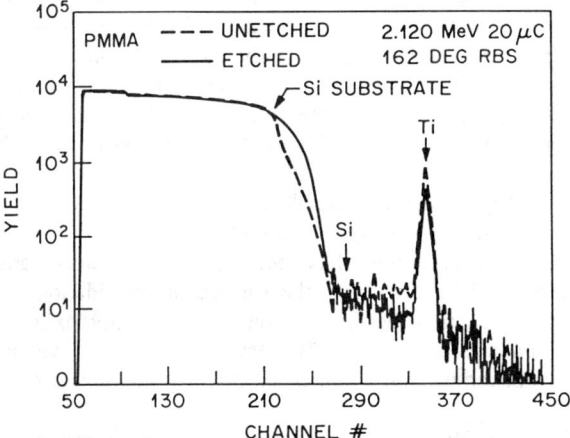

Figure 6. RBS spectra of a PMMA film treated with TiCl$_4$ for 2 mins. (dotted line) and then etched for 30 mins. by O$_2$ RIE (solid line).

hydrophilic and moderately hydrophilic polymers, the polar groups, such as C=O, OH and COOH, hydrogen bond to a surface water layer on the polymer. In the functionalization process, $TiCl_4$ is hydrolyzed at the polymer surface to form a surface TiO_2 layer that acts as an etch barrier during O_2 RIE. Our results indicate that at least 3 monolayers of Ti are necessary to form a tenacious etch mask and that etching selectivities as high as 3000 can be attained with 5-6 monolayers of Ti. Hydrophobic polymers such as PS lack hydrogen bonding functionalities and hence lack surface water, thus explaining the absence of a significant surface TiO_2 layer and poor etching selectivities. XPS analysis of PS samples corroborates this hypothesis (13).

Several substituted poly(styrene) polymers were treated with $TiCl_4$ to determine the effect of the substituent on the functionalization process (Figure 8). The polymer films were treated with $TiCl_4$ for 1 min. and were etched in an O_2 plasma for 30 mins. to determine the etching properties. The substituents that possess oxygen atoms e.g. OH, OAc, OMe, COOH, COOR, OCO_2R, COR and even SO_2 and NO_2 made the polymer films etch resistant to O_2 RIE indicating that the polymer surface is hydrophilic enough to sorb sufficient water required to form an etch resistant TiO_2 layer upon reaction with $TiCl_4$. Separate experiments with poly(styrene-allyl alcohol) copolymers revealed that 0.2-0.3 atomic % of hydroxyl oxygen is sufficient to give functionalized polymers after treatment with $TiCl_4$. Poly(styrene) with Cl, Br, CH_3 or CH_2Cl substituents did not functionalize after treatment with $TiCl_4$ as evidenced by the etching rates.

These results provided the basis for postulating a new imaging scheme using 1) photooxidation of hydrophobic polymers to give a hydrophilic polymer surface having C=O, OH and COOH groups, 2) selective sorption of water on the hydrophilic sites, 3) reaction of $TiCl_4$ with surface water to form TiO_2 on the hydrophilic exposed sites and 4) development by O_2 RIE to provide negative-tone images (Figure 9). Since this process only has utility at the polymer surface, optimum results should be obtained for very highly absorbing polymers. Owing to its very high absorption, PS or its derivatives should be most sensitive at 193 nm. Initial experiments were conducted with 193 nm radiation from an ArF laser while imaging experiments were performed using KrF excimer laser radiation (248 nm) from a deep-UV stepper since no reliable fine line imaging tool is available when 193 nm radiation is used as the exposure source.

Photooxidation of poly(styrene) has been a subject of considerable interest over the past 25 years (14,15). However, the surface photooxidation of poly(styrene), the aspects of which are most pertinent to the proposed photooxidative scheme, has only been examined recently (16,17). Free radical intermediates have been proposed to account for the formation of oxidized groups upon 254 nm irradiation of poly(styrene).

Sensitivity data for 193 nm exposures were obtained by imaging 1 mm to 1 cm diameter spots in a 1.2 μm thick poly(styrene) film using a Questek ArF excimer laser. Sensitivity was found to be a function of the fluence. For example, one pulse was sufficient to result in a full thickness image after treatment with $TiCl_4$ and O_2 RIE when the fluence was 6 mJ/cm^2/pulse (Figure 10). Considerably more dose (32 mJ/cm^2) was required to obtain the same result when the fluence was 1

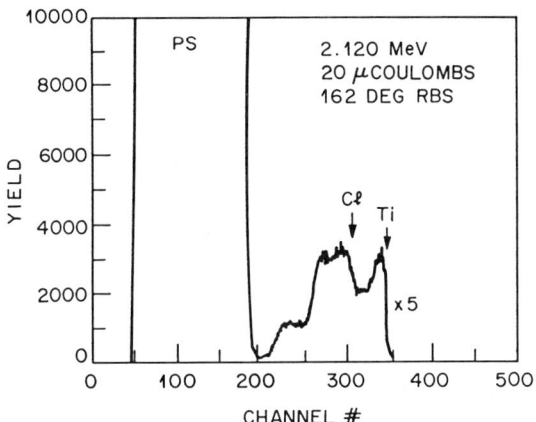

Figure 7. RBS spectra of a poly(styrene) film treated with $TiCl_4$ for 2 minutes.

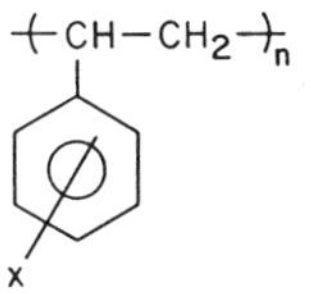

$$\{CH-CH_2\}_n$$

(phenyl ring with substituent X)

x = Cl, Br, I, CH_3, CH_2Cl, H
DID NOT FUNCTIONALIZE

x = OH, OAc, OMe, COOH,
COOR, OCO_2R, COR, NO_2, SO_2R
FUNCTIONALIZED

Figure 8. Structures and functionalization nature of poly(styrene) derivatives reacted with $TiCl_4$.

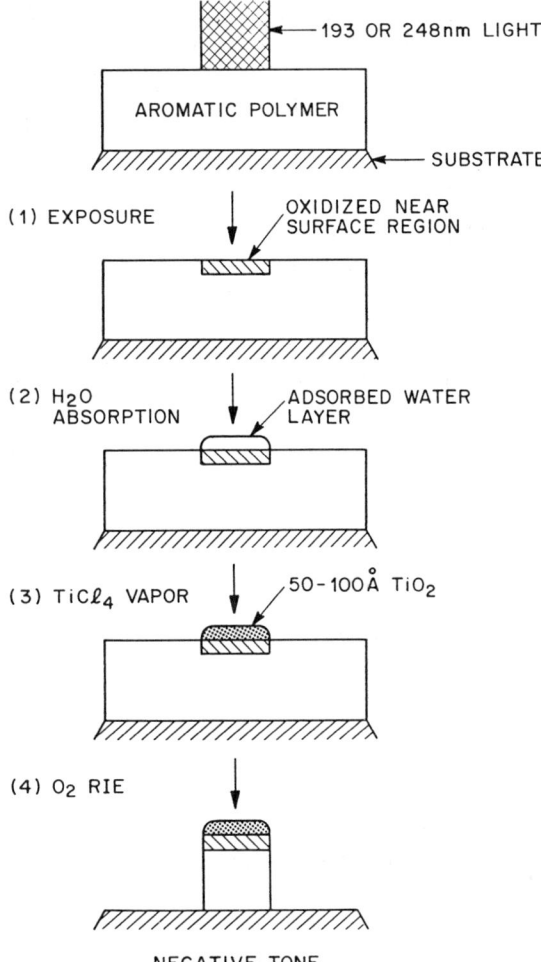

Figure 9. Photooxidative imaging scheme using the gas-phase functionalization of hydrophilic polymer regions by $TiCl_4$ treatment and O_2 RIE development.

mJ/cm^2/pulse. In the latter case, a rough value for the contrast could be determined as 2.2. Chlorinated poly(styrene) films were found to be slightly more sensitive; 3 mJ/cm^2 was sufficient to give full thickness image at the fluence of 1 mJ/cm^2/pulse.

Poly(styrene) exhibited only traces of patterns at doses ≤500 mJ/cm^2 when exposed at 248 nm on a KrF excimer laser stepper. However, Sub half-micron features (Figure 11) could be resolved with a sensitivity of 200 mJ/cm^2 and a contrast >2 (Figure 12) in 1-2 μm thick chlorinated poly(styrene) films. The sensitivity of 200 mJ/cm^2 is rather low, but is significant considering that the absorption of poly(styrene) is less than 0.1/μm at 248 nm (Figure 13). The sensitivity and contrast of chlorine-containing poly(styrenes) were found to be a function of the amount of Cl and the position of Cl on the polymer chain in addition to the relative humidity, TiCl$_4$ treatment time and etching time. Chlorine at the α-carbon imparted the highest selectivity (Table VI). Sensitivity also was found to be directly proportional to the amount of chlorine in the polymer.

Table VI. Lithographic Sensitivity as a Function of the Amount and Position of Chlorine in Poly(styrene) Polymers

Polymer	Cl/C	Lithographic Sensitivity
1. Chlorinated poly(styrene)[a]	0.143	200
2. Poly(2-chloro styrene)	0.125	>500
3. Poly(4-chloro styrene)	0.125	>500
4. Poly(4-chloromethyl styrene)	0.111	300

a. Chlorine is present both on the methine and aromatic ring carbons.

The patterns generated by this method have some line edge roughness and background residue resulting from the diffusion of TiCl$_4$ into the unexposed regions. Residue was also observed in control experiments where a chlorinated poly(styrene) film was treated with TiCl$_4$ without any exposure to light and was etched by O$_2$ RIE. Another factor contributing to the residue is the reflected light from the Si substrate.

The "residue" can be minimized using bilayer schemes and sensitivity can be increased by either using poly(vinyl biphenyl) derivatives that are more absorbing at 248 nm or by adding anthracene derivatives to chlorinated poly(styrene) polymer. The present formulations are not production worthy because of this "residue" and we are currently working on approaches that may eliminate this problem.

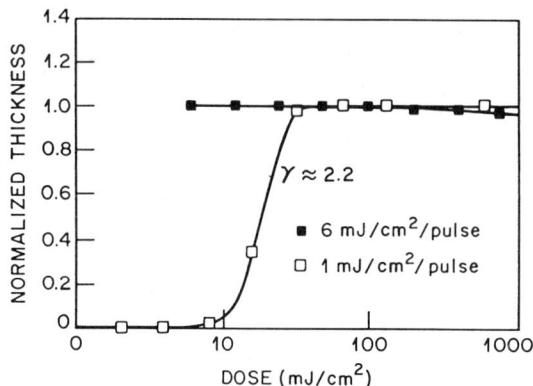

Figure 10. Sensitivity curves for 1.2 μm thick poly(styrene) films exposed at 193 nm at fluences of 1 and 6 mJ/cm^2/pulse and developed by O$_2$ RIE.

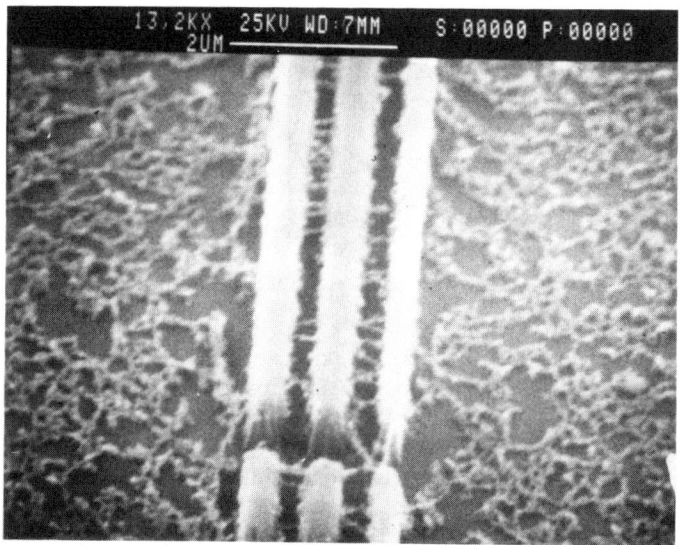

Figure 11. SEM of 0.4 μm line and space patterns in a 1.2 μm thick chlorinated poly(styrene) film exposed with 250 mJ/cm^2 of 248 nm light, treated with TiCl$_4$ and developed by O$_2$ RIE.

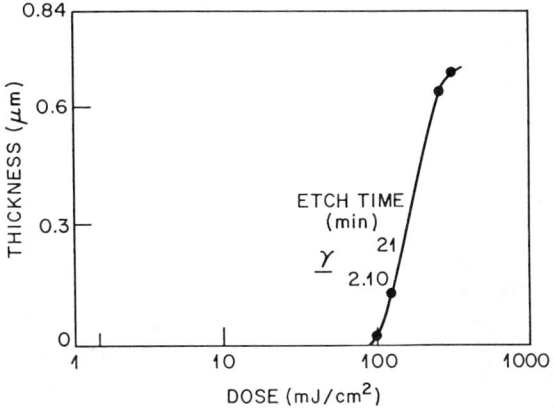

Figure 12. Sensitivity curve for 0.8 μm thick chlorinated poly(styrene) resist exposed at 248 nm and developed for 21 mins. by O_2 RIE.

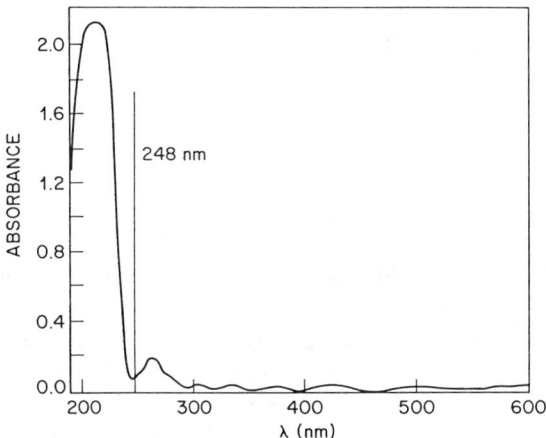

Figure 13. Absorption spectrum of a 1 μm thick poly(styrene) film on a quartz disc.

Conclusions

TiCl$_4$ readily functionalizes hydrophilic polymers such as poly(vinyl alcohol), m-cresol novolac and methacrylic acid copolymers as well as moderately hydrophobic polymers such as poly(methyl methacrylate), poly(vinyl acetate), poly(benzyl methacrylate) and fully acetylated m-cresol novolac. TiCl$_4$ did not react with poly(styrene) to form etch resistant films indicating that very hydrophobic films follow a different reaction pathway. RBS analysis revealed that Ti is present only on the surface of hydrophilic and moderately hydrophobic polymer films, whereas it was found diffused through the entire thickness of the poly(styrene) films. The reaction pathways of hydrophilic and hydrophobic polymers with TiCl$_4$ are different because TiCl$_4$ is hydrolysed by the surface water at the hydrophilic polymer surfaces to form an etch resistant TiO$_2$ layer. Lack of such surface water in hydrophobic polymers explains the absence of a surface TiO$_2$ layer and the poor etching selectivities.

A photooxidative scheme has been developed to pattern sub half-micron images in single layer resist schemes by photochemical generation of hydrophilic sites in hydrophobic polymers such as poly(styrene) and chlorinated poly(styrene) and by selective functionalization of these hydrophilic sites with TiCl$_4$ followed by O$_2$ RIE development. Sub half-micron features were resolved in 1-2 μm thick chlorinated poly(styrene) films with exposures at 248 nm on a KrF excimer laser stepper. The polymers are much more sensitive to 193 nm (sensitivity 3-32 mJ/cm^2) than to 248 nm radiation (sensitivity ~200 mJ/cm^2) because of their intense absorption at 193 nm.

Acknowledgments

We thank Larry Stillwagon for many helpful discussions, Norris Hobbins for assistance with XFS measurements, Vic McCrary and Vince Donnelly for exposures with the ArF laser and Ray Cirelli for deep-UV stepper exposures.

Literature Cited

1. Taylor, G. N.; Stillwagon, L.; Venkatesan, T. *J. Electrochem. Soc.* 1984, *131*, 1658.
2. Wolf, T. M.; Taylor. G. N.; Venkatesan, T.; Kraetsch, R. T. *ibid.* 1984, *131*, 1664.
3. MacDonald, S. A.; Ito, H.; Hiraoka, H.; Willson, C. G. *Mid-Hudson Section Soc. Plastic Engrs., Ellenville, NY, October 28-30*, 1985, p. 177.
4. Coopmans, F.; Roland, B. *Solid State Technol.* 1987, *30*, 6, 93.
5. Roland, B.; Vandendriessche, J.; Lombaerts, R.; Denturck, B.; Jakus, C. *Proc. SPIE* 1988, *920*, 120.
6. Venkatesan, T.; Taylor, G. N.; Wagner, B.; Wilkens, B.; Barr, D. *J. Vac. Sci. and Technol.* 1981, *19*, 1379.
7. Stillwagon, L. E.; Silverman, P. J.; Taylor, G. N. *Mid-Hudson Section, Soc. Plastic Engrs., Ellenville, NY, October 28-30*, 1985, p. 87.

8. Taylor, G. N.; Nalamasu, O.; Stillwagon, L. *Microcircuit Engg.* 1988, 7, (in Press).
9. Feit, E. D.; Stillwagon, L. E.; *Polym. Eng. and Sci.* 1980, 20, 1058.
10. Tarascon, R.; Hartney, M.; Bowden, M. J. In *Materials for Microlithography: Radiation sensitive Polymers*; Thompson, L. F.; Willson, C. G.; Frechet, J. M.; Eds.; ACS Symposium Series No. 266; American Chemical Society: Washington, DC, 1984; pp. 361-388.
11. Warshawsky, A.; Deshe, A. *J. Polym. Sci., Chem. Ed.* 1985, 23, 1839.
12. Pol, V.; Bennewitz, J. H.; Escher, G. C.; Feldman, F.; Firtion, V. A.; Jewell, T. E.; Wilcomb, B. E.; Clemens, J. *Proc. SPIE* 1986, 633, 6.
13. Taylor, G. N.; Stillwagon, L. E.; Baiocchi, F. A.; Vasile, M. J. *Microcircuit Engg.* 1987, 6, 381.
14. Grassie, N.; Weir, N. A.; *J. Appl. Polym. Sci.* 1965, 9, 963, 987, 999.
15. Geuskens, G.; Baeyens-Volant, D.; Delaernois, G.; Lu-Vinh, Q.; Piret, W.; David, C. *Eur. Polm. J.* 1978, 14, 291, 298.
16. Peeling, J.; Clark, D. T. *Polym. Degrad. and Stabil.* 1981, 3, 97.
17. Clark, D. T.; Munro, H. S. *ibid.* 1984, 8, 213; *ibid.* 1984, 9, 63, 185.

RECEIVED July 17, 1989

Chapter 13

Kinetics of Polymer Etching in an Oxygen Glow Discharge

Charles W. Jurgensen

AT&T Bell Laboratories, 600 Mountain Avenue, Murray Hill, NJ 07974

This paper is a critical review of the literature and a summary of my recent work on the etching kinetics of organic and organosilicon polymers in oxygen glow discharges. The most important application for etching polymers in oxygen glow discharges is the pattern transfer step in multi-layer lithography. Anisotropic etching is required and observed in this application; this suggests that bombardment-induced processes play a dominant role in the etching mechanism. The etching rate for a bombardment-induced mechanism is equal to the flux of bombarding particles times the average yield per particle. The simplest kinetic model for a bombardment-induced process assumes that the yield per bombarding particle depends only on its energy and its angle relative to the surface normal. This paper discusses etching results of organic and organosilicon polymers to determine the range of etching conditions where these results are consistent with this simplest kinetic assumption.

Multi-Layer Lithography. The etching kinetics of organic and organosilicon polymers in oxygen glow discharges is important because the oxygen "reactive ion etching" (O_2 RIE) behavior of these materials is the basis for the pattern transfer step in multi-layer lithography [1,2]. Single-layer optical lithography is now capable of half micron resolution on planar, nonreflective substrates; however, thickness variations and reflections off topography are severe problems for single-layer resists on reflective topographic substrates. These difficulties associated with device topography are eliminated in multi-layer lithography by coating the substrate with an organic planarizing layer that ideally provides a level, nonreflective surface on which to image. This simplifies the imaging step, but additional processing steps are required to transfer the pattern through the planarizing layer. Tri-layer schemes use a conventional resist to pattern an intermediate masking layer which is subsequently used to pattern the planarizing layer during the O_2 RIE pattern transfer step; the masking layer may be either an inorganic oxide [3], or an organosilicon polymer [4]. Silicon (or another oxide precursor) is incorporated into the imaging layer in bi-layer lithography [5,6,7] to enable it to function as the O_2 RIE mask. Surface functionalization schemes [8,9] achieve multi-layer performance in a single resist layer by selective incorporation of an O_2 RIE resistant species into the surface of the exposed resist film to allow O_2 RIE development of the latent image. The O_2 RIE pattern transfer step is a critical process in all these schemes; thus it is important to understand the factors controlling polymer etching rates, selectivity, uniformity, anisotropy, and process latitudes.

The Problem with Plasmas. Plasma processes are characterized by many independent degrees of freedom including the rf-voltage, rf-frequency, pressure, gas composition, gas flow rate, sample temperature, reactor geometry, magnetic field strength, and others. All these processing variables have an effect on etching rates, selectivity, uniformity, and anisotropy; but a fundamental interpretation of these effects requires that one understand the effect of the processing variables on the fundamental variables such as the radical concentration, ion flux, ion energy distribution, and others. Plasmas are not understood well enough to predict these fundamental variables as a function of the processing variables; thus one must use plasma diagnostics to determine the dependence of the fundamental variables on the processing variables. Unfortunately plasma diagnostics are themselves complex, and it is difficult to measure or estimate the radical flux, ion flux, ion energy distribution, ion angular distribution, energetic neutral flux, and other fundamental variables. Interpreting the results observed in empirical studies is difficult because one never has complete information on all these fundamental variables.

Bombardment-Induced Kinetics. The directionality observed in anisotropic pattern transfer processes results from the highly directional angular distribution of the energetic particles bombarding the surface being etched. In O_2 glow discharges, these particles are O_2^+ ions and the energetic neutral O_2 products of charge transfer collisions [10]. Highly selective and anisotropic pattern transfer is routinely achieved in multi-layer lithography; this indicates that chemical and physical processes are acting synergisticly because purely physical sputtering is not selective while purely chemical etching is not anisotropic. The etching rate for a bombardment-induced process is equal to the bombarding particle flux times the yield per bombarding particle; thus, to characterize the kinetics of a bombardment-induced chemically-assisted etching process, one must determine the yield per bombarding particle as a function of its mass, energy, angle relative to the surface normal, the flux ratio of bombarding to chemically assisting species, sample temperature, and other fundamental variables. If the chemically assisting species is ground state oxygen molecules then the bombarding to assisting flux ratio will be low at typical O_2 RIE pressures, so the pressure may have little effect on the yields under these conditions. The simplest bombardment-induced kinetic model assumes that the yield per bombarding particle only depends on its energy and its angle relative to the surface normal. This review will seek to determine the range of etching conditions where this simplest kinetic assumption is consistent with published etching results. The first section of this paper is a critical review of the literature on the etching kinetics of organic polymers in O_2 glow discharges. The second section of this paper reviews the oxidation kinetics of organosilicon polymers in O_2 glow discharges. Some important aspects of polymer etching kinetics can not be learned by etching planar samples. For example, the angle dependence of the yield can not be determined except by bombarding at off normal incidence or by etching topographic substrates. The angle dependence has not been directly studied in polymer etching, but it is particularly important for understanding etching profiles as discussed in the final section of this paper.

Organic Polymers

Chemical Etching Regime. Taylor and Wolf [11], studied the O_2 plasma etching rates of many organic polymers at high pressure (.5 to 1 torr) in a barrel-type etching system where the physical etching component is negligible. Pederson [12] conducted a similar study for CF_4 and CF_4/O_2 discharges. The etching rates observed in these studies were strong functions of temperature, and the relative rates of different polymers varied more than an order of magnitude. These large variations were rationalized in terms of chemical properties such as the chain scission yield. Cook and Benson [13,14] used electron paramagnetic resonance spectroscopy to study the oxidation of photoresist (novolac) downstream from a microwave discharge. They found that the etching rate is proportional to the O atom concentration, and shows an Arrhenius temperature dependence with an activation energy of 11 kcal/mole. Spenser et al. [15] found the same activation energy in a downstream stripping system.

Ion Beam Studies. Gokan et al. [16,17,18] studied Ar^+ and O_2^+ ion beam etching (IBE) rates of several organic polymers. Since carbon has the smallest sputtering yield of the atoms present in organic polymers, they expected that the relative Ar^+ sputtering rates would be inversely proportional to the mass density of carbon (density times mass fraction carbon) in the polymer. The relative etching rates varied by a factor of 3, and were in good agreement with the expected correlation. Furthermore, a better correlation was obtained when one carbon atom per oxygen atom in a monomer unit was neglected in calculating the carbon mass density. They concluded that C=O groups have a higher sputtering yield than carbon atoms; this is reasonable because CO is a stable molecule and one would expect its binding energy to the surface to be much smaller than that of a carbon atom. The

O_2^+ IBE rates were proportional to but 15 times higher than the Ar^+ IBE rates, and CO was the predominant carbon containing O_2 IBE product. The O_2^+ IBE yields [17] increased as the O_2 chamber pressure increased, but surprisingly the yield decreased with increasing bombardment energy. At the lowest pressure, the yields were 2 carbons per O_2^+, indicating that both O atoms reacted to form CO. This is an example of the reactive ion etching mechanism, but it is interesting to note that this mechanism was only observed at their lowest chamber pressure (5×10^{-5} torr). As the O_2 chamber pressure increased, the yields increased up to 6 carbons per O_2^+ at the highest pressure studied (0.4 mtorr). They noted that this chamber pressure was too low for charge transfer collisions to play an important role in increasing the flux of bombarding particles; thus they concluded that these yields resulted from an O_2 chemically-assisted, bombardment-induced etching mechanism. They were also able to explain the counterintuitive energy dependence of the yields in terms of this mechanism. Apparently their ion source was not equipped with accel-decel extraction electrodes to prevent target electrons from flowing back to the source plasma; thus they could not independently control acceleration voltage and beam current. The beam current increased rapidly with acceleration voltage, so the neutral-to-ion flux ratio (pressure to beam current ratio) decreased with acceleration voltage. They concluded that the bombardment-induced chemistry per ion depends directly on the neutral-to-ion flux ratio, and that this dependence dominated the direct energy dependence when the acceleration voltage was increased.

These studies [16,17,18] are important because they show that ground state O_2 can enhance polymer O_2^+ IBE yields, but they leave several important questions unanswered. What is the energy dependence of the yield at constant O_2 pressure and beam current? What happens when the neutral-to-ion flux ratio is increased by more than an order of magnitude to the range typical for O_2 RIE pattern transfer processes? Does the yield saturate with increasing pressure at constant ion flux and energy? Do the O_2 molecules enhance the yield by a thermal spike mechanism [19,20] or by an enhanced sputtering mechanism [21]? The chemically enhanced sputtering mechanism is a momentum transfer collision cascade mechanism where the product molecules leave the surface within a picosecond of the collision and have a non-Maxwellian energy distribution. The chemical enhancement in this mechanism results from the chemical stability of CO molecules which allows a CO molecule to be removed from the polymer chain with much less energy than would be required to remove a CH_X radical or C atom. The thermal spike mechanism assumes that the product molecules leave the surface on a time scale greater than a picosecond which is the time required for the surface to reach local thermal equilibrium [19]. In this mechanism, the product molecules have a modified Maxwellian energy distribution at the (time varying) local surface temperature [19]. A low energy (< 1000 eV) ion penetrates ≈ 40 Å into the surface; thus it deposits its energy into the ≈ 1000 atoms that are within 40 Å of the point where the ion hit. For a 500 eV ion, this results in a local temperature of ≈ 5000 °K (roughly the temperature at the surface of the sun) which persists for tens of picoseconds. The chemical enhancement in this mechanism results from thermally induced reactions with O_2 adsorbed to the surface and from thermal desorption of CO groups which may be formed when O_2 reacts with the radicals that may remain after a thermal spike has cooled. These mechanisms may be distinguished by measuring the product energy distribution [19,20,21,22], but such measurements have not been attempted in polymer etching. The most important distinction between these mechanisms is that energy is a scalar quantity while momentum is a vector quantity; thus yields may be angle dependent for a momentum transfer mechanism, but must be nearly independent of angle for a thermal spike mechanism. Physical sputtering yields are strongly angle dependent [22,23], but chemically-enhanced bombardment-induced etching yields are nearly independent of angle in the systems that have been studied to date [24,25,26]. The angle dependence was not determined in Gokan's O_2 IBE studies [16,17,18]; however, it is relevant to the simulation of pattern transfer processes as will be discussed in the final section of this paper.

Role of O Atoms in Pattern Transfer Regime. Oxygen atoms spontaneously etch polymers and are present under O_2 RIE conditions, so it is important to determine if they have a significant effect on polymer O_2 RIE rates under typical etching conditions. Hartney et al. [27] measured the effect of power density and pressure on organic polymer O_2 RIE rates while using mass spectrometric flux analysis to track the O atom partial pressure. Their organic polymer O_2 RIE rate increased with pressure, but the O atom partial pressure decreased with pressure. This experiment shows that O atoms can not explain the observed trend, and suggests that O atoms do not participate in the rate controlling step. Selwyn [28] used laser-induced fluorescence to measure the O atom concentration distribution over a bare electrode, and over a polyimide polymer sample in a low pressure O_2/Ar glow discharge. Etching rates increase with bias voltage, but Selwyn found that the O atom concentration near the polymer surface decreased with bias voltage. Selwyn concluded that the O atoms are not rate controlling, and suggested a dominant role for bombardment-induced processes. The O atom concentration gradient increased with bias voltage, which shows that the O atom consumption rate increases with the etching rate. Selwyn [28] suggests that bombardment-induced processes control the etching rate, which in turn controls the O atom consumption rate, and hence the concentration near the polymer surface. Three studies of O_2 RIE profiles [29,30,31] have reported that O atoms can result in lateral etching rates (undercutting) as high as 15% of the vertical etching rate; moreover, this lateral etching component is temperature sensitive (activation energy 2.1 kcal/mole [30]). This radical induced undercutting may be almost completely suppressed by holding the substrate at room temperature [29,30], or by adding hydrocarbons to the plasma to reduce the O atom concentration [31].

Role of Bombardment in Pattern Transfer Regime. Gokan et al. [18] measured the relative O_2 RIE rates of a series of polymers and found that the RIE rates scaled inversely with the mass density of carbon as in their IBE study (several materials deviated from the correlation). The relative RIE rates [18] varied by a factor of 3 and do not show the large variations characteristic of the radical-induced mechanism [10]. These results strongly suggest a bombardment-induced mechanism, but one can not directly apply the IBE results to RIE conditions because IBE neutral-to-ion flux ratios are orders of magnitude lower than under RIE conditions.

Paraszczak et al. [32] studied polyimide etching in a dual rf-microwave etching system. They used a Langmuir probe to monitor the plasma density, and used the 13.5 MHz rf power to control the sheath acceleration voltage while independently controlling the plasma density with the microwave power source. In one experiment, they varied the sheath acceleration voltage at constant plasma density, and observed that the etching rate increased as the square root of voltage over the range from 15 to 160 Volts. The ion flux is proportional to the plasma density [33], so this experiment is performed at constant ion flux; thus it determines the etching yield as a function of bombardment energy to within a proportionality factor. They [32] noted that the momentum per bombarding particle varies as the square root of its energy, and argued that the square root energy dependence implies that the rate controlling step is a physical sputtering (momentum transfer) process. I agree that these results reflect the energy dependence of the yield; however, I do not agree that these results imply that the rate controlling step is a physical sputtering process for the following reasons: (1) A straight line with a positive intercept on the rate axis fits this data as well as their square root energy dependence. (2) Physical sputtering yields do not scale as the square root of bombardment energy [22]. I will return to a discussion of the energy dependence after showing that the results presented by Paraszczak et al. [32] are consistent with the energy dependent yields reported by Jurgensen and Rammelsberg [34].

In a second experiment, Paraszczak et al. [32] observed that the etching rate increases linearly with the plasma density when the microwave power source is used to vary the plasma density at constant sheath acceleration voltage. They argued that the O atom concentration is proportional to the plasma density and interpreted the results of this experiment in terms of a radical-dominated mechanism. They did not measure the O atom concentration, and it is not clear why they abandoned the physical sputtering mechanism which they had used to explain the acceleration voltage dependence. The ion flux is proportional to the plasma density [33] (at constant electron temperature as reported in this experiment); thus a linear increase in etching rate with plasma density can also be interpreted in terms of a bombardment-induced mechanism. This experiment shows that the etching rate is proportional to the ion flux at constant acceleration voltage; thus it implies that the yield is independent of the neutral-to-ion flux ratio under these etching conditions.

Visser and de Vries [29] used an energy flux diagnostic to study the O_2 RIE rates of novolac, polystyrene and polymethylmethacrylate (PMMA) in a symmetric parallel plate etching system. They reported that the etching rates of the novolac and polystyrene are insensitive to temperature, but the etching rate of PMMA at 150 °C was three times its etching rate at 40°C. They determined that the etching reactions are exothermic for novolac, polystyrene, and for PMMA at 40 °C; however, the PMMA reaction becomes endothermic at temperatures higher than 100 °C. At low temperature, the PMMA etching rate in an Ar plasma was much lower than in an O_2 plasma, but at high temperature the PMMA etched at the same rate both plasmas. They [29] concluded that PMMA etches by a bombardment-induced oxidation mechanism at low temperatures, but at high temperatures it etches by chain scission followed by depolymerization and evaporation of the monomer.

Visser and de Vries [29] used in situ temperature measurements and an energy balance to determine the rate at which bombarding particles deliver energy to the sample. They determined etching rates and energy deposition rates on both the powered and grounded electrodes as a function of O_2 pressure over the range from 1.3 to 12.5 Pa. The novolac etching rate increased with pressure on the powered electrode while it decreased with pressure on the grounded electrode; however, it was proportional to the bombardment energy flux on both electrodes. This correlation is plotted as triangles on Fig. 1 where the lower three etching rates were obtained on the grounded electrode while the higher rates were obtained on the powered electrode. Their system is nominally an equal area system, but the powered electrode still developed a negative self bias; thus the difference in etching rates between the powered and grounded electrodes primarily reflects the difference in sheath acceleration voltages on these electrodes. This energy flux diagnostic results in a correlation that simultaneously accounts for the effect of pressure and acceleration voltage on polymer O_2 RIE rates. They concluded that the proportionality between etching rates and the bombardment energy flux implies that the rate controlling step is a bombardment-induced process.

Jurgensen [10] has shown that the sheath thickness is on the order of the mean free path for charge transfer collisions under typical O_2 RIE conditions. Thus the flux of energetic neutral products of charge transfer collisions is on the order of the ion flux, and charge transfer collisions control the ion energy distribution at the electrode. Jurgensen and Shaqfeh [35] have presented a theory which uses measurements of the pressure, sheath thickness, and sheath voltage drop to estimate the flux and average bombardment energy of O_2^+ ions, and of energetic neutral O_2 products of charge transfer collisions. Jurgensen and Rammelsberg [34] have applied this theory to study the O_2 RIE kinetics of a hard-baked organic novolac polymer. They studied the effect of pressure, sheath voltage drop, rf frequency, sample temperature, and O_2 flow rate on O_2 RIE rates (R), and on the energy flux delivered by bombarding particles (Q). Figure 2 shows the effect of pressure

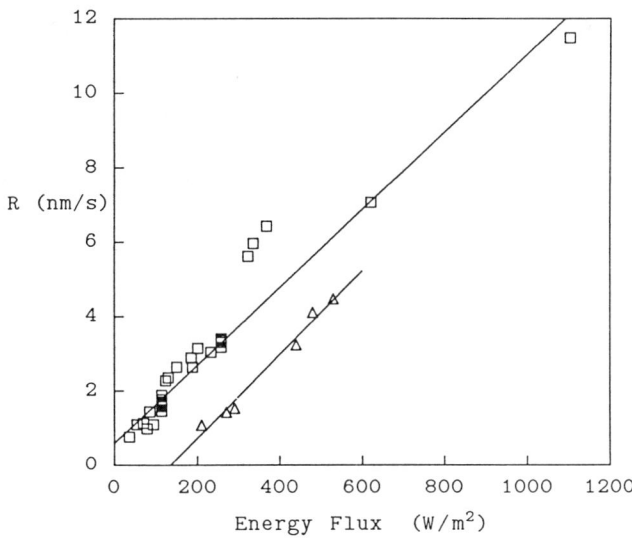

Fig. 1. The organic polymer etching rate is a linear function of the bombardment energy flux as determined by Visser and de Vries [29] (triangles) and by Jurgensen and Rammelsberg [34] (squares).

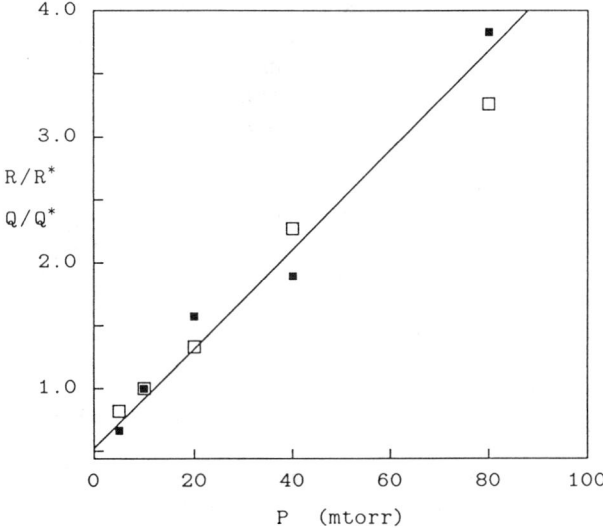

Fig. 2. The organic polymer O_2 RIE rate (filled points) tracks the bombardment energy flux (open points) as the pressure is varied at a constant 500 V self bias, and at 13.5 MHz.

on the O_2 RIE rate (filled points) and on the bombardment energy flux (open points) at 500 V self bias, 13.5 MHz, 20 SCCM O_2 flow, and 20 °C sample temperature. The etching rates and energy fluxes in Figs. 2 and 3 are ratioed to the results at 10 mtorr where the etching rate is 0.1 um/min and the estimated energy flux is 110 W/m^2. Note that the etching rate tracks the energy flux as reported by Visser and de Vries [29]. Jurgensen and Rammelsberg [34] also reported that the etching rate and bombardment energy flux both increased by a factor of 2.5 when the frequency was increased from 9 to 18 MHz at 10 mtorr, and 500 V self bias. Figure 3 shows the effect of self bias voltage on the relative etching rate (filled points) and bombardment energy flux (open points) at 13.5 MHz and 10 mtorr (squares) or 80 mtorr (triangles). This figure shows that the etching rate tracks the bombardment energy flux at both pressures. The flow rate was varied from 5 to 100 sccm, and the sample temperature was varied from 15 to 90 °C, but neither of these variables had a significant effect of either the etching rate or the bombardment energy flux. The squares in Fig. 1 show the correlation between etching rate and energy flux, as reported by Jurgensen and Rammelsberg [34]. These results span the following range of parameter space: pressure from 5 to 80 mtorr, self bias from 200 to 1000 volts, applied power density (not an independent variable) from 0.1 to 1.4 W/cm^2, frequency from 9 to 18 MHz, O_2 flow rate from 5 to 100 sccm, and sample temperature from 15 to 90 °C. Note that the etching rate is a linear function of the energy flux over this entire range of etching conditions.

Figure 1 shows that the results reported by Jurgensen and Rammelsberg [34] (squares) are in qualitative agreement with those reported by Visser and de Vries [29] (triangles), but the results of these studies do not overlay. Jurgensen and Rammelsberg [34] have discussed the differences in reactor geometry and in the methods used to estimate the bombardment energy flux in an attempt to understand what is responsible for the quantitative differences between these studies. They concluded that the relative energy flux estimates shown in Figs. 2 and 3 are accurate, but the absolute energy flux estimates (squares shown in Fig. 1) are low because the theory does not account for the effect of rf modulation on the electron density in the sheath [35]. The correction for this effect would shift the squares in Fig. 1 to the right towards the results presented by Visser and de Vries [29] (triangles). They also argued that energetic electrons may contribute to the energy flux measured in the symmetric system used by Visser and de Vries [29]. In particular energetic electrons are known [36] to bombard the counter-electrode where three of their results were obtained (lower 3 triangles in Fig. 1). Energetic electrons are expected to be less effective at inducing chemical reactions, so this could explain why these three data points are shifted to the right of the results presented by Jurgensen and Rammelsberg [34]. The above speculations may be tested by direct comparison between these methods of estimating the ion flux.

Bombardment-induced Yields in Pattern Transfer Regime. Visser and de Vries [29] estimated the sheath acceleration voltages in their system, and used the measured energy flux to estimate the ion flux. From the etching rate and the ion flux, they determined that the yield per ion was 3.6 monomer units for the novolac ($C_8H_9O_2$) at an estimated 500 V acceleration voltage. For this etching condition, the yields for novolac, polystyrene, and PMMA at 40 °C were all about 30 carbons per bombarding ion. This yield is 5 times larger than the largest O_2 IBE yield reported by Gokan and Esho [16], but this is consistent with Gokan's finding that the yield increases with the neutral-to-ion flux ratio. The results presented by Visser and de Vries [29] imply that the yield is independent of the neutral-to-ion flux ratio in the RIE pressure range, and this is consistent the results presented by Paraszczak et al. [32].

Figure 1 shows that the etching rate is nearly proportional to the energy flux delivered by bombarding particles; this implies that bombardment-induced processes control the etching rate, and that the yield per incident particle is nearly proportional to its energy. Figure 4 shows the energy dependence of the carbon atom yield (squares) as determined by Jurgensen and Rammelsberg [34] from the etching rate and estimated total flux of ions and energetic neutrals. The triangles and right hand axis on Fig. 4 are the etching rates reported by Paraszczak et al. [32] as function of sheath acceleration voltage at constant plasma density. The ion flux was constant in Paraszczak's [32] experiment, so his etching rates are proportional to the energy dependent yield and it should be possible to rescale his etching rates to superimpose them on Jurgensen's [34] yield trend. The line shown on Fig. 4 is a linear least squares fit to the yields reported by Jurgensen and Rammelsberg [34], but it is also a reasonable fit to the etching rates reported by Paraszczak et al. [32]. Thus the yields in these studies show nearly the same energy dependence. Jurgensen's yield extrapolates to a finite value of 3 $\pm$2 carbons per O_2 at zero bombardment energy. Subtracting 2 carbons per bombarding O_2 gives the induced yield per incident particle which extrapolates to zero at zero bombardment energy (within experimental error). For the rescaling factor chosen in Fig. 4, Paraszczak's [32] results extrapolate to a yield of 4.4 ($\pm$0.2) carbons per O_2^+ at zero energy. This implies that O atoms induce half the etching rate at the lowest acceleration voltage in Paraszczak's [32] study which agrees with the amount of undercutting observed on their etching profiles [32]. The etching profiles observed in Jurgensen's [37] system do not show any radical induced undercutting and will be discussed in the final section of this paper.

Under typical O_2 RIE conditions, the induced yield is much larger than the 2 carbons removed by the reactivity of an energetic O_2 ion or neutral. Thus the term "reactive ion etching" is actually a double misnomer. It implies that the bombarding particles are O_2^+ ions, but the flux of energetic neutral products of charge transfer collisions is roughly equal to the ion flux under typical O_2 RIE conditions [10]. In addition, it implies that the reactivity is supplied by the bombarding ion, but Fig. 4 shows that the induced yield is much larger than this reactive component. The mechanism implied by the name "reactive ion etching" was observed at the lowest pressures in

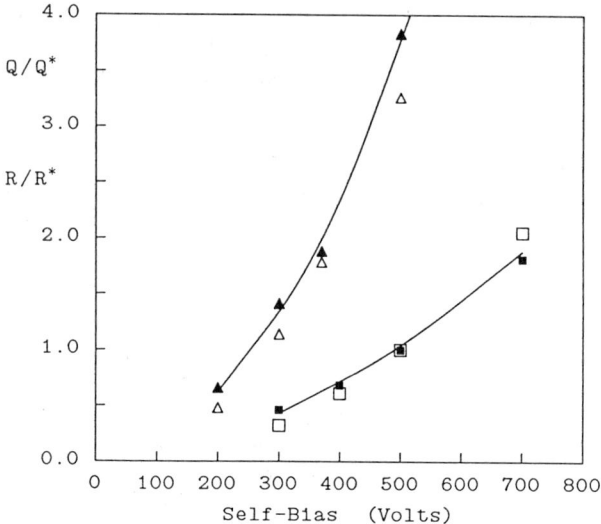

Fig. 3. The organic polymer O_2 RIE rate (filled points) tracks the energy deposition rate (open points) as the self bias voltage is varied at 13.5 MHz and 10 mtorr (squares) or 80 mtorr (triangles).

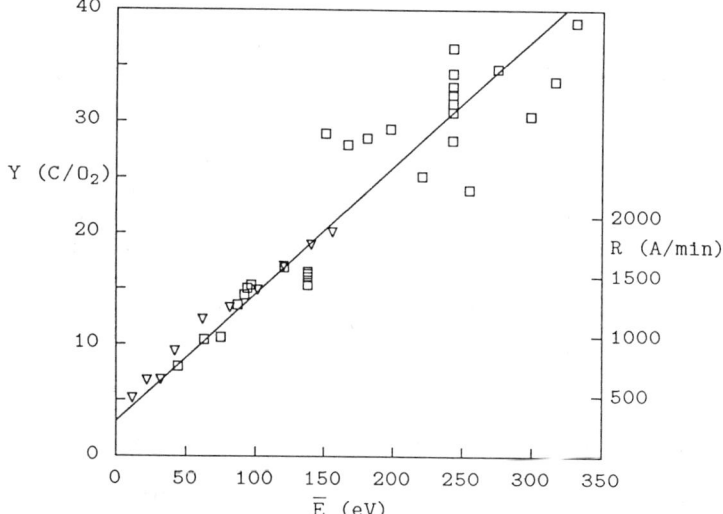

Fig. 4. The carbon atom yield (squares and left hand axis) reported by Jurgensen and Rammelsberg [34], is compared to the etching rate at constant plasma density (triangles and right hand axis) reported by Paraszczak et al. [32]. The line is a linear least squares fit to the Jurgensen's [34] results.

Gokan and Esho's [17] O_2^+ IBE yield study, but this mechanism does not apply under the "RIE" conditions. The term "reactive ion etching" not only refers to a particular (and incorrect) etching mechanism, but it is also used to refer to a reactor geometry where the powered electrode is much smaller than the grounded electrode and where the wafer sits on the powered electrode. Jurgensen's [34] results were obtained on such a system, but it is interesting to note that Visser and de Vries [29], and Paraszczak et al. [32] obtained their results on systems that are not traditionally called "reactive ion etching" systems. Nevertheless, the results obtained in these three studies clearly indicate that the same fundamental mechanism occurs in all three reactor configurations.

The bombardment-induced yields shown in Fig. 4 are so large that it seems unlikely that momentum transfer could play a dominant role in the etching mechanism. These yields are consistent with a thermal spike mechanism for the bombardment induced chemistry; in this mechanism, ground state oxygen molecules react with bombardment generated radical sites on the polymer to form carbonyl groups which are later thermally desorbed from the hot microregion created by a bombarding particle. This mechanism predicts that the yield increases with bombardment energy, and that the yield saturates with increasing pressure at constant ion energy and flux. This saturation effect was suggested in the O_2 IBE yield study presented by Gokan and Esho [17], and is consistent with the yield being independent of the neutral-to-ion flux ratio in the RIE regime as found in all the RIE studies [29,32,34]. The thermal spike mechanism also implies that the yield is nearly independent of angle which has important implications for modeling etching profiles.

Organosilicon Polymers

General Considerations. In 1980 Taylor and Wolf [11] noted that dimethylsiloxane has "no appreciable" etching rate in an O_2 plasma, and surmised that a thin <100 Å thick SiO_2 film formed on top of the polymer and protected it from further etching. The principle that silicon (or another oxide former) can drastically reduce a polymer's O_2 RIE rate is the basis for selectivity during the O_2 RIE pattern transfer step for many multi-layer lithographic processing schemes including "spin on glass" tri-layer schemes [4], bi-layer schemes [5,6,7], and surface functionalization schemes [8,9]. Line width loss caused by mask erosion is a serious concern for all these schemes, so it is important to understand organosilicon polymer O_2 RIE kinetics. More recently Butherus et al. [38] have reported another application for O_2 plasma etching of silicon containing polymers where the objective is to form a thick SiO_2 layer (>2000 Å) for use as a permanent dielectric between metal levels in multi-level metalization schemes. This application requires rapid oxidation of the organosilicon polymer, so it is also important to understand organosilicon polymer oxidation kinetics in O_2 glow discharges for this application. Note that the oxide layer must protect the organosilicon polymer from further oxidation in the masking application, while the entire layer must be rapidly oxidized in the dielectric application. Four etching regimes have been observed for organosilicon polymers in O_2 glow discharges. The transient regime occurs during the early stages of the etching or oxidation process while the oxide layer forms on the surface of the polymer. The diffusion-controlled regime [38] occurs under conditions where sputtering is absent; in this regime the oxide thickness and total thickness loss (initial thickness minus total thickness of the oxide and polymer) scale as the square root of the etching time. The steady-state regime [39,40] occurs when the sputtering rate balances the oxidation rate such that the oxide thickness is independent of time (after the initial transient), and the total thickness loss increases linearly with time (after the initial transient). Finally, the total thickness loss and the thickness of the partially oxidized layer increase linearly with time after an initial transient in the anomalous transport regime [40]. Before discussing the experimental results, it is useful to present a simple kinetic model which contains the transient, diffusion-controlled, and steady-state regimes as special cases.

Kinetics of Oxide Growth. The kinetic model derived in this section is an extension of the Deal-Grove [41] oxidation model to include a loss term that accounts for oxide loss by sputtering. This model assumes that the plasma conditions determine both the oxygen radical concentration on the surface of the oxide layer, and the rate at which oxide is lost by physical sputtering. Secondly, it assumes quasi-static diffusion of oxygen radicals across a uniform oxide layer with no recombination losses. Finally it assumes that the oxygen radicals react by first order kinetics at the polymer-oxide interface to form oxide and volatile oxidation products which diffuse back though the oxide to escape. In the quasi-static approximation, the flux of radicals across the oxide layer is set equal to their rate of consumption at the oxide-polymer interface [41] to obtain

$$F = kC_i = \frac{D(C_S - C_i)}{X} = \frac{kC_S}{(1 + Xk/D)} \quad (1)$$

where F is the O atom flux, k is the rate constant for the reaction between O atoms and polymer, C_i is the O atom concentration at the polymer-oxide interface, D is the O atom diffusion coefficient in the oxide, C_S is the O atom concentration on the surface of the oxide (plasma side), and X is the oxide thickness. The rate at which the polymer is consumed is proportional to the O atom flux which yields

$$\frac{dP}{dt} = -\nu F = \frac{-\nu k C_S}{(1 + Xk/D)} \quad (2)$$

where P is the remaining thickness of the organosilicon polymer layer, t is time, and ν is the volume of polymer removed per O atom. If the stoichiometry of the oxidation reaction is assumed to be

$$C_\alpha H_\beta Si_\eta O_\gamma + (2\alpha + \tfrac{1}{2}\beta + 2\eta - \gamma)O \rightarrow \alpha CO_2 + \tfrac{1}{2}\beta H_2O + \eta SiO_2 \quad (3)$$

then the volume of polymer consumed per O atom is given by

$$\nu = \frac{(12\alpha + \beta + 28\eta + 16\gamma)}{\rho_p N_0 (2\alpha + \frac{1}{2}\beta + 2\eta - \gamma)} \quad (4)$$

where ρ_p is the polymer density, and N_0 is Avogadro's number. Finally, a silicon material balance gives

$$\frac{dX}{dt} = -M\frac{dP}{dt} - S \quad (5)$$

where S is the SiO_2 sputtering rate, and M is the ratio of the Si mass density in the polymer to the Si mass density in the oxide. For the polymer composition shown in Eq. 3, M is given by

$$M = \frac{60\eta\rho_p}{(12\alpha + \beta + 28\eta + 16\gamma)\rho_{ox}} \quad (6)$$

where ρ_{ox} is the density of the oxide layer. Equations 2 and 5 may be combined to obtain a closed ordinary differential equation for the time dependence of the oxide thickness

$$\frac{dX}{dt} = \frac{M\nu k C_S}{1 + kX/D} - S \quad (7)$$

The transient solution to Eq. 7 will be discussed in the next section. Equation 5 can also be directly integrated to relate the remaining polymer thickness to the time dependent oxide thickness

$$P(t) = P_0 - St/M + (X_0 - X(t))/M \quad (8)$$

where X_0 and P_0 are the initial oxide and polymer thicknesses. Finally, one may define the total thickness loss $L(t)$ as the initial thickness minus the sum of the oxide and polymer thicknesses to obtain

$$L(t) = (X(t) - X_0)(1/M - 1) + St/M \quad (9)$$

It is important to note that Eqs. 5, 8, and 9 were derived entirely from a silicon material balance and the assumption that physical sputtering is the only silicon loss mechanism; thus these equations are independent of the kinetic assumptions incorporated into Eqs. 1, 2, and 7. This is an important point because several of these kinetic assumptions are questionable; for example, Eq. 2 assumes a radical dominated mechanism for $X = 0$, but bombardment-induced processes may dominate for small oxide thickness. Moreover, ballistic transport is not included in Eq. 1, but this may be the dominant transport mechanism through the first ≈ 40 Å of oxide. Finally, the first ≈ 40 Å of oxide may be annealed by the bombarding ions, so the diffusion coefficient may not be a constant throughout the oxide layer. In spite of these objections, Eq. 2 is a three parameter kinetic model (k, C_S, and D), and it should not be rejected until clear experimental evidence shows that a more complex kinetic scheme is required.

Transient Regime. Equation 7 is separable and may be integrated to obtain the time dependence of the oxide thickness. Assuming a finite sputtering rate, the result is

$$X(t) - X_0 = \frac{M\nu DC_S}{S} \ln\left[\frac{M\nu kDC_S - DS - SkX_0}{M\nu kDC_S - DS - SkX(t)}\right] - St \quad (10)$$

Watanabe and Ohnishi [39] have proposed another model for the polymer consumption rate (in place of Eq. 2) and have also integrated their model to obtain the time dependence of the oxide thickness. Time dependent oxide thickness measurement in the transient regime is the clearest way to test the kinetic assumptions in these models; however, neither model has been subjected to experimental verification in the transient regime. Equation 9 may be used to obtain time dependent oxide thickness estimates from the time dependence of the total thickness loss, but such results have not been published. Hartney et al. [42] have recently used variable angle XPS spectroscopy to determine the time dependence of the oxide thickness for two organosilicon polymers and several etching conditions. They did not present kinetic model fits to their results, nor did they compare their results to time dependent thickness estimates from the material balance (Eq. 9). More research on the transient regime is needed to determine the validity of Eq. 10 or the comparable result for the kinetic model presented by Watanabe and Ohnishi [39].

Diffusion Controlled Regime. Butherus et al. [38] studied plasma oxidation of polysiloxane in a barrel etcher, and were able to convert several thousand angstroms of polysiloxane to SiO_2 in ≈ 20 minutes. They observed that the oxide thickness scales as the square root of time, and concluded that the rate controlling step in this process is the diffusion of oxygen radicals through the oxide film. More recently Namatsu [43] has also observed that the oxide thickness scales as the square root of time under conditions where the ion bombardment energy is low. In terms of the kinetic model presented earlier, the diffusion-controlled regime is described by assuming $S = 0$ and $kX/D \gg 1$ in Eq. 7, and integrating to obtain

$$X^2 - X_0^2 = 2DM\nu C_S t \quad (11)$$

This equation predicts that the oxide thickness scales as the square root of the etching time (for $X \gg X_0$) as observed experimentally [38,43]. Note that D in Eq. 11 is the diffusion coefficient of oxygen radicals in the oxide layer, not in the polymer. The radical concentration at the surface of the oxide is determined by the plasma conditions. Thus the only polymer dependent property in Eq. 11 is the product $M\nu$ which is determined entirely by the density and stoichiometric composition of the polymer. This constant increases with the silicon content of the polymer such that $M\nu$ for a dimethylsiloxane is four times larger than for a diphenylsiloxane. Namatsu [43] observed that the oxide accumulates more rapidly on methylsiloxane than on phenylsiloxane as predicted by Eq. 11; however, he did not interpret his results in terms of this kinetic model. Namatsu [43] also found that the molecular weight and structure of several methylsiloxane polymers has a large effect on the oxidation rate in the diffusion-controlled regime even though the silicon and carbon contents of these polymers were similar. This result was not expected from Eq. 11 because $M\nu$ only depends on composition of the polymer and is independent of its molecular weight and detailed structure. One can, however, rationalize Namatsu's [43] results in terms of Eq. 11 if one assumes that the molecular weight and polymer structure can affect the structure of the oxide and hence the diffusion coefficient in the oxide. Namatsu [43] did not interpret his results in terms of Eq. 11, and has rejected the assumptions that lead to this model. He argues that the diffusion coefficient in the polymer itself is important, and that the trend with polymer structure reflects the trend in the diffusion coefficients in the polymer. Namatsu [43] did not present measurements of diffusion coefficients in his polymers, nor did he explain how the oxide thickness could scale as the square root of time if diffusion across the oxide layer is not the rate controlling step. Namatsu's [43] results are interesting regardless of how they are interpreted; they either indicate that the oxide remembers the structure of its polymer precursor (my interpretation), or that the kinetic assumptions leading to Eq. 11 must be abandoned in the regime where they seem most reasonable (Namatsu's interpretation).

Etching Rates in Steady-State Regime. Watanabe and Ohnishi [39] proposed a steady-state RIE model for organosilicon polymer etching; it assumes that an oxide forms on the surface of the polymer and reaches a steady-state thickness where the rate of silicon loss by physical sputtering is balanced by oxidation of the underlying organosilicon polymer. Their kinetic model was not based on a microscopic mechanism; however, Eq. 11 also predicts that the oxide thickness will reach a steady-state where the rate of oxidation balances the sputtering losses. Jurgensen et al. [40] pointed out that steady-state etching behavior is expected for any kinetic model where the polymer oxidation rate decreases as the oxide thickness increases. Such behavior is plausible because transport through the oxide layer is in series with oxidation of the polymer. The steady-state etching rate ($R = S/M$) is proportional to the sputtering rate of SiO_2 and inversely proportional to the mass density of silicon (density times mass fraction silicon) in the polymer. This result is a direct consequence of the silicon material balance (Eq. 9) together with the assumption that the oxide thickness is independent of time in the steady-state regime. Watanabe and Ohnishi [39] tested this model for several silyl-styrene polymers and found that it quantitatively predicts the etching rate of these materials under high bombardment energy O_2 RIE conditions; however they observed significant deviations from predicted behavior under lower bombardment energy O_2 RIE conditions. In a later study, Gokan et al. [44] found that the steady-state model quantitatively predicts the etching rate of 12 organosilicon polymers under O_2^+ IBE conditions for bombardment energies greater than 100 eV. In this same study, they observed significant deviations from the steady-state model under O_2 RIE conditions similar to those where Watanabe and Ohnishi [39] observed such deviations. Jurgensen et al. [40] studied the O_2 RIE behavior a silyl-novolac and two silyl-methacrylate polymers and observed quantitative agreement with the steady-state model for all three polymers under high bombardment energy ($>250\,eV$) etching conditions. At lower bombardment energies, the silyl-methacrylates exceeded the predicted etching rate, while the silyl-novolac continued to etch at the predicted rate down to 130 eV. Hartney et al. [42] recently found excellent agreement with the predicted steady-state etching rate for diphenylsiloxane based negative photoresist over a wide range of etching conditions; however, they observed steady-state etching rates twice as large as predicted for poly(trimethylsilylmethylstyrene) (PSMS) and a copolymer of this material with chloromethylstyrene. I will delay further discussion of the deviations from the steady-state model until after presenting surface analysis results.

Physical sputtering of the oxide layer is the rate controlling step in the steady-state regime, so it is important to study SiO_2 sputtering yields under O_2 RIE conditions to understand the effect of etching conditions on selectivity in multilayer lithography. Jurgensen and Rammelsberg [34] determined Si yields for a silyl novolac and for SiO_2 from the measured sputtering rate (for SiO_2) or steady-state etching rate (for the novolac) and from the estimated flux of bombarding ions and energetic neutrals. Figure 5 shows the Si yield of the silyl novolac (squares) and of SiO_2 (triangles) as a function of the average bombardment energy over the entire range of etching conditions discussed earlier. The SiO_2 sputtering yield falls on the same curve as the results for the silyl novolac as assumed in the steady-state model. The least squares line shown on Fig. 5 is a fit to the silyl novolac results and indicates an apparent sputtering threshold of 50 eV. Figure 6 shows the experimental (squares) and expected (curve) selectivity for etching the an organic novolac planarizing layer relative to the silyl novolac as a function of the average bombardment energy. The equation for the expected selectivity is

$$SEL = \frac{12(\chi_{Si}\rho)_{SP} \, Y_C}{28(\chi_C\rho)_{OP} \, Y_{Si}} \qquad (12)$$

where Y_C is the energy dependent carbon atom yield shown in Fig. 4, Y_{Si} is the energy dependent silicon atom yield shown in Fig. 5, $(\chi_{Si}\rho)_{SP}$ is the mass fraction of silicon times the density of the organosilicon polymer, and $(\chi_C\rho)_{OP}$ is the mass fraction of carbon times the density of the organic polymer. This equation applies to a wide range of etching conditions and to other polymer-organosilicon polymer systems because the O_2 RIE rate of most organic polymers scales inversely with the mass density of carbon [19,20,21], while the O_2 RIE rate of many organosilicon polymers scales inversely with the mass density of silicon [39,40,42]. These results show that one can improve selectivity by reducing the bombardment energy; however, this is only true as long as the organosilicon polymer continues to etch by the steady-state mechanism.

Steady-State Oxide Thickness. The steady-state etching rate ($R = S/M$) does not contain any of the kinetic parameters; thus it does not contain any information about the kinetics of the oxidation process. In contrast, the steady-state oxide thickness is determined by the kinetics of the transport and oxidation processes; thus one can learn about these processes by studying the steady-state oxide thickness. The silicon material balance (Eq. 9)

13. JURGENSEN *Polymer Etching in an Oxygen Glow Discharge* 227

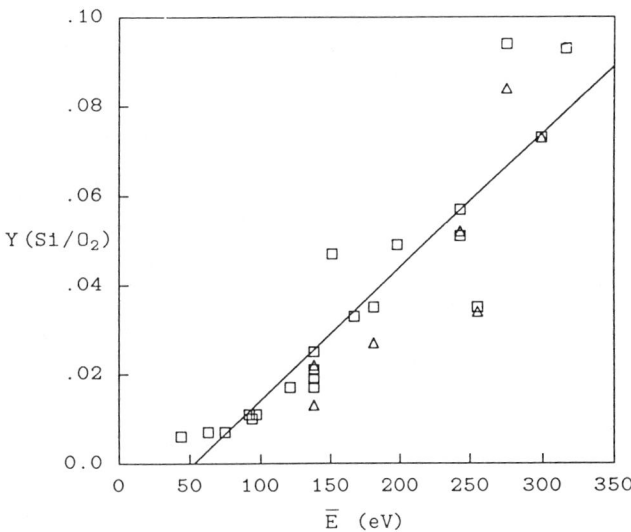

Fig. 5. The silicon atom yield of a silyl novolac (squares) and of SiO_2 (triangles) are shown as a function of the average bombardment energy. The line is a linear least squares fit to the silyl novolac results.

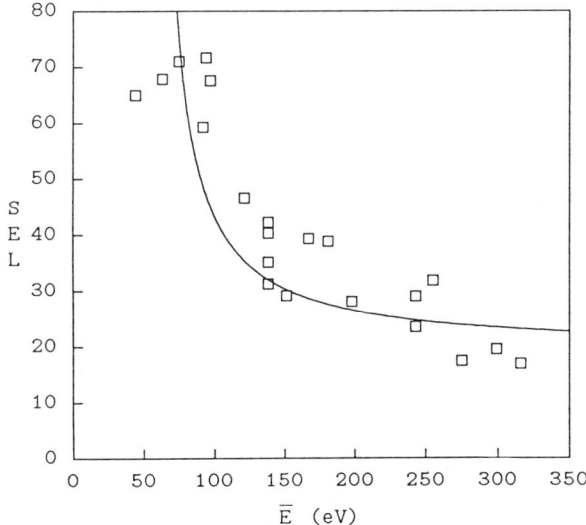

Fig. 6. The selectivity for etching the organic novolac relative to the silyl novolac is shown as a function of the average bombardment energy. The curve is based on the least squares fits to the yield trends.

may be used to relate the steady-state oxide thickness to the initial thickness loss during the transient regime. Jurgensen et al. [40] estimated the steady-state oxide thickness using this method and by depth profiles based on Ar ion milling with glancing angle X-ray photoelectron spectroscopy (XPS). Both methods gave oxide thickness estimates in the range from 30 to 50 Å under conditions where steady-state model predicted the etching rate. The observed steady-state oxide thickness was nearly independent of etching conditions and polymer silicon content. Hartney et al. [42] used variable angle XPS spectroscopy to measure the oxide thickness, and also obtained thicknesses in this range. These results [40,42] appear to conflict with the results obtained in the diffusion-controlled regime where oxide formation rates of several hundred Å/min are observed [38,43]. In particular, the observed oxide thickness would appear to require SiO_2 sputtering rates greater than 100 Å/min in the steady-state regime; however, this is much larger than the observed [40,42] SiO_2 sputtering rates which range from 2 to 30 Å/min depending on etching conditions. This discrepancy implies that energetic bombardment plays an important role in the steady-state regime in addition to sputtering the oxide layer. Bagley et al. [45] resolved this dilemma by showing that energetic ion bombardment renders these materials permanently resistant to oxidation. They propose that ion bombardment compacts the oxide surface, dramatically reducing diffusion coefficients in the oxide layer.

The results presented by Bagley et al. [45] imply that the oxide diffusion coefficient is much smaller in the steady-state regime than in the diffusion-controlled regime where physical bombardment is absent. It may be possible to account for this effect in terms of the diffusive transport model presented earlier by using a smaller oxide diffusion coefficient in the steady-state regime. To explore this possibility, one may set $dX/dt = 0$ in Eq. 7 to obtain

$$X_{SS} = \frac{D}{k}\left[\frac{M\nu k C_S}{S} - 1\right] \tag{13}$$

where X_{SS} is the steady-state oxide thickness and the first term in the brackets is the ratio of the initial etching rate to the steady-state etching rate and is much larger than one. This term equals the selectivity, so Eq. 13 predicts that X_{SS} is proportional to the selectivity which decreases with bombardment energy. Jurgensen et al. [40] reported that the steady-state oxide thickness is nearly independent of etching conditions, while Hartney et al. [42] reported that the oxide thickness increases with bombardment energy. Equation 13 also predicts that the steady-state oxide thickness increases with the polymer's silicon content; however, Jurgensen et al. [40] found that the steady-state oxide thickness is nearly independent of the silicon content. These failures show that the diffusive transport model is not valid when the sample is subjected to energetic O_2^+ bombardment; this suggests that ballistic transport is the dominant transport mechanism in the steady-state regime. The steady-state oxide thickness for a ballistic transport mechanism is equal to the penetration depth of the implanted oxygen [42]. This mechanism predicts that the steady-state oxide thickness is independent of the silicon content of the polymer and increases with bombardment energy as found experimentally [40,42].

Anomalous Transport Regime. The characteristic signature of the steady-state etching regime is a rapid initial etching rate (while the oxide layer forms) followed by a much slower constant etching rate. This etching behavior is always observed for high silicon content organosilicon polymers under O_2 RIE conditions; however, the predicted steady-state etching rate appears to be a lower bound while rates that exceed this prediction are often observed under moderate bombardment energy etching conditions [39,40,42,44]. The predicted steady-state etching rate was derived entirely from a silicon material balance and three assumptions: (1) The oxide thickness is constant (after the transient regime). (2) Physical sputtering is the only silicon loss mechanism (after the transient regime). (3) The physical sputtering yield is identical to that for SiO_2. At least one of these assumptions must fail when an organosilicon polymers etch faster than predicted by the steady-state model. Jurgensen et al. [40] and Hartney et al. [42] have both used surface analysis to determine which of these assumptions is failing. Jurgensen et al. [40] used ion milling and glancing angle XPS to depth profile silylmethacrylate samples that had been etched at for various times under both high and low bombardment energy etching conditions. At high bombardment energy, the sample etched at the rate predicted by the steady-state model, and after the initial transient the oxide thickness was constant and equal to the value expected from the material balance (Eq. 9). After the initial transient at low bombardment energy, the sample etched at twice the predicted rate and the oxide thickness continued to increase. The accumulated oxide thickness was in excellent agreement with the thickness predicted from the total thickness loss and the material balance (Eq. 9). They concluded that the assumption 1 was violated under conditions where this organosilicon polymer exceeded the predicted etching rate. Hartney et al. [42] used variable angle XPS to determine the oxide thickness on the silylstyrene polymer under conditions where it etched at twice the predicted rate. They found that the oxide thickness was constant after the initial transient and concluded that the polymer structure can affect the oxide structure and its sputtering yield (assumption 3 violated). This conclusion appears to conflict with Gokan's [44] finding that all organosilicon polymers etch at the predicted steady-state etching rate under O_2^+ IBE conditions.

The results obtained in these studies [40,42] apparently conflict and lead to opposite conclusions; however, the differences may result from different surface analysis methods, not from differences in the samples studied. Jurgensen et al. [42] determined the silicon concentration as a function of depth by integrating under the total silicon XPS peak including both oxygen bound and carbon bound silicon. This procedure was used because bonds are broken and reform in the ion milling process, so the chemical bonding information may not be reliable in ion milling-XPS depth profiles. They [40] also used this procedure to determine the carbon and oxygen concentration profiles. After a long low energy etch, the silicon and oxygen concentrations were elevated to a depth of 240 Å while the carbon concentration was depressed to this depth [40] (compared to an unetched but Ar^+ sputtered sample); however, the XPS spectra also indicated that at 240 Å, the silicon and oxygen were both bound to carbon. Thus carbon is selectively oxidized and removed from the polymer which concentrates the silicon that remains, but this silicon may still be bound to carbon. In contrast, Hartney et al. [42] used variable angle XPS to determine the thickness of the layer where silicon was bound to oxygen. This method would not detect the presence of a silicon enriched layer unless the silicon was present as an oxide. Thus these studies do not necessarily conflict with each other, but may be giving two distinct pieces of information.

The above interpretation implies that carbon bound silicon is accumulating under the oxide layer when the etching rate exceeds the steady-state prediction. The sputtering XPS depth profiles [40], silicon material balance [40] and angle resolved XPS thickness measurement [42] all agree that 30 to 50 Å of oxide are formed during the initial transient. The etching rate in the anomalous transport regime is constant after the initial transient which implies that the accumulating silicon does not contribute additional mass transfer resistance. This is consistent with the conclusion that physical bombardment can dramaticly reduce a material's diffusivity [45], and with the interpretation that the penetration depth is ≈ 40 Å. This leads to the question of what determines organosilicon polymer etching rates under moderate bombardment energy etching conditions. A reasonable working hypothesis assumes that the mass transfer resistance in the diffusion-controlled regime (D in Eq. 11) is equal to the mass transfer resistance of the material which accumulates under the first ≈ 40 Å of oxide. Organosilicon polymers which form good diffusion barriers in the diffusion-controlled regime (low D in Eq. 11) are expected to etch by the steady-state mechanism in moderate bombardment energy etching conditions, while materials which form poor diffusion barriers in the diffusion controlled regime (high D in Eq. 11) are expected to etch by the anomalous transport mechanism in moderate bombardment energy etching conditions. This hypothesis has not been carefully tested, but it appears to be consistent with the results presented in several studies [39,40,42,43,44]. This mechanism differs from those proposed by Jurgensen et al. [40] (which involved k in Eq. 1), Hartney et al. [42] (which involved Y_{Si} in Eq. 12), and Namatsu [43] (which involved D for the polymer); however it is similar to the mechanism proposed by Gokan et al. [44].

Bombardment Induced Pattern Transfer Models

Jurgensen and Shaqfeh [46] have formulated a kinetic theory of interface evolution to describe the time evolution of etching profiles in bombardment-induced glow discharge etching processes. This theory assumes that an axisymmetric angular distribution of energetic particles is incident on the surface being etched, and that the yield per incident particle is a function of its energy and angle relative to the surface normal. The evolution equation that results from these assumptions is intractable except in two special cases where it reduces to a partial differential equation. Fortunately one of these special cases applies to the planarizing layer in multilayer lithography, while the other applies to the masking layer. One of these simplifications is the case where the yield is independent of angle as expected for the thermal spike mechanism of bombardment-induced etching. Shaqfeh and Jurgensen [47,37] have assumed that this simplification applies to the planarizing layer in multi-layer lithography and have found that the predicted etching profiles are in good agreement with those observed experimentally for a tri-layer pattern transfer process. Visser [unpublished results] has also obtained some evidence that O_2 RIE yields are independent of angle by etching photoresist patterns with different wall angles. The rate controlling step for organosilicon polymers in the steady-state regime is a physical sputtering process, and physical sputtering yields are strongly angle dependent [22,23]; therefore the above simplifying approximation does not apply to the masking layer in bi-layer lithography. Fortunately, the masking layer is not shadowed by remote portions of the interface (it is raised and convex), and this allows our evolution equation [46] to be reduced to the hyperbolic conservation law discussed by Ross [48]. Analysis of this case [48] shows that facet edges (slope discontinuities) will spontaneously develop when the yield is angle dependent. Dr A. Tanaka at NTT has many beautiful (but unpublished) SEMs showing the formation and time evolution of facets in the masking layer of a bi-layer system; however, the clearest published SEM showing such facets in bi-layer lithography was presented by Saito et al. [49] (fig 10). This phenomenon has important implications for bi-layer pattern transfer processes because it implies that some line width loss will occur even with a perfectly monodirectional angular distribution and a perfect 90 degree wall angle in the organosilicon resist layer.

Conclusions

Physical bombardment plays a dominant role in the O_2 reactive ion etching pattern transfer step in multi-layer lithography. The results of organic polymer O_2 RIE and O_2 IBE etching studies are consistent with a thermal spike mechanism for the bombardment-induced chemistry. The silicon atom yield of organosilicon polymers in the steady-state regime is equal to the SiO_2 sputtering yield for the same etching condition as assumed in the steady-state etching model. These yields show threshold behavior at low bombardment energy, but this behavior is often obscured because many organosilicon polymers do not etch by the steady-state mechanism at low and intermediate bombardment energies. These etching results are being incorporated into pattern transfer models [37,46,47] that predict etching profiles and process latitudes in multi-layer lithography.

Acknowledgments

I thank Mark Hartney, Robert Visser, Akinobu Tanaka, Eric Shaqfeh, Mike Vasile, Elsa Reichmanis, and Gary Taylor for stimulating discussions.

Literature Cited

1. B. J. Lin, in *Introduction to Microlithography*, edited by L. F. Thompson, C. G. Wilson, and M. J. Bowden Amer. Chem. Soc. Symp. Ser. **219**, (ACS, Washington, D. C., 1983), p. 287.
2. M. A. Hartney, D. W. Hess, and D. S. Soane, *J. Vac. Sci. Technol.* **B 7**, 1 (1989).
3. J. M. Moran and D. Maydan, *J. Vac. Sci. Technol.* **16**, 1620 (1979).
4. J. R. Havas, *Electrochem. Soc. Extended Abstracts* **76**, 743 (1976).
5. Y. Ohnishi, M. Suzuki, K. Saigo, Y. Saotome, and H. Gokan, **SPIE Vol. 539** *Advances in Resist Technology and Processing* II, 62 (1985).
6. E. Reichmanis, G. Smolinsky, and C. W. Wilkins, Jr., *Solid State Technology* **28**(8), 130 (1985).
7. E. Babich, J. Paraszczak, M. Hatzakis, and J. Shaw, *Microelectronic Engineering* **3**, 279 (1985).
8. T. M. Wolf, G. N. Taylor, T. Venkatesan, and R. T. Kraetsch, *J. Electrochem. Soc.* **131**, 1664 (1984).
9. F. Coopmans, and B. Roland, *Solid State Technology* **30**(6), 93 (1987).
10. C. W. Jurgensen, *J. Appl. Phys.* **64**, 590, (1988).
11. G. N. Taylor, and T. M. Wolf, *Polym. Eng. and Sci.* **20**, 1087, (1980).
12. L. A Pederson, *J. Electrochem. Soc.* **129**, 205 (1982).
13. J. M. Cook, and B. W. Benson, *J. Electrochem. Soc.* **130**, 2459 (1983).
14. J. M. Cook, *Solid State Technology* **30**(4), 147 (1987).
15. J. E. Spencer, R. A. Borel, and A. Hoff, *J. Electrochem. Soc.* **133**, 1922 (1986).
16. H. Gokan, S. Esho, and Y. Ohnishi, *J. Electrochem. Soc.* **130**, 143 (1983).
17. H. Gokan, and S. Esho, *J. Electrochem. Soc.* **131**, 1105 (1984).
18. H. Gokan, K. Tanigaki, and Y. Ohnishi, *Solid State Technol.* **28**(5), 163 (1985).
19. M. W. Thompson, and R. S. Nelson, *Phil. Mag.* **7**, 2015 (1962).
20. S. C. McNevin, *J. Vac. Sci. Technol.* **B 4**, 1203 (1986).
21. F. H. M. Sanders, A. W. Kolfschoten, J. Dieleman, R. A. Haring, A Haring, and A. E. de Vries, *J. Vac. Sci. Technol. A* **2**, 487 (1984).
22. H. Oechsner, *Appl. Phys.* **8**, 185 (1975).
23. R. E. Lee, *J. Vac. Sci. Technol.* **16**, 164 (1979).
24. H. Okano, and Y. Horiike, *Electrochem. Soc. Extended Abstracts* **80-1**, 291, (1980).
25. H. Okano, and Y. Horiike, *Jpn. J. Appl. Phys.* **20**, 2429, (1981).
26. T. M. Mayer, R. A. Barker, L. J. Whitman, *J. Vac. Sci. Technol.* **18**, 349, (1981).
27. M. A. Hartney, W. M. Greene, D. S. Soane, and D. W. Hess, **SPIE Vol. 920** *Advances in Resist Technology and Processing* V, 108 (1988).
28. G. S. Selwyn, *J. Appl. Phys.* **60**, 2771, (1986).
29. R. J. Visser, and C. A. M. de Vries, *Proceedings 8th International Symposium on Plasma Chemistry* edited by K. Akashi and A. Kinbara (IUPAC, Tokyo, 1987), p.1029

30. H. Namatsu, Y. Ozaki, and K. Hirata, *J. Vac. Sci. Technol.* **21**, 672 (1982).
31. H. Namatsu, Y. Ozaki, and K. Hirata, *J. Electrochem. Soc.* **130**, 523 (1983).
32. J. Paraszczak, E. Babich, J. Heidenreich, R. McGouey, L. Ferreiro N. Chou, and M. Hatzakis, **SPIE Vol. 920** *Advances in Resist Technology and Processing V*, 242 (1988).
33. J. E. Allen, in *Plasma Physics*, Institute of Physics Conference Series **20** (London Institute of Physics, London, 1974), p. 131.
34. C. W. Jurgensen, and A. Rammelsberg, accepted *J. Vac. Sci. Technol. A*
35. C. W. Jurgensen, and E. S. G. Shaqfeh, *J. Appl. Phys.* **64**, 6200, (1988).
36. B. Chapman, *Glow Discharge Processes* (Wiley, New York, 1980), pp. 209-214.
37. C. W. Jurgensen, and E. S. G. Shaqfeh, **SPIE Vol. 1086** *Advances in Resist Technology and Processing VI*, (1989).
38. A. D. Butherus, T. W. Hou, C. J. Mogab, and H. Schonhorn, *J. Vac. Sci. Technol. B* **3**, 1352 (1985).
39. F. Watanabe, and Y. Ohnishi, *J. Vac. Sci. Technol. B* **4**, 422 (1986).
40. C. W. Jurgensen, A. Shugard, N. Dudash, E. Reichmanis, and M. J. Vasile, *J. Vac. Sci. Technol. A.* **6**, 2938 (1988).
41. B. E. Deal, and A. S. Grove, *J. Appl. Phys.* **36**, 3770 (1965).
42. M. A. Hartney, J. N. Chiang, D. S. Soane, D. W. Hess, and R. D. Allen, **SPIE Vol. 1086** *Advances in Resist Technology and Processing VI*, (1989).
43. H. Namatsu, to be published in *J. Electrochem. Soc.*
44. H. Gokan, Y. Saotome, K. Saigo, F. Watanabe, and Y. Ohnishi, in *Polymers for High Technology*, edited by M. J. Bowden, and S. R. Turner, Amer. Chem. Soc. Symp. Ser. **346**, (ACS, Washington, D. C., 1987), p. 358.
45. B. G. Bagley, W. E. Quinn, C. J. Mogab, and M. J. Vasile, *Materials Letters* **4**, 154 (1986).
46. C. W. Jurgensen, and Shaqfeh, submitted to *J. Vac. Sci. Technol. B*
47. E. S. G. Shaqfeh, and C. W. Jurgensen, submitted to *J. Appl. Phys.*
48. D. S. Ross, *J. Electrochem. Soc.* **135**, 1235 (1988).
49. K. Saito, S. Shiba, Y. Kawasaki, K. Watanabe, and Y. Yoneda **SPIE Vol. 920** *Advances in Resist Technology and Processing V*, 198 (1988).

RECEIVED July 28, 1989

Chapter 14

Quantitative Analysis of a Laser Interferometer Waveform Obtained During Oxygen Reactive-Ion Etching of Thin Polymer Films

B. C. Dems, P. D. Krasicky, and F. Rodriguez

School of Chemical Engineering, Olin Hall, Cornell University, Ithaca, NY 14853

> A rigorous analysis of a laser interferometer waveform, obtained during oxygen Reactive-Ion Etching (RIE) of thin polymer films, was developed and used to measure the polymer refractive index to within 2-3%. This in turn allowed the *in situ* etch rate to be measured without any prior knowledge of the polymer's physical properties. Film "roughening", imparted to the film surface during RIE, was assessed semi-quantitatively through an amplitude reduction factor, which accounts for any diffuse character at the plasma/polymer interface. Amplitude reduction factors measured for films etched at 5mTorr were near unity (smooth surface) and invariant with respect to incident RF power over the range 0.125-0.75 Watts/cm^2. However, amplitude reduction factors decreased with incident RF power at 35mTorr, indicating greater roughening. Scanning Electron Micrographs showed roughness correlation lengths of etched films to be on the order of 0.1-0.2μm.

Laser interferometry is used extensively in the integrated circuit and thin film processing industries. It is well-suited for end-point detection and *in situ* etch rate monitoring during the plasma processing of thin polymer films. The qualitative features of the laser interferogram can also shed light on physical changes occurring at any particular time in the process. This paper describes a rigorous analysis of an interferogram which allows the polymer's refractive index to be calculated directly from the waveform. This in turn allows *in situ* etch rate determination without prior knowledge of the polymer film's physical properties. The analysis also allows the degree of surface roughening imparted to the film during plasma processing to be assessed on a relative basis. The analysis is restricted to (a) process etch gases which do not contain film-forming pre-cursors and (b) non-absorbing polymer films which remain homogeneous at all times.

Background

A model for the type of thin film system realized in Reactive-Ion Etching (RIE) is shown in Figure 1. The polymer film is labeled as region #2, the substrate as #3, and the plasma environment as #1. The polymer-substrate (2-3) interface is assumed to be sharp and smooth, but the plasma-polymer (1-2) interface may be diffuse or rough. Unpolarized light of vacuum wavelength λ is incident upon the system from region #1 at an angle ϑ_1 relative to the normal. The system's overall reflectance R is given by (1),

$$R = \frac{I}{I_o} = \left(\frac{r_{23}^2 + 2fr_{12}r_{23}\cos\phi + f^2r_{12}^2}{1 + 2fr_{12}r_{23}\cos\phi + f^2r_{12}^2r_{23}^2} \right) \quad (1)$$

where I_o and I are the incident and reflected light intensities, respectively. The r's with subscripts are Fresnel reflection coefficients for sharp, smooth interfaces between the regions denoted, while f is a reduction factor for reflection from the plasma-polymer (1-2) interface and accounts for spreading of the interface into a transition region. R is a periodic function of the film thickness d through the phase function $\phi = \phi_2 + \phi_t$, which includes the phase delay (2),

$$\phi_2 = (4\pi d/\lambda)(n_2^2 - n_1^2\sin^2\vartheta_1)^{½} \quad (2)$$

associated with passage of light through the film to the substrate and back, as well as any additional phase shift ϕ_t introduced by the transition region. When the 1-2 interface is perfectly sharp and smooth, then $f = 1$, $\phi_t = 0$, and the maximum value of the periodically oscillating reflectance equals the reflectance of the bare substrate after the film is completely removed. The optical effect of $f < 1$ is a reduction in amplitude of oscillation of R from what it would be for an otherwise sharp, smooth interface.

For a sharp, smooth film with $\phi_t = 0$, the final reflectance maximum occurs just as film removal is completed. With a transition region, f and ϕ_t are constant if this region maintains fixed size and shape, but are expected to vary once the region begins to collapse into the substrate near the endpoint of film removal. An observable effect of f and ϕ_t varying is the end-point according to the laser interferogram occurring slightly before the final extrapolated reflectance maximum. The shape of the R versus time curve near the endpoint can provide clues to the shape of the refractive index profile through the transition region (Figure 2). In RIE, film roughening is expected from ion bombardment of the exposed surface. Roughening of the 1-2 interface can produce an amplitude reduction in reflectance oscillations similar to that due to diffuseness provided the laser beam is sufficiently wide compared to the roughness correlation length. But scattering from the rough interface also deflects energy out of the specular directions and introduces additional corrections into R. As a first approximation, these corrections can be neglected and the interferometer waveform for RIE analyzed using a modified version of diffuse interface transition layer theory.

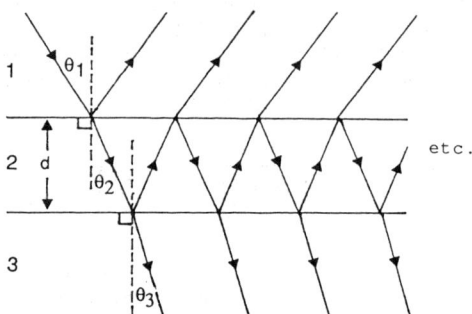

Figure 1: Basic Film Model

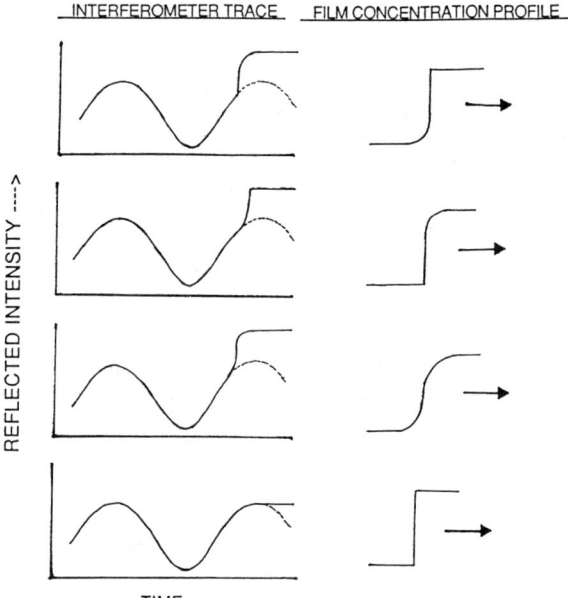

Figure 2: Waveform Dependence on Concentration Profile Near the End-Point

When $|r_{12}| \ll 1$, as for most polymer films dissolving in solvents, R is well-approximated by a power series in r_{12} to first order:

$$R = r_{23}^2 + 2fr_{12}r_{23}(1-r_{23}^2)\cos\phi \tag{3}$$

Physically, this amounts to considering emerging rays whose histories involve no more than one reflection from the 1-2 interface (figure 1). R is then a purely sinusoidal function of ϕ (and of film thickness d), oscillating about the value $R_o = r_{23}^2$. The factor f reduces the oscillation amplitude symmetrically about $R = R_o$, facilitating straightforward calculation of polymer refractive index from quantities measured directly from the waveform (3). When $|r_{12}|$ is not small, as in the plasma etching of thin polymer films, the first order power series approximation is inadequate. For example, for a plasma/poly(methylmethacrylate)/silicon system, $r_{12} = -0.196$ and $r_{23} = -0.442$. The waveform for a uniformly etching film is no longer purely sinusoidal in time but contains other harmonic components. In addition, amplitude reduction through the f factor does not preserve the vertical median R_o making the film refractive index calculation non-trivial.

Figure 3 is a sketch of a typical waveform obtained during an oxygen/RIE process. Referring to this figure, let $R_\pm$ be the extreme values of R when the transition layer is present. Let $R_\pm^o$ be the corresponding values if the interface were sharp (f=1). These values can be identified by setting $\cos\phi = \pm 1$ in equation 1 which gives,

$$R_\pm = \left(\frac{r_{23} \pm fr_{12}}{1 \pm fr_{12}r_{23}}\right)^2 \tag{4}$$

$$R_\pm^o = \left(\frac{r_{23} \pm r_{12}}{1 \pm r_{12}r_{23}}\right)^2 \tag{5}$$

Remember that $R_+^o = r_{13}^2$, the reflectance of the bare substrate medium. As before, let R_o be the value of R for which $\cos\phi=0$ in equation 1,

$$R_o = \left(\frac{r_{23}^2 + f^2 r_{12}^2}{1 + f^2 r_{12}^2 r_{23}^2}\right) \tag{6}$$

Note that the points on the waveform for which $R=R_o$ are equally spaced along ϕ for uniformly decreasing film thickness, making them identifiable. The plasma itself contributes to the measured intensity. This contribution, defined as R_p, must be subtracted from the reflectance values before any calculations are made. R_p is measured by extinguishing the plasma and measuring the corresponding reduction in intensity.

When f=1, the directly measurable ratio $x^o = (R_-/R_+^o)^{\frac{1}{2}}$ for normal or near normal incidence can be used to determine n_2, from the known value of n_1 and n_3. At normal incidence,

$$R_+^o = \left(\frac{n_3 - n_1}{n_3 + n_1}\right)^2 \tag{7a}$$

$$R_-^o = \left(\frac{n_3 n_1 - n_2^2}{n_3 n_1 + n_2^2}\right)^2 \tag{7b}$$

Solving for n_2 in terms of x^o yields,

$$n_2 = (n_3 n_1)^{1/2} \left[\frac{\left(\dfrac{n_3 + n_1}{n_3 - n_1}\right) - x^o}{\left(\dfrac{n_3 + n_1}{n_3 - n_1}\right) + x^o}\right]^{1/2} \tag{8}$$

For unpolarized light, this relation is a very good approximation for incidence angles $\vartheta_1 \leq 15°$. When $f < 1$, the observed ratio is $x = (R_-/R_+)^{1/2} < x^o$. If r_{12} were small, then the fact that f would compress the waveform symmetrically about the value R_o could be used to find x^o, and hence n_2. Note that this analysis would also immediately determine f because equation 3 would apply, and f would be the amplitude reduction factor for the reflectance oscillations. But if r_{12} is not small, then a more careful analysis is needed.

Consider the quantities $\Delta R_\pm^{1/2} \equiv |R_\pm^{1/2} - R_\pm^{o1/2}|$, which denote the shifts in the extreme values of $R^{1/2}$ due to spreading of the interface. Using equations 4 and 5, these can be written as

$$\Delta R_\pm^{1/2} = \left|\frac{(1-f)r_{12}(1-r_{23}^2)}{(1 \pm fr_{12}r_{23})(1 \pm r_{12}r_{23})}\right| \tag{9}$$

One then finds

$$\left(\frac{\Delta R_-^{1/2}}{\Delta R_+^{1/2}}\right) = \left(\frac{1 + fr_{12}r_{23}}{1 - fr_{12}r_{23}}\right)\left(\frac{1 + r_{12}r_{23}}{1 - r_{12}r_{23}}\right) \tag{10}$$

If the right side of equation 10 were known or could be estimated reasonably well, the equation could then be used to find R_-^o from normalized values of R_+, R_+^o, and R_-, which are available experimentally. If $(1-f)$ were small, then the fractional shifts $\Delta R_\pm/R_\pm^o$ would also be small, and vice-versa, so that the ratio of shifts could be written as

$$\left(\frac{\Delta R_-}{\Delta R_+}\right) \approx \left(\frac{R_-}{R_+}\right)^{1/2}\left(\frac{\Delta R_-^{1/2}}{\Delta R_+^{1/2}}\right) \tag{11}$$

But the right side of equation 10 is not known *a priori*. A crude approximation might be to set it equal to unity, and equation 11 would then give

$$(\Delta R_-/\Delta R_+) \approx (R_-/R_+)^{1/2} = x \tag{12}$$

This crude approximation gives a rather simple result but is probably not accurate enough for most purposes. However, it turns out that the quantities on the right side of equation 10 can be obtained from a more careful analysis of the interferometer waveform. And once these quantities are known, n_2 can then be found directly from them without recourse to equations 11 or 12.

Consider the general form of the reflectance R from equation 1, namely

$$R = \left(\frac{a + c(\cos\phi)}{b + c(\cos\phi)}\right) \quad (13)$$

where
$$a = r_{23}^2 + f^2 r_{12}^2 \quad (14a)$$
$$b = 1 + f^2 r_{12}^2 r_{23}^2 \quad (14b)$$
$$c = |2 f r_{12} r_{23}| \quad (14c)$$

In terms of these one can write previously defined quantities as

$$R_\pm = \left(\frac{a \pm c}{b \pm c}\right) \quad (15)$$

$$R_o = a/b \quad (16)$$

Now define the directly measurable ratios

$$Q_\pm \equiv R_\pm/R_o = \left(\frac{1 \pm c/a}{1 \pm c/b}\right) \quad (17)$$

Letting $A = a/c$ and $B = b/c$, and solving equations 17 for A and B, one obtains

$$A = \left(\frac{Q_+ - Q_-}{Q_+ + Q_- - 2 Q_+ Q_-}\right) \quad (18a)$$

$$B = \left(\frac{Q_+ - Q_-}{2 - Q_+ - Q_-}\right) \quad (18b)$$

Letting $y = |f r_{12}/r_{23}|$ and $z = |f r_{12} r_{23}|$, one finds that $2A = y + 1/y$ and $2B = z + 1/z$, from which y and z are found to be

$$y = |A - (A^2 - 1)^{\frac{1}{2}}| \quad (19a)$$
$$z = |B - (B^2 - 1)^{\frac{1}{2}}| \quad (19b)$$

Equation 19b is valid in general, whereas equation 19a holds only if $y \leq 1$. If $y > 1$, then the opposite sign preceding the square root term in equation 19a should be used. Once A and B are known, it follows that

$$|f r_{12}| = (zy)^{\frac{1}{2}} \quad (20a)$$
$$|r_{23}| = (z/y)^{\frac{1}{2}} \quad (20b)$$

Now the task is to determine f.

First define the directly measurable quantity $w \equiv (R_+/R_+^o)^{1/2}$. Using equations 4, 5, and 20, w can be written in terms of y, z, and f as

$$w = \left[\frac{(1 + y)(1 + z/f)}{(1 + z)(1 + y/f)}\right] \quad (21)$$

which can then be solved for f to give

$$f = \left[\frac{y - z - y(1 + z)(1 - w)}{y - z + (1 + z)(1 - w)}\right] \quad (22)$$

Equations 21 and 22 involve the assumption that r_{12} and r_{23} are both of the same algebraic sign, so that either $n_1 < n_2 < n_3$ or $n_3 < n_2 < n_1$. Otherwise, appropriate sign adjustments must be made. Knowing f and $|fr_{12}|$, one may then determine $|r_{12}|$.

Finally, assuming that $n_1 < n_2 < n_3$, a condition typical in RIE, one notes that

$$|r_{12}| = \left(\frac{n_2 - n_1}{n_2 + n_1}\right) \quad (23a)$$

$$|r_{23}| = \left(\frac{n_3 - n_2}{n_3 + n_2}\right) \quad (23b)$$

which leaves two ways to solve for the polymer refractive index, n_2,

$$n_2 = \left(\frac{1 + |r_{12}|}{1 - |r_{12}|}\right) n_1 \quad (24a)$$

$$n_2 = \left(\frac{1 - |r_{23}|}{1 + |r_{23}|}\right) n_3 \quad (24b)$$

A local etch rate can be calculated once n_2 is determined. The change in thickness corresponding to one complete oscillation is known as the thickness period, d_p,

$$d_p = \frac{\lambda}{2\,[n_2^2 - n_1^2 \sin^2\vartheta_1]^{1/2}} \quad (25)$$

which is constant as long as the film properties are not changing during the process. The local etch rate is related to d_p and the time period of oscillation, T (figure 3),

$$\text{local etch rate} = d_p/T \quad (26)$$

Experimental

All waveforms were collected using a custom-built, asymmetric, capacitively coupled (13.56MHz), reactive-ion etching apparatus. A circulating bath maintains the aluminum powered electrode temperature at 24.0 ±0.5°C. The apparatus is equipped with a He-Ne laser (λ=632.8nm) interferometer as depicted in Figure 4. The laser beam was aimed at the substrate surface at an angle of 10° off the normal. Laser intensity was measured by a photodiode whose

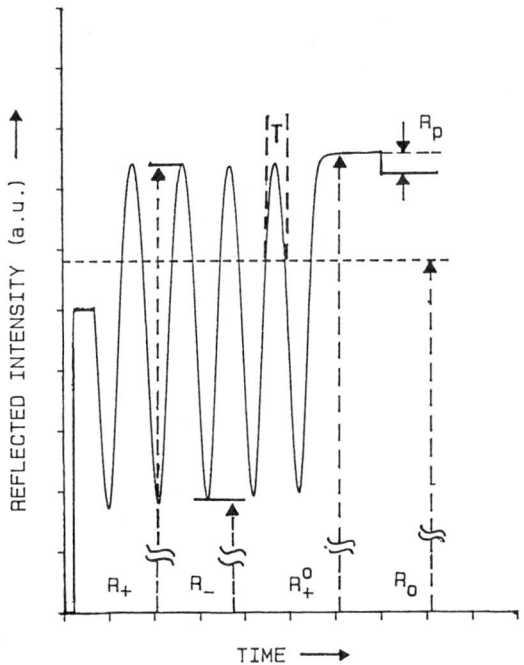

Figure 3: Typical Oxygen RIE Interferogram

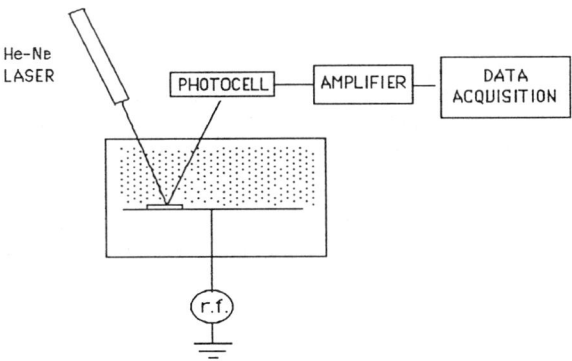

Figure 4: RIE/Laser Interferometer

signal was passed through a transimpedance amplifier/low pass filter (cutoff frequency = 10Hz) circuit and digitally recorded by an IBM personal computer. Dry oxygen was used exclusively as the etch gas. Films of poly(methylmethacrylate), PMMA (KTI Chemicals,950k), and VMCH, a commercially available poly(vinyl-chloride-vinyl acetate-maleic anhydride) terpolymer were spun from 6% solutions in chlorobenzene onto 3" silicon wafers at 1500 rpm and pre-baked at 160°C for 1 hour yielding film thicknesses of 1.0 ± 0.1µm. The film coated wafers were placed on the powered electrode without any thermal bond in between unless otherwise specified.

Surface roughness was measured according to ANSI standard 1346.1-(1978) using a Tencor Instruments Alpha-Step 200 stylus profilometer located at the National Nanofabrication Facility of Cornell. Five surface roughness measurements were made for each sample and their average values recorded. Details of the experimental apparatus set-up and its operation are given elsewhere (Dems, B. C.; et. al. Intl. Polym. Proc., in press.).

Data Analysis

The interferometric data were digitally smoothed (attenuation range = 0.9995-1.005) and differentiated so that the local minima ($=R_-$) and maxima ($=R_+$) of the interferogram could be located and averaged over a given run. An average value of the medium/substrate reflectance, R_+^o ($=r_{13}^2$), was also calculated over a ten second period after the end-point had been reached. The plasma intensity was then determined by extinguishing the plasma. The baseline measurement for the background intensity was measured by closing the shutter on the laser for about five seconds immediately before each run. Initial results showed that it was very difficult to determine R_o experimentally due to small increases (≈5%/min.) in etch rate with time. Therefore, an initial guess of $R_o = \frac{1}{2}(R_+ + R_-)$ was used. These values were input to a computer program which computed the polymer's refractive index ($=n_2$) by the following iterative scheme.

After subtracting the baseline and plasma intensity, experimental R_-, R_+, R_+^o, and R_o values were substituted in equations 17-20 and equation 22 and solved for $|r_{12}|$ and $|r_{23}|$. The polymer refractive index was then calculated with equations 24a and 24b and the results between the two compared. The value of R_o was then appropriately adjusted and the process repeated using the new R_o until the n_2-values calculated by equations 24a and 24b agreed to within 0.1%. Singularities in the expressions for parameters A and B in equations 18a and 18b were avoided by making a 5% change in R_o when necessary. Convergence was usually achieved in less than ten iterations.

Results

Results of refractive index measurements for PMMA and VMCH during oxygen RIE at 35mTorr and various incident RF power densities are tabulated in Table I. The values in the first column are those obtained using the full reflectance relation (equation 1) while those in the second column were calculated assuming $r_{12} \ll 1$. The calculated n_2 values are invariant with respect to power level to within experimental error. The more rigorous analysis makes a 2-4 percent correction to the $r_{12} \ll 1$ case and yields results that are in better agreement with values cited in the literature or measured by other methods.

Table I. Calculation of Polymer Refractive Index from Laser Interferometer Waveform during Oxygen RIE*

Power Density (Watts/cm^2)	Self-Bias (-VDC)	PMMA		VMCH	
		$r_{12} \approx r_{23}$	$r_{12} \ll r_{23}$	$r_{12} \approx r_{23}$	$r_{12} \ll r_{23}$
0.125	250	1.461	1.444	1.508	1.481
0.25	350	1.481	1.442	1.512	1.462
0.50	450	1.482	1.446	1.501	1.456
0.75	520	1.466	1.435	1.528	1.484
0.875	550	1.486	1.444	1.511	1.454
Standard deviation of measurement		±0.010		±0.011	
Accepted value		1.489		1.535**	

* RIE conditions: Flow=20 SCCM O_2, pressure=35mTorr
** Measured experimentally with Rudolph Instruments ellipsometer

The laser interferometer appears to be sensitive enough to detect small changes in surface structure. Figure 5 is a plot of amplitude reduction factor versus incident power density obtained for PMMA and VMCH films at 35mTorr. It is seen that f decreases with RF power suggesting that more surface roughening occurs at higher ion bombardment energies. Extrapolation of the curves to zero power does not necessarily go through f=1 (i.e. sharp interface), particularly for the PMMA curve. This is consistent with the fact that scattering of light by pre-etched PMMA films was always observed during the experiments whereas scattering by pre-etched (post-bake) VMCH films, while present to a small extent, was noticeably less. Roughness amplitudes estimated from f using transition layer theory (3) range from 60 to 130 Å for PMMA and from 80 to 140 Å for VMCH.

Table II summarizes surface roughness values measured for PMMA and VMCH samples etched for 1.0 minute at 35mTorr at various power densities. Although the measured values of 80 - 105 Å fall within the ranges obtained from the interferometer and transition layer theory, there is no significant variation with power density. Differences in surface roughness between pre-etched films of PMMA and VMCH are also negligible according to the stylus measurements.

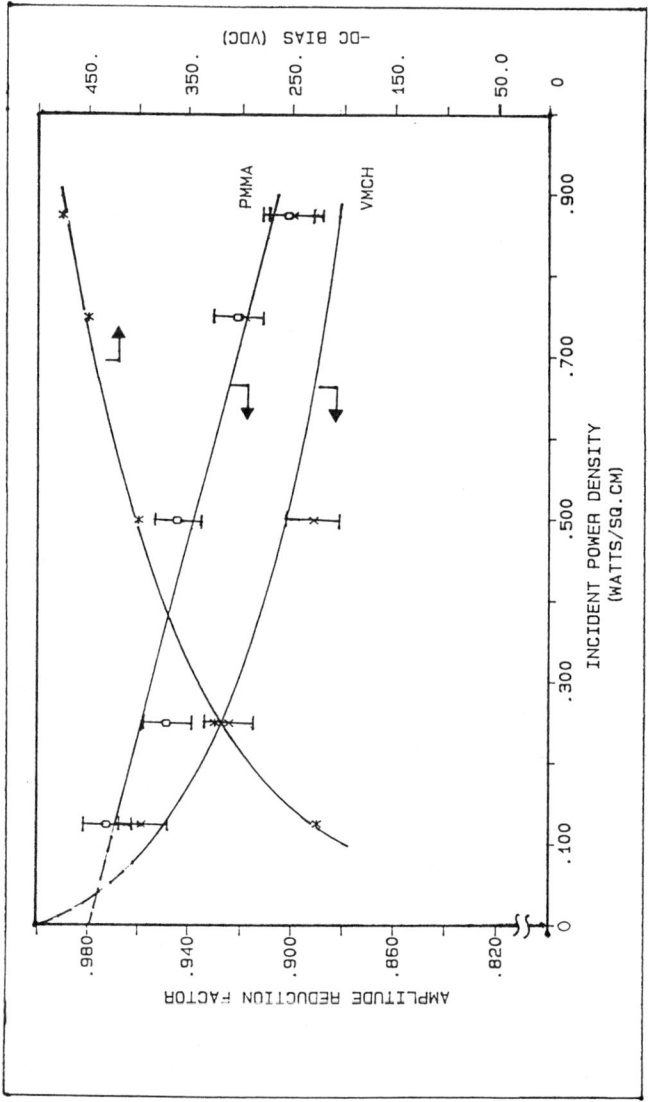

Figure 5: Amplitude Reduction Factor and Self-Bias Potential Versus Power Density at 35mTorr

Table II. Average Surface Roughness of Etched PMMA and VMCH Films

POWER DENSITY (Watts/cm^2)	SELF-BIAS (-VDC)	SURFACE ROUGHNESS*(Å) PMMA	VMCH
unetched		40	35
0.125	250	100	85
0.250	350	90	95
0.500	450	80	105
0.750	520	90	80
1.050	575	85	90

* Surface Roughness measured with Tencor Instruments Alpha-Step 200 stylus profilometer per ANSI Standard 1346.1-(1978)

O_2/RIE Conditions: flow=20 SCCM, chamber pressure=35mTorr
Etch time = 1.0 minute

Table III is a summary of results obtained for O_2/RIE of PMMA at 0.125 Watts/cm^2 and 0.75 Watts/cm^2 and 5mTorr and 35mTorr. The amplitude reduction factor is approximately equal to unity (sharp/smooth surface) at 5mTorr for both power levels while values less than unity occur at 35mTorr, indicating greater roughening. The etch rates at constant power are 2-3 times lower at 5mTorr than the corresponding rates at 35mTorr. In addition, the dark space width, measured with a cathetometer, increased from 1.1cm at 35mTorr to 3.0cm at 5mTorr while self-bias potentials were only modestly higher at 5mTorr. Non-local transport theories (4) predict energy deposition rates that track closely with observed etch rates. However, estimated ion fluxes at 5mTorr for both power levels are less than those at 35mTorr. Thus, it appears that surface roughness scales with ion flux to the surface. Thermal heating effects, which are known to cause the etch rate of PMMA to accelerate with time (5), and could influence surface roughness, were investigated by measuring the etch rate of samples that were thermally bonded (Apiezon Type N vacuum grease) to the electrode and comparing them to unbonded samples. The thermally bonded samples etched at virtually the same rate as the unbonded samples for the etch times investigated (< 1.5 minutes).

Table III. Amplitude Reduction Factor and Oxygen RIE Rate of PMMA

Power Density (Watts/cm^2)	Pressure (mTorr)	Self-Bias (-VDC)	f	Etch Rate* (nm/min)
0.125	5	250	0.993 ±0.015	135
0.750	5	600	0.999	435
0.125	35	230	0.966 ±0.015	245
0.750	35	500	0.945	1160

* Time-weighted average

SEM photomicrographs of etched films confirm these findings. Figures (6a)-(6d) are SEM photomicrographs taken of PMMA films etched for one minute under the following O$_2$/RIE conditions: (6a) baked/unetched, (6b) 0.125 Watts/cm^2, 35mTorr, no thermal bond; (6c) 0.75 Watts/cm^2, 35mTorr, no thermal bond; (6d) 0.75 Watts/cm^2, 35mTorr, with thermal bond. It is clear that the film surfaces etched at 35mTorr become progressively rougher with increasing power density while thermal bonding had no effect. Similar SEM analyses showed the films etched at 5mTorr to remain smooth after identical etching intervals. The characteristic wavelength of the corrugated surface folds of the roughened films appears to increase from less than 0.1μm at low power to about 0.2μm at high power. These wavelengths are less than wavelength of the probing laser beam and insure that the beam is indeed optically averaging the surface roughness. Had the surface corrugations been of sufficiently long wavelength, then the reflected beam signal would not average over corrugations, thereby changing interpretation of results.

In these oxygen/RIE experiments, the effect of the transition layer phase shift ϕ_t is such that the endpoint begins close to the final extrapolated reflectance maximum, perhaps anticipating it slightly as expected for a rough film (Figure 7). The approach to completion of etch removal appears to be more gradual than is ordinarily seen in film dissolution experiments, with no evidence of acceleration or steep inflection in the concentration profile within the transition region. The concentration profile for a simple rough interface is not expected to show any sharp inflection unless some intrinsic layering structure (i.e. residual etch products) were present, and none is evident from the interferogram. Furthermore, there is no reason to expect acceleration of etch removal near the endpoint since etch rate is limited by generation of active etchant species (atomic oxygen and O$_2^+$ ions) in the gas phase (6-9). The observations are consistent with this expectation.

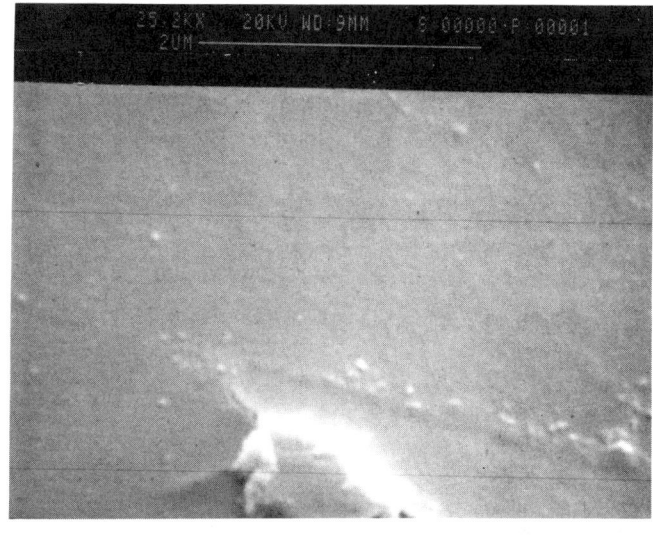

(6a) 1μm ─────

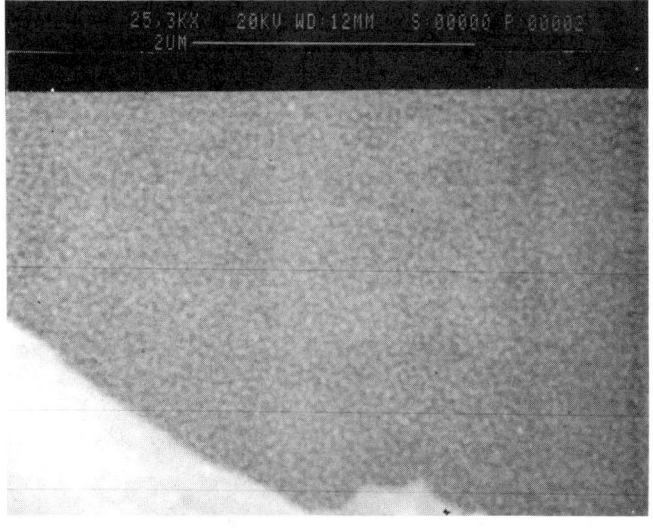

(6b) 1μm ─────

Figure 6. Scanning electron micrographs of hard-baked PMMA films before and after a 1.0-min oxygen-RIE treatment at the following conditions: a, unetched; and b, etched, 0.125 W/cm^2, 35 mTorr, -230 VDC, no thermal grease. Continued on next page.

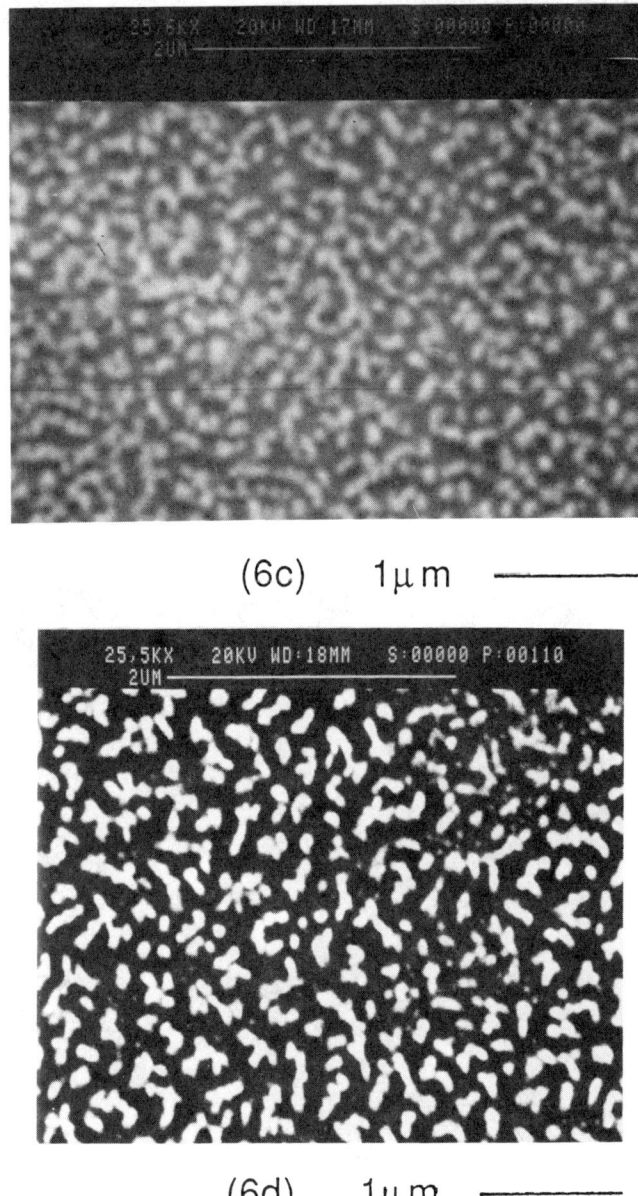

(6c) 1μm

(6d) 1μm

Figure 6. Continued. c, Etched, 0.75 W/cm^2, 35 mTorr, -500 VDC, no thermal grease; and d, same as in c except with thermal grease.

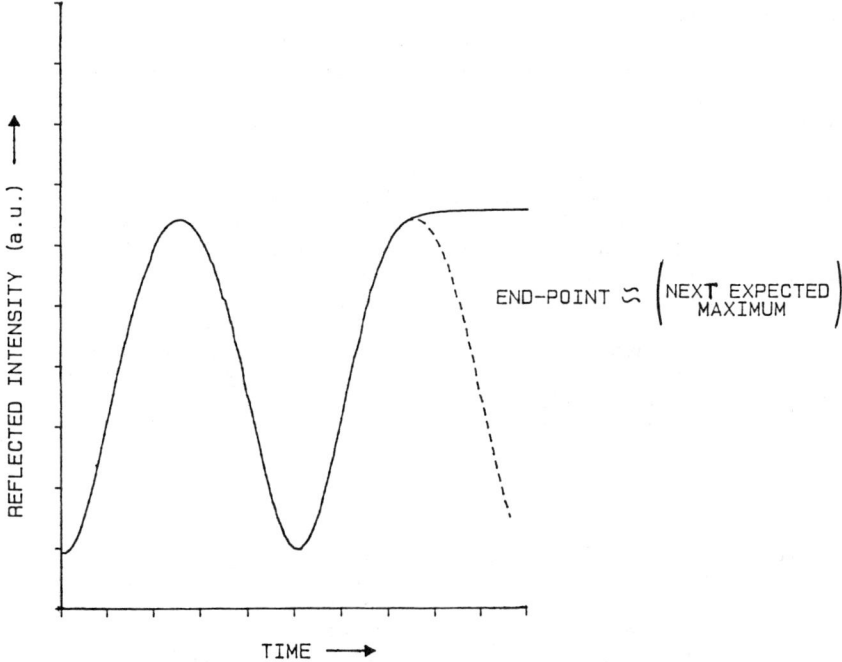

Figure 7: End-Point Behavior of PMMA During Oxygen/RIE

Discussion

Refractive index measurements made by this technique are of sufficient accuracy to make this a valuable research tool in studying and characterizing the dry etch behavior of novel polymer resist materials. In situ etch rate can be measured and analyzed quickly without any prior knowledge of the polymer's physical properties.

Surface "roughness" measurements of reactive-ion etched samples suggest that laser interferometry may be more sensitive than mechanical stylus measurements. This apparent difference in sensitivity is due to the fact that surface roughness correlation lengths in this study are less than $0.2 \mu m$. This is less than laser beam's width, which insures that the probing beam is optically averaging the surface roughness and makes it intrinsically sensitive enough to use as a RIE characterization tool. On the other hand, lateral surface roughness dimensions of this order are beyond the sensitivity of a 5-12μm diameter mechanical stylus tip. Similar conclusions were drawn in the ellipsometric surface study by Nee (10).

The key factor in obtaining precise information from the oxygen/RIE laser interferogram is careful process control which minimizes signal noise. The calculations are sensitive to small changes in reflected beam signal. In many cases, the amplitudes of waveform oscillations can vary by a few percent during a run which can lead to proportionate changes in calculated n_2 and f values. Estimates of the plasma intensity, R_p, are a function of the time at which it is measured and the surface area of exposed etchable material during the process. The plasma intensity is typically a few percent greater during the etch when etchable material is still present. This was confirmed by measuring the plasma intensity periodically throughout an etch cycle.

Conclusions

The full expression for the reflected intensity of a laser interferometer in a plasma/polymer/silicon system can be used to measure the polymer refractive index to within about 3 percent. This leads to an efficient and non-destructive in situ etch rate measurement technique which does not require any prior knowledge of film properties. The analysis can also be used to semi-quantitatively gauge the degree of "roughening" imparted to a polymer film surface during RIE processing.

Acknowledgments

This work was supported by the Office of Naval Research. This work was performed in part at the National Nanofabrication Facility at Cornell University which is partially supported by the National Science Foundation.

Literature Cited

1. Krasicky, P. D.; Groele, R. J.; Rodriguez, F. Chem. Eng. Comm. 1987, 54, 279.
2. Vašíček A. In Optics of Thin Films; North-Holland: Amsterdam, 1960; Chapter 3.
3. Krasicky, P. D.; Groele, R. J.; Rodriguez, F. J.Appl.Polymer Science 1988, 35, 641.
4. Jurgenson, C. W.; Shaqfeh, E. S. G. J. Appl. Phys., 1988, 64, 6200.
5. Visser, R. J.; de Vries, C. A. M. ISPC-8 Tokyo, 1987, p 1029.
6. Battey J. F. IEEE Trans. on Elec. Dev., 1977, 24(2), 140.
7. Hartney, M. A.; Greene, W. M.; Soane, D. S.; Hess, D. W. SPIE Proceedings -Adv. in Resist Tech. and Proc. V, 1988, Vol.920, p 108.
8. Gokan, H.; Esho, S. J. Electrochem. Soc., 1984, 131, 1106.
9. Soller, B. R.; Shuman, R. F. J. Electrochem. Soc., 1984, 131, 1353.
10. Nee, S. F. Applied Optics, 1988, 27(14), 2819.

RECEIVED July 7, 1989

Chapter 15

Evaluation of Several Organic Materials as Planarizing Layers for Lithographic and Etchback Processing

L. E. Stillwagon and Gary N. Taylor

AT&T Bell Laboratories, 600 Mountain Avenue, Murray Hill, NJ 07974

> Several materials were surveyed as topographic substrate planarizing layers for use in optical lithography or in etchback processes that are used in the fabrication of multilevel metal-insulator structures. Two types of materials were examined: resins that were applied by spin coating and baked to enhance planarization, and liquid monomers that were applied by spin coating and hardened photochemically after an appropriate leveling period. The planarizing properties of the materials were compared by measuring their abilities to level 20-500 μm wide isolated square holes on Si substrates. Other relevant properties, such as etching resistance, uv absorbance and glass transition temperature, are reported and discussed.

As optical lithography is pushed to its resolution limit, the depth of focus of the exposure tools will decrease and planarization, that is leveling, of substrate topography may be necessary to properly pattern topographic substrates (1). A single layer that planarizes substrate topography in addition to serving as the imaging layer may be used, or a multilayer-resist structure may be used with the bottom layer serving as the planarizing layer. A more immediate need for substrate planarization is in etchback processing that is used in the fabrication of multilevel metal-insulator structures (2-4).

Figure 1 shows two processes for leveling substrate topography. In the first a low molecular weight polymer (resin) is applied to the topographic substrate by spin coating from a concentrated solution of the resin (4-9). During spin coating the film profiles over isolated features that are narrower than about 50-100 μm are at least partially planar while the profiles over features wider than this are conformal (10). Thus, it is necessary to bake the coated substrate after spin coating to lower the film viscosity and enhance leveling. In the second process a low viscosity, liquid monomer is applied by spin coating and the film is hardened (cured) after an appropriate leveling or flow period (10). Here this process is done at room temperature and the film is hardened by uv irradiation.

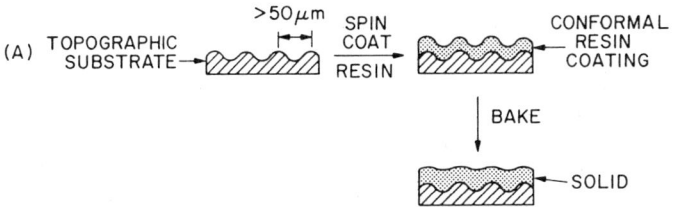

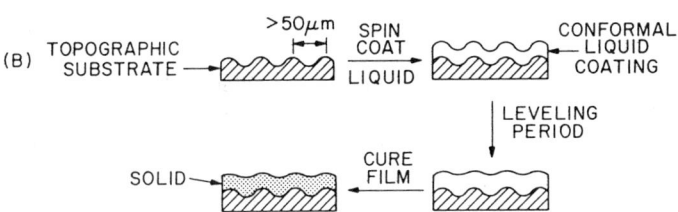

Figure 1. Schematic drawing of two planarization processes using (A) a low molecular weight polymeric resin and (B) a low viscosity, liquid monomer.

In addition to having adequate planarizing properties, materials used in optical lithography should be resistant to plasmas used to etch typical substrates; the most severe are plasmas that are used to etch aluminum substrates. The materials should be thermally stable and should contain small amounts of metal impurities that adversely affect device performance. Materials used as bottom layers in multilayer-resist structures should be highly absorbing at the exposure wavelength to eliminate substrate reflection of light that degrades resolution, and should be highly crosslinked so that the bottom layer is resistant to spinning solvents used to apply the top layers. The absorption requirement can be met by adding small amounts of appropriate dyes. For etchback processing, the material must be resistant to O_2 reactive-ion etching and should contain low amounts of harmful metal impurities.

In this paper the planarizing properties of some commercially available resins and monomers are evaluated. Other important properties such as etching resistance, film absorbance and glass transition temperature T_g are reported and discussed. Some of the materials that we evaluated are not marketed for use in the microelectronics industry. Consequently, they are not available as filtered spin coating solutions and may contain high levels of metal impurities that adversely affect device performance.

Experimental

Silicon substrates with isolated square holes that were fabricated using standard photolithographic and plasma etching techniques were used to evaluate the planarizing properties of the materials. The hole widths were 20, 50, 100, 200, 300, 400 and 500 μm and the hole depth was 1 μm. Figure 2 shows a schematic drawing of a typical topographic substrate. Planarization was determined either by film thickness measurements using an automated instrument (Nanospec/AFT Model no. 010-0180, Nanometrics, Inc.) or by measuring film profiles over the topographic features with a Dektak IIA stylus profilometer (Sloan Technology) equipped with a 12.5 μm radius stylus. In the former case, the film thickness was measured near the center of the square hole ($h(0)$ in Figure 3) and in the topographically higher region far away (several hole widths) from the center of the hole. This latter thickness was assigned to the initial film thickness h_0. Planarization was determined using

$$\text{Planarization } (\%) = 100 \frac{h(0) - h_0}{d} \qquad (1)$$

where d is the depth of the hole on the uncoated substrate. This depth was measured using the profilometer. Planarization was determined from film profile measurements by

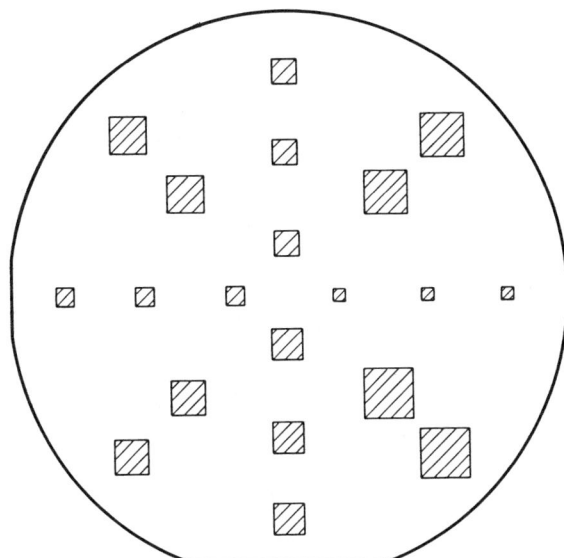

Figure 2. Schematic drawing of a typical topographic substrate.

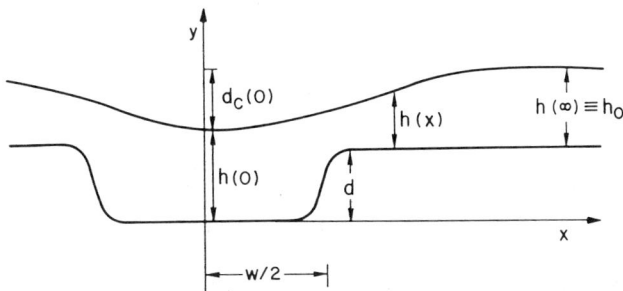

Figure 3. Drawing of a film profile over a hole.

$$\text{Planarization } (\%) = 100\frac{d - d_c(0)}{d} \qquad (2)$$

where $d_c(0)$ (Figure 3) is the depth at the center of the hole on the coated substrate.

Table I describes the resins that were evaluated. Films of the resins were applied to topographic substrates by spin coating from concentrated solutions of the resins in volatile spinning solvents using a Headway Research Model EC101 spin coater. After 2 minutes of spinning the coated substrates were transferred to a hot plate for baking. The temperature of the hotplate was measured with a surface thermometer.

Table I. Resin Information

Trade Name	Material Type	Softening Point (°C)
HPR-206 Positive Photoresist (Olin-Hunt)	Mixed Isomer Novolac + Diazonaphthoquinone Photoactive Compounds	120-140
AZ Protective Coating (Hoechst)	Mixed Isomer Novolac	120-140
PC1-1500D (Futurrex)	Polyester	-
Varcum 29-801 (BTL Speciality Corp.)	Ortho-Cresol Novolac	90-100 (8)
Kristalex 3085 (Hercules)	Poly(α-methylstyrene)	85

Liquid monomer films were also evaluated as planarizing layers. 1-Naphthyl acrylate was prepared by reacting 1-naphthol with acryloyl chloride in the presence of a tertiary amine. Ethoxylated bisphenol-A dimethacrylate and t-butylphenyl glycidyl ether were obtained from ARCO Speciality Chemicals and Wilmington Chemical Co., respectively. Viscosities of the monomers were measured with calibrated capillary viscometers (Cannon Instrument Co.) or were determined from the thicknesses of spin-coated films using the equation of Emslie, et al (11). The low viscosity monomers were applied to the substrates by spin coating and after an appropriate leveling period were hardened by exposure to light from an Optical Associates, Inc. deep-uv light source, Model no. 29D. Small amounts of Irgacure 651 (Ciba-Geigy Corp.) or triarylsulfonium hexafluoroantimonate (UVE 1014,

General Electric Co.) were added to the acrylic and epoxy monomers, respectively, to initiate photopolymerization. The acrylic monomers that polymerized by a free-radical mechanism were irradiated under nitrogen. The output of the lamp measured at the substrate was about 4 mW/cm^2 at 260 nm and about 11 mW/cm^2 at 310 nm.

Glass transition temperatures of the uv-hardened films were measured with a Perkin Elmer Model DSC-4 differential scanning calorimeter (DSC) that was calibrated with an indium standard. The films were scraped from silicon substrates and placed in DSC sample pans. Temperature scans were run from -40 to 100-200°C at a rate of 20°C/min and the temperature at the midpoint of the transition was assigned to T_g.

Oxygen reactive-ion etching (O_2 RIE) rates of the films were measured using a single-wafer etcher. Typical etching conditions were 8-10 sccm of O_2, 5 mtorr pressure, -400 V bias and a power density of about 0.4 watts/cm^2. Reactive-ion etching rates in a plasma used to etch aluminum substrates were measured using an Applied Materials AME 8110 Hexode Reactor. The plasma was formed in a gas mixture of BCl_3, Cl_2 and CHF_3, and the bias and power density were -230 V and 0.1 watts/cm^2.

Results and Discussion

If we neglect film shrinkage the degree of planarization achieved during the leveling period is determined by the type of geometry and the value of the dimensionless parameter $t\gamma h_0^3/\eta w^4$ where t is the length of the leveling period, γ is the film surface tension, η is the film viscosity and w is the width of the topographic feature ([12]). The type of geometry and topographic widths are determined by integrated circuit designers. The film thickness is limited to about 2 μm by etching considerations, and the surface tension of most organic materials is about 30 dyn/cm. Thus, the only properties considered when surveying candidates as planarizing materials were viscosity and the capability to have long leveling periods. The best materials will be those capable of long, low viscosity leveling periods.

Resins. The first three materials listed in Table I are marketed for use in the microelectronics industry. The last two materials have lower softening points than mixed isomer novolac resins and should have lower viscosities when baked at high temperatures. Positive photoresist is widely used as a planarizing layer, but in the future, materials with better planarizing properties will be required. Mixed isomer novolacs by themselves have been reported to have better planarizing properties than positive photoresist ([9]). We evaluated two other commercially available mixed isomer novolacs in addition to the material from Hoechst, Inc. All three had similar planarizing properties. Pampalone and coworkers ([8]) have reported that ortho-cresol novolacs had much better planarizing properties than positive photoresist. The poly(α-methylstyrene) material used here is a mixture of oligomers and, unlike the other materials, did not crosslink during baking at high temperatures. Table II shows the planarization of 20 - 200 μm wide holes achieved by *unbaked* films of the materials.

Table II. Planarization Achieved by Unbaked Films

Material	Film Thickness (μm)	Planarization (%)			
		20	50	100	200
Positive Photoresist	2.8	45	35	13	-6
Mixed Isomer Novolac	2.9	45	36	10	-9
Polyester	2.7	64	38	16	-5
Ortho-Cresol Novolac	3.6	-	40	-2	-9
Poly(α-methyl styrene)	2.1	57	32	-1	-5

Spinning solution concentrations were adjusted so that the final film thicknesses after spin coating and baking were about 2 μm. Only the film profiles over holes narrower than 50-100 μm were leveled to some degree during the spin coating process. The planarity of the film profile decreased with increasing hole width and leveling is just beginning for the 100 and 200 μm wide holes. These observations agree with a recent analysis of the relationship between spin coating and planarization (10). Negative values for the degree of planarization were observed for the wider holes. It has been shown that during the early stages of leveling the planarity of the film profiles in the center of the holes worsens before improving (12).

Table III lists the planarization achieved by unbaked and baked films of positive photoresist, mixed isomer novolac and the polyester material. Three baking temperatures (200, 250 and 300°C) were tried for the photoresist and mixed isomer novolac, while the polyester was baked at 200°C as recommended by the supplier. All films were baked for 10 min. During the early stages of baking the film viscosity is lowered and flow (leveling) is enhanced, but loss of volatile species from the film leads to a film thickness decrease that slows leveling and also degrades some of the leveling that may have occurred earlier (9,10). These two competing processes occur simultaneously during the baking and their interplay determines the planarity of film profiles over underlying substrate topography. The three materials undergo reactions at higher temperatures that drastically increase the film viscosity and eventually lead to a highly crosslinked, insoluble film. Films of these materials became insoluble after baking for less than a minute at temperatures of 200°C and above.

Table III. Planarization Achieved by Unbaked and Baked Films

Material	Bake Conditions	Film Thickness (μm)	Planarization (%)			
			20	50	100	200
Positive Photoresist	Unbaked	2.8	45	35	13	-6
Positive Photoresist	210°C	2.0	70	30	4	-1
Positive Photoresist	260°C	2.0	78	33	8	-3
Positive Photoresist	305°C	1.9	74	32	4	-3
Mixed Isomer Novolac	Unbaked	2.9	45	36	10	-9
Mixed Isomer Novolac	205°C	2.1	84	56	9	-8
Mixed Isomer Novolac	255°C	2.1	96	70	24	-15
Mixed Isomer Novolac	310°C	2.0	84	65	25	-12
Polyester	Unbaked	2.7	64	38	16	-5
Polyester	200°C	2.2	66	52	29	-6

For the positive photoresist, baking at 200 - 300°C enhanced leveling of the profiles over 20 μm wide holes, but degraded the planarity of the profiles over 50 and 100 μm wide holes. Apparently significant flow occurred only over distances of roughly 10 μm before thermal reactions greatly increased the film viscosity. Thus, thermal flow was able to offset the degrading effects of film shrinkage for only the film profiles over the 20 μm wide holes. Wilson and Piacente (7) examined the planarizing properties of several different positive photoresist films baked at temperatures between 160 - 200°C and found similar results.

Baking films of the mixed isomer novolac at 205°C improved the planarity of the film profiles over the 20 and 50 μm wide holes. The degree of planarization was superior to that achieved by positive photoresist. The diazonaphthoquinone photoactive compounds in the positive photoresist apparently caused crosslinking (hardening) to occur at lower temperatures and the photoresist had a shorter flow period than the mixed isomer novolac by itself (9). Baking at 255°C improved the planarity of the film profiles over the 20, 50 and 100 μm wide holes. These profiles are more planar than those obtained after baking at 205°C, while baking at 310°C yielded roughly the same results.

Baking the polyester film at 200°C improved the planarity of the film profiles over the 20, 50 and 100 μm wide holes. These profiles were leveled better than those for the baked positive photoresist film. During baking, the polyester film shrinks less than the positive photoresist film and may also have either a lower viscosity and/or a longer, low viscosity flow period.

The thermal reactions that occur during baking at high temperatures limit the low viscosity leveling period for the three materials listed in Table III. Baking at lower temperatures to avoid or slow the thermal reactions did not provide as good planarization as that achieved at 200°C.

Table IV lists the planarization of 20 - 400 μm wide holes achieved by unbaked and baked films of positive photoresist, ortho-cresol novolac and poly(α-methylstyrene).

Table IV. Planarization Achieved by Unbaked and Baked Films

Material	Bake Conditions	Film Thickness (μm)	Planarization (%)					
			20	50	100	200	300	400
Positive Photoresist	Unbaked	2.8	45	35	13	-6	-	-
Positive Photoresist	210°C/10 min	2.0	70	30	4	-1	-	-
Ortho-Cresol Novolac	Unbaked	3.6	-	40	-2	-9	-	-
Ortho-Cresol Novolac	225°C/2 min	1.9	88	84	68	28	16	-14
Ortho-Cresol Novolac	225°C/5 min	1.8	90	86	72	37	21	-13
Ortho-Cresol Novolac	225°C/10 min	1.8	100	89	70	42	30	-11
Ortho-Cresol Novolac	225°C/15 min	1.9	98	91	69	40	32	-13
Poly(α-methylstyrene)	Unbaked	2.1	57	32	-1	-5	-	-
Poly(α-methylstyrene)	225°C/2 min	1.7	93	86	66	41	14	-3
Poly(α-methylstyrene)	225°C/10 min	1.6	97	88	75	51	24	4
Poly(α-methylstyrene)	225°C/30 min	1.5	99	88	76	59	29	8

The ortho-cresol novolac has a lower softening temperature and reacts (crosslinks) more slowly during baking than the mixed isomer novolac resins typically used in positive photoresists (8,13). The molecular weight of the resin, and therefore the viscosity, begin to increase after about 2 min of baking at 225°C (8); after about 6 min the films are insoluble indicating that they are highly crosslinked. Thus, films of this material had longer low viscosity flow periods than the three materials previously discussed and exhibited better planarizing properties in spite of the large film thickness decrease that occurred during baking. Improvement in the planarity of film profiles over holes as wide as 300 μm was observed. At 225°C the film thickness remained roughly constant for baking times greater than 2 min and the optimum baking time was between 5 and 10 min. No improvement in leveling occurred after 10 min since the resin was highly crosslinked at this point. Baking at 155°C for 30 min gave inferior planarization to that achieved at 225°C for 2 min.

The planarizing properties of the oligomeric poly(α-methylstyrene) were similar to that of the ortho-cresol novolac. The film remained soluble after baking at 225°C, and the film thickness decreased with increasing baking time. Chromatographic analysis of baked films showed that the lower molecular weight oligomers (MW = 100-200 g/mol) evaporated from the film during baking. Unlike the ortho-cresol novolac, baking for more than 10 min at 225°C continued to improve the planarity of the film profiles.

The leveling rate can be increased by increasing the film thickness (10,12). Table V shows the degree of leveling achieved by ortho-cresol novolac and poly(α-methylstyrene) when the final film thickness after baking was about 4 μm. The films were baked at 225°C for 10 min.

Table V. Planarization Achieved by 4 μm Thick Films

Material	Film Thickness (μm)	Planarization (%)						
		20	50	100	200	300	400	500
Ortho-Cresol Novolac	4.2	95	93	90	79	59	49	31
Poly(α-methylstyrene)	4.3	96	95	89	82	70	56	40

As expected, the degree of leveling was superior to that achieved by 2 μm thick films, and the poly(α-methylstyrene) film achieved better planarization over the wider holes. The improvement in the planarity of the film profiles was extended to holes as wide as 500 μm by doubling the film thickness.

The polyester film was etched about 1.5 times faster than the positive photoresist film during RIE in a plasma that is used to etch Al substrates. The ortho-cresol novolac and poly(α-methylstyrene) materials have high aromatic carbon content and should etch at roughly the same rate as positive photoresist, although we have not measured their etching rates. After baking at 200°C, 2 μm thick films of the positive photoresist, polyester and ortho-cresol novolac were highly absorbing at 436 and 366 nm and were highly crosslinked. Films of poly(α-methylstyrene) absorb strongly only at wavelengths less than about 220 nm and are not crosslinked by baking at 200°C. To be used as the bottom layer in a multilayer-resist structure a non-volatile dye that strongly absorbs at the exposure wavelength would have to be added to this material and the film would have to be crosslinked by some means. Gokan and coworkers (14) recently reported that films of low molecular weight polystyrene exhibited good planarizing properties when baked in a nitrogen atmosphere. They also investigated copolymers of styrene with chloromethylstyrene that could be crosslinked after baking by uv irradiation. Ortho-cresol novolac and poly(α-methylstyrene) etched at roughly the same rate as positive photoresist during low-pressure, high-bias O_2 reactive-ion etching, while the polyester etched at roughly twice the positive photoresist rate.

Monomers. We surveyed several candidates for the planarization process that uses low viscosity, liquid monomers at room temperature (Figure 1B). A low viscosity is required for good leveling properties and a high C/O ratio (15) is required for etching resistance. 1-Naphthyl acrylate (NA) had a viscosity of 30 cs and a C/O ratio of 6.5, and had the best combination of properties of the acrylate and methacrylate monomers that were surveyed. Materials with very low viscosities, for example benzyl methacrylate (2 cs), evaporated during spin coating. Naphthyl acrylate is monofunctional and formed soft films during uv irradiation. Ethoxylated bisphenol-A dimethacrylate EBDMA, that is difunctional and has a moderate C/O ratio of 4.5, was added to NA to act as a crosslinking agent. Hard films were formed when the mixtures were irradiated.

Table VI shows a comparison of the leveling properties of the ortho-cresol novolac film that was baked at 225°C for 15 min with those for a uv-hardened film of a 7:3 mixture of NA and EBDMA. The mixture had a viscosity of 75 cs and a 10 min leveling period was used before uv hardening. Both films had a final film thickness of about 2 μm.

Table VI. Planarization by 2 μm Thick Films

Material	RIE Rate		Planarization (%)						
	Al	O_2	20	50	100	200	300	400	500
Ortho-Cresol Novolac (225°C/15 min)	1.0	1.0	98	91	69	40	32	-13	-
7/3 NA-EBDMA (75 cs, 10 min)	1.3	1.1	100	95	85	75	64	46	29

The ortho-cresol novolac film partially planarized holes as wide as 300 μm, while the film of the monomer mixture partially planarized holes as wide as

500 μm and exhibited superior planarization for holes wider than 100 μm. The hardened NA-EBDMA film etched 30 % faster than the ortho-cresol novolac film in a plasma that is used for etching Al substrates and etched 10 % faster during O_2 reactive-ion etching.

Para-t-butylphenyl glycidyl ether BPGE had a similar viscosity and C/O ratio as those of NA and had the best properties of the photocurable epoxies that were surveyed, but this monomer dewetted from Si substrates immediately after spin coating and formed a puddle at the substrate center. Other monofunctional epoxies exhibited the same behavior. Mixtures of BPGE with multifunctional aromatic epoxies wetted Si substrates and could be used as planarizing layers.

Films of the NA-EBDMA mixtures that were cured by uv irradiation at room temperature had glass transition temperatures that were around room temperature. Films with glass transition temperatures around 100°C or greater are required for lithography. Baking the films after uv hardening increased the glass transition temperatures of the films, but film shrinkage also occurred that degraded planarization. Table VII lists the glass transition temperatures and film shrinkage for films of the 7:3 NA-EBDMA mixture that were irradiated for 5 min and then baked at 100, 150 and 200°C for 5 min.

Table VII. Glass Transition Temperatures and Film Shrinkage of UV-Hardened and Baked 7/3 NA-EBDMA Films

Baking Temperature (°C)	T_g (°C)	Film Shrinkage (%)
None	28	5
100	84	25
150	95	27
200	107	25

The unbaked films had glass transition temperatures in the 25-30°C range and the T_g increased with increasing baking temperature. The films shrank by 5 % during uv irradiation and an additional 20 % during baking. Baking at 150 or 200°C raised the T_g to about 100°C, but caused a 20 % degradation in planarization. In general the glass transition temperature of an uv-hardened film is not much greater than the irradiation temperature unless very long irradiation periods are used. Although we have not tried it, irradiating the films while they are being heated should yield films with high glass transition temperatures and eliminate the need for the post-exposure bake and the consequent degradation of planarization due to film shrinkage. This approach may require monomer mixtures with higher boiling points than those used here to avoid film loss due to monomer evaporation during the high temperature irradiations.

Films of the NA-EBDMA mixture absorbed light strongly only at wavelengths below about 300 nm and appropriate dyes would have to be added if films of these materials were used as bottom layers in multilayer-resist structures to be patterned at 366 or 436 nm.

Conclusions

The planarizing properties of the ortho-cresol novolac and poly(α-methylstyrene) materials are clearly superior to those of the positive photoresist, mixed isomer novolac and polyester materials used in these studies. The ortho-cresol novolac becomes highly crosslinked after baking for 5-10 min at 225 °C and further baking does not improve planarization. Poly(α-methylstyrene), on the other hand, remains soluble and continues to flow after baking for 30 min at 225 °C, thus affording improved leveling at long baking times. The ortho-cresol novolac appears to have the best overall properties as either a bottom layer in a multilayer-resist structure or as a planarizing layer in an etchback process, although poly(α-methylstyrene) may also be useful in the latter process. Neither of these materials are marketed for use in the microelectronics industry, but we have examined some experimental materials from Futurrex Inc. (Ting C. H., Pai P. and Sobczack Z., Paper to be presented at VLSI Multilevel Interconnection Conf. in Santa Clara, Ca., June, 1989) that have comparable planarizing properties to the ortho-cresol novolac and poly(α-methylstyrene) materials. These materials may be available to the microelectronic industry in the future.

Films of 7:3 mixtures of 1-naphthyl acrylate and ethoxylated bisphenol-A dimethacrylate had better planarizing properties than any of the resins that were examined and may be useful as layers for etchback processing. For use as bottom layers in multilayer-resist structures it will be necessary to bake the films after uv hardening to increase the T_g, and if the exposure wavelength is above 300 nm, an appropriate dye must be added to eliminate substrate reflections that degrade resolution.

Acknowledgments

We thank John Frackoviak, Avi Kornblit, Nick Ciampa and Hans Stocker for fabrication of the topographic substrates and for performing some of the reactive-ion etching experiments. We also thank Molly Hellman for doing chromatographic analysis of some of the resins, and Ken Takahashi, Frank Ventrice and Reddy Raju for contact angle measurements, assistance with T_g measurements and helpful discussions.

Literature Cited

1. Burggraaf, P., Semicond. Internat. 1986, 9(3), 55.
2. A. N. Saxena and D. Pramanik, Solid State Tech., Oct. 1986, 95.
3. Sato K., Harada S., Saiki A., Kimmura T., Okubo T. and Mulai K., IEEE Trans. Parts, Hybrids, Packag. 1973, php-9, 176.
4. Adams A. C. and Capio C. D., J. Electrochem.Soc. 1981, 128, 423.
5. Rothman L. B., ibid. 1980, 127, 2216.
6. White L. K., ibid. 1983, 130, 1543.
7. Wilson R. H. and Piacente P. A., ibid. 1986, 133, 981.
8. Pampalone T. R., DiPiazza J. J. and Kanen D. P., ibid. 1986, 133, 2394.
9. Schiltz A., Abraham P. and Dechenaux E., ibid. 1987, 134, 190.
10. Stillwagon L. E., Larson R. G. and Taylor G. N., ibid. 1987, 134, 2030.
11. Emslie A. G., Bonner F. T. and Peck L. G., J. Appl. Phys. 1958, 29, 858.
12. Stillwagon L. E. and Larson R. G., ibid. 1988, 63, 5251.
13. Pampalone T. R., Solid State Technol. 1984, 27(6), 115.
14. Gokan H., Mukainaru M. and Endo N., J. Electrochem. Soc. 1988, 135, 1019.
15. Gokan H., Esho S., Ohnishi Y., ibid. 1983, 130, 143.

RECEIVED July 18, 1989

NOVEL CHEMISTRY AND PROCESSES FOR MICROLITHOGRAPHY

NOVEL CHEMISTRY AND PROCESSES FOR MICROLITHOGRAPHY

The previous sections have dealt with new materials or new processes that pertain to reasonably well established microlithographic techniques. Of course, it was not all that long ago that these same techniques were considered to be novel, and highly speculative. For example, an ACS microlithography symposium held in 1982, would probably have placed a paper titled "High Resolution Chemically Amplified Resist" into a section on novel resist chemistry. With this is mind, it will be interesting to see how future ACS symposia treat the topics presented in the following section.

The chapters in this section fall into two broad areas. One group describes the use of unconventional materials to design new resist systems. The other group consists of fundamental scientific studies aimed at understanding the events which occur within a resist film. The combination of these two areas not only provides information on todays resist systems, but also provides the basic understanding required to rationally design future materials and processes.

Scott A. MacDonald
IBM Almaden Research Center
650 Harry Road
San Jose, CA 95120–6099

Chapter 16

New Negative Deep-UV Resist for KrF Excimer Laser Lithography

Masayuki Endo, Yoshiyuki Tani, Masaru Sasago, and Noboru Nomura

Semiconductor Research Center, Matsushita Electric Industrial Company Ltd., 3–15, Yagumo-nakamachi, Moriguchi, Osaka 570, Japan

> A photosensitive composition, consisting of an aromatic azide compound (4,4'-diazidodiphenyl methane) and a resin matrix (poly(styrene-co-maleic acid half ester)), has been developed and evaluated as a negative deep UV resist for high resolution KrF excimer laser lithography. Solubility of this resist in aqueous alkaline developer decreases upon exposure to KrF excimer laser irradiation. The alkaline developer removes the unexposed areas of this resist. No swelling-induced pattern deformation occurs and high aspect ratio sub-half-micron patterns in 1 micron film thickness are obtained with high sensitivity.

1. Introduction

KrF excimer laser lithography that utilizes shorter wavelength has become of great interest as a means of fabricating 0.3-0.5 micron patterns in semiconductors (1-3).
One problem with KrF excimer laser lithography is the lack of high resolution resist. Many attempts to obtain suitable resist for KrF excimer laser has been reported. Co- and terpolymers of PMMA with indenone (4), oximinobutanone methacrylate (5), and methacrylonitrile (6) have been used as high resolution positive deep UV resists. However, their sensitivity to KrF excimer laser irradiation and resolution are not sufficient for use in practical KrF excimer laser lithography process (7). Naphthoquinonediazide-based positive resists have high optical density, so the profiles of their pattern are degraded (2,3,7). Recently, Willson et al. have reported on 1,3-diacyl-2-diazo linkage derivative dissolution inhibition system with high sensitivity as a positive deep UV resist. However, the resolution and pattern

profiles of the resist have not been fully described (8).
Orvek et al. have presented organosilicon positive photoresist for KrF excimer laser. Half-micron patterns were obtained using a KrF excimer laser contact printing system, while the complicated bilayer process is required for the use of the resist (9).

As for negative deep UV resist, O'Toole et al. have exhibited half-micron pattern resolution in 0.5 micron film thickness using the new resist and PIE process (10). The pattern profiles, however, were re-entrant, due to the large photo absorption and the applications to single-layer-resist system have not been presented (11).

We have developed a new negative deep UV resist for KrF excimer laser lithography. The resist is composed of 4,4'-diazidodiphenyl methane ((a) in Figure 1) as a photosensitive azide compound and poly(styrene-co-maleic acid half ester) ((b) in Figure 1) as a resin matrix.

Azide-phenolic resin photoresists have been reported by workers at Hitachi. They are used for i-line (12) or for deep UV light (13), and the applications to KrF excimer laser lithography have not been demonstrated.

We have found the combination of the azide compound and the styrene resin is well suited for achieving high resolution and high aspect ratio patterns using KrF excimer laser stepper system, because of the absence of swelling-induced pattern deformation during alkaline development and the suitable optical density at 248 nm in terms of sensitivity.

In this paper, the material characteristics and lithographic evaluation of this new resist are demonstrated. The resist meets the requirements for KrF excimer laser lithography, which exhibits high sensitivity, high resolution and high aspect ratio pattern profiles.

2. Experimental

2.1. Resist preparation

The photosensitive azide compound was 4,4'-diazidodiphenyl methane (m.p. 44.0 ℃). The poly(styrene-co-maleic acid half ester) was used as a resin matrix.

The azide compound was mixed with the styrene resin in the range of 10 to 40 wt%, and dissolved in 2-methoxyethyl acetate.

2.2. Spectroscopic characterization

The IR spectra of this new resist films on silicon substrates were measured with a Shimadzu FTIR-4000 Fourier transform spectrometer. The UV spectra of 4,4'-diazidodiphenyl methane in a quartz cell and the films of poly(styrene-co-maleic acid half ester) and the new resist on quartz substrates were measured with a Shimadzu UV-265FS double-beam spectrometer.

2.3. Lithographic evaluation

The new resist was spin-coated on a silicon substrate and baked for 20 min. at 80 °C in a convection oven. After exposure, the resist film was developed with a 60s immersion in tetramethyl ammonium hydroxide (TMAH) aqueous solution.

The film thickness of the resist was 1.0 micron. The exposure was done with a 5X KrF excimer laser stepper system (N.A 0.36) we manufactured (14).

Sensitivity is defined as the exposure energy necessary for 50% resist thickness remaining in the exposed areas. Contrast values are assessed by measuring the slope of the linear portion of the curve obtained by plotting the thickness of the relief image as a function of the logarithm of the exposure energy (15). The film thickness was measured with a Nanospec AFT film thickness monitor (Nanometrics).

The resist pattern profiles were evaluated using a JEOL JSM-T200 scanning electron microscope.

3. Results and Discussion

3.1. Optimization of the resist composition

Figure 2 shows the exposure characteristics for azide-styrene resin resist film with an azide concentration from 10 to 40 wt% (based on the styrene resin weight) and Figure 3 shows the contrast of the resist films as a function of the azide concentration. Development was done with a 60s immersion in 0.83% TMAH solution. The styrene resin matrix alone has been found to be a negative deep UV resist. However, rather low contrast (1.48) and low sensitivity (2.5 J/cm^2) are observed. The contrast and the sensitivity of the styrene resin is remarkedly increased by adding the azide, as shown in Figures 2 and 3.

When the concentration of the azide exceeds 30 wt%, sensitivity decreases (Figure 2) and the contrast becomes worse (Figure 3). This is due to the increase of the optical density of the resist. Large optical density prevents the light from penetrating into the resist (3,11). Also, the resist thickness remainig after development is maximum at the 30 wt% azide concentration (Figure 2). From these results, it was concluded that the azide-styrene resin resist which contains 30 wt% 4,4'-diazidodiphenyl methane based on poly(styrene-co-maleic acid half ester) would be most suited for KrF excimer laser lithography. The contrast (4.72) was excellent and the sensitivity (30 mJ/cm^2) was in the desired range. This resist composition was subjected to spectroscopic characterization and lithographic evaluation.

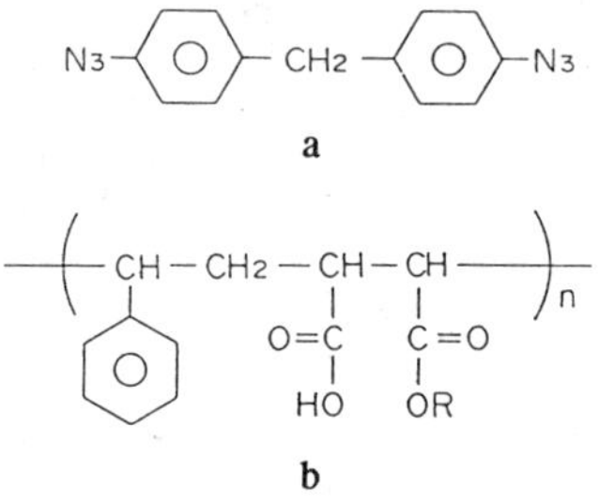

Figure 1. Chemical structures of (a) 4,4'-diazidodiphenyl methane and (b) poly(styrene-co-maleic acid half ester).

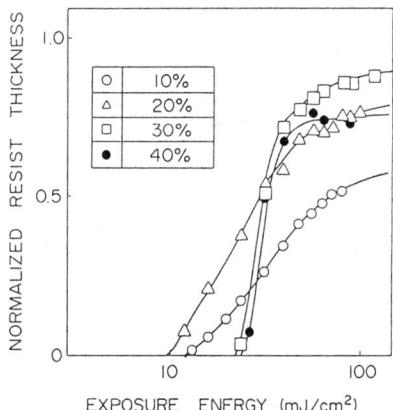

Figure 2. Effect of azide concentration on exposure characteristics for azide-styrene resin resist of 1.0 micron film thickness.

3.2. Spectroscopic characteristics

Figure 4 demonstrates a FT-IR spectra of this new resist before and after 80 °C prebaking, and after exposure to KrF excimer laser irradiation for 100 mJ/cm^2. The strong absorption at 2100 cm^{-1} due to the azide group stretching vibration in this resist is clearly present before prebaking. It can be seen that the characteristic bond at 2100 cm^{-1} of the azide decreased to half after prebaking and disappeared after exposure. No other significant changes are observed.

This indicates that the prebaking temperature higher than the melting point of the azide decomposes the azide (50%) and it totally decomposes upto 100 mJ/cm^2 irradiation. It is possible that subsequent reactions of the nitrene, generated from the azide thermolysis and photolysis, with the styrene resin could be responsible for solubility modulation of this type resist (16).

The UV spectra for the azide in a diethylene glycol dimethyl ether solution and for the styrene resin film with 1.0 micron thickness are shown in Figure 5. The azide has an intense absorption at around 248 nm (molar extinction coefficient at 248 nm = 3.0x10^4 1/M·cm). The syrene resin used as matrix polymer exhibits a significant transparency at 248 nm (70%).

The UV spectra for this resist film, before and after exposure to KrF excimer laser irradiation for 100 mJ/cm^2, are shown in Figure 6. The absorbance of the azide renders the reist film of 1.0 micron thickness essentially opaque at 248 nm. After exposure of 100 mJ/cm^2, the absorbance bleaches from 0.5 to 6.0% at 248 nm. Intense absorption by this resist at 248 nm closely relates to the pattern profile of the resist, which will be discussed in the last section.

3.3. Dissolution kinetics

In order to determine the appropriate development conditions, we examined dissolution characteristics for resist films in the aqueous alkaline developers by measuring film thickness as a function of development time. In Figure 7, dissolution characteristics for the new resist before and after exposure to KrF excimer laser are compared with those for the styrene resin matrix. The exposure energy was 100 mJ/cm^2 and the alkaline concentration in TMAH solution was 0.83%. Large differences in solubility are observed between these three films.

This indicates that the thermally or photochemically decomposed azide (Figure 4) inhibits the dissolution of the styrene resin into the alkaline developer. The inhibition may be due to the increase of the molecular weight of the styrene resin in the presence of the decomposed azide. Hydrogen abstraction from the polymer by nitrene of the decomposed azide and subsequent polymer radical recombination result in a increase in the molecular weight of the polymer (17).

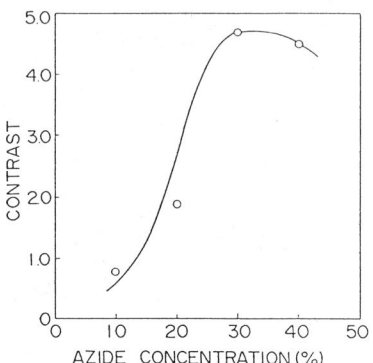

Figure 3. Effect of azide concentration on contrast for azide-styrene resin resist of 1.0 micron film thickness.

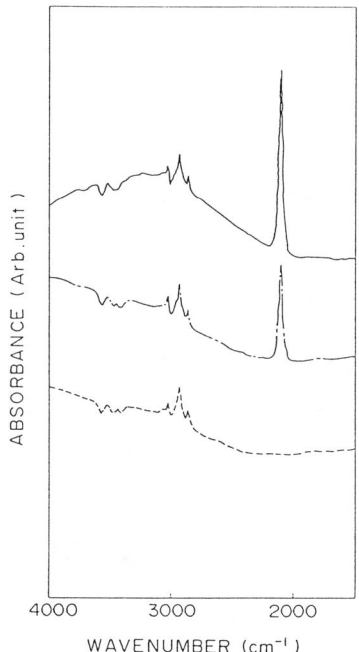

Figure 4. FT-IR spectra for the new resist film with 1.0 micron thickness, (———) before prebaking, (—-—) after prebaking and (-----) after exposure to KrF excimer laser irradiation.

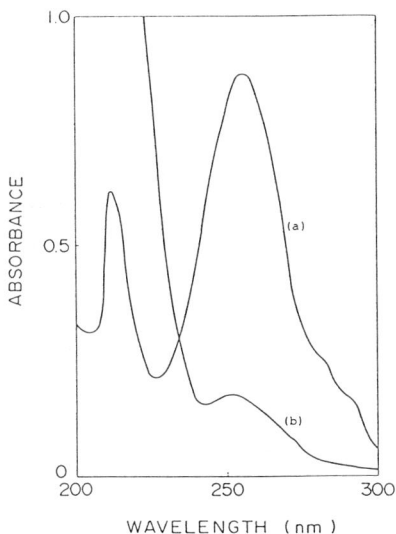

Figure 5. UV spectra for (a) 4,4'-diazidodiphenyl methane in a diethylene glycol dimethyl ether (7.2×10^{-5} mole/l) and (b) poly(styrene-co-maleic acid half ester) (1.0 micron film thickness).

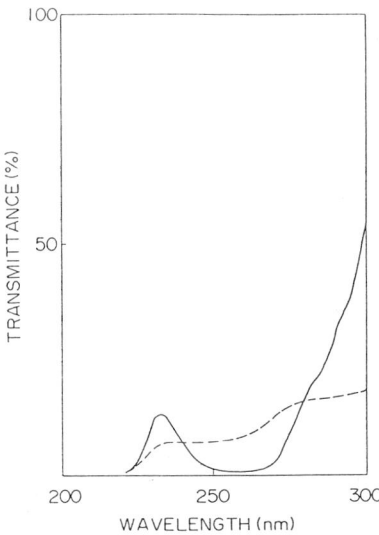

Figure 6. UV spectra for this new resist film with 1.0 micron thickness (———) before and (-----) after exposure to KrF excimer laser irradiation.

Consequently, the exposed resist film did not dissolve into 0.83% TMAH solution for at least 1 min., while the unexposed resist film dissolved away in 48s. This dissolution kinetics study demonstrates the high sensitivity and high contrast potentials of the new resist.

Figure 8 shows the effect of the alkaline concentration in TMAH solutions on the contrast and sensitivity of the new resist. Sensitivity of the resist increases as the alkaline concentration increases, however, the contrast is maxima (4.72) at 0.83% TMAH solution. This means that the higher concentration over 0.83% cannot distinguish the difference of the dissolution rate between the unexposed and exposed resist film. For instance, the higher concentrated developer also attacks the exposed areas and the loss of resist thickness occurs. The alkaline concentration in TMAH solution, therefore, is optimized at 0.83%. This developer concentration was subjected to the following lithographic evaluation.

3.4. Lithographic evaluation

Figure 9 shows an SEM photograph of 0.6 micron down to 0.45 micron line-and-space patterns of the new resist in 1.0 micon film thickness exposed with KrF excimer laser stepper system (N.A. 0.36). The energy required for the pattern fabrication was only 50 mJ/cm^2, and the development was done with a 60s immersion in 0.83% TMAH solution. High aspect ratio patterns of such thick resist films were successfully obtained using this resist.

As shown in Figure 6, the resist film strongly absorbs KrF excimer laser light. In the KrF excimer laser exposure of the resist, photon energy absorption is highest at the top of the resist film and lowest at the interface between the resist and substrate. This is due to the attenuation of the irradiation in the resist layer. Decreases in solubility followed by such photochemical reaction occur to a much greater extent in the vicinity of the resist film surface. Moreover, the thermally decomposed azide decreases solubility of the unexposed and exposed resist film (Figure 7).

Therefore, this resist development can be regarded as an etching-type process. The insoluble surface layer of the exposed area acts as an etching mask while the alkaline developer removes unexposed areas except for the insoluble layer of the thermally reactioned regions. The developer could not remove the bottom thermally reactioned layer because of the decrease of the alkaline strength during the development. Finally, slightly concave patterns, without swelling-induced pattern deformation, were successfully fabricated. More rectangular patterns could be attained by varying the development conditions.

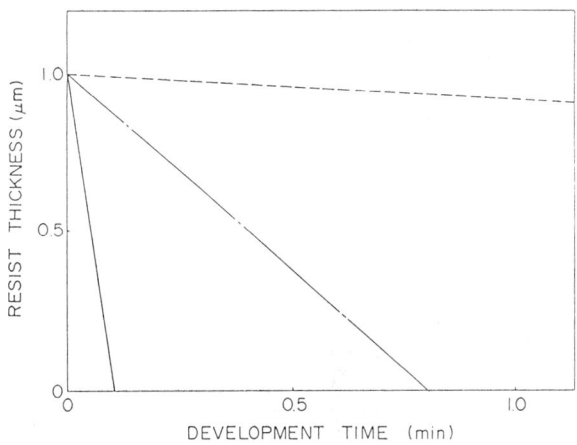

Figure 7. Dissolution characteristics for (———) unexposed poly(styrene-co-maleic acid half ester) film, (—-—) unexposed this new resist film and (-----) exposed this new resist film to KrF excimer laser irradiation.

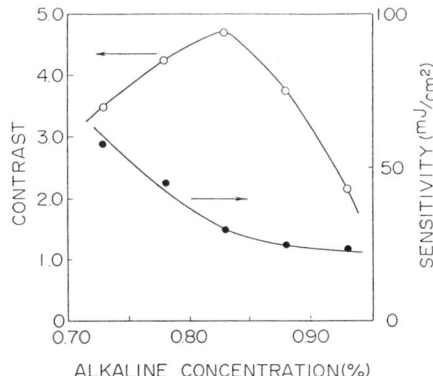

Figure 8. Effect of alkaline concentration in TMAH solution on contrast and sensitivity of the new resist with 1.0 micron film thickness.

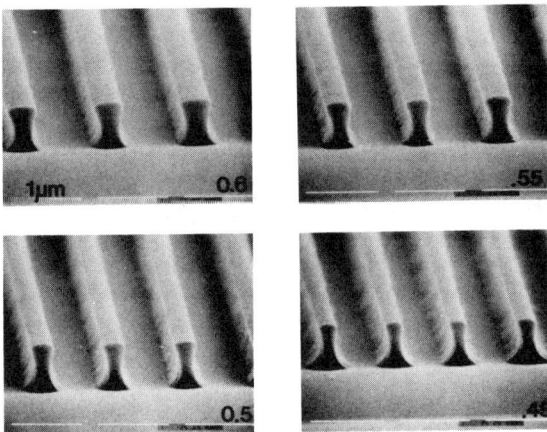

Figure 9. SEM photographs of 0.6 micron down to 0.45 micron line-and-space patterns of this new resist with 1.0 micron film thickness.

4. Conclusions

A negative deep UV resist, consisting of a photosensitive 4,4'-diazidodiphenyl methane and a poly(styrene-co-maleic acid half ester) resin, has been found to meet the requirements needed for KrF excimer laser lithography.

The resist has a high sensitivity and high resolution capability. The suitable photo absorption, excellent dissolution kinetics and no swelling during development contribute to such superior characteristics of this resist.

We achieved high aspect ratio sub-half-micron pattern fabrication in 1.0 micron film thickness using this new resist. We are convinced that this new resist could make possible simple and efficient single-layer-resist system for KrF excimer laser lithography.

Literature Cited

1. Jain,K.; Willson,C.G.; Lin,B.J. IBM J. Res. Develop. 1982, 26, 151.
2. Pol,V.; Bennewitz,J.H.; Escher,G.C.; Feldman,M.; Firtion,V.A.; Jewell,T.E.; Wilcomb,B.E.; Clemens,J.T. Proc. SPIE 1986, 633, 6.
3. Endo,M.; Sasago,M.; Hirai,Y.; Ogawa,K.; Ishihara,T. Proc. SPIE 1987, 774, 138.
4. Hartless,R.L.; Chandross,E.A. J. Vac. Sci. Technol. 1981, 19, 1333.
5. Wilkens,Jr.,C.W.; Reichmanis,E.; Chandross,E.A. J. Electrochem. Soc. 1980, 127, 2510; 1980, 127, 2514.
6. Reichmanis,E.; Wilkens,C.W. In Polymer Materials for Electronic Applications; American Chemical Society Symposium Series: Washington, DC, 1983; Vol.184, p.30.
7. Wolf,T.M.; Hartless,R.L.; Shugard,A.; Taylor,G.N. J. Vac. Sci. Technol. 1987, B5, 396.
8. Willson,C.G.; Miller,R.D.; McKean,D.R.; Pederson,L.A.; Regitz,M. Proc. SPIE 1987, 771, 2.
9. Orvek,K.J.; Cunningham,Jr.,W.C.; Mcfarland,J.C. Microcircuit Engineering 1987, 6, 393.
10. Sheats,J.R.; O'Toole,M.M.; Hargreaves,J.S. Proc. SPIE 1986, 631, 171.
11. O'Toole,M.M.; Grandpre,M.P.; Feely,W.E. J. Electrochem. Soc. 1988, 135, 1026.
12. Iwayanagi,T.; Hashimoto,M.; Nonogaki,S.; Koibuchi,S.; Makino,D. Polym. Eng. Sci. 1983, 23, 935.
13. Koibuchi,S.; Isobe,A.; Makino,D.; Iwayanagi,T.; Hashimoto,M.; Nonogaki,S. Proc. SPIE 1985, 539, 182.
14. Nakagawa,H.; Sasago,M.; Endo,M.; Hirai,Y.; Ogawa,K. Proc. SPIE 1988, 922, 400.
15. Thompson,L.F. Solid State Technol. 1974, 17(7), 27.
16. Tsunoda,T.; Yamaoka,T.; Osabe,Y.; Hata,Y. Photogr. Sci. Eng. 1976, 20, 188.
17. Nonogaki,S.; Hashimoto,M.; Iwayanagi,T.; Shiraishi,H. Proc. SPIE 1985, 539, 189.

RECEIVED June 14, 1989

Chapter 17

Characterization of a Thiosulfate Functionalized Polymer

A Water-Soluble Photosensitive Zwitterion

C. E. Hoyle, D. E. Hutchens, and S. F. Thames

Department of Polymer Science, University of Southern Mississippi, Hattiesburg, MS 49406-0076

> The characterization of a water-soluble zwitterionic polymer derived from aminoethane thiosulfuric acid (AETSA) and a diglicydyl ether of bisphenol A (DGEBA) has been accomplished. The polymer has been shown to form an associative network in water at high concentrations. The gel network disassociates at higher temperatures or under a shearing stress. Upon heat treatment above 165 °C, the polymer sustains a measurable weight loss and is crosslinked to form a water-insoluble film presumably with disulfide formation. Photolysis of coalesced films, either direct or sensitized, results in sulfur-sulfur bond cleavage with subsequent disulfide formation and loss of water sensitivity. Initial imaging studies demonstrate the potential of the thiosulfate functionalized polymer as a photoresist.

A large number of polymeric materials have been developed over the past two decades which are photochemically reactive. In many cases, such polymers are initially soluble in organic solvents prior to exposure with insolubilization accompanying ultraviolet radiation. This often presents a problem in practical applications where handling of organic solvents is objectionable or expensive. A need exists to develop functional polymers which are both water soluble and photochemically labile.

A number of reports in the literature describe the use of alkyl thiosulfates to modify reactive vinyl type monomers and/or preformed polymers with the expressed goals of producing polymers with enhanced water solubility ([1-6]). The alkylthiosulfate modified polymers have been shown to be thermally and photochemically reactive and capable of producing crosslinked films with varying degrees of stability ([5]).

This paper describes the synthesis and characterization of a new zwitterionic water-soluble thiosulfate polymer (Poly[γ-(amino β-thiosulfate) ether]-PATE) via chemical reaction of a diglicydyl ether of bisphenol A (DGEBA) with aminoethane thiosulfuric acid (AETSA) as a reactive nucleophile. Due to the presence of the ionizable thiosulfate moiety, the resulting polymer is water soluble. Further, the ability of the water soluble polymer PATE to form an inner zwitterionic salt allows for significant stability in water at near neutral pH. Since AETSA itself, as well as cured DGEBA resins, are known to be of negligible toxicity, such polymers are also most likely non-toxic.

The associative, thermochemical, and photochemical properties of PATE polymer will be described and evidence will be provided supporting the thesis of labile sulfur-sulfur bonds. Photolysis of a model compound and a polymer similar to PATE but lacking thiosulfate functional groups has been conducted to provide supportive evidence for the proposed sulfur-sulfur bond cleavage reactions. Finally, a preliminary investigation of the imaging characteristics of the PATE polymer is presented in order to place the photochemical investigation in perspective.

Experimental

Materials. Reagent grade solvents, dimethyl formamide (DMF), dimethyl acetamide (DMAC), dimethyl sulfoxide (DMSO) and methanol were purchased from Baker, stored over molecular sieves once opened, and used without further purification. Aminoethane thiosulfuric acid (AETSA) purchased from Kodak, and Taurine, purchased from Alfa were purified by recrystallization. Each was thrice recrystallized from hot, deionized water. The crystalline precipitate was dried (48 hours at 40 °C) *in-vacuo* and subsequently stored in a desiccator. Benzophenone (BP) was purchased from Aldrich Chemical Company. QUANTACURE BTC (BTC), (4-benzolybenzyl) trimethylammonium chloride, was used as supplied by Aceto, Inc., Flushing, New York. Phenyl glycidyl ether (PGE) was purchased from MCB, distilled *in-vacuo*, and stored at -15 °C. Epon 828® was used as supplied by Shell Chemical Company. The epoxy equivalent weight (EEW) for Epon 828® determined by an appropriate titration, was found to be 187.7.

Elemental Analysis. All samples submitted for elemental analysis were dried at 40 °C in a vacuum oven at less than 1 torr pressure for 24 hours and then sealed in ampoules. Elemental analyses were performed by MHW Laboratories of Phoenix, Arizona.

Instrumentation. Fourier transform infrared (FTIR) spectra were recorded on a Nicolet 5DX using standard techniques. Spectra were measured from various sample supports, including KBR pellets, free polymer films and films cast on NaCl windows. Spectra for quantitative analysis were recorded in the absorbance mode. The height of the 639 cm^{-1} absorbance was measured after the spectrum was expanded or contracted such that the 829 cm^{-1} absorbance was a constant height. In some spectra an artifact due to instrumental response appeared near 2300 cm^{-1}.

Proton-decoupled ^{13}C NMR spectra were obtained using a JEOL FX90Q spectrometer. Polymer solutions for analysis were 5 to 15 weight percent. All chemical shifts are referenced externally to trimethyl silane. Ultraviolet spectroscopy was performed with a Perkin-Elmer model 330 spectrophotometer, using double-beam, background-cancelling techniques. HPLC analysis was performed using an LDC Minipump, a Rheodyne 7125 sample injector and a Perkin-Elmer LC-75 variable wavelength UV detector operating at 245 nm. The detector used air as a reference with offset background cancelling. The mobile phases were mixtures of acetonitrile and water, most commonly 85 parts water and 15 parts acetonitrile by volume. The columns were a Waters microbondapack CN alone or in series with a Waters C-18 column (3.9 mm i.d. X 30 cm.). The flow rate varied between 0.9 and 1.0 mL/min to generate a pressure of less than 2000 psi at the pump exit.

Synthesis of Poly[γ-(amino β-thiosulfate) ether] (PATE), Poly[γ-amino β-sulfonic acid] (PASE), and Hydroxy-3-aminoethane thiosulfuric acid (AETSAPPE). Details of the synthesis of these three compounds are given elsewhere (7).

Preparative Photolysis. The preparative photolysis of an aqueous solution (pH=8.5) of AETSAPPE (2.5 M) was conducted in a 1-inch diameter quartz test tube in a Rayonet Reactor (Southern New England Radiation Co.) fitted with 254 nm lamps. Within two hours the solution gelled and the reaction was terminated. Upon acidification the solution cleared, and the product could be re-precipitated by addition of base. This indicates loss of the thiosulfate functionality. The product was dissolved in dilute HCl, precipitated with acetone, and filtered. This process was repeated three times, and the final precipitate was washed with water. The product (20 to 30 mg) was dried *in vacuo* for 24 hours and stored in a dessicator until use. Comparison of the ^{13}C NMR spectrum of the product with the starting AETSAPPE ^{13}C NMR spectrum clearly shows that the thiosulfate methylene peak shifted upfield, from 39 ppm to 35 ppm. The complete ^{13}C NMR and IR analysis of the product were consistent with the disulfide product. Further, elemental analysis of the product confirmed that the product was the desired disulfide product 2-amino (2-hydroxy 3-(phenyl ether) propyl) ethyl disulfide (AHPEPED): Expected C: 58.39, H: 7.08, N: 6.20, S: 14.18; actual C: 58.26, H: 7.22, N: 6.06, S: 14.28.

Quantitative Photolysis. Photolysis experiments were performed by exposing samples to a 450 Watt medium pressure mercury lamp. A shutter was placed between the samples and the lamp so that the exposure time could be accurately controlled. Unless otherwise stated, the samples were placed 4 inches from the lens. Filters (254 nm, 280 nm and 366 nm) when used, were placed immediately in front of the shutter. Sample holders were available for 1" x 1" quartz plates, NaCl windows and quartz UV cuvettes, and samples of each type were utilized.

UV intensity measurements were made with an International Light 700A Research Radiometer. The measuring head was tightly covered with aluminum foil for zeroing, and then exposed to the lamp output under exactly the same conditions as the actual samples (i.e., same distance, angle, elevation, etc.). The results of these experiments were used to evaluate the quantum yield or efficiency of the photochemical process. Specifically, photolysis of AETSAPPE to yield AHPEPED was followed quantitatively by HPLC (Waters) using a 50:50, water:acetonitrile mobile phase and a C-18/cyano column combination in series. AETSAPPE eluted at a retention volume of 5.9 mL and AHPEPED at an elution time of 3.2 mL. Three unknown products of relatively low amounts also eluted with retention volumes of 3.0 mL, 5.3 mL, and 7.9 mL. For the sensitized photolysis, a 85:15 water:acetonitrile mobile phase was used to separate AHPEPED and BTC from AETSAPPE.

Photolysis of PATE Films. PATE films (5 μ to 20 μ thick) were obtained by casting on glass 10-20% solids aqueous solutions containing 0.02% wetting agent. The resultant films, heated in a drying oven at 100 °C for four minutes, were quite water soluble.

For IR studies, films of 5 to 10 microns were applied to NaCl disks by injecting known, small volumes of dilute solutions of PATE in DMSO. The solutions were spread to cover a 1-cm diameter area on the NaCl disk, and subsequently dried 10 minutes at 105 °C followed by photolysis.

Results and Discussion

Synthesis of Poly [γ-(amino β-thiosulfate) ether] (PATE), Poly [γ-(amino β-sulfonic acid) ether] (PASE), and Hydroxy-3-(aminoethane thiosulfuric acid) (AETSAPPE). The general scheme for synthesis of both the PATE and PASE polymers is shown in Scheme I. In the case of the PATE polymer, the mild conditions employed (40-60° C) insure that crosslinking of the resultant polymer by disulfide linkages is minimized. Specific details of the synthesis are given elsewhere (7). Depending on the exact reaction conditions as well as the nature (value of n may vary but is close to 0.1 in Scheme Ia) of the starting diglycidal ether of bisphenol A, the molecular weight (determined by viscometry in DMSO and the amino equivalent weight) of the PATE polymer varied between about 3000 and 9000. In the polymer synthesis employed the primary amine of aminoethane thiosulfate (AETSA) reacts as a difunctional monomer, since it possesses two active hydrogen atoms and reacts with two epoxy groups. AETSA can also act as a monofunctional end-capping agent, and the combined effect of these two reactions is the formation of the PATE polymer. The value of such a synthetic scheme is that the desired thiosulfate-functional polymer is formed directly from stable monomeric materials. It is therefore unnecessary to subject a thiosulfate-functional vinyl monomer to polymerization conditions (i.e. anionic, cationic or radicals) which could result in premature degradation of the thiosulfate bond, with a subsequent high probability of disulfide crosslinks. Additionally, this synthesis does not rely on the availability of a linear (uncrosslinked) prepolymer and therefore represents an entirely new method for introducing the thiosulfate (Bunte salt) functionality into a polymer molecule.

The synthesis of hydroxy-3-aminoethane thiosulfuric acid (AETSAPPE) is shown in Scheme II. The same basic conditions used for the polymer synthesis were employed to synthesize the model compound (AETSAPPE) although the work-up conditions were less stringent. The structure was confirmed by carbon-13 NMR and elemental analysis.

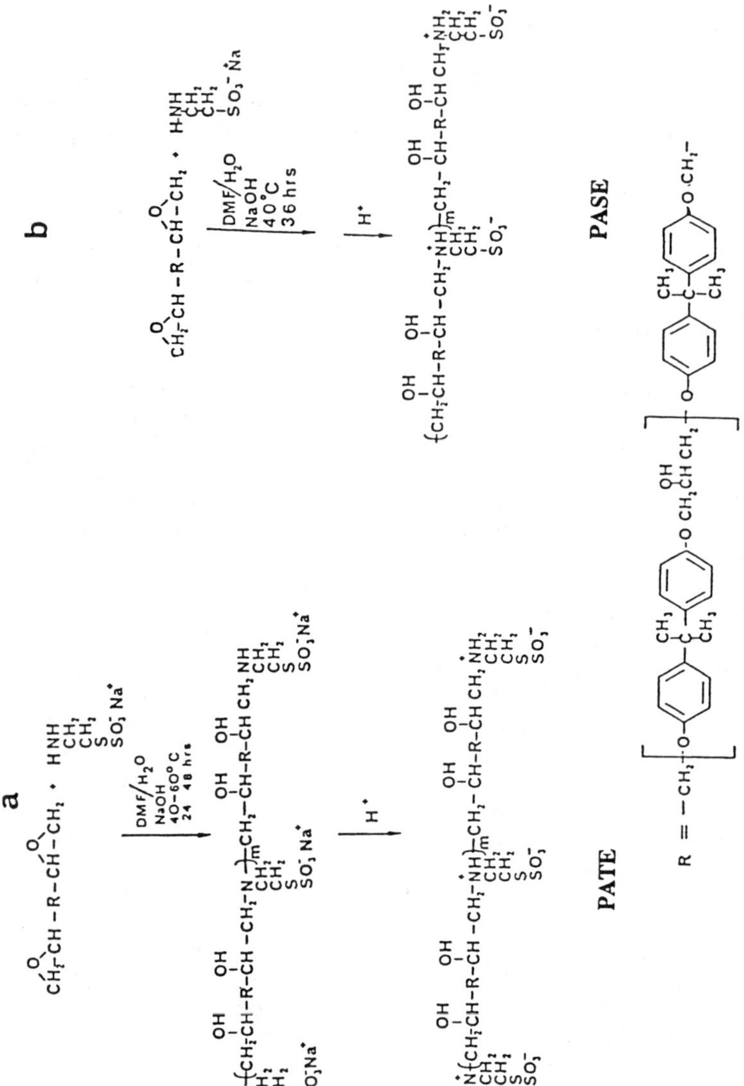

SCHEME I

Titration of PATE. The titration curve (pH vs. acid/base concentration) of the PATE polymer (Figure 1) shows a large increase in pH near the isoprotic point with addition of a small amount of sodium hydroxide, confirming the zwitterionic nature of PATE near the isoprotic point.

Rheological Behavior of PATE. Given the reported tendency of AETSA zwitterions to associate in the crystal lattice, and the potentiometric evidence that AETSA-functional polymers are zwitterions, it is reasonable to expect a tendency toward association. Such association of AETSA functional ends would result in a network-like structure. However, such ionic aggregation is far weaker than covalent bonding, and would be expected to dissipate at high temperatures or shear fields. Viscosifying effects consistent with AETSA association have indeed been noted. Specifically, 15% and higher solids aqueous solutions of PATE resins are opaque solid gels at room temperature. When heated above 45 °C the solutions become clear and free flowing. However upon cooling, the solutions reform their gel-like structure within a few minutes, and opacify within a few hours. These phenomena imply a form of chain-chain association with the gel visualized as a loose associative matrix structure.

The apparent viscosity measured by a Contraves Rheometer illustrates the effect of heat on the solution viscosity. The zero-shear apparent viscosity was evaluated for aqueous PATE solutions of 1 g/dL to 15 g/dL over a temperature range of 25 °C to 60 °C (Figure 2). The results demonstrate that at room temperature increasing the concentration of polymer 15-fold increases the viscosity by a factor of 13. However, at higher temperatures, the increase in viscosity with concentration over the same range is only about a factor of 2. These data are consistent with formation of weak electrostatic interactions which are overcome by heating. In another experiment, a Gardner viscosity tube filled with a 17.1% aqueous PATE solution was heated to 80 °C and allowed to cool to room temperature. As the sample reaches room temperature a large increase in the viscosity marks the onset of a gel-like state (Figure 3).

While the Gardner tube experiment illustrates the initial formation of a network as a function of viscosity, the associations apparently continue to form for some hours. Visual observation of the solution indicates that maximum opacity occurs after about 20 to 30 hours after heat removal: this increase in opacity is likely the result of the formation of progressively larger aggregates. Direct evaluation of the opacification of the solution was measured as a function of optical density. A quartz cuvette was filled with the 17.1% aqueous solution of PATE heated to 80 °C. Complete opacity (0% transmission of light) was achieved in 28 hours (Figure 4). The initial, rapid decrease in transmission corresponds to the gelation of the solution. However, a much longer time frame is required to form a large matrix capable of efficient light scattering.

In summary, the rheological studies of PATE are consistent with a proposed molecular association model for PATE solutions. Kinematic viscosity evaluation shows that at concentrations of 15% to 20% solids, a gelatinous solution results. The apparent viscosity measurements illustrate that network formation can be overcome by heating indicating that the association is electrostatic in nature.

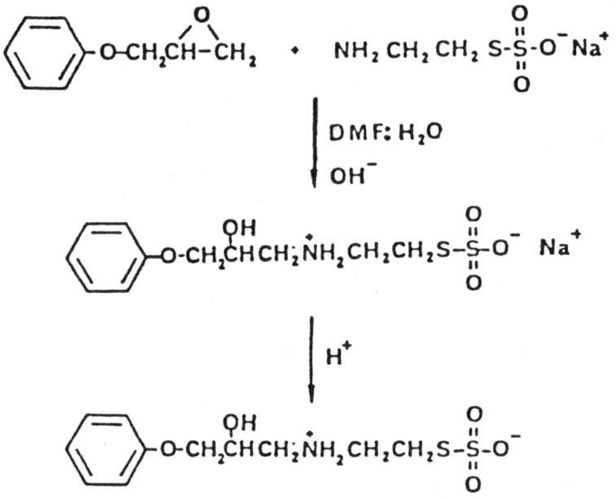

SCHEME II

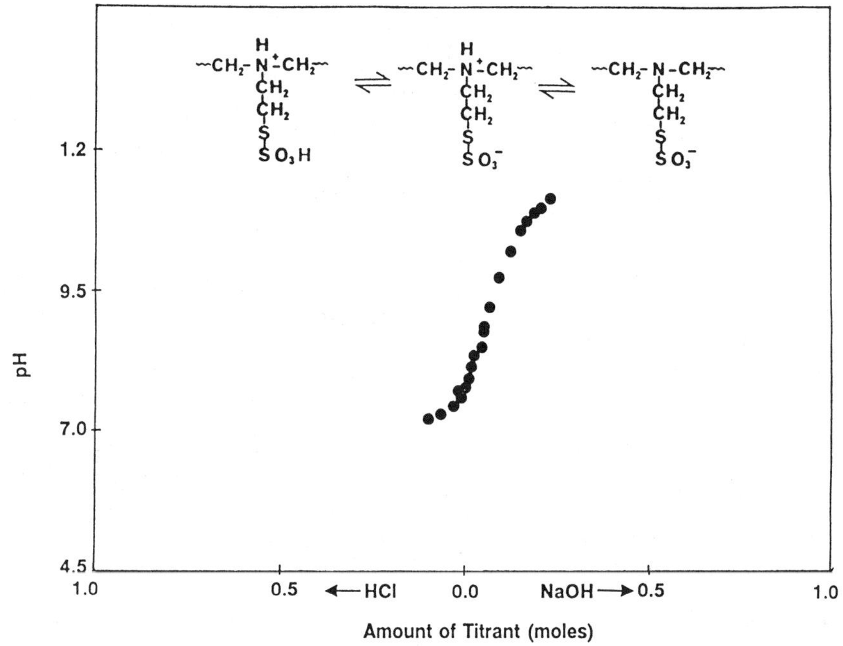

Figure 1. Titration curve of PATE illustrating zwitterion formation near isoprotic point.

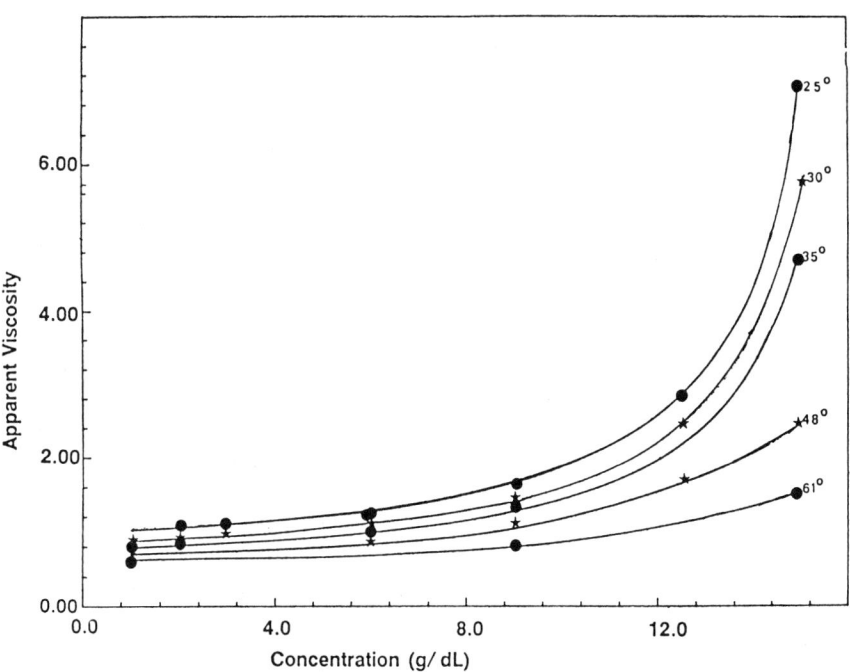

Figure 2. Concentration dependence of aqueous solutions of PATE on the apparent viscosity.

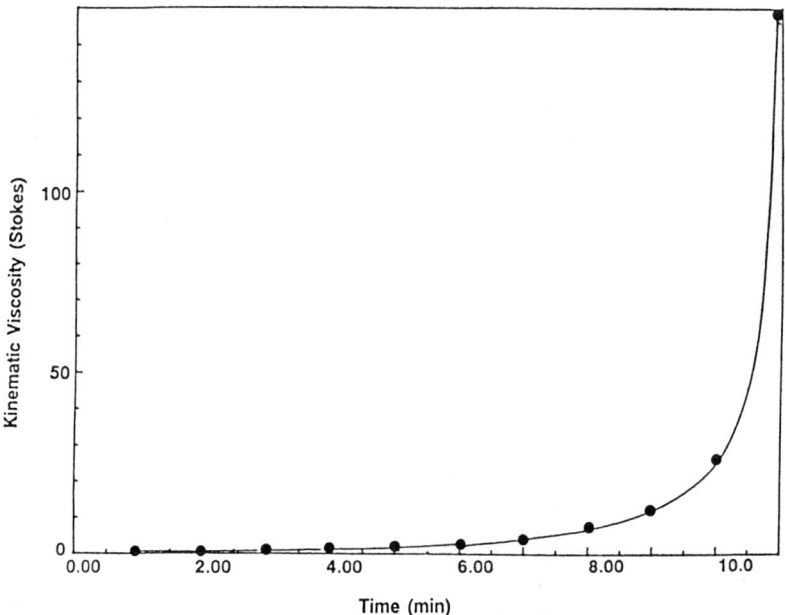

Figure 3. Dependency of kinematic viscosity on time since removal of heat (80 °C) from a 17.1% aqueous solution of PATE.

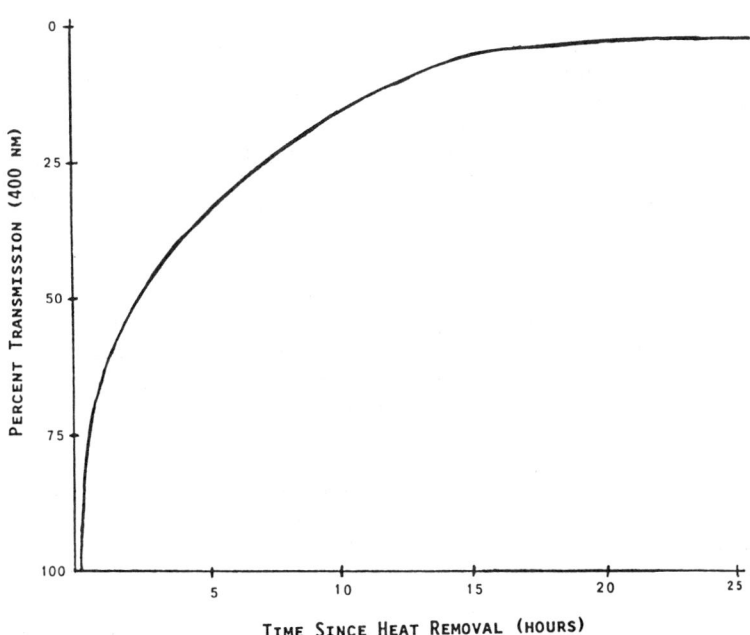

Figure 4. Increase of percent transmission upon time since removal of heat from a 17.1% aqueous PATE solution.

Thermal Polymerization of PATE Films. Films of 0.008 to 0.010 inch, wet film thickness were prepared from 15% to 20% aqueous PATE solutions. All resulting films were coalesced by oven drying for 4 minutes at 100 °C, or at room temperature with the addition of a small amount of a high boiling solvent such as DMF, methoxy ethanol or butoxy ethanol. After drying, the films could easily be washed from the glass or steel surface with water and redissolved. However, after thermolysis at 170 °C (for 10-15 minutes) the films became insoluble in water and organic solvents, confirming the presence of a crosslinking, water-insolubilizing reaction. The films were light tan colored, possessed excellent adhesion, were quite brittle, and harder than comparably heat cured amine/epoxy systems. Due to their brittle nature free PATE films could not be obtained intact by removing the films from glass or steel. However, soaking in water resulted in slightly hazy and softened films suggesting incomplete decomposition of the thiosulfate moiety. Infrared spectral analysis of thermally cured films confirm this conclusion, showing that the thiosulfuric acid stretching bands at 639, 1033 and 1170-1230 cm^{-1}, although significantly reduced, were still present to some extent after 20 minutes of heating (Figure 5). Unfortunately, the absorption characteristic of disulfides (S-S stretching at 500-400 cm^{-1}) is too weak to be of significant value; however, no other changes were noted in the IR spectrum which implies a crosslinking mechanism other than disulfide bond formation.

To gain additional insight into the necessary structural features for the crosslinking reaction, the PASE polymer, which does not contain the labile S-S bond, was subjected to the same curing process as the PATE polymer. Thus, if crosslinking occurs as a result of thiosulfate degradation, the PASE polymers would not crosslink, since the sulfonic acid group lacks the sulfur-sulfur bond. Indeed, after prolonged heating, no crosslinking was noted. Further, IR analysis of the films, before and after thermal curing, showed no spectral changes indicating that the labile S-S bond must be present for crosslinking to occur.

Any crosslinking mechanism involving S-S bond rupture should result in a detectable and quantifiable weight loss. Figure 6 shows the results of an experiment in which a PATE sample was slowly heated. Substantial weight loss occurs above 165 °C. Figure 7 depicts the quantitative weight loss of a sample heated rapidly to about 174 °C and held isothermally, simulating a thermal curing process. The recorded loss of 7.2% is less than the theoretical value of 11.3%. Nonetheless, these data are consistent with the proposal that the crosslinking reaction proceeds via a mechanism involving S-S bond cleavage and weight loss. Non-quantitative thermal degradation of thiosulfate with concomitant disulfide formation accommodates such a proposed mechanism: since the weight loss is consistently less than expected for complete reaction, some thiosulfate functionality must remain in the film thereby providing for the residual water sensitivity previously noted.

In summary, the presumption from thermal curing data is that heating above 165 °C results in sulfur-sulfur bond rupture and weight loss. Since the degradation is not quantitative, the residual thiosulfate functionality renders the "cured" films somewhat water sensitive, resulting in swelling and weakening of the polymer upon exposure to water.

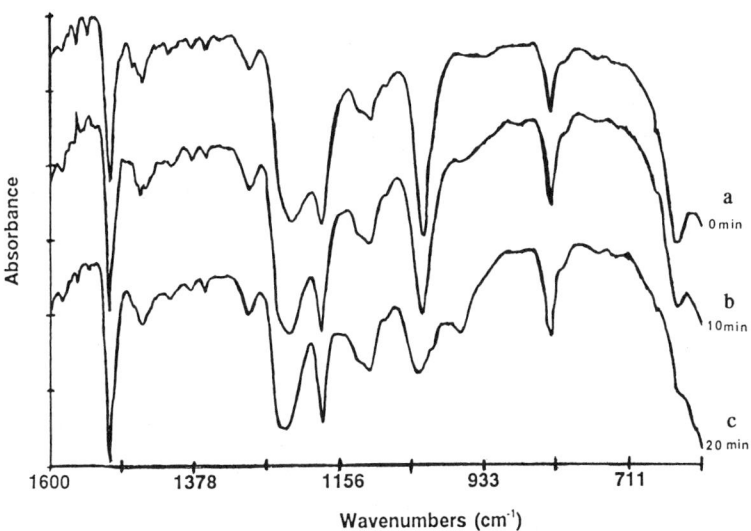

Figure 5. FT-IR of PATE film (a) before and after (b) 10 minutes and (c) 20 minutes thermolysis at 170 °C.

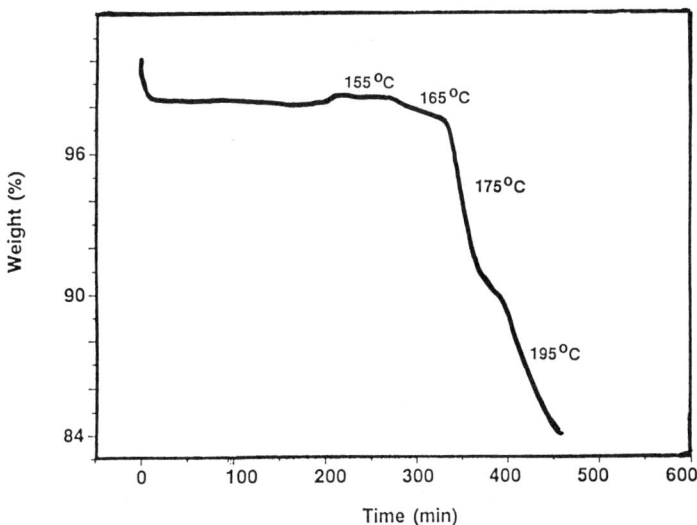

Figure 6. Weight loss of PATE film as a function of temperature.

Photolysis of PATE Films. Photolysis of 2-8 micron films of PATE on glass or steel under the full arc of a focused 450 Watt medium pressure mercury lamp for 10 minutes yields totally insoluble films in water and a variety of organic solvents. These include ethyl ether, MEK, MIBK, THF, carbon tetrachloride, cyclohexane, pyridine, methylene chloride, methoxy ethanol, benzene, xylenes and acetone. DMSO alone swelled the film. Upon soaking in warm water for 10 minutes, the films could be removed intact.

FTIR analysis of the photolyzed PATE film confirmed loss of the thiosulfate group (exemplified by reduction of the 639 cm^{-1} absorption--Figure 8). As in the thermal case, no other changes were noted in the FTIR spectrum. This provides further evidence that the thiosulfate moiety decomposes and crosslinks via disulfide formation upon exposure to UV light as shown in Scheme III. Storage of the photolyzed films in aqueous solution for a period of greater than 3 years resulted in no visible signs of opaqueness, degradation, or swelling. Figure 9 shows the loss of the 639 cm^{-1} thiosulfate absorbance band upon photolysis of a 5-10 micron film through a 254 nm line filter, using the 829 cm^{-1} aromatic C-H out-of-plane-bend absorption as a reference. The 639 cm^{-1} thiosulfate band is the only absorbance band characteristic of the thiosulfate group which is not obscured by some other group associated with the polymer backbone. Clearly, the rate of loss of thiosulfate moiety is greatest early in the photolysis, and slows later in the reaction, resulting in a total loss of 53% of the thiosulfate group. The decrease in the reaction rate is consistent with formation of a chromophore, i.e. a disulfide, which competes with the thiosulfate for UV radiation. Similar results have been noted upon photolysis at 280 nm.

In an attempt to sensitize the thiosulfate bond cleavage, benzophenone (10% by weight) was incorporated into the polymer film. Upon photolysis at 366 nm, the 639 cm^{-1} thiosulfate band was reduced (Figure 10) as in the case of direct photolysis at 254 nm and 280 nm. Since benzophenone is a known triplet sensitizer it is likely that the S-S bond cleavage in the thiosulfate group occurs from a triplet excited state in the sensitized reaction. Incidentally photolysis of a PATE film at 366 nm in the absence of benzophenone resulted in no loss of the 639 cm^{-1} IR peak. Unfortunately due to the film thickness, we were unable to obtain accurate quantum yields for either the direct or sensitized photolysis. Finally it should be noted that no chemical evidence has been presented to confirm disulfide formation. Results from the photolysis of a PATE-type model compound will be offered to substantiate the claim of disulfide formation as well as quantitate the primary photolysis step. But first, we consider photolysis of a PASE polymer film.

Photolysis of PASE Films. Given the overwhelming evidence that PATE resins crosslink by degradation of the S-S bond in the thiosulfate with concomitant disulfide formation, an additional study was undertaken to confirm this thesis using PASE, a polymer that is identical to PATE except for its lack of a sulfur-sulfur bond: this excludes degradation by the proposed sulfur-sulfur bond cleavage mechanism. After 120 minutes exposure to the full output of a medium pressure mercury lamp, PASE films remained water soluble, indicating that no crosslinking had occurred. In addition, no changes in the IR spectrum could be

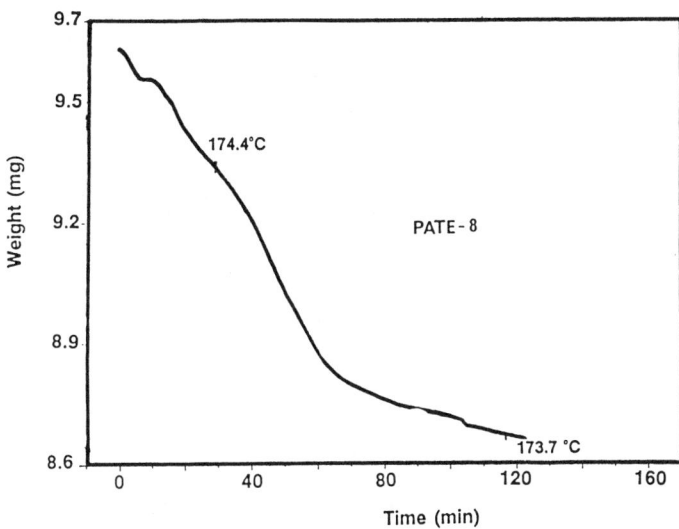

Figure 7. Isothermal (174 °C) weight loss of PATE film.

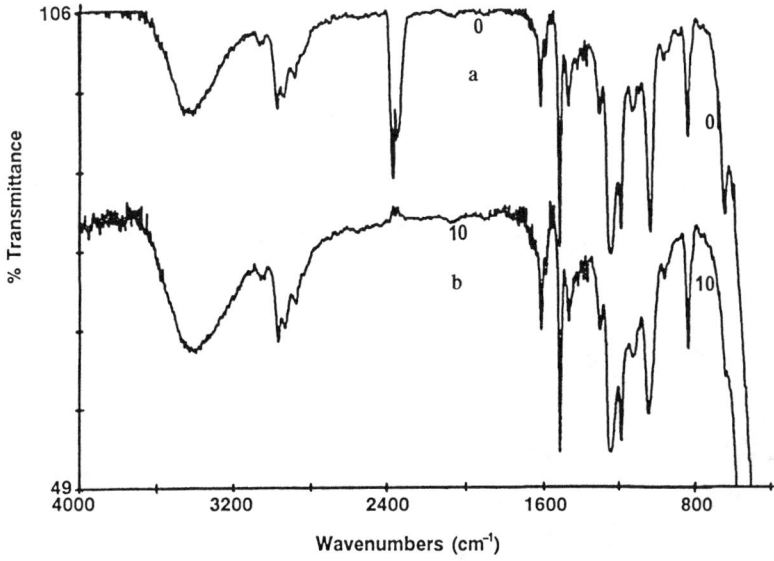

Figure 8. FT-IR of PATE film (a) before and (b) after 10 min photolysis with unfiltered 450 Watt medium-pressure mercury lamp.

294 POLYMERS IN MICROLITHOGRAPHY

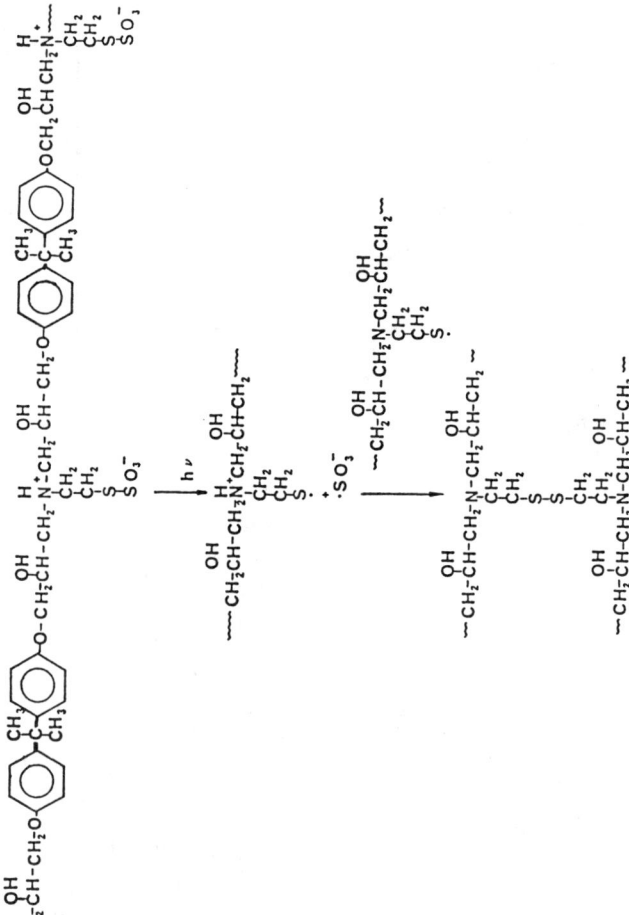

SCHEME III

Photolysis of Thiosulfates to Disulfides

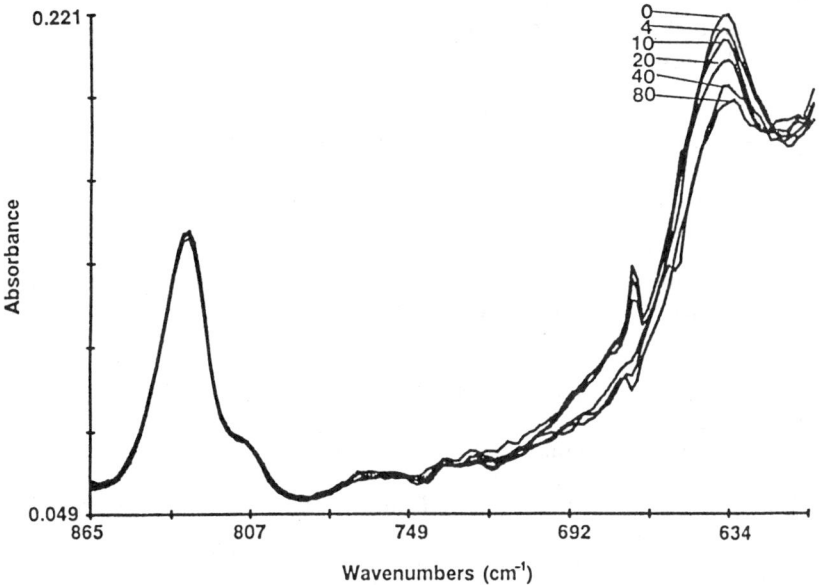

Figure 9. FT-IR of PATE film photolyzed for time periods of 0, 4, 10, 20, 40, and 80 min with isolated 254 nm line of 450 Watt medium-pressure mercury lamp.

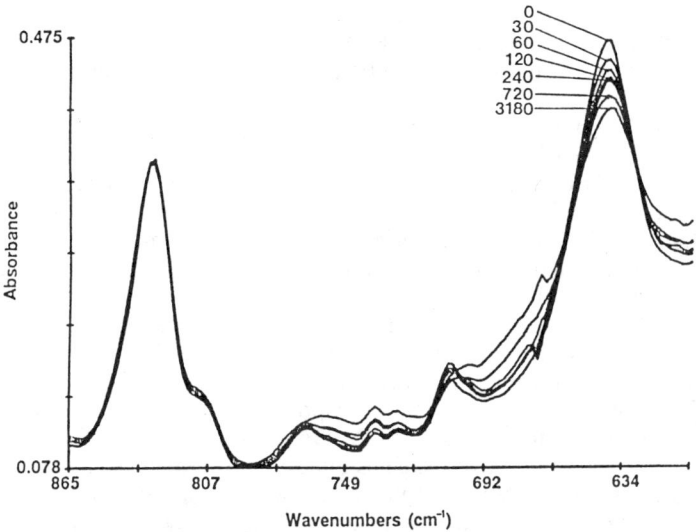

Figure 10. FT-IR of PATE film containing 10% benzophenone (BP) photolyzed for time periods of 0, 30, 60, 120, 240, 720, and 3180 min with isolated 366 nm line of 450 Watt medium-pressure mercury lamp.

detected, even after 120 minutes (Figure 11). These results are consistent with the thesis that the presence of a thiosulfate group is a requirement for crosslinking in the PATE polymer.

Photolysis of Hydroxy-3-(aminoethane thiosulfuric acid) propyl phenyl ether (AETSAPPE). AETSAPPE possesses identical functionality to the thiosulfate group in PATE, but being only monofunctional, AETSAPPE is incapable of forming insoluble polymer upon photolysis. As a result, the reaction products should remain soluble in typical organic solvents, making separation and identification of products possible.

Preparative photolysis of AETSAPPE (0.25 M aqueous solution) at 254 nm (Rayonet reactor) resulted in the formation of the disulfide product 2-amino(2-hydroxy-3-(phenyl ether) propyl) ether disulfide (AHPEPED) as the primary photoproduct. Photolysis of AETSAPPE at 254 nm (isolated line of medium pressure mercury lamp) resulted in rapid initial loss of starting material accompanied by formation (analyzed by HPLC) of AHPEPED (Figure 12a and 12b) (Scheme IV). Similar results were obtained for photolysis at 280 nm. Quantum yields for disappearance of AETSAPPE and formation of AHPEPED at 254 nm and 280 nm are given in Table I. The photolytic decomposition of AETSAPPE in water was also accomplished by sensitization (λ_{ex} =366 nm) with (4-benzoylbenzyl) trimethylammonium chloride (BTC), a water soluble benzophenone type triplet sensitizer. The quantum yield for the sensitized disappearance (Table I) is comparable to the results for direct photolysis (unfortunately, due to experimental complications we did not measure the quantum yield for AHPEPED formation). These results indicate that direct photolysis of AETSAPPE probably proceeds from a triplet state.

The data for the model compound study strongly supports the tenet of a photolytic bond cleavage of the sulfur-sulfur bond in PATE polymers (Scheme IV). Combined with the results for PASE, we feel quite confident in postulating the formation of a disulfide linkage upon photolysis of PATE films (Scheme III).

Preliminary Examination of PATE Imaging Characteristics. The data herein have shown that PATE resins are easily photolyzed in the deep UV region to form crosslinked films which are of sufficient integrity for photoresists. However, in addition to these film performance properties, a potential resist material must meet other equally important criteria. For example, the masked (unphotolyzed) portion of the resist film must be removed prior to etching, without damage to the cured film. Also, the cured films must withstand an etchant bath. Therefore, since PATE resins seem to meet the necessary requirements of solubility and solvent resistance, investigation of performance under crude simulated processing conditions was undertaken.

Films of PATE were cast on copper plated circuit boards. Typical solutions contain from 8% to 17% PATE solids and 0.02% Flourad FC 135 surfactant. No other additives, such as adhesion promoters, sensitizers, viscosity modifiers or stabilizers, were employed in the formulations. Films of 0.001 to 0.008 inches wet film thickness were cast on abrasively cleaned circuit boards and then oven dried for 5 minutes at 105 °C. Thus, in each case the

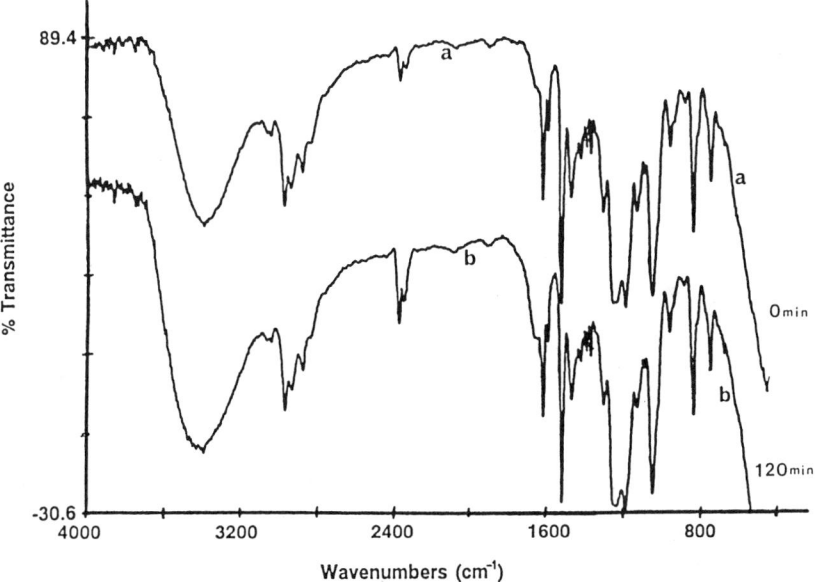

Figure 11. FT-IR of PASE film (a) before and (b) after 120 min photolysis with unfiltered 450 Watt medium-pressure mercury lamp.

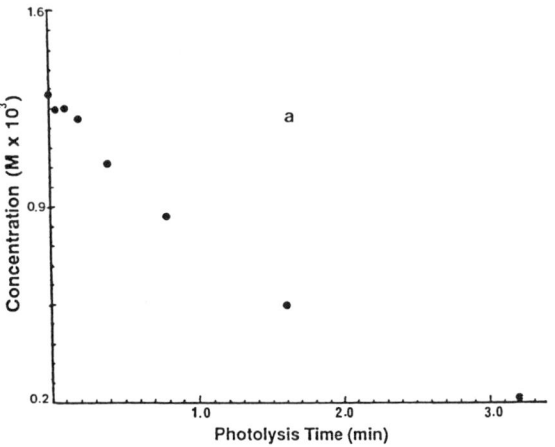

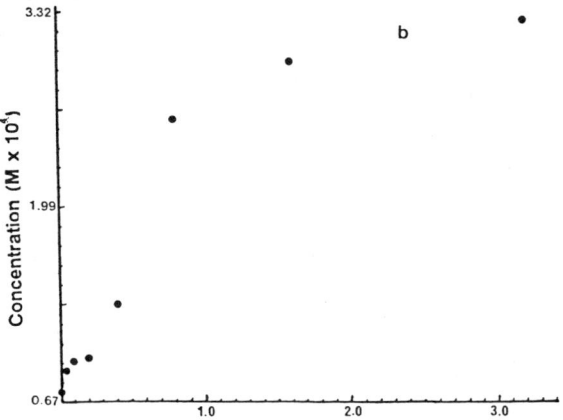

Figure 12. (a) Plot of concentration of AETSAPPE versus photolysis time with the 254-nm isolated line of a 450 Watt medium-pressure mercury lamp (b) Plot of concentration of AHPEPED versus photolysis time of an aqueous AETSPPE solution with the 254-nm isolated line of a 450 Watt medium-pressure mercury lamp.

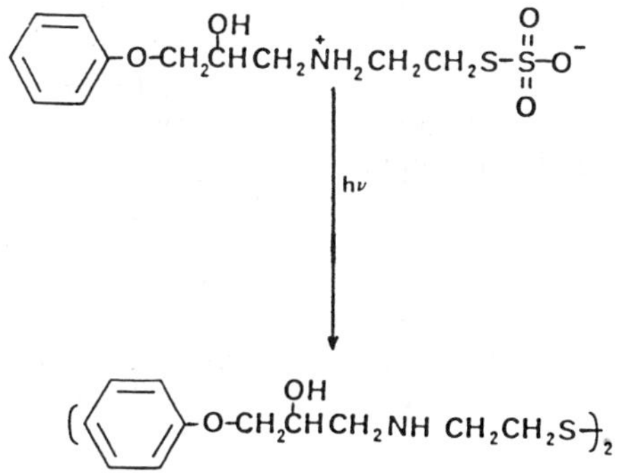

SCHEME IV

Table I. Quantum Yields for Loss of AETSAPPE (ϕ_{LOSS}) and formation of AHPEPED ($\phi_{AHPEPED}$)

λ_{ex}	ϕ_{LOSS}	$\phi_{AHPEPED}$	Sensitizer
254 nm	0.23	0.30	None
280 nm	0.22	0.15	None
360 nm	0.00	0.00	None
360 nm	0.23	---*	Yes (OD=0.84 at 366 nm)

*See text for explanation

approximate dry film thickness was less than 25 microns. However, it was found that thicker films (10 to 25 microns) were more difficult to develop, and gave poor line resolution, so efforts were confined to films less than 10 microns. Typically, 8 to 15 minutes of photolysis under the full arc of a 450 W medium pressure mercury lamp from a distance of 4 inches were required to crosslink the films. A thin metallic strip machined as a support for silicon microchips was clamped onto the surface of the circuit board for masking purposes. An initial attempt was made to etch the copper immediately after photolysis using a Kepro^{tm} BTE-202 bench-top etcher containing a strong $FeCl_3$ solution (3.03 M). Unfortunately, the etchant bath crosslinked the unphotolyzed resin, so a developing step in neutral pH water was required. Photocured PATE resins were thus developed for 10 to 30 seconds under warm tap water: the boards were immediately placed in the etchant for 2 minutes, then washed and dried. Figure 13 shows a section of one of the circuit boards produced by this technique.

In order to establish the ultimate resolution possible for the PATE polymer, aqueous solutions of PATE (a different batch from that used in the quantitative photolysis investigation) were spin cast to give thin films (less than 1 micron) on a silicon wafer, exposed to a 240-260 nm source (PE 600; Scanspeed 50,000; Aperture 3; UV-2, 240-260 nm) and subsequently developed by rinsing with a neutral pH water stream. Figures 14a and 14b show the resultant electron micrographs corresponding to mask line resolutions of 2.5 and 1.5 microns. At 1.5 micron resolution, a "snaking" or swelling of the pattern is noted. Attempts to generate higher resolution patterns resulted in an increased tendency to swelling, therefore defining the ultimate resolution of the PATE system tested at about 1.5-2.0 microns.

Figure 13. Image of mask produced by exposure of a PATE film followed by appropriate rinsing/etching procedure on copper board.

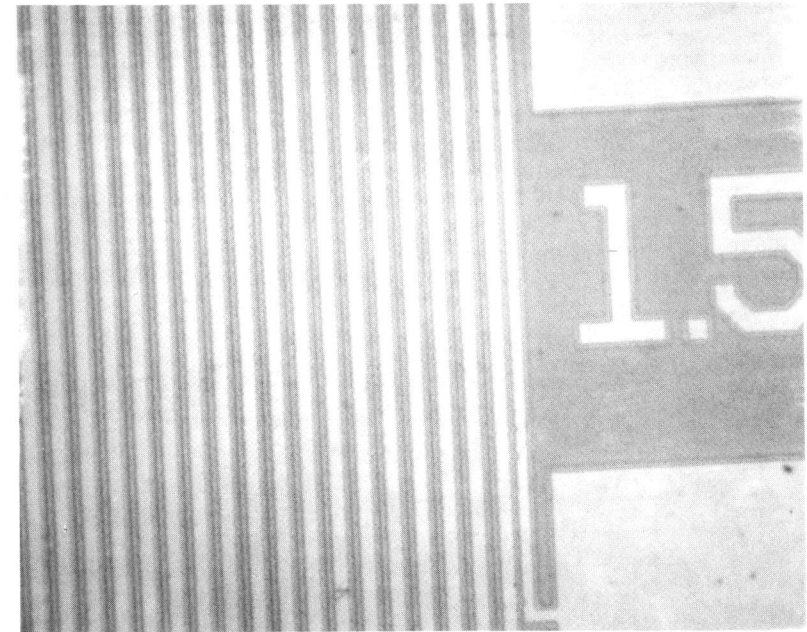

Figure 14. Image generated by exposure of spin watered PATE film on silicon wafer: (a) 2.5 micron line resolution (b) 1.5 micron line resolution.

Conclusions

This paper describes the successful synthesis and examination of poly[γ-(amino β-thiosulfate) ether] (PATE), a water soluble photolabile polymer. Evidence has been presented that the PATE polymer is zwitterionic and forms weak associations in aqueous solutions. Heat treatment of PATE films result in extensive crosslinking, presumably through a disulfide bond. This work presents strong evidence that PATE is activated by deep UV radiation, and that a disulfide crosslink is formed. Sensitization experiments demonstrate that the crosslinking reaction can be induced by a triplet sensitizer. Finally, preliminary results point out the potential for application of PATE films as active photoimaging systems.

Acknowledgments

We thank C. G. Willson for help with the imaging investigations which were performed at IBM. We also gratefully acknowledge the efforts of E. Bernardo and S. Buckley for their assistance in preparation of certain PATE samples.

Literature Cited

1. Beerman, C. German Patent 1 143 330, 1963.
2. Feldstein, R., Bunte Salt Polymers: Synthesis, Reactivity and Properties. Ph.D. Dissertation, The American University, 1971.
3. Harris, J. R., Investigation of Bunte Salts as Potential Polymeric Emulsifying and Crosslinking Agents, Ph.D. Dissertation, The University of Southern Mississippi, 1986.
4. Vandenberg, E., U. S. Patent 3,706,706, 1972.
5. Okawara, M. and Ochiai, Y., ACS Symposium Series 121, 41 (1980).
6. Stewart, M. and Dawson, J., U.K. Patent 2 050 438 A, 1979.
7. Hoyle, C. E.; Hutchens, D. E.; and Thames, S. F., Macromolecules (1989).

RECEIVED July 31, 1989

Chapter 18

Pyrimidine Derivatives as Lithographic Materials

Yoshiaki Inaki, Minoo Jalili Moghaddam, and Kiichi Takemoto

Department of Applied Fine Chemistry, Faculty of Engineering, Osaka University, Suita, Osaka 565, Japan

Pyrimidine derivatives were found to be applicable to both negative and positive type deep-UV photoresists. Intermolecular photodimerization of pyrimidine bases in the side chain of various polymeric and dimeric compounds upon irradiation of UV light (270 nm) led to the photocrosslinking of the polymer chains or photopolymerization of the dimeric compounds. This in turn allows the use of these materials as deep-UV negative type photoresists. Photolithographic evaluation of a typical polymer showed very high sensitivity with good resolution. On the other hand, the polymers containing a thymine photodimer in the main chain underwent dissociation of the thymine photodimers upon irradiation to UV light (250 nm), leading to breakage of the polymer chains. These polymers could be used as positive type photoresists and high resolution (0.3 μm) was demonstrated.

It is well known that pyrimidine bases convert to photodimers upon irradiation to UV light near the λ max(> 270 nm). This photochemical reaction has a lethal effect in biological systems due to the photochemical transformation of pyrimidine bases of nucleic acids. However the photodimerization is a reversible reaction and the photodimers split to afford the original monomers very efficiently upon irradiation at a shorter wavelengths as shown in Scheme 1 (1).

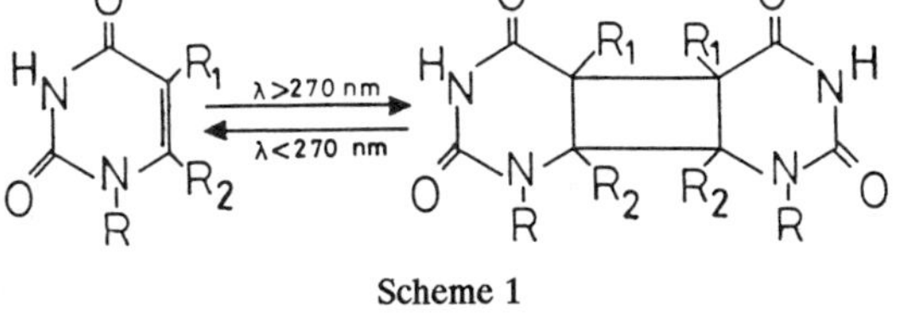

Scheme 1

This knowledge regarding the efficient photochemical dimerization of the pyrimidine bases, led us to study, in detail, both the intramolecular

0097–6156/89/0412–0303$06.00/0

photodimerization of pyrimidine bases grafted onto the side chain of a polymer and dimeric model compounds (2-13), as well as the intermolecular photodimerization. The latter reaction leads to photocrosslinking of the oligomers and polymers thus allowing them to be applied as negative type photoresists.

On the other hand, due to the efficient photoreversal reaction of the pyrimidine photodimers at shorter wavelength, the design of a new type of polymers containing thymine photodimers in the main chain seemed to be of interest. Photolysis of these polymers causes cleavage of the photodimers in the polymer chain and reduces the molecular weight of the polymer. This decrease in molecular weight would be expected to increase the solubility of the polymers in the irradiated regions allowing their use as positive photoresists. These polymers might be obtained by photopolymerization of the corresponding dimeric models containing pyrimidine bases at the ends of their molecules.

The object of this study is to develop new photoresists for deep-UV lithography, by using the reversible photoreaction of pyrimidine bases (17-19). Applicability of pyrimidine containing polymers to both negative and positive type photoresists is due to this photoreversible reaction in which cyclobutane dimers are either formed or cleaved depending on the exposure wavelength (Scheme 2).

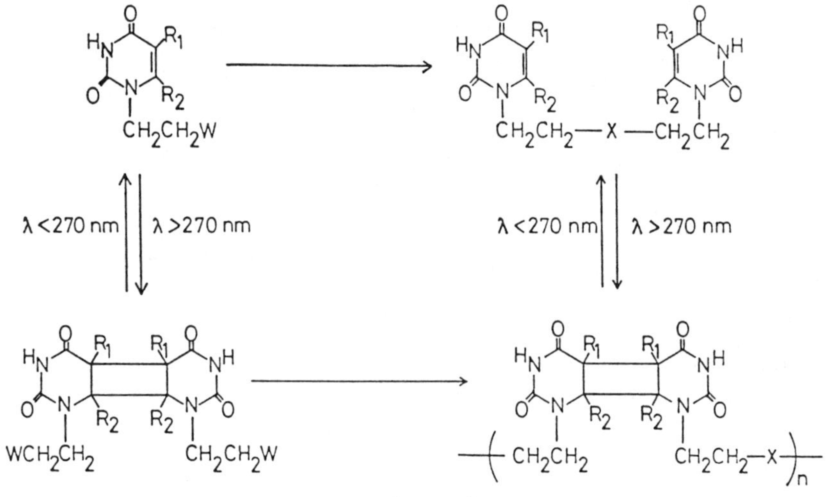

Scheme 2

EXPERIMENTAL

Material

Negative Type photoresists:
Polymethacrylates and copolymers of butadiene and methacrylate having various pyrimidine derivatives (Figures 1, and 2) were prepared by free radical polymerization of the methacrylate monomers (14-16). In the case of the poly(MAOT[1]-alt-MAOT[3]Me[1]), the polymer was obtained by the reaction of the polymethacrylic anhydride with the hydroxyethyl

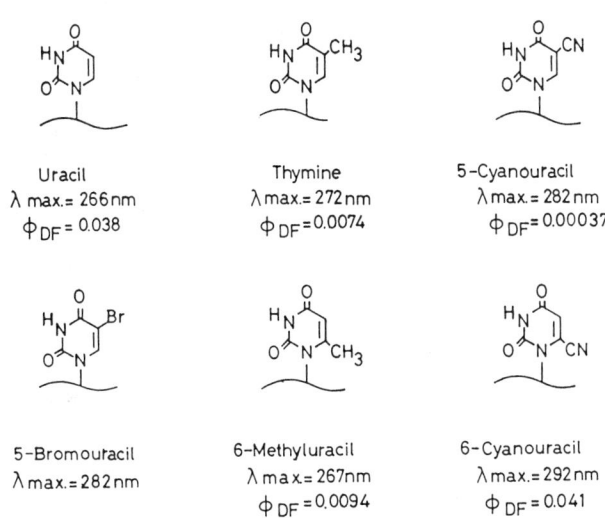

Figure 1. Polymethacrylates containing pyrimidine bases.

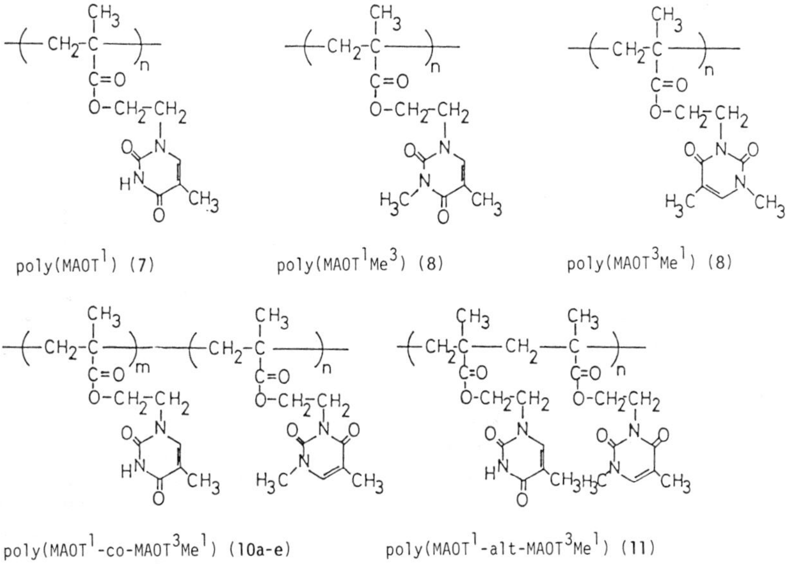

poly(MAOT^1) (7)   poly(MAOT^{1}Me^3) (8)   poly(MAOT^{3}Me^1) (8)

poly(MAOT^1-co-MAOT^{3}Me^1) (10a-e)   poly(MAOT^1-alt-MAOT^{3}Me^1) (11)

Figure 2. Polymethacrylates containing thymine derivatives.

derivatives of thymine followed by the reaction of another cyclic type of thymine derivatives (18).

Bis-pyrimidine derivatives were prepared by the reaction of the hydroxyethyl derivatives of pyrimidine with various bifunctional compounds such as dichloro-siloxane derivatives (Figure 3) (20).

Positive Type Photoresists:
Polyamides containing thymine photodimer units in the main chain (17a,b) were prepared by polycondensation of thymine photodimer derivatives (15a,b), which were obtained by the photochemical reaction of the monomeric compound, and various diamines by the activated ester method (Figures 4 and 5) (17, 19).

Instrumentation

The photosensitivity spectra were recorded on a Nihon spectrophotograph, CT-40 and a 500 W Xe-short arc lamp was chosen as the deep-UV source. A Canon mask aligner (PLA-521F) with cold mirror (CM250, in the case of resolution evaluation of the positive photoresist, and CM290 in the case of negative type photoresist) was used as the UV irradiation instrument in which a Xe-Hg lamp was chosen as the light source, and irradiation of the silicon wafers were carried out with contact printing through a mask. Ultraviolet spectra were measured with a Nihon-Bunko (UVIDEC-660) spectrophotometer. Glass transition temperature(Tg) of the polymers were measured with a Seiko differential scanning calorimeter(DSC-20). The molecular weight distribution of the polymers were determined by GPC method using Toyo Soda HLC-CP8000 with a thermostated column TSK gel G4000HT, and a UV detector operating at 270 nm with dimethylformamide as the eluent.

RESULTS AND DISCUSSION

Photodimerization of the Pyrimidines

Photochemical reactions of the pyrimidine polymers in solution were studied to determine the quantum yields of the intramolecular photodimerization of the pyrimidine units along the polymer chains. Photoreactions of the polymers were carried out in very dilute solutions to avoid an intermolecular(interchain) photodimerization. Quantum yields determined at 280 nm for the polymers (1-6 in Figure 1) are listed in Table I. The quantum yield of the 5-bromouracil polymer [poly(MAOU-5Br)] could not be determined because of side reactions of the base during the irradiation.

Among the examples shown in Table I, the 5-cyanouracil derivative [poly(MAOU-5CN)] had the lowest photochemical reactivity, while the 6-cyanouracil analog [poly(MAOU-6CN)] had the highest reactivity. The highest photodimerization reactivity of the 6-cyanouracil derivative might be caused by the capto-dative substituent effect; the electron donating group is the N-1 group, and the electron acceptor group is the cyano group (12).

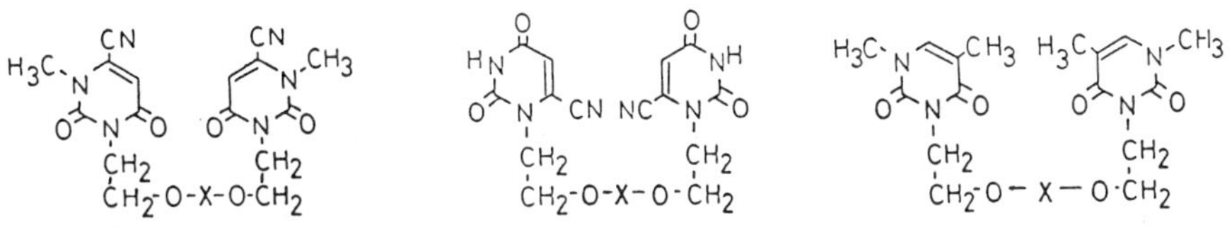

(18) (19) (20)

-X-	
a :	$-Si(CH_3)_2-O-Si(CH_3)_2-O-Si(CH_3)_2-$
b :	$-Si(CH_3)_2-O-Si(CH_3)_2-$
c :	$-SiPh_2-O-SiPh_2-$

Figure 3. Bis-pyrimidine derivatives.

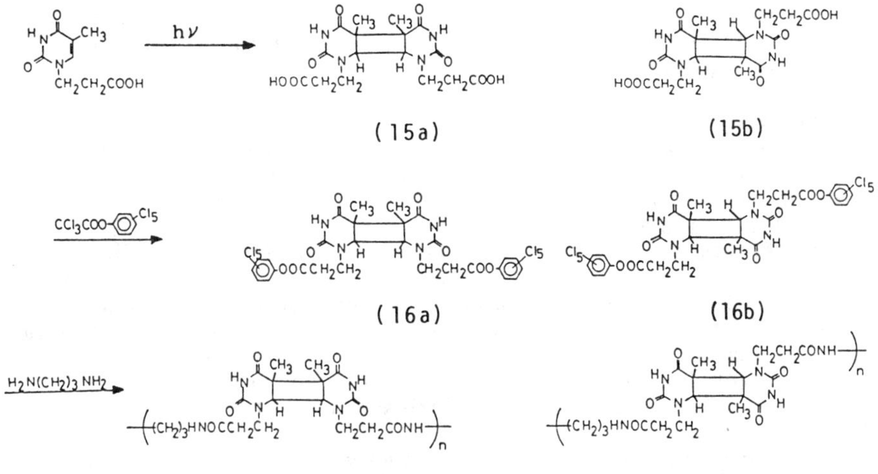

Figure 4. Isomers of polyamides containing thymine photodimer.

Table I. Quantum Yields in Solution and Lithographic Evaluation of Polymethacrylates with Pendant Pyrimidine bases

Pyrimidines	Polymers	Φ_d[a]	λ[b] (nm)	λ max[c] (nm)	E_0[d] (mJ/cm^2)
Uracil	Poly(MAOU)	0.035	--	--	--
5-Methyluracil	Poly(MAOT^1)	0.0095	230-300	280	34.6
6-Methyluracil	Poly(MAOU-6Me)	0.0081	245-305	280	95.0
5-Cyanouracil	Poly(MAOU-5CN)	0.00037	255-320	290	76.9
6-Cyanouracil	Poly(MAOU-6CN)	0.041	255-355	305	19.5
5-Bromouracil	Poly(MAOU-5Br)	--	245-325	295	19.8

a) Quantum yields of photodimerization in dimethyl sulfoxide solution.
b) Photosensitive wavelength range.
c) Maximal photosensitive wavelength.
d) Minimum required energy for photocrosslinking at the maximal wavelength.

The quantum yields for the polymeric thymine derivatives, 7-11 in Figure 2, determined at 280 nm in dimethyl sulfoxide solution are plotted against the molar composition of the polymers as shown in Figure 6a. Poly(MAOT^1-co-MAOT^{3}Me^1)s have lower quantum yields for intramolecular photodimerization than poly(MAOT^1) and poly(MAOT^{3}Me^1). Among the copolymers, poly(MAOT^1-alt- MAOT^{3}Me^1), in which the T^1 and T^{3}Me^1 base units are strictly alternating, has a slightly lower quantum yield than random copolymer, poly(MAOT^1-co-MAOT^{3}Me^1) of the the same composition. These provided that the photodimerization reaction of the adjacent T^1 and T^{3}Me^1 bases imparts more strain to the polymer chain than the equivalent reaction of T^1 units or T^{3}Me^1 units.

Photopolymerization:
Photopolymerizations of the bis-pyrimidine derivatives were carried out in solid film by irradiation with UV light from a spectroirradiator. The rate of the photopolymerization was found to depend on the wavelength of light. The highest reactivity was observed at ~ 300 nm (Figure 7a), while the highest reactivity for photoreversal of the photodimer was obtained by irradiation at ~ 240 nm (Figure 7b). The polymers obtained the photopolymerization were not soluble in ethyl acetate. Therefore, sensitivity curves for insolubilization of the bis-pyrimidine derivatives via photopolymerization were obtained by development with ethyl acetate/toluene (Figure 8). The photopolymerization using an eximer laser (XeCl: 308 nm) source gave essentially the same result.

Photoreversal of the Pyrimidine Photodimers

Photodissociation of the polyamides (Figures 4 and 5) was carried out in the solid films. Polymer films (1-2 μm) were cast onto quartz substrates and were exposed to monochromatic 250 nm light from a spectroirradiator. The dissociation of the thymine photodimers was followed by monitoring the absorbance at 270 nm. The results obtained for the polyamide prepared from the reaction of propane diamine and the isomers of the thymine photodimer (cis-syn(17a), cis-anti(17b), and

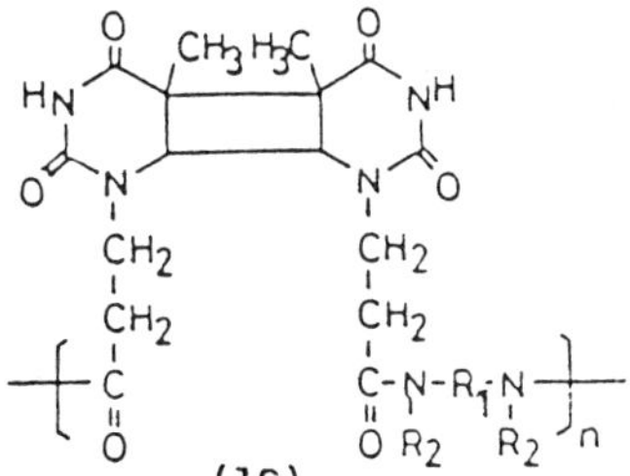

(18)

	R_1	R_2	T_g (°C)	T_m (°C)
a:	$-N\langle\rangle-(CH_2)_3-\langle\rangle N-$		26.7	212-215
b:	$-(CH_2)_3-N\langle\rangle N-(CH_2)_3-$	H	—	175-178
c:	$-C_6H_4-CH_2-C_6H_4-$	H	72.0	258-261
d:	$-C_6H_4-O-C_6H_4-$	H	52.5	243-245
e:	$-C_6H_4-O-Si(Me)_2-O-C_6H_4-$	H	—	196-198
f:	$-(CH_2)_3-Si(Me)_2-O-Si(Me)_2-(CH_2)_3-$	H	—	150-153

Figure 5. Polyamides containing thymine photodimer.

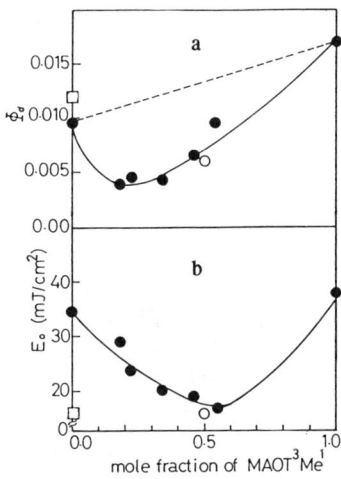

Figure 6. (a) Quantum yields for intramolecular reaction and (b) photosensitivities vs MAOT³Me¹ units in polymers for homopolymers and copolymers in Fig. 2. ●: Poly(MAOT¹), Poly(MAOT³Me¹), and Poly(MAOT¹-co-MAOT³Me¹). ○: Poly(MAOT¹-alt-MAOT³Me¹). □: Poly(MAOT¹Me³)

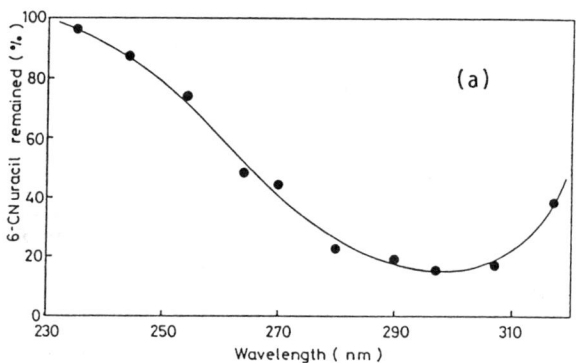

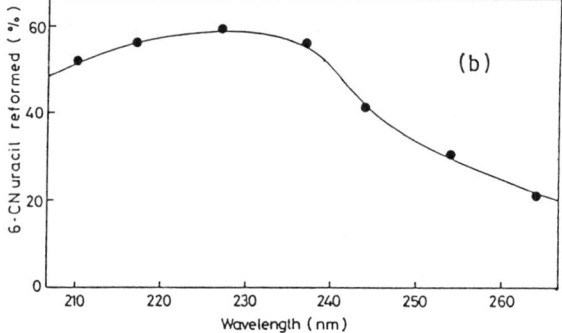

Figure 7. (a) Wavelength dependency of photopolymerization of the 6-cyanouracil derivatives. (b) Wavelength dependency of photoreversal of the photodimer.

a mixture of both isomers(17a+b)) are shown in Figure 9. From this figure, it is perceived that, regardless of the structure of the thymine photodimers, polymer chain scission occurres at the same rate. Furthermore, to investigate the effects of diamine structure in the polymer chains on the dissociation rate of the polymers, the photoreactions of different polyamides (18a-f) were compared in Figure 10. From this figure it is seen the polymers with a more flexible framework (lower Tm and Tg), 18b, 18e, and 18f, dissociated faster than polymers 18c, and 18d that possess a more rigid framework (higher Tm and Tg). However, the maximum dissociation conversion is the same for all the polymers (the same G values). This result might be related to the fact that the photoreactions were carried out at room temperature, and therefore, the polyamides with a Tg lower than room temperature had more conformational freedom of the polymer chains, and hence the photodissociation reactivity would be higher than for those polymers with high Tg.

The molecular weights of the polymers before and after irradiation were followed by GPC to determine changes in the molecular weights of the polyamides. It was found that photolysis of the polymer resulted in polymer chain scission, leading to the appearance of oligomers containing thymine bases at the end of the molecules.

Lithographic Sensitivity

Negative type photoresist:
The wavelength range for which each polymer undergoes photodimerization, the wavelength at which each polymer displays maximal sensitivity, and the sensitivity (E; the minimum incident input energy per unit area required to produce an insolubilized film of the same thickness as the the initial film) are listed in Table I. Polymer sensitivities were determined for the wavelength where each polymer displays maximum sensitivity. All formulations employed the same spinning solvent and the same developer.

Poly(MAOU-6CN) which showed the highest quantum yield for intramolecular dimerization in solution, showed the highest photosensitivity value. The photoresist sensitivities of poly(MAOU-6Me) and poly(MAOU-5CN), which have low reactivity for photodimerization in solution, are low. A very interesting result is the high sensitivity of poly(MAOU-5Br) which has similar E value as poly(MAOU-6CN). In this case there is evidence for debromination in addition to the photodimerization in solution; the UV maximum at 282 nm shift to 266 nm, which is assigned to uracil base, accompanied by decrease of absorbance. This debromination should form a free radical, and afford enhanced sensitivity by crosslinking.

From the photosensitivity data (Table II) of the thymine substituted polymethacrylates (Figure 2), the minimum energy required to effect crosslinking at the wavelength where the polymers display maximum sensitivity is plotted also against the composition of thymine units in Figure 6b. Comparison of Figure 6a and Figure 6b shows that increasing the concentration of T^3Me^1 thymine units in the copolymer, which reduces the favorable stacking conformation between the adjacent thymine units, effected more compatibility for intermolecular photodimerization in the film state. Therefore, polymers with lower

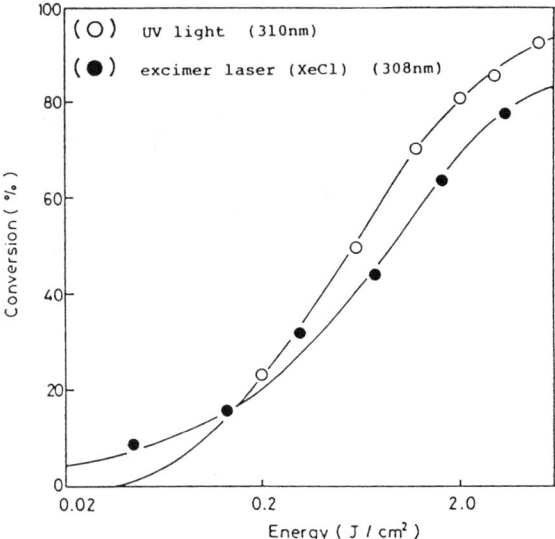

Figure 8. Photopolymerization of the 6-cyanouracil derivative.

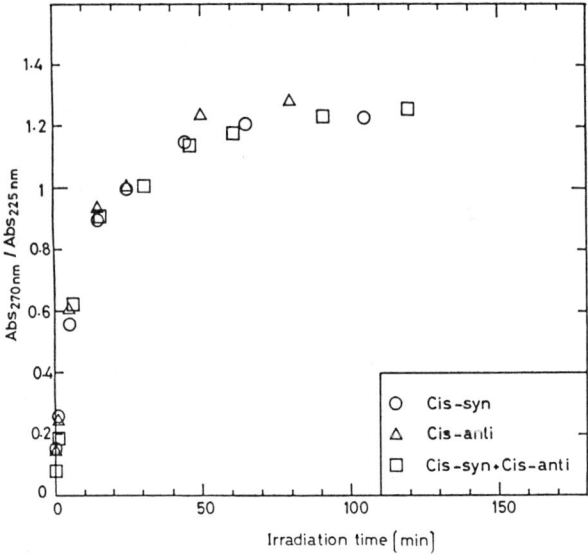

Figure 9. Photoreversal of the polymers containing isomers of thymine photodimer.

quantum yields of intramolecular photodimerization in solution showed higher photosensitivity values. This was confirmed by comparing the quantum yields of the polymers in the film state tabulated for some of the polymers in Table III, and the photosensitivity values in Table II. These results show that suppressing intramolecular, and intensifying intermolecular photodimerization, polymers of higher photosensitivity values can be obtained.

Table II. Lithographic Evaluation of Polymethacrylates with Pendant Thymine bases

Polymer	[MAOT^{3}Me^1][a] (unit %)	λ [b] (nm)	λ max[c] (nm)	E_0 [d] (mJ/cm^2)
Poly(MAOT^{3}Me^1)	100	230-305	280	38.0
Poly(MAOT^1)	0	230-300	280	34.6
Poly(MAOT^{1}Me^3)	0	230-305	280	16.7
Poly(MAOT^1-co-MAOT^{3}Me^1)-1	18	250-305	280	28.0
Poly(MAOT^1-co-MAOT^{3}Me^1)-2	22	230-305	280	23.4
Poly(MAOT^1-co-MAOT^{3}Me^1)-3	34	230-310	280	20.3
Poly(MAOT^1-co-MAOT^{3}Me^1)-4	46	230-310	280	18.8
Poly(MAOT^1-co-MAOT^{3}Me^1)-5	56	230-310	280	16.7
Poly(MAOT^1-alt-MAOT^{3}Me^1)	50	200-315	270	15.8

a) MAOT^{3}Me^1 units in copolymer.
b) Photosensitive wavelength range.
c) Maximal photosensitive wavelength.
d) Minimum required energy for photocrosslinking at the maximal wavelength.

Table III. Quantum Yields and Maximum Photodimerization Conversion of Thymine Bases in Polymethacrylates with Pendant Thymine Bases in the Film State

Polymer	Φ d Solution[a]	Film	Conv.max (%) Film	E_0 [b] (mJ/cm^2)
Poly(MAOT^1)	0.0095	0.086	40	34.6
Poly(MAOT^{3}Me^1)	0.017	0.080	37	38.0
Poly(MAOT^1-co-MAOT^{3}Me^1)-3[c]	0.0044	0.159	42	20.3

a) Quantum yield for intramolecular dimerization in solution.
b) Maximum required energy for photocrosslinking (Table II).
c) MAOT^{3}Me^1 unit in copolymer: 34 unit%.

Positive type photoresist:
The photosensitivity tests were carried out only for the polyamides containing different isomers of thymine photodimers to determine the effect of isomer structure on photosensitivity.

The polymers 17b and 17a+b were dissolved in dimethylformamide by 10%(w/w) and were spin coated onto a silicon wafer. The thickness of the film was found to be 0.3 μ m. The minimum required energy for complete removal of the photoresist layer after development, versus the wavelength of UV irradiation is shown in Figure 11. The spectral sensitivities of the polymers prepared with different isomers were approximately the same at each wavelength throughout the deep-UV region. Moreover, the highest sensitivity was observed at 250 nm.

Resolution Evaluation

Negative type:
Polymethacrylates containing 6-cyanouracil or 5-bromouracil units in the side chain of the polymer had the highest photosensitivity. However, since these polymers were tacky in nature, a copolymer of butadiene and the methacrylate monomer with pendant 6-cyanouracil was synthesized. This polymer showed resolution down to 1 μm (Figure 12).

Positive type:
A typical pattern obtained with polymer 17b is shown in Figure 13. An exposure time of approximately 0.9 min was determined to be the best irradiation time after development of the images in a solution of methanol and isopropanol (v/v 8/2). The light energy corresponding to this exposure time was 345.6 mJ/cm^2. The width of the bars are in the range of 0.5 - 0.9 μm. High resolution was observed down to 0.5 μm, which was the resolution limit of the line and space patterns in the mask. The thickness of the film after development and rinsing was 0.28 μm, corresponding to 93% thickness remaining. The resolution evaluation for other polymers was carried out under the same conditions except that the smallest line and space features on the mask were 0.1 μm. An imaged pattern for the polymer 18c is shown in Figure 14. Here it was determined that polyamides containing thymine photodimers in the main chain are inherently capable of very high resolution down to 0.3 μm.

Conclusion
Polymethacrylates containing 6-cyanouracil or 5-bromouracil units in the side chain of the polymer displayed the highest photosensitivity. Copolymers of butadiene with the methacrylate monomer with pendant 6-cyanouracil are capable of resolving 1 μm features and behaved as negative photoresists.

Polyamides containing thymine photodimer units in the main chains showed excellent resolution values (0.3 μm) and behaved as positive photoresists. It is concluded that polymers containing pyrimidine bases displayed high resolution and high sensitivity when used in both negative and positive photoresists formulations.

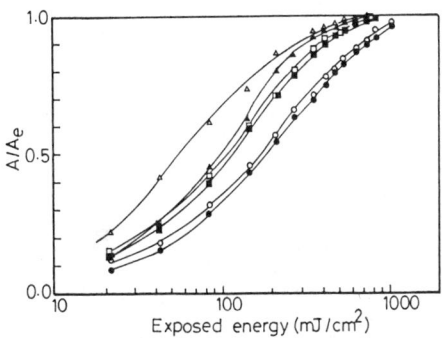

Figure 10. Photoreversal of the polyamides containing thymine photodimer. ■: 18a, □: 18b, ●: 18c, ○: 18d, ▲: 18e, △: 18f

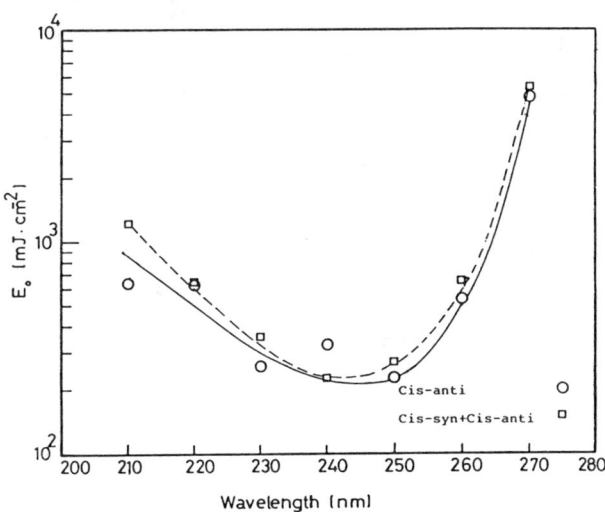

Figure 11. Photosensitivity spectra of polyamides containing thymine photodimer.

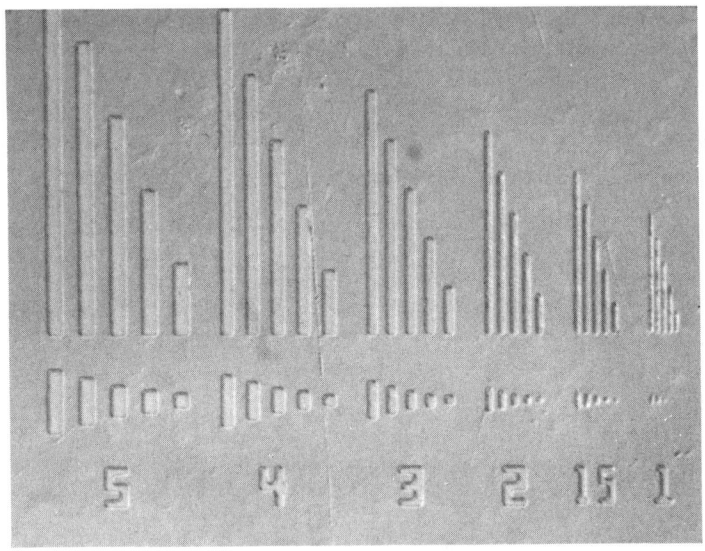

Figure 12. Resolution pattern of polymethacrylate containing 6-cyanouracil.

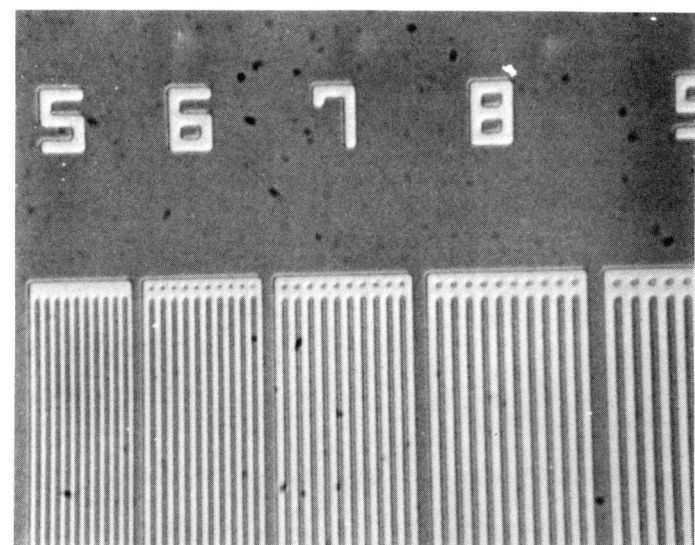

Figure 13. Resolution pattern of polyamide 17a.

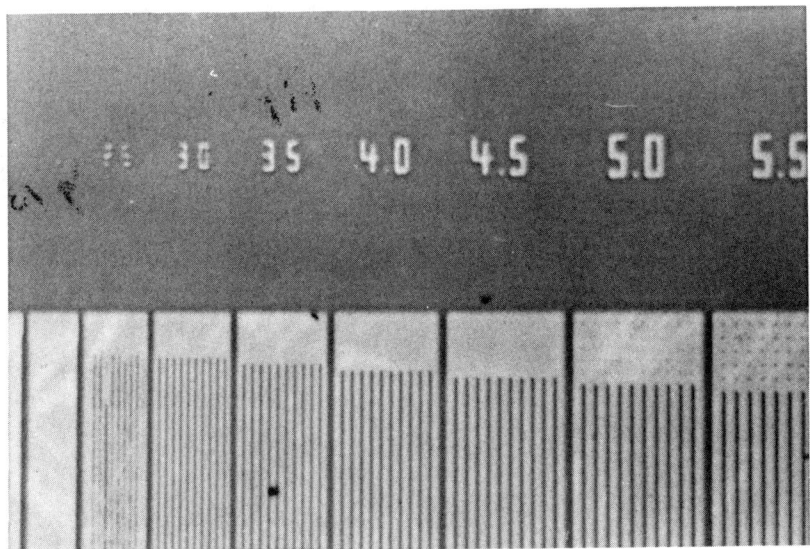

Figure 14. Resolution pattern of polyamide 18c.

Acknowledgment

The authors would like to express grateful acknowledgment to Japan Synthetic Rubber Co., for the photosensitivity evaluation of the pyrimidine polymers.

References

1. Wang, S. Y. Ed.; *Photochemistry and Photobiology of Nucleic Acids*; Academic: New York, 1976; Vol. I and II.
2. Kita, Y.; Uno, Y.; Inaki, Y.; Takemoto, K. *J. Polym. Sci. Polym. Chem. Ed.* 1980, 18, 427.
3. Kita, Y.; Uno, Y.; Inaki, Y.; Takemoto, K. *ibid.*, 1981, 19, 477.
4. Kita, Y.; Uno, Y.; Inaki, Y.; Takemoto, K. *ibid.*, 1981, 19, 1733.
5. Kita, Y.; Uno, Y.; Inaki, Y.; Takemoto, K. *ibid.*, 1981, 19, 2347.
6. Inaki, Y.; Suda, Y.; Kita, Y.; Takemoto, K. *ibid.*, 1981, 19, 2519.
7. Kita, Y.; Uno, Y.; Inaki, Y.; Takemoto, K. *ibid.*, 1981, 19, 3315.
8. Suda, Y.; Inaki, Y.; Takemoto, K. *ibid.*, 1983, 21, 2813.
9. Suda, Y.; Inaki, Y.; Takemoto, K. *ibid.*, 1983, 22, 623.
10. Suda, Y.; Kono, M.; Inaki, Y.; Takemoto, K. *ibid.*, 1984, 22, 2427.
11. Suda, Y.; Inaki, Y.; Takemoto, K. *Polymer J*, 1984, 16, 303.
12. Inaki, Y.; Fukunaga, S.; Suda Y.; Takemoto, K. *J. Polym. Sci. Polym. Chem. Ed.*, 1985, 23, 119.
13. Inaki, Y.; Takemoto, K. in *Current Topics in Polymer Science*; Ottenbrite, R. M.; Utracki, L. A.; Inoue, S. Eds.; Hanser Publisher: Munich, 1987; Vol. 1, p 79.
14. Takemoto, K.; Inaki, K. in *Advances in Polymer Science*; Springer-Verlag: Berlin Heidelberg, 1981; Vol. 41, p 1.
15. Takemoto, K.; Inaki, Y. in *Functional Monomers and Polymers*; Takemoto, K.; Inaki, Y.; Ottenbrite, R. M. Eds; Marcel Dekker: New York, 1987; p 149.
16. Kita, Y.; Futagawa, H.; Inaki, Y.; Takemoto, K. *Polymer Bull.*, 1980, 2, 195.
17. Moghaddam, M. J.; Hozumi, S.; Inaki, Y.; Takemoto, K. *J. Polym. Sci. Polym. Chem. Ed.*, 26, 1988, 27, 3297.
18. Moghaddam, M. J.; Hozumi, S.; Inaki, Y.; Takemoto, K. *Polym. J.*, in press.
19. Inaki, Y.; Moghaddam, M. J.; Kanbara, K.; Takemoto, K. *J. Photopolym. Sci. Technol.*, 1988, 1 , 28.
20. Moghaddam, M. J.; Dr. of Engineering Thesis, Osaka University, Osaka, 1989.

RECEIVED June 14, 1989

Chapter 19

Synthesis of New Metal-Free Diazonium Salts and Their Applications to Microlithography

Shou-ichi Uchino, Michiaki Hashimoto[1], and Takao Iwayanagi

Central Research Laboratory, Hitachi Ltd., Kokubunji, Tokyo 185, Japan

> New metal-free diazonium salts have been synthesized and applied to microlithography as photobleachable dyes for a contrast enhancing layer (CEL), negative working photoactive compounds, and photoacid generators. 4-N,N-dimethylaminobenzenediazonium trifluoromethanesulfonate (D1) is the most suitable photobleachable dye. Upon i-line (365nm) exposure, 0.4-μm line-and-space patterns were obtained using a positive resist in conjunction with the D1-CEL. D1 also shows good properties as a negative working sensitizer for a two-layer resist system formed by means of a "doping" process, in which the diazonium salt is distributed in the top and bottom layers. 4-Methoxybenzenediazonium trifluoromethanesulfonate can be used as a photoacid generator for acid catalyzed cross-linking. Submicron resolution can be achieved with high sensitivity.

In recent years, demands on microlithographic techniques have become much more severe along with the reduced sizes of semiconductor devices. Various techniques to enhance the performance of microlithographic resist systems have been developed. Contrast enhanced lithography (CEL) is one technique which improves resist resolution (1, 2). The CEL process involves spin-coating a CEL layer containing a photobleachable dye on top of a conventional photoresist. This CEL layer is opaque before exposure, but during exposure, the most highly exposed regions bleach first while the least exposed regions bleach later. Therefore, optical images that are degraded by the lens system of the exposure apparatus are sharpened by passing through the CEL layer. The photobleachable dyes used in contrast enhancing materials can be classified into four categories: nitrones (1, 2), polysilanes (3), styrylpyridiniums (4), and diazonium salts (5-9).

[1]Current address: Yamazaki Works, Hitachi Chemical Company Ltd., Hitachi, Ibaraki 317, Japan

The relationship between resist contrast and optical properties (extinction coefficient and bleaching quantum yield) of photobleachable dyes was previously described by a simple model(9), and water-soluble diazonium salts with good optical properties were developed for the CEL process(9).

Diazonium salts are also useful as a photosensitive material in a photobleachable two-layer resist system based on a "doping process"(10). High-resolution resist patterns were obtained using this two-layer resist scheme and an i-line reduction projection aligner.

As mentioned above, the conventional diazonium salts have good optical properties as CEL dyes and negative working sensitizers for the two-layer resist system. However, almost all diazonium salts are stabilized with metal-containing compounds such as zinc chloride, tetrafluoroborate, hexafluoroantimonate, hexafluoroarsenate, or hexafluorophosphate, which may not be desirable in semiconductor fabrication because of potential device contamination. To alleviate the potential problem, new metal-free materials have been sought for.

In this paper we report on the use of trifluoromethanesulfonates (Table 1) of 4-N,N-dimethylaminobenzenediazonium (D1) and 4-methoxybenzene-diazonium (D2) as CEL dyes, negative working sensitizers, and photoacid generators for chemical amplification resist systems(11).

Experimental
Synthesis of diazonium salts.
The details of the synthesis of 4-N,N-dimethylaminobenzenediazonium trifluoromethanesulfonate (D1) are described below. N,N-Dimethyl-p-phenylenediamine (27.7 g, 0.2 mole) was dissolved in acetic acid (150 ml). Trifluoromethanesulfonic acid (30.6 g, 0.2 mole) was added to the solution in an atmosphere of nitrogen. The solution was kept at room temperature while isopentyl nitrite (26 g, 0.22 mole) was added dropwise. The solution was allowed to react for half an hour. After completion of diazotization, tetrahydrofuran (400 ml) was added to the solution and the precipitated D1 (55 g) was filtered. D1 (20 g) was recrystallized from acetone (90 ml)/tetrahydrofuran (260 ml). Precipitated D1 crystals (14 g) were filtered and dried. The melting point of D1 is 127°C and the maximum absorption in water appears at 376 nm. 4-Methoxybenzenediazonium trifluoromethanesulfonate (D2) was synthesized similarly.

Lithographic evaluation.
CEL dye.
A CEL solution was obtained by dissolving poly(N-vinylpyrrolidone) (PVP) (7 g) and D1 (5.8 g) in 50 wt% aqueous acetic acid. (87.2 g). The CEL layer was spin-coated onto a photoresist, RI-7000P (Hitachi Chemical Co.), and baked at 80°C for 20 minutes. Exposure was performed with an in-house i-line reduction projection aligner. The resist was developed in a 2.38 wt% tetramethylammonium hydroxide aqueous solution. The film thickness was measured with an Alpha-step 200 (Tencor)

profilometer. Spectroscopic measurements were performed on a Hitachi 340 UV spectrophotometer and a Hitachi 260-10 IR spectrophotometer.

Negative two-layer resist. A cresol novolac resin (Alnovol PN-430) was spin-coated on a silicon wafer and baked at 80℃ for one minute on a hot plate. The silicon wafer was set on a spin-coater and a photosensitive solution consisting of D1 (3 wt%), PVP (5 wt%), acetic acid (46 wt%), and water (46 wt%) was deposited so as to contact the phenolic resin film on the silicon wafer. The solution on the phenolic resin film stood for 2 minutes. The residual solution on the phenolic resin film was spun to form a D1-PVP top layer and baked at 80℃ for two minutes. By this process, a two-layer resist was formed (10).

The resist was exposed with an in-house i-line reduction projection aligner or a 600-W Xe-Hg lamp (Cannard Hanovia) in conjunction with a UVD2 band pass filter (Toshiba glass Co.). After removing the top layer by rinsing in water, the resist was developed with a 2.38 wt% tetramethylammonium hydroxide aqueous solution.

Photoacid generator. D1 (4 wt%) was mixed with poly(glycidyl methacrylate) (PGMA) (20 wt%) in ethyl cellosolve acetate. The mixture was spin-coated on a silicon wafer and baked at 80℃ for 1 minute. Exposure was performed with a 600-W Xe-Hg lamp in conjunction with a UVD2 filter. The resist was developed in a mixture of methyl ethyl ketone to ethanol (7/1 w/w).

In another application, the diazonium salt (D1 or D2) (2.5 wt%) was dissolved in a mixture of cyclohexanone and acetic acid containing 12.5 wt% poly(4-hydroxystyrene) and 2.5 wt% Methylone resin (GE 75108). The resist was exposed with a 600-W Xe-Hg lamp through a 313-nm interference filter. After exposure the resist was baked at 80℃ for 3 minutes and developed in a 1 wt% tetramethylammonium hydroxide aqueous solution.

Results and discussion
Synthesis of diazonium salts. The conventional diazotization of amino compounds is performed as described below. An amino compound is dissolved in acidic water, then aqueous sodium nitrite is added dropwise to the solution, while the solution temperature is kept below 0℃. After completion of diazotization, a diazonium ion stabilizer such as zinc chloride, sodium tetrafluoroborate, or sodium hexafluorophosphate is added to the reaction solution. Therefore, diazonium salts that are synthesized by the conventional process contain metal ions that may contaminate semiconductor devices. In addition, because some diazonium salts are extremely soluble in water, it is often difficult to isolate them from the aqueous reaction solution.

To avoid the above mentioned problems, diazonium salts were synthesized by diazotizing amino compounds with isopentyl nitrite in acetic acid containing a strong acid such as trifluoromethanesulfonic acid. After completion of diazotization, the reaction solution was poured into an organic solvent. The precipitated crystals were filtered and dried.

This process makes it possible to remove metal ions from the diazonium salts and also provides easy isolation of the diazonium salts that are highly soluble in water.

Thermal stability of D1. The thermal stability of the D1 in 50 wt% acetic acid aqueous solution was evaluated by kinetic analysis. The thermal decomposition of diazonium salts in an aqueous solution is a first-order reaction. The thermal decomposition constants of D1 at 25, 75, and 90°C are 9.1×10^{-5}, 2.9×10^{-2}, 1.3×10^{-1}/hr, respectively. The Arrhenius plot of D1 in acetic acid solution is shown in Figure 1. The activation energy of D1 (23.0 kcal/mol) and the decomposition rate constant at 5°C were obtained from Figure 1. This decomposition rate constant ($k_{5°C} = 3.7 \times 10^{-7}$/hr) indicates that 1% of the D1 in the acetic acid aqueous solution will decompose in about 3 years at 5°C.

Photoproduct of D1 in solid state. Major photoproduct from D1 in the solid state was isolated by silica gel column chromatography and analyzed by IR and NMR spectroscopy (Figures 2 and 3). NMR spectrum shows that a photoproduct of D1 has the absorption band of the N,N-dimethylamino group (2.86 ppm) and the symmetric absorption band of aromatic protons (6.57-7.10 ppm). The strong IR absorption of the sulfonyl group (1210 and 1420 cm^{-1}) can be seen in Figure 3. From NMR and IR spectra, the main photoproduct of the D1 was identified as trifluoromethanesulfonic acid 4-N,N-dimethylaminophenyl ester.

Application to a CEL dye. The UV spectra of D1 in a PVP matrix before and after bleaching are shown in Figure 4. The absorption maximum of D1 lies at 376 nm. The photobleaching curve of the D1-CEL layer consisting of D1 and PVP is shown in Figure 5. The transmittance of the unbleached layer is less than 1% and that of the completely bleached layer is about 90%. Contrast and resolution of the photoresist are expected to be greatly improved with the use of the D1-CEL.

The exposure curves of the positive photoresist Hitachi Chemical RI-7000P with and without the D1-CEL are shown in Figure 6. The resist contrast with the D1-CEL ($\gamma=3.9$) is three times higher than that without the D1-CEL ($\gamma=1.3$). Scanning electron microphotographs of submicron resist patterns printed with and without the D1-CEL are shown in Figure 7. The photographs show that the CEL layer containing D1 enables imaging of 0.4-μm line-space patterns on an i-line reduction projection aligner.

Application to the two-layer resist system. Photobleachable resist systems that have a strong absorption before exposure and that bleach completely upon UV exposure alleviate the light reflection from the substrate. A photobleachable resist system formed by means of the "doping process" was reported in our previous paper (9). This resist system consists of two layers in which a diazonium salt is distributed in both the top and bottom layers. When exposed to i-line, the diazonium salt in

Table 1. Newly synthesized diazonium salts and chemical structure of Methylone used in this experiment

NAME	CHEMICAL STRUCTURE	λ_{max} (nm)	ε_{max} x10^4
D 1	$H_3C \diagdown N - \bigcirc - N_2SO_3CF_3$ $H_3C \diagup$	376	3.37
D 2	$H_3CO - \bigcirc - N_2SO_3CF_3$	312	2.37

$$O-CH_2CH=CH_2$$
$$\bigcirc -(CH_2OH)_m$$
METHYLONE $m = 1 - 3$

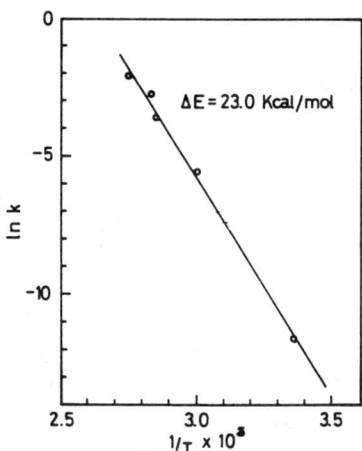

Figure 1. Arrhenius plot of D1 thermal decomposition in a 50 wt% acetic acid aqueous solution.

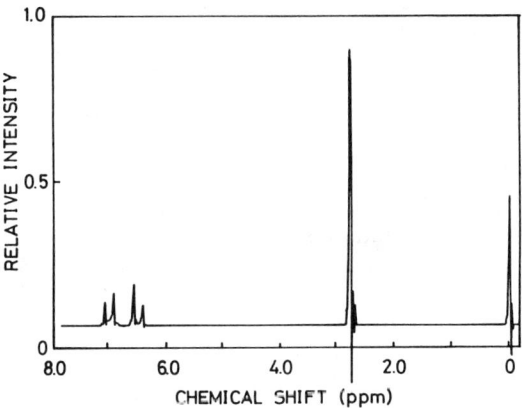

Figure 2. NMR spectrum of the main photoproduct from D1.

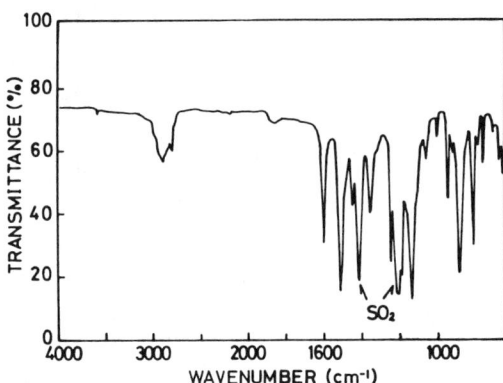

Figure 3. IR spectrum of the main photoproduct from D1.

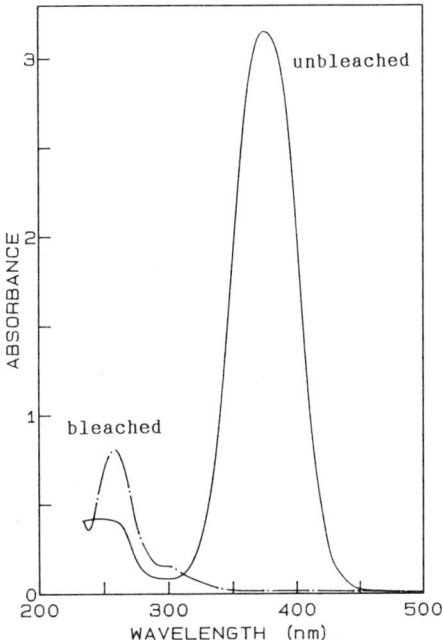

Figure 4. UV absorption spectra of a D1-CEL layer consisting of 4-N,N-dimethylaminobenzenediazonium trifluoromethanesulfonate and PVP. The solid line is the unbleached layer, and the broken line is the completely bleached layer.

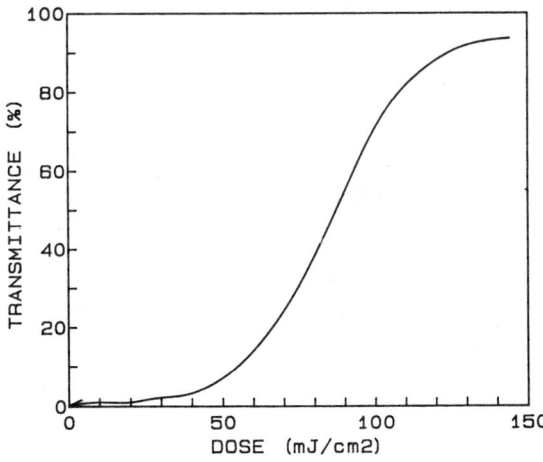

Figure 5. Photobleaching curve of the D1-CEL layer exposed at the i-line (365 nm).

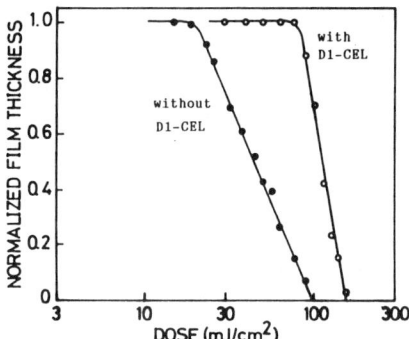

Figure 6. Exposure curves of positive resist Hitachi Chemical RI-7000P with and without the D1-CEL. The contrast (γ-value) of the D1-CEL resist is three times of that of the non-CEL resist.

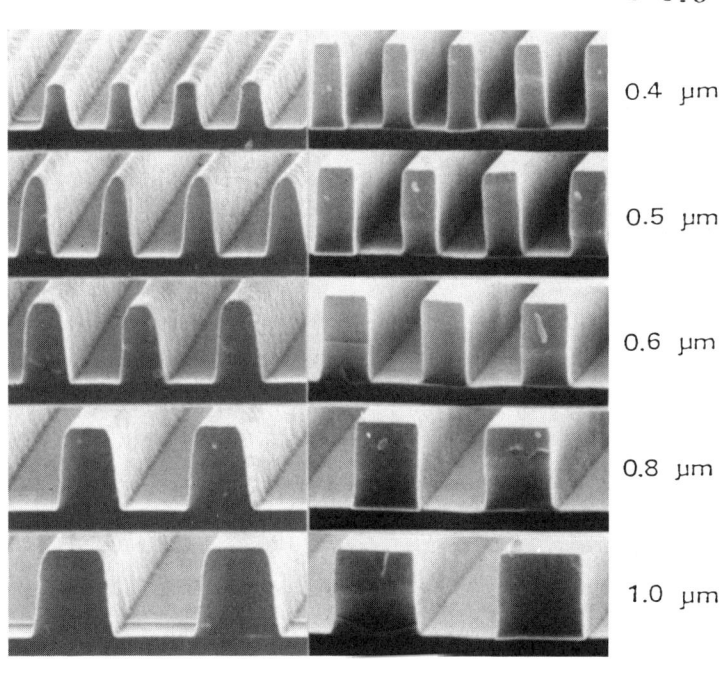

Figure 7. SEM photographs of resist patterns printed with and without the D1-CEL on commercially available resist (Hitachi Chemical RI-7000P).

the top layer bleaches to act as a contrast enhancing material, while the diazonium salt in the bottom layer decomposes to cause insolubilization of the phenolic resin in the base developer.

The exposure curve of the two-layer resist based on the doping process is shown in Figure 8. The two-layer resist system has a high contrast and high resolution capability. Submicron line-and-space patterns are obtained using this two-layer resist system (Figure 9).

Application to photoacid generators. Recently much attention has been focused on chemically amplified resist systems using onium salts(11), halogen compounds(12), or nitrobenzyl ester(13) as photoacid generators. Since the newly synthesized diazonium salts have trifluoromethanesulfonic acid anion as the counter ion, they generate a strong protonic acid upon UV exposure. D1 and poly(glycidyl mathacrylate) (PGMA) were mixed and exposed to i-line. The exposure curve of the D1-PGMA resist system is shown in Figure 10. Here it can be seen that the D1-PGMA resist works as a highly sensitive negative resist. The newly synthesized diazonium salts might be useful as photoacid generators for chemical amplification resist systems. However, the D1-PGMA resist system did not show excellent image quality because of swelling during the development process.

Another acid-catalyzed cross-linking type resist system devoid of swelling was evaluated, which consisted of a phenolic resin, Methylone resin, and the diazonium salt. Both D1 and D2 were used. The exposure curve of the resist system containing D2 as the photoacid generator is shown in Figure 11. The sensitivity and contrast of the resist dramatically improved with post exposure baking. An SEM photograph of the submicron pattern of the acid catalyzed resist system using D2 as the photoacid generator is shown in Figure 12. The resist system which contains D1 does not show high sensitivity and high contrast characteristics. It is probable that the photogenerated acid is wasted by the protonation of the N,N-dimethylamino group of D1. The details of the reaction mechanism are under investigation.

Conclusion

New metal-free diazonium salts were synthesized and applied to microlithography. A CEL layer consisting of D1 and PVP has good optical properties for i-line exposure and resolves 0.4-μm line-and-space positive resist patterns. D1 is also a useful material as a negative working sensitizer for phenolic resins. 4-Methoxybenzenediazonium trifluoromethanesulfonate can be used as a photoacid generator for the design of mid-UV resist system based on acid-catalyzed cross-linking. Submicron resist patterns are resolved using this highly sensitive resist system.

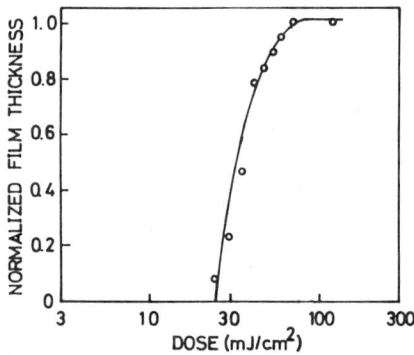

Figure 8. Exposure curve for a negative two-layer resist system formed by "doping." Exposure was performed at the i-line(365nm).

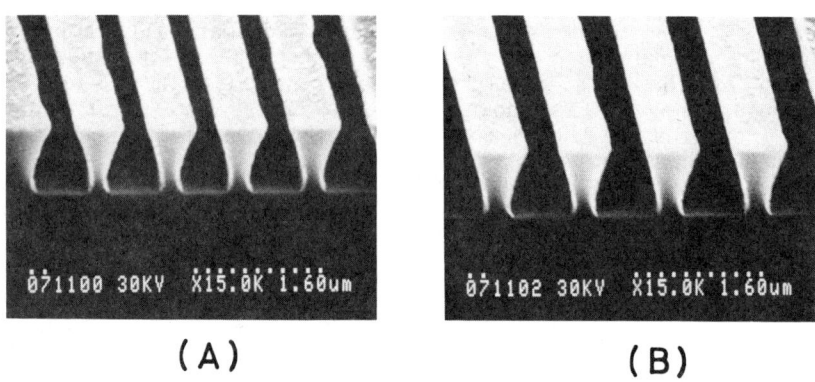

Figure 9. SEM photograph of the negative two-layer resist formed by means of "doping." Exposure was performed at the i-line. (A) 0.5-μm L&S, (B) 0.6-μm L&S.

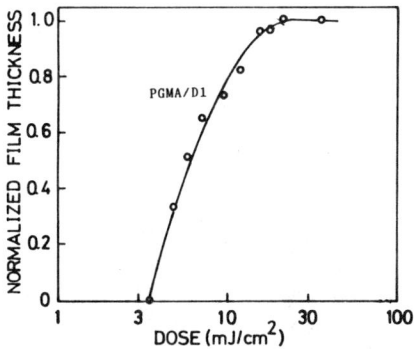

Figure 10. Exposure curve of PGMA sensitized with D1. Exposure was performed at the i-line.

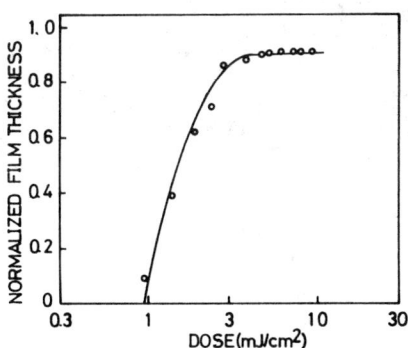

Figure 11. Exposure curve of the acid-catalyzed resist system consisting of poly(4-hydroxystyrene), D2, and Methylone. Exposure was performed using a Xe-Hg lamp equipped with a 313-nm interference filter.

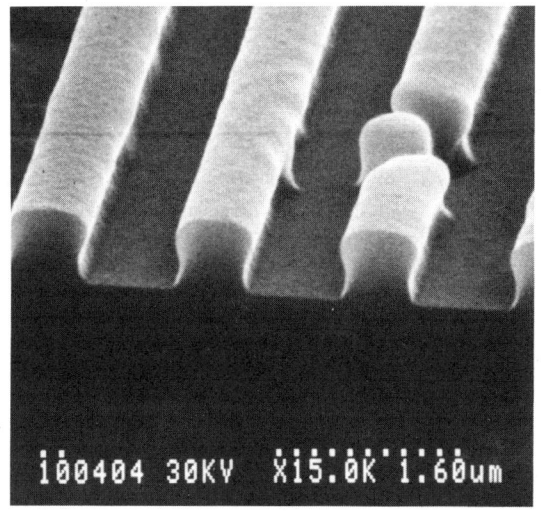

0.7 μm L & S

Figure 12. SEM photograph of the acid-catalyzed resist system exposed at 313 nm with an exposure energy of 5 mJ/cm^2.

Acknowledgments

The authors would like to thank Mr. Tsuneo Terasawa and Mr. Toshihiko Tanaka for performing the optical exposure experiments and Dr. M. Toriumi for performing dissolution rate measurement using laser interferometory.

Literature Cited

1. Griffing, B. F.; West, P. R., Polym. Eng. Sci., 1983, 23, 947.
2. Griffing, B. F.; West, P. R., IEEE Electron Device Lett., 1983, EDL-4, 14.
3. Hofer, D. C.; Miller, R. D.; Willson, C. G.; Neureuther, A. R., Proc. SPIE, 1984, 469, 108.
4. Yonezawa, T.; Kikuchi, H.; Hayashi, K.; Tochizawa, N.; Endo, N.; Fukuzawa, S.; Sugito, S.; Ichimura, K., Conf. "Photopolymers: Principles, Processes, and Materials," SPE, Mid Hudson Section, Ellenvill, New York, 1988, 183.
5. Halle, L. F.; J. Vac. Sci. Technol., 1985, B3, 323.
6. Nakase, M., Proc. SPIE, 1985, 537, 60.
7. Sasago, M.; Endo, M.; Hirai, Y.; Ogawa, K.; Ishihara, T., Proc. SPIE, 1986, 631, 321.
8. Tomo, Y.; Jinbo, H.; Yamashita, Y.; Ohno, S.; Asano, T.; Nishibu, S.; Umehara, H., Conf. "Photopolymers: Principles, Processes, and Materials," SPE, Mid Hudson Section, Ellenvill, New York, 1988, 195.
9. Ueno, T.; Uchino, S.; Iwayanagi, T.; Nonogaki, S.; Tanaka, T.; Shirai, S.; and Moriuchi, N., J. Imaging Sci., 1988, 32, 144.
10. Uchino, S.; Iwayanagi, T.; Hashimoto, M., Proc. SPIE, 1988, 920, 100.
11. Ito, H.; Willson, C. G., ACS Symp. Ser., 1984, 242, 11.
12. deGrandpre, M.; Graziano, K.; Thompson, S. D., Proc. SPIE, 1988, 923, 158.
13. Houlihan, F. M.; Shugard, A.; Gooden, R.; Reichmanis, E., Proc. SPIE, 1988, 920, 67.

RECEIVED June 14, 1989

Chapter 20

Photobleaching Chemistry of Polymers Containing Anthracenes

James R. Sheats

Hewlett Packard Laboratories, Palo Alto, CA 94304

We describe a dye system that is particularly useful for resolution enhancement in optical lithography because of the possibility of complete separation of dye and resist exposures, and because of its utility in the deep UV spectral region. Substantial variations in reactivity occur in connection with the substituents of the anthracene nucleus, the type of polymer matrix, the presence of sensitizers, and binding of the anthracene to the polymer backbone. The best results are obtained with small alkyl substituents and an external sensitizer. Variations are also observed in deep UV behavior. Novel reactions occurring in the absence of oxygen are described; these are intensity and wavelength dependent and presumably involve highly excited states. It is probable that these reactions lead to the crosslinking of some anthracene-containing polymers.

Despite the steadily shrinking dimensions of VLSI circuits, which are now well into the submicron regime for advanced production, optical lithography continues to be the patterning method of choice (1). Although electron beam direct writing can produce extremely small features, it is unlikely that its throughput will allow it to be competitive in high volume production. The excellent intrinsic resolution of x-ray lithography must be considered along with the cost of synchrotron sources and the formidable problems of mask making. For these reasons optical methods will be vigorously pursued as long as they are viable.

There are in general two aspects to resolution in optical lithography: the imaging system and the resist. Current activities involving shorter wavelength and higher numerical aperture (N.A.) imaging are described in ref 1. Single-layer resist materials have also been improved considerably in recent years (2); however it is not easy to satisfy all the requirements in a single material; thus multilayer approaches, despite the disadvantages of a greater number of processing steps, may play a role in wringing the maximum possible performance from a given exposure tool (3).

Two general types of multilayer process may be distinguished, depending on whether they use oxygen reactive ion etching (RIE) (4) or optical exposure (5-11) to transfer the pattern into the resist. Optical pattern transfer may suffer from some limitation due to substrate reflections, but has an advantage in the simplicity of the equipment compared to RIE.

This paper will describe the chemistry underlying one such process, which we have called Photochemical Image Enhancement (9,11) because it uses photobleaching to improve the image incident on the resist. It is related to, but distinct from, other photobleachable dye processes such as Contrast Enhanced Lithography (CEL) (6) and Built on Mask (BOM) (8,12). It involves no inherent throughput penalty (as does CEL), and the processing is simpler than the Portable Conformable Mask (PCM) that has been successfully used in production (5); it is quite substantially simpler than RIE-based methods. A resolution of 0.5 μm with between 1 and 2 μm total depth of focus has been demonstrated using 436 nm, 0.42 N.A. imaging (13); similar results were also obtained by Hargreaves, et al. with i-line exposure (14).

A schematic of PIE is given in refs. 9 and 11. Briefly, a layer of polymer containing a dye that is photobleachable by the imaging radiation is applied over the resist. Image-wise exposure of the dye creates a latent image of dye concentration (while not affecting the resist), and this image is transferred to the resist by a flood exposure at a wavelength that is strongly absorbed by the dye and to which the resist is sensitive; the dye is nonreactive during the flood exposure. The image quality is enhanced by the exponential dependence of transmittance on dye concentration arising from the coupling of Beer's law to the photobleaching reaction: regions of high [dye] transmit much less light than those with lower [dye]. More extensive theoretical analysis is given in refs. 9, 13, and 15. CEL similarly uses a dye over the resist, but the dye and resist are simultaneously exposed in the imaging instrument. The optimum contrast enhancement is thus only transiently obtained.

The chemistry involves the photooxidation of 9,10-substituted anthracenes, which must be embedded in a polymer of high oxygen permeability. Previous publications have described the preliminary applications to lithography (13) and the issue of oxygen permeability (15); the present report concentrates on photobleaching kinetics, sensitizer concentration effects, and deep UV exposure characteristics. Data concerning the effect of having the anthracene bound to a hydrocarbon polymer chain are presented and compared to observations on the photochemistry of anthracene/polymer mixtures under nitrogen, with various wavelengths and sensitizers.

EXPERIMENTAL

The apparatus for recording photobleaching has been described in previous publications (10,15,16). Briefly, an argon ion laser (364, 476, 514nm, or all visible lines) or a He-Cd laser (442 nm) with greatly expanded beam illuminated the sample (fused silica wafer) while its spectrum was recorded by an HP8450A spectrometer.

Some sources of error are discussed in ref. 15; in addition, it should be noted that these exposures were started with a manual shutter while simultaneously pressing the "start" button of the spectrometer, which leads to timing uncertainty of around +/-1 sec. 248 nm irradiation was from a Lumonics excimer laser, with a Laser Precision Rj-7200 energy meter for dosimetry. Pulse lengths are nominally 35 nsec (manufacturer's specification). Oxygen was excluded by a cell with fused silica windows and Viton o-rings, and a flow of N_2 (from liq. N_2 boil-off) at a few cc/sec. (A few tests with Ar ($[O_2]$ ~ 0.4 ppm by Spectra-Gases assay) gave the same results.) Dose measurement for the cw laser is estimated to be uncertain by as much as +/-10-20% due to spatial positioning uncertainty, although the photodiodes were calibrated by ferrioxalate actinometry (17,18). The data in Figures 4,5 and 7 were obtained with the laser setup described in ref. 16, which is more accurate (+/-3%). The exposure in Figure 6 used an Hg lamp with bandpass filter for 365 nm (10 nm FWHM); dose accuracy is similar to the spectrophotometer system. The film in Figure 8 was exposed in the gas cell and then removed for spectral analysis. The dose quoted for the all-lines exposure is somewhat less accurate than the others, because an approximate average photodiode responsivity was used; thus probably ~+/-25%.

Hydrocarbon polymers were purchased from Aldrich Chemical Co., and siloxanes from Petrarch Systems. The ketocoumarin Kc450, from Kodak, is 3,3'-carbonylbis(7-diethylaminocoumarin). Diphenylanthracene (DPA), dimethylanthracene (DMA) and eosin were purchased from Aldrich; 1,2-bis(10-(trimethylsiloxy)-9-anthryl)ethane (DSAE) was provided by Professor H.-D. Becker of the University of Goteborg, Sweden. The three copolymers were provided by Dr. J.S. Hargreaves of Hewlett Packard Co., who has published synthetic procedures elsewhere (19). They are abbreviated as follows: 1:2 P(MAMMA:PMDSMA), 1:2 P(VDPA:PMDSMA), and 1:2 P(VPA:PMDSMA); where MAMMA = (10-methyl-9-anthryl)methyl methacrylate; VDPA = vinyldiphenylanthracene, or 9-(p-ethenylphenyl)-10-phenylanthracene; VPA = 9-vinylphenylanthracene and PMDSMA = 3-methacryloxypropylpentamethyldisiloxane. All chemicals were used as received; films were spun from chlorobenzene solution. The films were in general not baked since baking caused no apparent difference in photochemical behavior; the films in Figures 4-7 received a 90^0C, 30 min. oven bake. Further details are given in ref. 15.

BLEACHING KINETICS

Figures 1-3 show deep UV transmittance (248 or 260 nm) for several different anthracene derivatives in poly(phenylsilsesquioxane) (PPSQ), with and without an external triplet sensitizer (Kc450, or "Kc"). It is clear that there are substantial differences in the bleaching contrast.

By "bleaching contrast" is meant the ratio $R_c=[(D_2/D_1)/E_2/E_1)]$, where D refers to transmitted dose, E to incident dose, and the points 1 and 2 are two points relatively close to each other on the curve (E_2/E_1 ~ 1.1). We have argued elsewhere (15) that this is the most appropriate figure of merit with which to evaluate photobleachable materials for image enhancement (including for CEL and BOM); the larger R_c is, the better. R_c will vary along the bleaching curve, and the greatest image resolution enhancement should be obtained if the maximum of R_c occurs approximately at the nominal line edge.

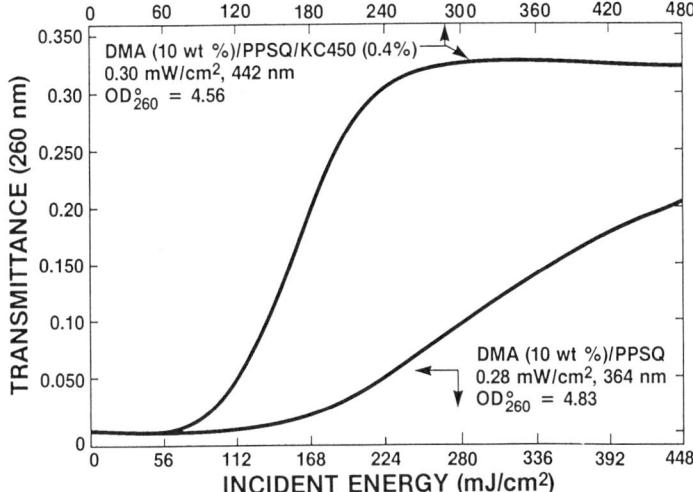

Figure 1. Transmittance T (260 nm) vs. incident energy E (364 or 442 nm) for DMA/PPSQ films as indicated (film irradiated at 442 nm contains 0.4 wt.% Kc450)

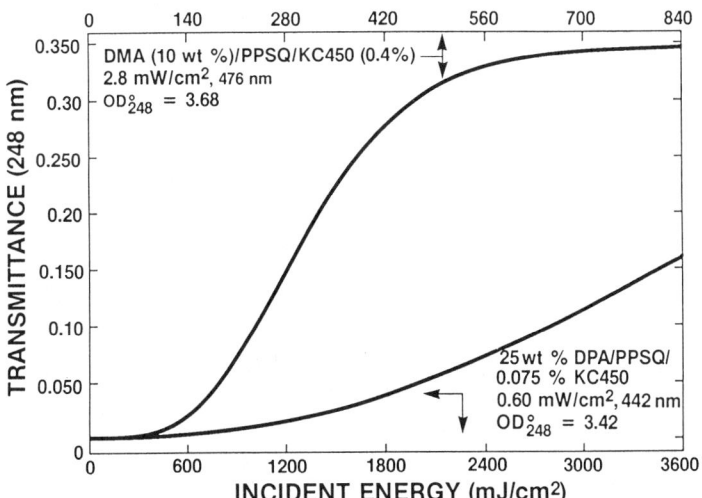

Figure 2. T (248 nm) vs. E (442 or 476 nm) for 10 wt.% DMA/PPSQ/0.4% Kc450 and 25% DPA/PPSQ/0.075% Kc450 as indicated.

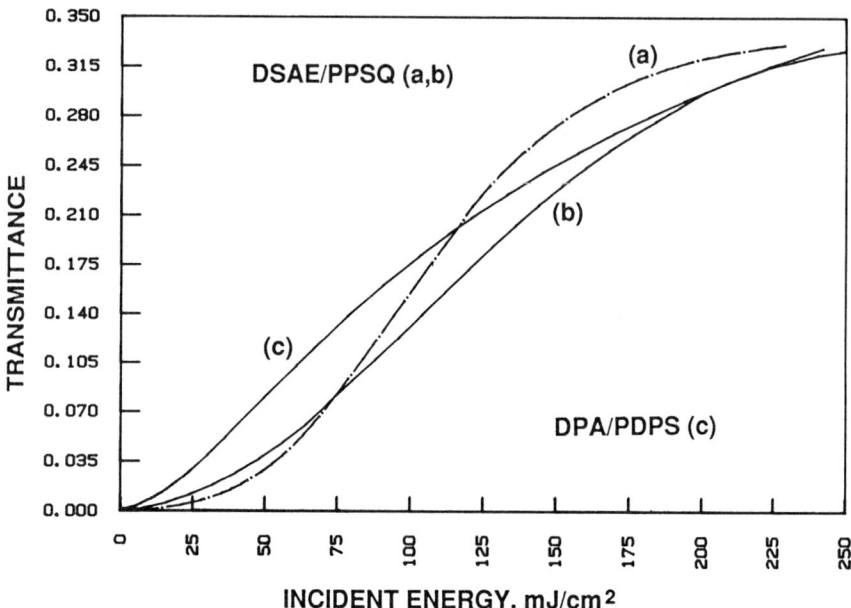

Figure 3. T vs. E for films with (a) 28% DSAE/PPSQ/4.8% Kc450, 140 μm/cm², 442 nm, T_{248}, $OD^0_{248}=3.4$; (b) 29.1% DSAE/PPSQ/0.146% Kc450, 160 μm/cm², 442 nm, T_{248}, $OD^0=2.84$; and (c) 40% DPA/PDPS, 6.46 mW/cm², 364 nm, T_{260}, $OD^0_{260}=3.15$. The energy scale for (b) is 10x greater than shown, and for (c) it is 32X greater.

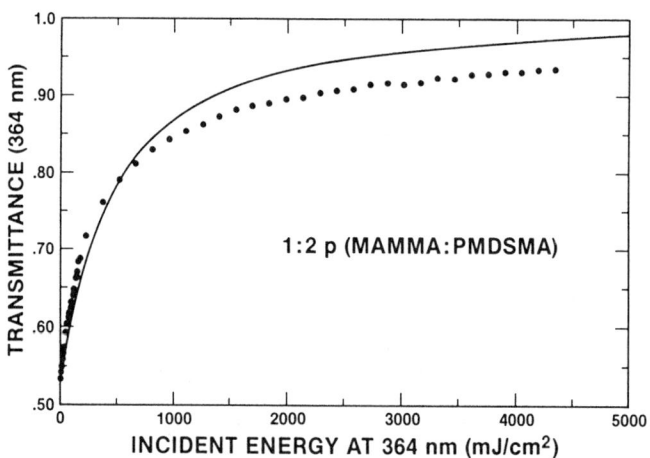

Figure 4. T vs. E for 1:2 P(MAMMA:PMDSMA), 200 mW/cm², 364 nm. Solid line: calculation for second order kinetics.

20. SHEATS *Photobleaching Chemistry of Polymers Containing Anthracenes* 337

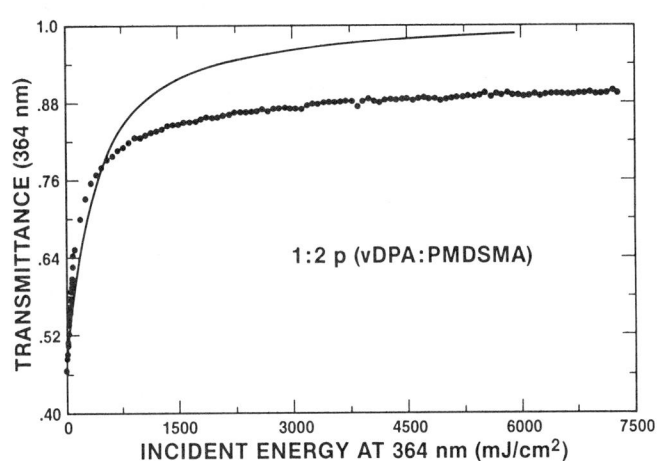

Figure 5. T vs. E for 1:2 P(VDPA:PMDSMA), 100 mW/cm^2, 364 nm. Solid line: calculation for second order kinetics.

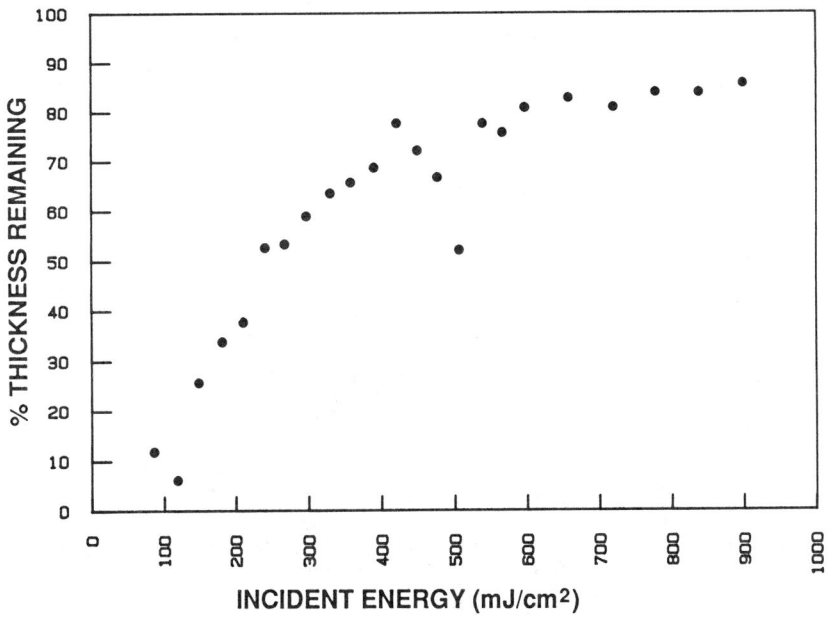

Figure 6. % thickness remaining vs. incident energy for 1:2 P(MAMMA:PMDSMA) exposed over hardbaked Hunt HPR 206 on a Si wafer; developed 60 sec. in butyl acetate. Initial thickness 0.34 μm.

338 POLYMERS IN MICROLITHOGRAPHY

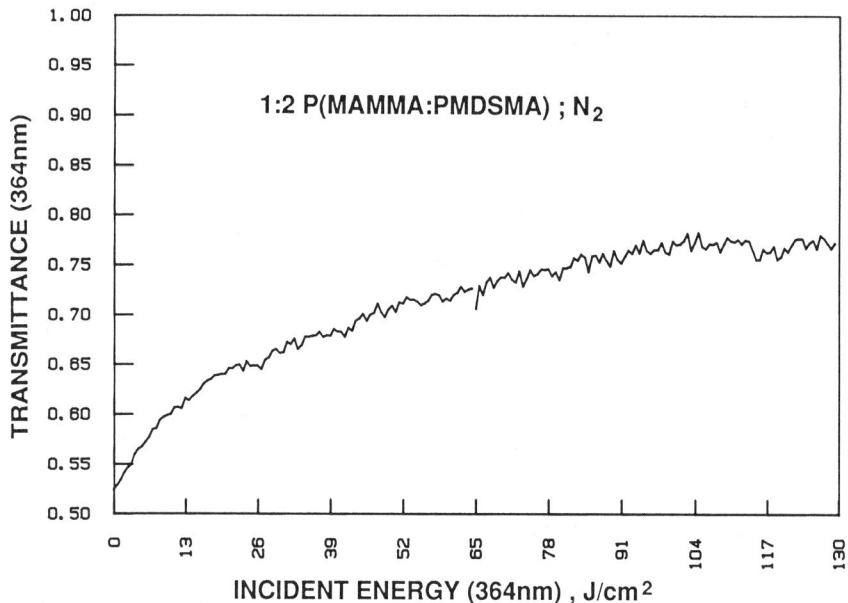

Figure 7. T vs. E for 1:2 P(MAMMA:PMDSMA) irradiated under N_2 at 364 nm, 130 mW/cm².

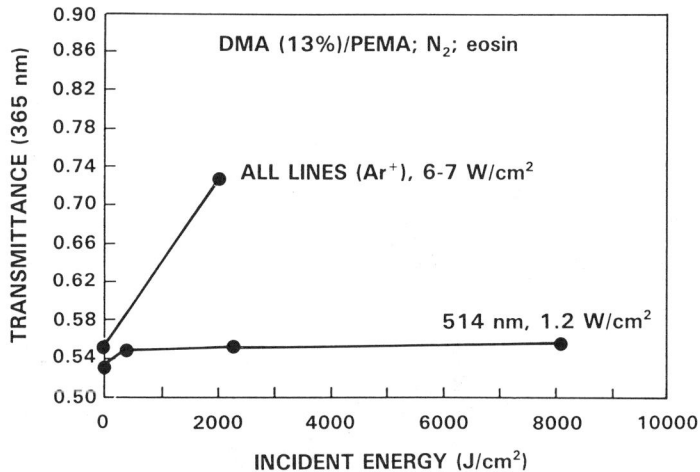

Figure 8. T (365 nm) vs. E (as indicated) for 13% DMA/PEMA/1% eosin, irradiated under N_2. Note that the all-lines dose includes ~50% 514 nm light, which is demonstrated in this figure to be non-actinic; therefore the efficiency of the blue-light reaction is ~2x greater than shown.

R_c will also depend on the initial absorbance (greater initial absorbance implies larger R_c (15)), and so all of the curves in the three figures cannot be strictly compared to one another. However, several pairwise comparisons are precisely valid, and the absorbances are close enough that a common comparison among all of them is useful. The following points are evident:

1) DPA with Kc is very similar to DMA without Kc (the small superiority of DMA shown here is likely due to the higher initial absorbance).
2) DSAE with a high [Kc] has a higher contrast than DPA with Kc, but lower than DMA with Kc.
3) DSAE with Kc has lower contrast if [Kc] is low than if it is high. (The data are not shown, but DPA and DMA contrast do not vary with [Kc].)
4) DPA without Kc has the lowest contrast, and R_c is just barely greater than 1 at its maximum.
5) DMA with sensitizer has the highest contrast.

A thorough theoretical analysis of the bleaching kinetics for DMA and DPA in several polymers, using transmittance at 365 nm to avoid any complications due to possible absorbance in the medium or photoproducts, is being published elsewhere (J.R. Sheats, J. Phys. Chem., 1989). The reactions known from solution phase studies (20) lead (13) to the following rate equation for anthracene (A) (same for oxygen):

$$dA/dt = -\alpha\theta IA[\beta A/(\beta A + 1)] \qquad (1)$$

where θ is a function of oxygen concentration and various rate constants; for the present purpose it can be taken to be the triplet yield (13). β is the ratio of the rate constant for the oxidation step ($^1\Delta$ O_2 + A) to the rate constant for decay of singlet oxygen. (α is the base e extinction coefficient, and I = intensity.) When a sensitizer is used, the first factor of A is replaced by the sensitizer concentration, and α and θ pertain to the sensitizer rather than anthracene.

This analysis leads to the conclusion that equation 1 is obeyed well by certain anthracene/polymer combinations, but there are significant deviations for others. DMA/PEMA (poly(ethyl methacrylate)) and DPA/PBMA (poly(butyl methacrylate)) fit equation 1 exactly, while DPA/PEMA (both with and without sensitizer) and DPA/PDPS (poly(diphenylsiloxane)) deviate; the actual reaction proceeds more slowly with dose than predicted, and the deviation increases as the dose increases.

The parameter β is related to the contrast. If $\beta A >> 1$, equation 1 reduces to that of a simple first order reaction (such as CEL materials are usually assumed to follow (6)). If $\beta A << 1$, the reaction becomes second order in A. In a similar manner, the sensitized reaction varies between zero order and first order. For the anthracene loadings required by the PIE process (13,15), A is close to 1M, so $\beta >> 1$ is required for first order unsensitized kinetics. Although in solution, β for DMA is ~500, and ~25 for DPA (20), we have found $\beta = 3$ for DMA/PEMA, and $\beta = 1$ for DPA/PBMA. Thus although the chemical trends are in the same direction in the polymer as in solution, the numbers are quite different, indicating a substantial restrictive effect of the polymer on the reaction. That is, we can conclude that steric effects, or "microviscosity" effects will have a substantial effect on the reaction kinetics. Although, as mentioned, the model does not quantitatively describe DPA/PEMA (which is similar to DPA/PPSQ), the discrepancy is not large, and the behavior is approximately second order; likewise that for DPA/PEMA (or DPA/PPSQ) with Kc450 is essentially first order (13). Such effects of free volume and microscopic mobility have been discussed some time ago by Somersall, et. al (21) and by others (22,23).

The trends seen here, such as the greater R_c of DMA relative to DPA, and the approximate equality of DMA (unsensitized) to DPA (sensitized), are thus consistent with expectation. DSAE has substantial absorbance at 442 nm (0.143 at 436 nm in 0.68μm thickness, 29.1 wt.%) (cf. ref. 15), so the low [Kc] case in Figure 3 can be considered as essentially unsensitized. It is noteworthy that DSAE behaves more like DPA than DMA, even though its substituents are aliphatic. In solution (20), β is determined by electronic structure: disubstituted systems react more efficiently than monosubstituted ones, and aliphatic substituents more than aromatic ones. In polymers, however, the hindering effect of the polymer lends greater importance to steric characteristics. It seems plausible, then, that the greater bulk, or lower conformational mobility, of DSAE causes it to react less efficiently than DMA, even though both are aliphatic. Its contrast is still better than DPA.

Figure 3 shows that a second order system (DPA/PDPS) is essentially just a "linear image transfer agent"; at this initial absorbance it gives neither enhancement nor degradation of contrast. This places a lower limit on the kinetic order that one can utilize in a process such as PIE. It is worthwhile to consider, however, that such a linear system could still have value in PIE (although it would be worthless for CEL). One might image with the KrF laser stepper at 248 nm (where available resists have undesirably strong unbleachable absorbance (3)), and then transfer the pattern at 260-265 nm, where the anthracene absorbance is highest and where some potential resists have relatively lower absorbance (24,25). (Exposure of the resist during imaging would be avoided by adjusting its sensitivity relative to that of the anthracene.)

Figures 4 and 5 present the kinetic behavior of two copolymers of an anthracene-containing monomer (methacrylate or styryl) with a siloxane-containing methacrylate; the anthracene moiety is related to DMA and DPA respectively. Although the transmittance at 364 nm rather than 248 or 260 is recorded, comparison to the data of Figures 1-3 is made possible by the theoretical curves, which are the best possible fits for a second-order kinetic law (26). It can be seen that there is a major deviation; the bleaching proceeds more slowly than second order, and gets progressively slower as the reaction proceeds. The deviation is more pronounced for the phenyl-substituted case than for the methyl. It is virtually impossible to completely bleach either film at 364 nm. This is paralleled by an insolubilization process (14,19), as shown in Figure 6. The appearance of fully insoluble material is at approximately the same dose as that at which the deviation from expected kinetics (i.e., for DMA) begins to be noticeable. The deviation can not be related to oxygen permeability since reciprocity failure due to insufficient oxygen is not observed for intensities 2.5x higher.

It might be thought that the insolubilization is caused by anthracene dimerization (27,28), although DPA has never been observed to dimerize (29) due to steric hindrance from the non-planar phenyl rings. However, it was found that a film of P(VPA:PMDSMA) with $\sim 10^{-2}$M Kc450, extensively irradiated at 436 nm, also becomes insoluble. This reaction is extremely unlikely to be dimerization, since it has been well verified in solution phase studies that dimerization proceeds via the singlet state only (28-33); even for dianthrylethanes (where the excited triplet is held in close proximity to a ground state partner for its entire lifetime), dimerization is found only for the singlet except for some carbonyl-substituted species (30).

Another possible explanation is that singlet O_2 somehow leads to crosslinking. The reactions of $^1\Delta$ O_2 have been extensively studied (34), and do not appear relevant to these copolymers. The only functionality that could conceivably react with singlet O_2 is a vinyl chain termination, which could produce a hydroperoxide that might then participate in crosslinking. However, in a study of free radical polymerized PMMA (35), the maximum fraction of polymer chains with vinyl ends was found to be 0.36, for bulk polymerized material; in benzene solution the fraction was 0-3%. This result, plus the fact that the insolubilization occurs immediately during photolysis at room temperature, makes it very unlikely that such hydroperoxides are involved.

PHOTOREACTIONS UNDER INERT GAS

In an attempt to clarify these phenomena, mixtures of anthracenes and polymers were irradiated under N_2 or Ar, both at 364 nm (direct excitation) and with triplet sensitization, using the blue lines of an Ar^+ laser as well as 514 nm. 1:2 P(MAMMA:PMDSMA) was also irradiated at 364 nm (Figure 7). Some data for DMA/PEMA irradiated at 364 nm are given in ref. 15, where it is shown that bleaching does occur in the absence of O_2 (transmittance rises from 0.63 to 0.80 with ~ 5 J/cm^2, for a 13.6 wt.% film); the bleaching is similar to that of the copolymer. DPA also undergoes such bleaching, albeit with substantially lower efficiency. The effect can also be seen with triplet sensitization: a DMA/PEMA film with eosin, irradiated with all lines of the Ar^+ laser, is slowly bleached (Figure 8). However, when 514 nm alone is used, no reaction can be detected after 8 kJ/cm^2 (Figure 8). The excitation of eosin triplets is demonstrated by the fact that 514 nm irradiation in air very quickly (with a few hundred mJ/cm^2) bleaches the DMA. The triplet energy of eosin is high enough to be transferred to anthracene (eosin is 43 kcal/mol (36), and anthracene 42 (37)). The absence of reaction (under N_2) at 514 nm demonstrates that eosin itself has no effect. Therefore the bleaching reaction must proceed through the anthracene triplet. The bleaching is accompanied by a change in polymer molecular weight.

The most probable explanation of these results is found upon examining the absorption spectrum of the first triplet state T_1 of anthracene, which is strong ($\epsilon > 4 \times 10^4$ l/mol cm) and maximized in the blue ($\sim 425 - 450$ nm, depending on substituent) (38,39). Its absorbance at 514 nm is negligible. Thus the unexpected bleaching very likely results from absorbance by T_1 to produce a highly excited triplet, from which novel photochemistry may well occur. (Blue light is present directly during sensitized irradiation; with 364 nm irradiation the T_n excitation may be either by "trivial" (emission-absorption) energy transfer or by a Forster-Dexter mechanism.) The nature of this reaction is unknown, but it appears to be a relatively specific, bimolecular reaction from a highly excited (hence short-lived) state; the state is bound (39) and so simple dissociation (as seen elsewhere (40)) is ruled out. Energy transfer from T_2 is known for a variety of cases (41), and a rearrangement reaction from a highly excited triplet in the pleiadene family (possibly T_7) was observed in a 77K rigid glass (42). The latter reaction was observed only via direct two-photon excitation. Bimolecular reactions of upper excited states are quite rare (41). In the present case, where Birks (43) shows the state as ~T_7, electron transfer or hydrogen atom abstraction by anthracene are plausible candidates for the process; the formation of radicals is likely.

An anthracene radical adjacent to another anthracene could couple (44) to produce crosslinked material in the case of the copolymers; the concentration in the 1:2 copolymer is high enough for such juxtaposition to be common. The resonance stabilization of the anthryl radical should decrease the rate of alternative decay pathways. According to the picture of the photooxidation kinetics as described above, such crosslinking should further restrict the conformational mobility of the reacting anthracene rings and thus hinder the reaction (effectively reducing β). At the 100-300 mJ/cm^2 dose at which crosslinking is found (Figure 6), Figure 7 shows that a few % of the anthracenes have been bleached in the non-oxidative reaction, which is enough to crosslink chains of average molecular weight ~5×10^4. However, 1 J/cm (365 nm) does not cause crosslinking of 1:2 P(MAMMA:PMDSMA) under N_2. ~200 J/cm^2 (+/-50%) at 355-375 nm under N_2 produces thorough crosslinking (which is not reversed by baking at 105^{0}C for 50 min., and is therefore not dimerization (45)); the dose requirements for this reaction have not been more precisely determined. Thus, although O_2 is not required for crosslinking, it greatly enhances it, possibly by forming peroxide crosslinks.

Hargreaves has suggested that the insolubilization of some closely related polymers is due to photolytic homolysis of the endoperoxide O-O bond and subsequent generation of carbon-centered radicals from the O radicals (19). There are several facts that make this an extremely unlikely explanation for the data described here; these include the quantitative insufficiency of the maximum amount of endoperoxide reaction obtainable with a few hundred mJ/cm^2 dose (homolysis quantum yield <0.5 (46), and extinction coefficient ~1 (M cm)$^{-1}$ (47)), and the synthetic utility of such homolysis reactions in related molecules in the presence of good hydrogen atom donors (implying facile epoxide formation) (48). Clearly the crosslinking observed under N_2 is not accounted for by this mechanism.

Further understanding of this system will require excited state spectroscopy to identify in real time the species present. The present hypothesis, however, suggests ways of overcoming the impediment to the use of anthracene-containing copolymers in PIE. For example, lowering the molecular weight of the polymer would require more crosslinks before a gel is formed. The bleaching (oxidative) reaction could then be finished before the kinetics are significantly affected. Similarly, increasing the efficiency of the oxidation reaction relative to the crosslinking reaction (for example, by providing a more fluid environment) would accomplish the same result. Finally, using a polymer with less readily abstractable hydrogen atoms is predicted to reduce the rate of the crosslinking reaction.

SENSITIVITY AND SENSITIZER CONCENTRATION EFFECTS

Figure 9 shows the bleaching of 25 wt.% DPA/PPSQ with three different Kc450 concentrations. Equation 1 predicts that the rate of the reaction should scale linearly with the product of intensity and sensitizer concentration (at the same wavelength). (This formula is only valid if at least one member of the donor/acceptor pair can diffuse freely, or if the pair members are close enough for electron exchange energy transfer to occur readily. These conditions are satisfied since the [A] is ~1M, and Kc can transfer energy either to anthracene or directly to O_2 which is mobile.) One cannot expect exact agreement with this formula because of the effects of different absorbance (26). However, the data show that the prediction is a semiquantitatively useful rule. It also shows that the anthracene photobleaching reaction can be quite sensitive at reasonable sensitizer loadings, which bodes well for practical utility. (Note that the data are at 476 nm, where Kc450 absorbs ~2x less than at 436 nm.) The exposure time for PIE is thus limited by oxygen permeability (15) rather than intrinsic sensitivity.

Some sensitivities with 364 nm irradiation are as follows: DMA/PEMA, 0.0059; DPA/PEMA, 0.0057; DPA/PPSQ, 0.0092 cm^2/mJ. (This value is the initial rate dA/dt divided by the intensity; see ref. 15.) A comparable datum for DMA/PPSQ is not available, but by comparing the doses required to reach a specified absorbance, one finds that DMA/PPSQ is slightly faster than DMA/PEMA; the ratio is about 1.5 (+/-20%). These numbers are a little smaller than those for conventional positive resist (49). With deep UV irradiation, however, they will be ~7-20 times larger due to the larger extinction coefficient, which helps fulfill one of the key prerequisites of a CEL (or PIE) material for the deep UV.

DEEP UV EXPOSURE

A central aspect of the PIE concept is that the dye and resist exposures should be completely separated (9,15); any departure from this condition makes the system more like CEL and lowers its performance. Further bleaching of the anthracene is of course eliminated simply by removing oxygen; the bleaching reactions described above (e.g. Figure 8) are much lower in efficiency and therefore irrelevant insofar as pattern transfer is concerned. However, "antibleaching" due to the loss of oxygen from the anthracene peroxide upon DUV irradiation (47) is a concern; it constitutes a dose-dependent loss of contrast.

Although this reaction is well known, it was not expected to cause serious problems for PIE for two reasons (15). First, when the oxygen is regenerated, it is in the excited singlet state, and so should re-react efficiently with a nearby anthracene (the anthracene concentration is never lower than about 1/3 of original in even the fully exposed regions). Second, those oxygen molecules that do decay should be effectively re-excited by the high concentration of anthracene triplets that are generated by the deep UV light (the state excited at 260 nm decays on a subnanosecond time scale to the same state that is excited at 365 nm) (37). Since anthracene endoperoxide derivatives also possess a second, irreversible DUV photoreaction (involving breakage of the O-O bond and formation of ring epoxides and other products) (46,50), there should be little if any loss of contrast due to these processes.

Nevertheless, Figure 10 shows that 25 wt.% DPA/PPSQ/Kc450, previously bleached in O_2 by visible light, does undergo an antibleaching reaction. The extent of the reaction varies with the initial bleach: the greater the initial bleach, the greater the relative change. However, after a point of no further reaction is reached, the ordering of the curves is still the same; i.e., the greatest original extent of bleach is still the greatest. It is shown in ref. 15 that the final state curve is still useful ($R_c > 1$) despite this loss of contrast.

Similar results have been reported for DMA/PEMA (14). Figure 11 shows data for DSAE/PPSQ. Although the same effect is present, it is quantitatively different. The reduction in T is about a factor of 3 for a film initially bleached to T=32%, while for DPA about the same reduction is seen with T(initial) = 8%; thus it is apparently less severe for DSAE. The most interesting result is that DMA/PEMA, when irradiated under N_2 at 260 nm (+/-8nm FWHM bandwidth) by an Hg-Xe lamp, shows absolutely no change in transmittance with a dose of 100 J/cm^2 (15). Thus the antibleaching is an intensity dependent effect that is absent at low intensities and occurs only with the excimer laser (typically ~0.1-1 mJ/cm^2 in ~35 nsec).

Real-time excited state monitoring will be required to understand the mechanism of the antibleaching reaction. It is possible that the excimer laser pulse produces such high concentrations of singlet oxygen and triplet anthracene that bimolecular quenching processes greatly reduce the available concentrations (39). However, no data are available at this time to allow a quantitative analysis. From the lithographic point of view, the important result is that low intensity deep UV imaging sources such as the Perkin-Elmer Micrascan 1 can be used without any process degradation by this effect. The effectiveness of anthracene as a PIE or CEL material with KrF excimer laser sources can only be determined by further experiment, although it appears likely that quite useful enhancement factors can be gotten even in the presence of the degradation shown in the figures (cf. ref. 15).

CONCLUSIONS

The photobleaching kinetics of anthracene photooxidation have been investigated. Bleaching characteristics satisfactory for PIE (with Hg g-line and i-line as well as 248-265 nm deep UV imaging) and CEL (deep UV only) are observed (as long as the intensity does not exceed limits imposed by oxygen permeability of the film (15)). The kinetics depend on the anthracene substitution; small, mobile aliphatic substituents appear to be the best, although aromatics and larger aliphatic groups also work well with external sensitizers. Photobleaching under inert atmosphere (with much lower efficiency than under oxygen) has also been investigated, and ascribed to sequential multiphoton processes; it is hypothesized that these processes are involved also in producing the difficulties encountered when anthracene is bound to a methacrylate or methacrylate-co-styryl polymer chain. Deep UV stability data are presented indicating that anthracene derivatives can be fruitfully applied to deep UV PIE or CEL with low intensity (lamp) sources; further research is necessary to assess applicability with KrF laser sources, although at least some degree of utility at minimum is probable.

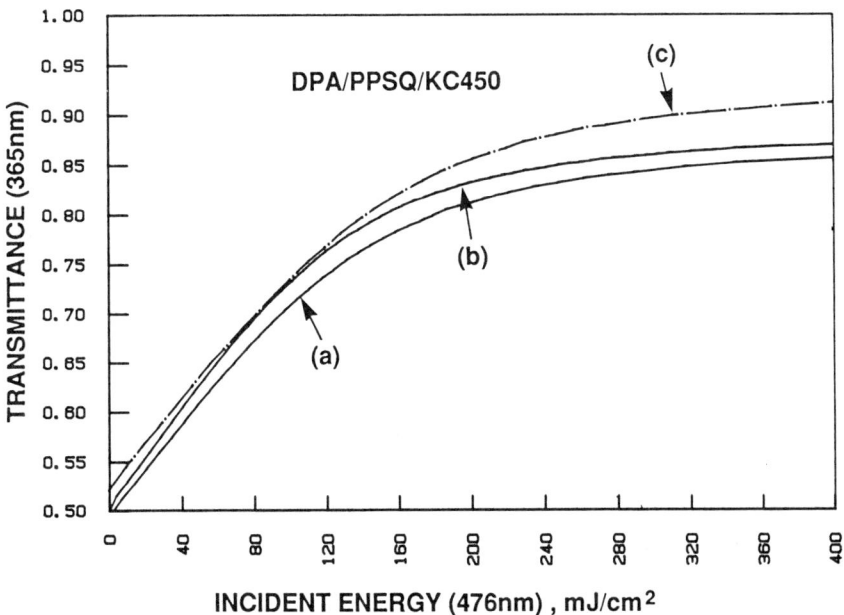

Figure 9. T (365 nm) vs. E (476 nm) for 25% DPA/PPSQ/Kc450, for 3 different [Kc450]: (a) 3.12%, 4.59 mW/cm^2, (b) 2.15%, 5.88 mW/cm^2, and (c) 0.356%, 53.4 mW/cm^2. The energy scale is correct for (a); (b) and (c) have been scaled by the [Kc450], so the true maximum dose is 580 and 3506 mJ/cm^2, respectively.

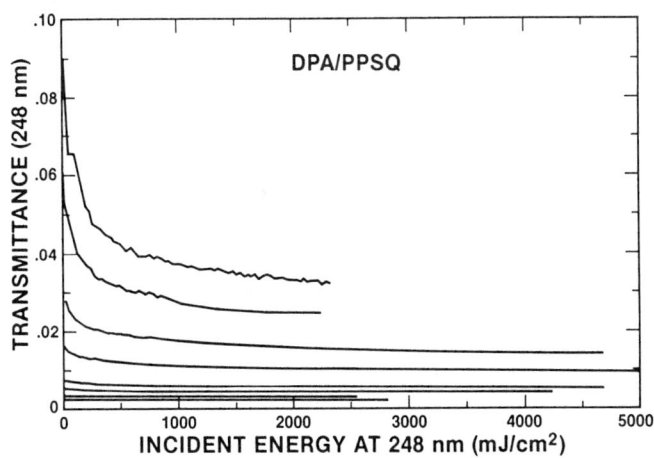

Figure 10. T (248 nm) vs. E (248 nm, KrF laser) under N_2, for 25% DPA/PPSQ/0.075% Kc450, previously exposed under O_2 at 476 nm to the T_{248} shown. Excimer laser pulse rep rate 25 pps, delivered in bursts of 10 or 100.

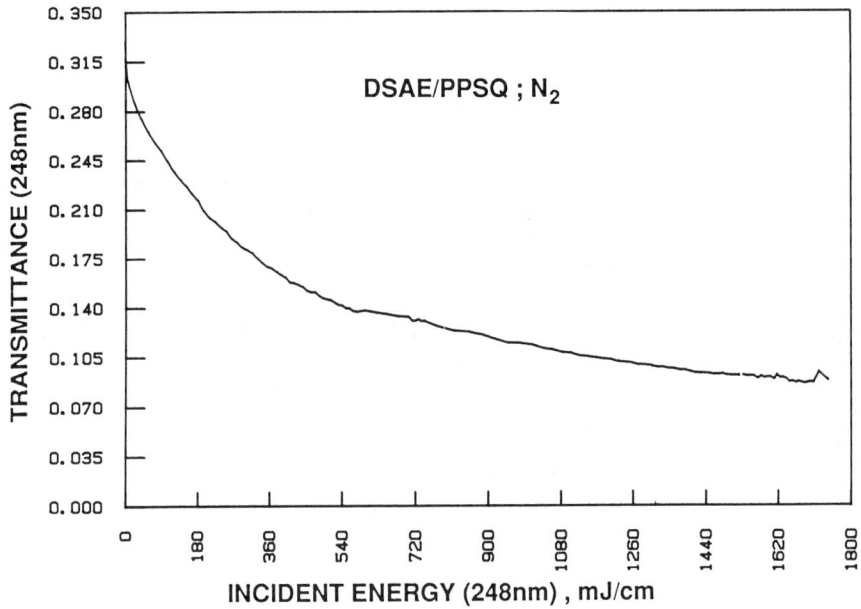

Figure 11. As in Figure 10, for 29.1% DSAE/PPSQ/Kc450 (film used in Figure 3).

ACKNOWLEDGMENTS

I am grateful to Professor Becker and Dr. Hargreaves for the materials they provided, and to Gerry Owen and Bob Gleason for support of this work.

REFERENCES

1. *SPIE Proc.* **1988**, *922*.
2. P. Trefonas III; B.K. Daniels, *SPIE Proc.* **1987**, *771*, 194.
3. J.R. Sheats, *Solid State Technology* **1989**, June issue.
4. M.A. Hartney; D.W. Hess; D.S. Soane, *J. Vac. Sci. Techn. B* **1989**, 7, 1.
5. K. Bartlett; G. Hillis; M. Chen; R. Trutna; M. Watts, *SPIE Proc.* **1983**, *394*, 49.
6. P.R. West; B.F. Griffing, *SPIE Proc.* **1983**, *394*, 33.
7. J.R. Sheats, *Appl. Phys. Lett.* **1984**, *44*, 1016.
8. F.A. Vollenbroek; W.P.M. Nijssen; H.J.J. Kroon; B. Yilmaz, *Microcircuit Eng.* **1985**, *3*, 245.
9. J.R. Sheats; M.M. O'Toole; J.S. Hargreaves, *SPIE Proc.* **1986**, *631*, 171.
10. J.R. Sheats; J.S. Hargreaves, *ACS Symp. Ser.* **1987**, *346*, 224.
11. J.R. Sheats, U.S. Patent 4,705,729, **1987**.
12. F.A. Vollenbroek; W.P.M. Nijssen; H.J.J. Kroon; B. Yilmaz, *SPE Preprints, Reg. Tech. Conf., "Photopolymers: Principles - Processes and Materials"*, Ellenville, N.Y., **1985**, 309.
13. J.R. Sheats, *SPE Preprints, Reg. Tech. Conf., "Photopolymers: Principles - Processes and Materials"*, Ellenville, N.Y., **1988**, 319; *Polym. Eng. Sci.* **1989**, *29*, 965.
14. J.S. Hargreaves; M.M. O'Toole; D. Burriesci, *J. Electrochem. Soc.* **1989**, *136*, 225.
15. J.R. Sheats, *SPIE Proc.* **1989**, *1086*, (paper #1086-46).
16. J.R. Sheats, *IEEE Trans. Elec. Dev.* **1988**, *ED-35*, 129.
17. C.G. Hatchard; C.A. Parker, *Proc. Roy. Soc., Ser. A* **1956**, *235*, 518.
18. J.N. Demas; W.D. Bowman; E.F. Zalewski; R.A. Velapoldi, *J. Phys. Chem.* **1981**, *85*, 2766.
19. J.S. Hargreaves, *J. Polym. Chem. A, Polym. Chem. Ed.* **1989**, *27*, 203.
20. B. Stevens; S.R. Perez; J.A. Ors, *J. Am. Chem. Soc.* **1974**, *96*, 6846.
21. A.C. Somersall; E. Dan; J.E. Guillet, *Macromolecules* **1974**, *7*, 233.
22. F.-D. Tsay; S.D. Hong; J. Moacanin; A. Gupta, *J. Polym. Sci., Polym. Phys. Ed.* **1982**, *20*, 763.
23. R. Richert, *Macromolecules* **1988**, *21*, 923.
24. S.R. Turner; K.D. Ahn; C.G. Willson, *ACS Symp. Ser.* **1987**, *346*, 200.
25. C.E. Osuch; K. Brahim; F.R. Hopf; M.J. McFarland; A. Mooring; C.J. Wu, *SPIE Proc.* **1986**, *631*, 68.
26. J.R. Sheats; J.J. Diamond; J.M. Smith, *J. Phys. Chem.* **1988**, *92*, 4922.
27. C. Linebarger, Amer. Chem. J. 1892, 14, 597.
28. H. Bouas-Laurent; A. Castellan; J.-P. Desvergne, *Pure Appl. Chem.* **1980**, *52*, 2633.
29. J. Bendig; W. Buchwitz; J. Fischer; D. Kreysig, *J. f. prakt. Chemie* **1981**, *323*, 485.
30. H.-D. Becker, *Pure Appl. Chem.* **1982**, *54*, 1589.
31. E.A. Chandross; J. Ferguson, *J. Chem. Phys.* **1966**, *45*, 3564.
32. J. Ferguson; A.W.H. Mau, *Mol. Phys.* **1976**, *27*, 377.

33. M.D. Cohen; A. Ludmer; V. Yakhot, *Chem. Phys. Lett.* **1976**, *38*, 398.
34. *J. Photochemistry* **1984**, *25*, 99-553 (Proc. of COSMO 84 Conf. on Singlet Molecular Oxygen, Clearwater Beach, FL, 1984)
35. T. Kashiwagi; A. Inaba; J.E. Brown; K. Hatada; T. Kitayama; E. Masuda, *Macromolecules* **1986**, *19*, 2160.
36. L.L. Costanzo; U. Chiacchio; S. Giuffrida; G. Condorelli, *J. Photochemistry* **1980**, *14*, 125.
37. N.J. Turro, *Modern Molecular Photochemistry*; Benjamin-Cummings, Menlo Park, CA, 1978.
38. J. Szczepanski; J. Heldt, *Z. Naturforsch.* **1985**, *40a*, 849.
39. G.N.R. Tripathi; M.R. Fisher, *Chem. Phys. Lett.* **1984**, *104*, 297.
40. D.M. Burland, *Acc. Chem. Res.* **1983**, *16*, 218.
41. N.J. Turro; V. Ramamurthy, W. Cherry; W. Farneth, *Chem. Rev.* **1978**, *78*, 125.
42. J. Kolc; J. Michl, *J. Am. Chem. Soc.* **1973**, *95*, 7391.
43. J. Birks, *Photophysics of Aromatic Molecules*; Wiley, New York, 1975.
44. K. Tokumara; N. Miukami; M. Udagawa; M. Itoh, *J. Phys. Chem.* **1986**, *90*, 3873.
45. H.-D. Becker; T. Elebring; K. Sandross, *J. Org. Chem.* **1982**, *47*, 1064.
46. R. Schmidt; H.-D. Brauer, *J. Photochem.* **1986**, *34*, 1.
47. W. Drews; R. Schmidt; H.-D. Brauer, *Chem. Phys. Lett.* **1980**, *70*, 84.
48. W. Adam, *Angew. Chemie I.E.* **1974**, *13*, 619.
49. F.H. Dill; W.P. Hornberger; P.S. Hauge; J.M. Shaw, *IEEE Trans. Electron Dev.* **1975**, *ED-22*, 445.
50. J. Rigaudy; C. Breliere; P. Scribe, *Tetrahedron Lett.* **1978**, 687.

RECEIVED August 2, 1989

Chapter 21

Lithography and Spectroscopy of Ultrathin Langmuir–Blodgett Polymer Films

S. W. J. Kuan[1], P. S. Martin[1], L. L. Kosbar[2], C. W. Frank[1], and R. F. W. Pease[3]

[1]Department of Chemical Engineering, Stanford University, Stanford, CA 94305
[2]Department of Chemistry, Stanford University, Stanford, CA 94305
[3]Department of Electrical Engineering, Stanford University, Stanford, CA 94305

> Ultrathin (0.9 - 15.3 nm) poly(methylmethacrylate) (PMMA) and (30 - 40 nm) novolac/diazoquinone films prepared by the Langmuir-Blodgett (LB) technique have been explored as high-resolution electron beam resists and photoresists, respectively. One-eighth micron line-and-space patterns have been achieved in PMMA using a Perkin Elmer MEBES I pattern generation system as the exposure tool. The etch resistance of PMMA films with thicknesses greater than 4.5 nm is sufficient to allow patterning of chromium film suitable for photomask fabrication. One micron line-and-space patterns have been fabricated by optical lithography in 30 nm thick novolac/diazoquinone films, and etched into 50 nm of chromium. Monolayer PMMA films containing 5 mol% pyrenedodecanoic acid (PDA) as a probe were prepared by transfer to a quartz substrate at different surface pressures and characterized by fluorescence spectroscopy. The ratio of excimer to monomer emission intensity (Ie/Im) has a maximum value at ~ 10 dyn/cm, which is suggestive of a structural rearrangement occurring in the Langmuir film at that surface pressure.

As the dimensions of integrated circuits keep shrinking, the desired resolution will soon be beyond the limit of conventional ultraviolet (UV) lithography. Deep UV and X- radiation, high and low energy electron beams and scanning tunneling microscopy (STM) have been proposed (1) as possible exposure systems for the next generation high resolution lithography. In optical lithography (UV or DUV) the resolution is limited by resist absorption, light diffraction, and rheological effects related to the resist development process. In electron beam lithography the major limitation on the resolution is imposed by electron scattering (proximity effect), which causes a nonuniform incident exposure in the pattern area. These resolution limiting effects generally become more serious with increasing resist thickness. Therefore, to improve the resolution in both optical and electron beam lithography, the use of ultrathin resists (with thicknesses < 200 nm) has been proposed (2-8).

Ultrathin resists have many technological advantages: In optical lithography they offer improved exposure and focus latitude (9), and alleviate the problem of absorption in conventional resists, such as novolac, especially for deep UV exposure. In electron-beam lithography the use of ultrathin resists will reduce electron scattering within the resists and thus make the proximity effect correction schemes (10) easier to implement. The most attractive advantage of an ultrathin resist is that it allows for electron penetration when the scanning tunneling microscope (STM) is used as a very low voltage exposure tool (1). Because the STM is capable of creating patterns with extremely high resolution (< 10 nm) and potentially at very high speed, it may become an important lithographic tool in the near future with the use of ultrathin polymer films as resist materials.

To prepare ultrathin polymer films on the surface of wafers, especially those of large diameter (6 or 8 inch), uniformity and defect density become important factors in determining the resist quality. The conventional spin coating method has been reported to introduce interference striations (11) and high defect densities (2,3) when used to prepare ultrathin polymer films. As an alternative approach, the LB technique has been proposed as being suited to the preparation of more uniform ultrathin polymer films (2). Using this technique monolayer polymer films can be transferred layer by layer to the surface of a solid substrate from the water surface. An important feature of the LB technique is that the accumulation of monolayer films allows the thickness of the built-up film to be controlled in a precise manner. Consequently, extremely uniform and ultrathin polymer films can be prepared.

In order to establish a basis for rationalizing the lithographic performance of ultrathin polymer films, a better understanding of their structure at the molecular level is required. However, to date there has been very little effort in this area (12-14). Ultrathin polymer films prepared using either the spin casting or LB technique not only have important technological applications, but are also interesting from a scientific point of view in that the resulting polymer chain configurations are expected to be different in films prepared by these two methods. In spin-cast films, due to the force exerted along the radial direction during spinning, the polymer chains may under certain circumstances be frozen in a nonequilibrium state along that direction as the solvent evaporates (15). In LB films, due to the interaction between the subphase (water) and the hydrophilic groups of the polymer, the polymer chains are expected to exist in a partially oriented geometry. For single component ultrathin polymer systems, both spin-cast and LB films can be used to study the polymer chain configurations in constrained geometries. In addition, interchain diffusion, chain relaxation and chain-substrate interaction upon annealing are areas for potential investigation.

Most commonly used positive photoresists are composed of several components, including a polymer resin and small molecule photo-active compound (PAC) as well as the casting solvent. The chemical compatibility of the resin and the PAC is often rather poor, and aggregation of the two materials during casting is of potential concern. Aggregation of the resin or PAC could cause variations in local development rates, which would be most obvious at line edges. We have previously (15) investigated the effects of the casting solvent on the homogeneity of the spin-cast film. LB film formation allows us to investigate the interactions of these materials in a slightly different manner. The molecules are inherently more ordered due to the hydrophilic and hydrophobic interactions at the air/water interface. This ordering may affect the homogeneity of mixed films of novolac and PAC. The fact that we can control the novolac and PAC concentration in each monolayer will also allow us to investigate other film parameters such as the interdiffusion of the PAC and the polymer upon prebake and the "sphere of influence" of the PAC molecules in inhibiting dissolution.

One of the major obstacles to investigating ultrathin polymer films is the small amount of detectable sample material and, as a result, high instrument sensitivity is crucial. Although polarized Fourier Transform Infrared Spectroscopy (13,14) has been used to study LB poly(octadecyl methacrylate) and poly(octadecyl acrylate) films, it was found that the signal-to-noise ratio decreases dramatically as the film thickness decreases from ~ 12 nm (6 layers) to less than 2 nm (one monolayer). Fluorescence spectroscopy has been used in the present study to investigate the microstructure of monolayer polymer films. With this technique, we are able to obtain the necessary signal-to-noise ratio to perform accurate measurements. Extrinsic fluorescence is capable of acting as a very sensitive probe of the polymer structure. A particular example is the use of pyrene excimer fluorescence, which has been employed in this study. The requirement for pyrene excimer formation is that two pyrene groups face each other in a sandwich arrangement at a separation distance of between 0.3 to 0.4 nm. Any structure changes on this scale will perturb the excimer forming sites and thus be detected.

Experimental

Materials. The nearly monodisperse atactic PMMA, which was used for the electron beam lithography and fluorescence spectroscopy studies, was obtained from Pressure Chemical. It has a weight average molecular weight (Mw) of 188,100 and Mw/Mn < 1.08. Pyrenedodecanoic acid (PDA) used in the fluorescence studies was obtained from Molecular Probes and used as supplied. Spectroscopic grade benzene purchased from J.T. Baker was used as the spreading solvent in the PMMA and PMMA/PDA solutions.

The novolac sample, which was provided by Kodak, was synthesized from pure meta-cresol and formaldehyde. It has a weight average molecular weight of 13,000 and a very broad polydispersity of 8.5. The polymer was purified by two precipitations from tetrahydrofuran into hexane. The PAC was a naphthoquinone-1,2-(diazide-2-)-5-sulfonyl ester provided by Fairmount Chemical (Positive Sensitizer 1010). A hydroxyl substituted benzophenone is attached to the sulfonyl ester. The spreading solvent was isopropyl acetate, which was obtained from Aldrich Chemicals and used as received.

Substrates and LB film Preparation. The substrates used in the electron beam lithography studies consisted of 50-nm chromium (Cr) films evaporated over 100-nm thermally grown silicon oxide formed onto 4-inch silicon wafers. The Cr and oxide layers provide an excellent contrast for evaluating etched Cr patterns with both optical and scanning electron microscopes (SEM). The substrates for the optical exposures were 3-inch silicon wafers upon which 50-nm chromium films had been evaporated. Three-inch quartz wafers obtained from Shin-Etsu Chem. Co. were used as substrates for the fluorescence measurements. The substrates were cleaned by immersion in a 9/1- H_2SO_4/H_2O_2 solution at 120 °C for 20 min, followed by six cycles of deionized water rinsing and a final spin drying under a nitrogen ambient.

The LB film depositions were performed using a Joyce-Loebl Langmuir Trough IV equipped with a microbalance for measurement of the surface pressure by the Wilhelmy plate method. Filtered deionized water with a pH of 7 was used for the subphase. For the electron beam lithography study, PMMA was spread on the water surface from a dilute benzene solution (~ 10 mg PMMA in 20 ml benzene). The novolac/PAC mixtures were spread from solutions (~ 20 mg solids in 10 ml solvent) of isopropyl acetate. For the fluorescence studies, the PMMA/PDA mixture was spread on the water surface from a dilute benzene solution (1.75 mg PDA and 8.33 mg PMMA in 20 ml benzene). Prior to compression, a 20 min interval was allowed for solvent evaporation. The Langmuir film was compressed to the desired transfer pressure at a rate of 50 cm^2/min, followed by a 20 minute equilibration period. The Cr-coated silicon wafers and quartz wafers were immersed into the subphase before the PMMA was spread on the water surface. The first monolayer of PMMA was transferred during the first upstroke of the substrate, at the speed of 2 mm/min.

Electron Beam Exposure. Single component LB PMMA films transferred at 5, 11, 15, and 17 dyn/cm, with thicknesses 0.9 nm (1 layer) to 15.3-nm (17 layers), have been prepared and investigated as high resolution electron beam resists by exposure with a modified Perkin Elmer MEBES I pattern generation system. The MEBES exposures were performed at a 20 MHz address rate, 10 kV accelerating voltage, 1/8 μm beam diameter and address size, and a 6 nA beam current resulting in a dose of 2 $\mu C/cm^2$ per scan. Equal line-space patterns with nominal feature sizes from 1.25 μm down to 0.125 μm were written. The dose ranges for this study were 1 - 200 $\mu C/cm^2$. After exposure PMMA was developed in a 3:7 2-ethoxyethanol:methanol solution for 13 sec.. Prebaking and postbaking conditions were 100°C for 2 hours and 90°C for 30 min, respectively. Following the postbaking process the samples were put in a Cr etching solution (Cyantek CR-14) for 30 sec. to transfer the resist pattern to Cr. The samples were then examined with a scanning electron microscope (SEM).

Optical Exposure. Multicomponent LB films were prepared from solutions of novolac/PAC varying in concentration from 5 - 50 wt% PAC, and transferred at 2.5 - 10 dyn/cm. The films were composed of 15 - 20 monolayers, with an average film thickness of 30 nm, as measured by ellipsometry. Exposures were performed with a Canon FP-141 4:1 stepper (primarily g-line exposure) at an exposure setting of 5.2 and with a fine line test reticle that contains line/space patterns from 20 to 1 μm (40 to 2 μm pitch). They then were then developed in 0.1 - 0.2 M KOH, depending on the PAC content. The wafers received a 20 min 120°C post development bake to improve adhesion to the Cr. Finally, the Cr was etched in Cyantek CR-14 chromium etchant, and the resist and Cr images were examined by SEM.

Fluorescence Measurement. Fluorescence spectra were measured on a Spex Fluorolog 212 spectrofluorometer equipped with a 450 W xenon arc lamp and a Spex DM1B data acquisition station. Spectra were recorded in the front-face illumination mode using 343 nm as the excitation wavelength. Single scans were performed using a slit width of 1.0 mm. PDA fluorescence emission spectra were recorded from 360 to 600 nm, with the monomer and excimer fluorescence measured at 376.5 and 485 nm, respectively. Monomer and excimer peak heights were used in the calculation of the ratio of excimer to monomer emission intensities (Ie/Im). Excitation spectra were recorded from 300 nm to 360 nm and monitored at 376.5 and 500 nm for the monomer and excimer excitation, respectively.

Results

Electron Beam Lithography. LB PMMA films with thicknesses greater than 6.3 nm withstood the 50-nm Cr etching and allowed the patterns to be transferred from the resists into Cr films. Fig. 1 is an example of the patterns in Cr employing a 8.1 nm (9 layer) LB PMMA film, transferred at 15 dyn/cm, as a positive electron beam resist. Resist films prepared at different transfer pressures did not appear to differ substantially in their lithographic performance. Those films thinner than 4.5 nm (5 layers) proved unsuitable for withstanding the chromium etch once they had been cycled through MEBES (even though nominally unexposed).

Optical Lithography. For the novolac/PAC films, the PAC concentration was varied rather than the film thickness. Films with 5 % PAC could not withstand the Cr

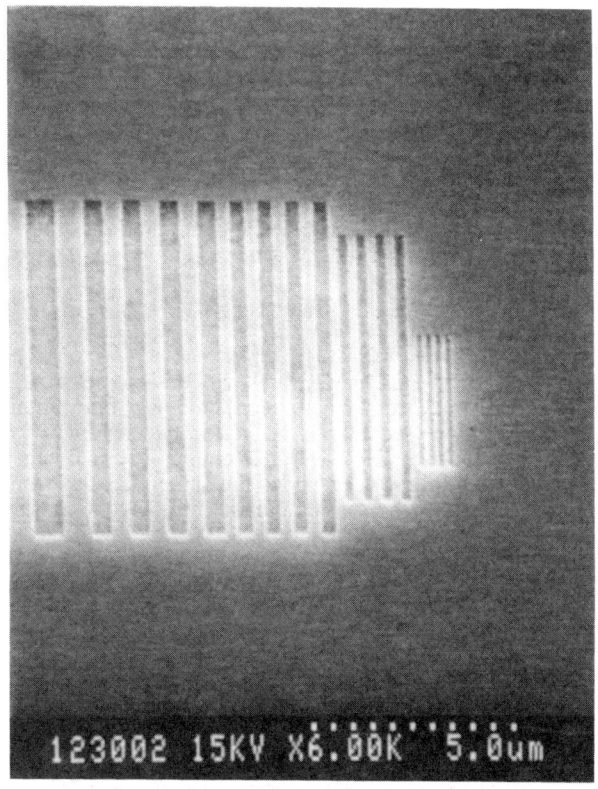

Figure 1. Patterns in 50 nm Cr film employing a 8.1 nm (9 layers) LB PMMA film as resist, exposed with a dose of 80 µC/cm^2. The linewidths of the patterns are 1/8, 1/4, 3/8, and 1/2 µm, respectively.

etchant. Films with 10 and 20 % PAC withstood the etchant and produced resolvable 1x1 um arrays (Fig. 2). The dose and development conditions would have to be optimized to achieve lines of the proper dimensions. Development of the films with 40 - 50 % PAC yielded observable resist images, but these images could not be replicated by Cr etching. An organic layer apprently remained in the bottom of the images and inhibited dissolution. There are also small defects in the films with high PAC concentrations (40 - 50 %), which do not appear in the films with lower PAC concentrations (Fig. 3). Films prepared at various transfer pressures did not exhibit significant lithographic differences.

Fluorescence Measurements. LB monolayer PMMA films doped with 5 mol% PDA have been prepared at surface transfer pressures ranging from 1 to 17 dyn/cm. Examples of PDA emission and excitation spectra are given in Figs. 4 and 5 respectively. The fluorescence measurements on these films have been plotted in Fig. 6 as I_e/I_m vs. surface pressure. I_e/I_m increases as the surface pressure increases, and decreases as the pressure passes some critical value ($\sim$ 10 dyn/cm). A summary of the results of the fluorescence excitation spectra is presented in Table I. The excitation spectrum, which is analogous to the absorption spectrum, is generated by monitoring the emission intensity at a specific wavelength as a function of excitation wavelength. From the excitation spectrum we can analyze the configurations of different absorbed species. All the excimer excitation spectra were red shifted ($\sim$ 4 nm) and broadened relative to that of the monomer.

Discussion

Electron Beam Lithography. LB PMMA films with thicknesses of 6.3 nm (7 layers) are sufficient for patterning a Cr film suitable for photomask fabrication. For ultrathin PMMA films the resolution (see Fig. 1) is limited by the smallest spot diameter available on MEBES I (1/8 μm). However, it is not possible to obtain this resolution if a thicker resist (> 100 nm) is used under the same exposure and development conditions, which demonstrates that ultrathin resists are able to minimize the proximity effect. Also, since the radius of gyration of 188,100 Mw PMMA is about 10 nm in the bulk, and the thickness of the 7 layer film (6.3 nm) is less than 10 nm, it is reasonable to assume there must be an alteration of chain configuration in the ultrathin films. This will be particularly true when the post-deposition baking temperature of the multilayer films is less than the glass transition temperature (115°C), as is the case for the present experiments. In such a case, interdiffusion of PMMA chains between the deposited layers may not result in chain configurations characteristic of the bulk.

Optical Lithography. Resist films can be prepared from LB films of polymer/small molecule mixtures. Fig. 7 gives a schematic representation of a Langmuir film of novolac and diazoquinone mixture. Single layers of novolac are substantially thicker ($\sim$ 2 nm) than PMMA monolayers ($\sim$ 1 nm) suggesting that it is not possible for the novolac to have all of the phenolic groups in contact with the water surface due to the polymer's branched structure. Images in the novolac/diazoquinone films could be resolved image down to about the resolution limits of the optical exposure tool used ($\sim$ 1 μm). Future electron-beam and deep UV exposures will allow us to evaluate the resolution of submicron images.

The homogeneity of the components in the LB films may affect the uniformity of dissolution in the exposed regions; however, aggregation of the PAC or novolac would cause regions with slower or faster dissolution, respectively. If the aggregated regions reached $\sim$ 0.2 μm or larger, they could create observable roughness on the edges of the developed and etched images. In general, the edges of the images appear to be quite smooth for PAC concentrations at or below 20 wt%, although at high

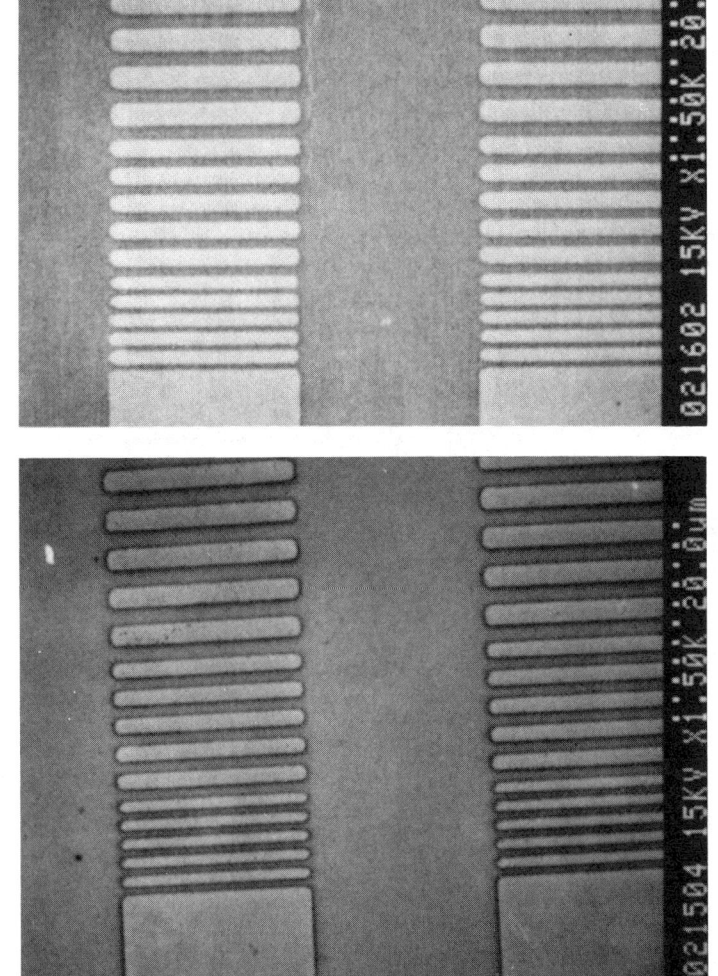

Figure 2. Patterns in 50 nm Cr films employing 15 layers (~ 30 nm) of LB novolac/diazoquinone resist, with (a) 10 wt% and (b) 20 wt% diazoquinone. The smallest arrays are nominally 1 μm lines-and-spaces.

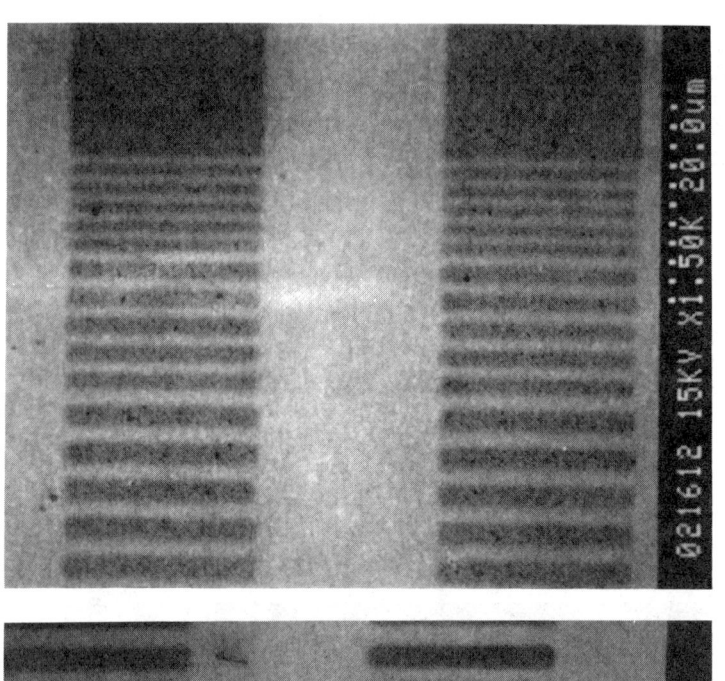

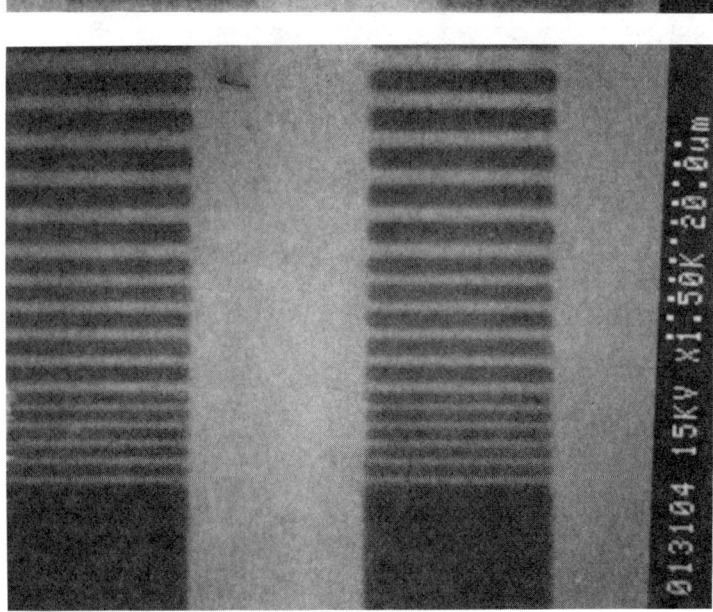

Figure 3. Developed patterns in 15 layers of LB novolac/diazoquinone resist with (a) 20 wt% and (b) 50 wt% diazoquinone.

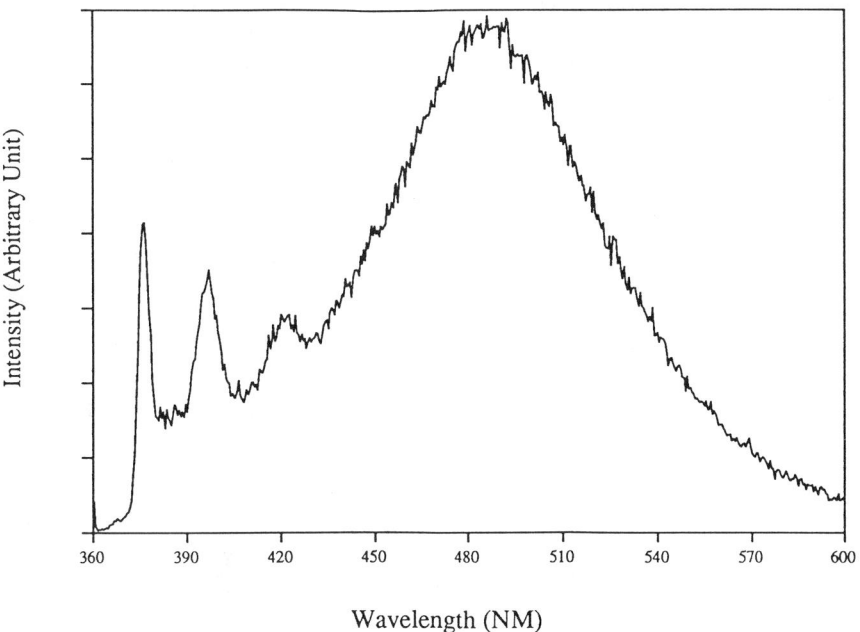

Figure 4. Fluorescence emission spectrum of 5 mol% PDA in monolayer LB PMMA film transferred at 7 dyn/cm. The excitation wavelength is 343 nm.

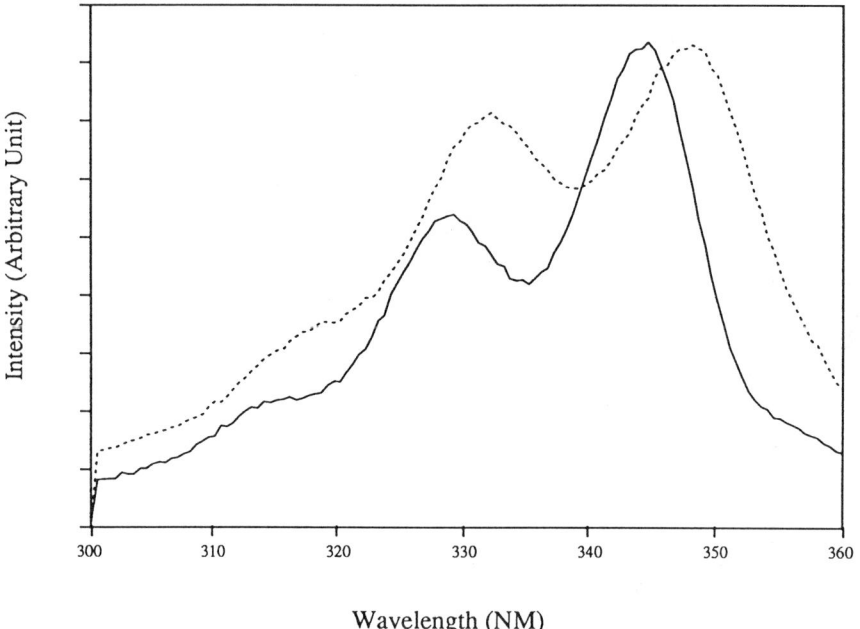

Figure 5. Fluorescence excitation spectra of 5 mol% PDA in monolayer LB PMMA film transferred at 7 dyn/cm. The excimer excitation (dashed curve) was monitored at 500 nm and monomer excitation (solid curve) was monitored at 376.5 nm.

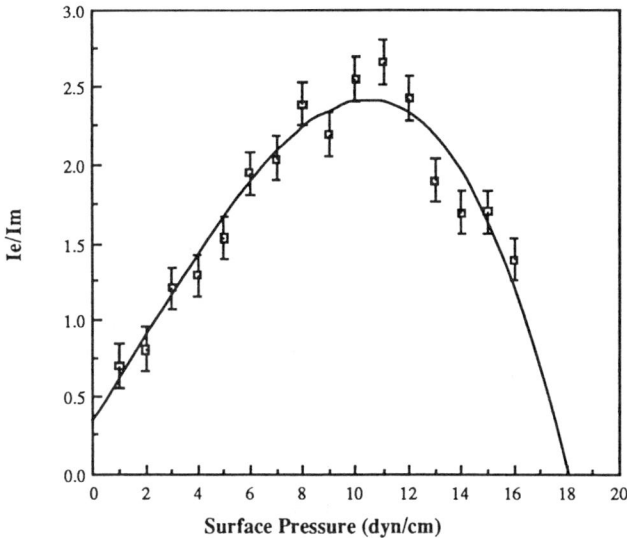

Figure 6. Ratio of excimer (Ie) to monomer (Im) fluorescence intensities of 5 mol% PDA in monolayer LB PMMA film as a function of transfer pressure.

Table I. Summary of the Results of PDA Fluorescence Excitation Spectra (188K Mw PMMA)

Surface Pressure (dyn/cm)	λ_{Max} (nm)[a]		Peak Broadening[b]	
	Excimer Exc.	Monomer Exc.	Excimer Exc.	Monomer Exc.
2	348	344	1.18	1.69
4	348	344	1.26	1.73
6	347.5	344	1.45	2.06
8	347.5	344	1.42	1.95
10	348	344	1.49	2.14
12	347.5	344	1.38	1.95
14	347.5	344.5	1.50	2.17
16	347.5	344.5	1.51	2.26

a. λ_{Max} is the wavelength at which the global maximum peak intensity of the PDA excitation spectrum occurs.
b. Peak Broadening is defined as the ratio of the intensity of the maximum peak of the PDA excitation spectrum to the intensity of its adjacent local minimum.

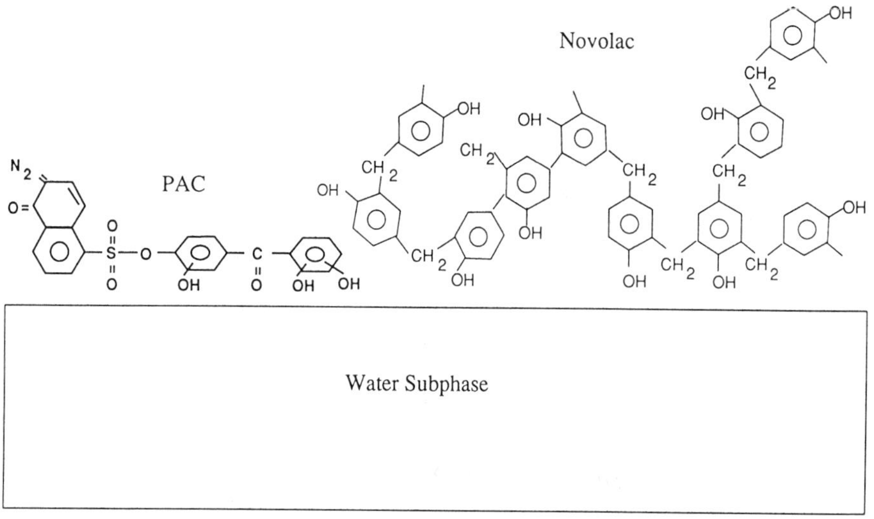

Figure 7. Schematic of Langmuir film of mixtures of novolac and diazoquinone (PAC).

concentrations (40 - 50 wt%) small (0.1 - 1 μm) variations are observable in the films. These may be related to the aggregation of the film components. Larger defects up to several hundred microns and thickness variations also occur in all of the films. It is not clear if these are directly related to aggregation, as thickness variations occur in films of pure novolac also.

Fluorescence Measurements. When the PMMA/PDA mixture forms a Langmuir film, the hydrophilic C=O groups of PMMA are expected to be directed toward the water phase with the chain backbone lying generally parallel to the water surface. The COOH groups of PDA are also expected to be directed toward the water with the long aliphatic chain and pyrene groups directed upward. Upon compression of the Langmuir film, the PMMA chains presumably become more compact and finally collapse. The collapse pressure of PMMA has always been a controversial issue in the literature. This has arisen because of different definitions and methods of determining collapse pressure, and different stereoregularity of the PMMA. Stroeve et al. (16) provides a detailed discussion of these points. They suggest that for Langmuir layers of syndiotactic PMMA (Mw ~ 100,000) there are two important regimes: a regime where reversible loss becomes important, at about 15 dyn/cm, and a regime where there is an irreversible collapse of the Langmuir layer, at surface pressures greater than or equal 34 dyn/cm. While the reversible loss mechanism must be associated with some of the polymer being forced out of the monolayer, this layer must be available to be drawn back into the interface upon expansion of the Langmuir layer. For the atactic-PMMA (Mw ~ 188,100) used in this study, these two regimes were observed from the pressure-area measurements.

5 mol% PDA was used as a probe to detect structural changes in the PMMA matrix. As the pressure increases, the two dimensional surface concentration of PDA increases and thus the probability of intermolecular excimer formation will be enhanced, causing Ie/Im to rise. However, once the pressure exceeds 10 dyn/cm, Ie/Im starts to decrease indicating that the structure of the excimer has been influenced by microstructural changes in the polymer matrix. The excitation spectra of PDA at different surface pressures indicate that the extent of the red shift (excimer vs. monomer), which is due to the ground state interaction of the dyes, is not influenced by the surface pressure in the range 0 - 17 dyn/cm. This suggests that the probe molecules are not unduly constricted as the inter/intra chain voids shrink in size during compression. In other words, as the pressure increases, the voids become smaller and consequently Ie/Im increases. However, despite the fact that the probes are forced to occupy a smaller area, they are not being constrained to the extent that they experience a change in ground state interaction. As the pressure surpasses ~10 dyn/cm, the polymer chain may start to buckle and form loops that extend above the monolayer, thereby being disruptive to excimer formation in this region. No clear picture on the molecular level is presently available. However, the spirit of our concept is contained in the highly schematic representation of Fig. 8.

Summary

Ultrathin LB PMMA and novolac/diazoquinone films have been demonstrated to act as high resolution electron beam and optical resists, respectively. Structural rearrangements in the LB PMMA films have been observed by using fluorescence spectroscopy. However, this rearrangement did not appear to influence the lithographic performance when seven or more layers of LB PMMA films were used as the resist. A more comprehensive study of the relationship between lithographic performance and LB film structure is currently underway.

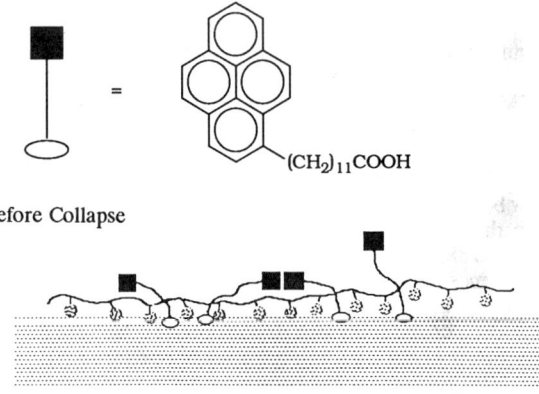
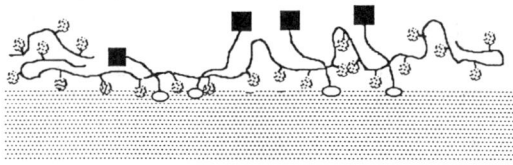

Figure 8. Schematic of PDA dyes in monolayer Langmuir PMMA film on the water surface before and after the film collapse.

Acknowledgments

The authors are grateful to Paul Jerabek for his help with the MEBES exposure. This study was supported by the Chemistry Division of the Office of Naval Research under Contract N00014-87-K-0426. P.S.M. would like to thank the Natural Sciences and Engineering Research Counsel of Canada for a post-doctoral fellowship. L.L.K. would like to thank IBM for support through their Resident Study Program.

References

1. McCord, M. A.; Pease, R. F. W., J. Vac. Sci. Technol. 1986, B(4), 86-88.
2. Kuan, S. W. J.; Frank, C. W.; Fu, C. C.; Allee, D. R.; Maccagno, P.; Pease, R. F. W., J. Vac. Sci. Technol. 1988, B6 (6), 2274-2279.
3. Jones, R.; Winter, C. S.; Tredgold, R. H.; Hodge, P.; Hoorfar, A., Polymer 1987, 28, 1619.
4. Barraud, A.; Rosilio, C.; Ruaudel-Teixier, A., Solid State Technol. 1979, Aug., 120.
5. Fariss, G.; Lando; J., Rickert, S., Thin Solid Films 1983, 99, 305.
6. Miyashita, T.; Yoshida, H.; Matsuda, M., Thin Solid Films 1987, 155, L11.
7. Broers, A. N.; Pomerantz, M., Thin Solid Films 1983, 99, 323.
8. Baraud, A.; Rosilio, C.; Ruaudel-Teixier, A., Thin Solid Films 1980, 68, 91.
9. Huynh, B., M.S. thesis University of California, Berkeley, 1988.
10. Owen, G.; Rissman, P., J. Appl. Phys. 1983, 54, 3573-3581.
11. Oagawa, K., Jap. J. Appl. Phys. 1988, 27, 855-860.
12. Hodge, P.; Khoshdel, E.; Tredgold, R. H.; Vickers, A. J.; Winter, C. S., Br. Polym. J. 1985, 17, 368.
13. Mumby, S. J.; Rabolt, J. F.; Swalen, J. D., Thin Solid Films 1985, 133, 161.
14. Mumby, S. J.; Swalen, J. D.; Rabolt, J. F., Macromolecules, 1985, 19, 1054.
15. Kosbar, L. L.; Kuan, S. W. J.; Frank, C. W.; Pease, R. F. W. Radiation Chemistry of High Technology Polymers; Reichmanis, E.; O'Donnell, J. H., Ed.; ACS Symposium Series , 1988; 381, 95-111.
16. Stroeve, P.; Srinivasan, M. P.; Higgins, B. G.; Kowel, S. T., Thin Solid Films, 1987, 146, 209-220.

RECEIVED June 29, 1989

Chapter 22

Dissolution of Phenolic Resins and Their Blends

J. P. Huang, E. M. Pearce, A. Reiser, and T. K. Kwei

Polymer Research Institute, Polytechnic University, Brooklyn, NY 11201

This work dealt with the kinetics of the dissolution of phenolic resin/polymeric inhibitor (sulfone, or methacrylate polymers) systems in alkali solutions in order to gain some insight into the dissolution mechanism. The effects of alkali concentration, size of the cation, salt addition and segmental mobility of resist on dissolution were studied. In addition, the relation between the chemical structure of a polymeric inhibitor and its effectiveness in retarding dissolution was explored in terms of the hydrophobicity/hydrophilicity. Based on these results, a model is proposed to account for most of the salient features of novolac dissolution.

Novolac resins, as the oldest synthetic polymers, have played an important role in microelectronic industry as positive photoresists. Studies of novolac dissolution have populated the literature; a recent survey shows that the rate of dissolution is influenced by the concentration of the alkali, size of the cation, addition of salt, and the presence of dissolution inhibitors (1-6). The voluminous experimental results, however, have not led to a clear understanding of the dissolution phenomena. Arcus (3) proposed an ion-permeable "membrane" model while Szmanda (4) and Hanabata (6) emphasized the importance of secondary structures of novolac molecules, for instance, inter- or intramolecular hydrogen bonding and the various isomeric configurations of the resins. These important contributions nevertheless point to a need for additional studies of the mechanism of dissolution.

This paper presents new data on dissolution kinetics. The effects of alkali concentration, size of the cation, and salt addition were studied. The influence of segmental mobility on dissolution was elucidated by measuring the temperature coefficients of the dissolution rates. Experiments were also carried out to study the relation between the chemical structure of a polymeric inhibitor and its effectiveness in retarding dissolution. Based on these results, a model for novolac dissolution is proposed to account for most of the salient features of novolac dissolution.

Experimental

Materials and Purifications. 2-Methyl pentene-1 (Aldrich) was distilled under normal pressure at 62°C after refluxing for one hour in the presence of $AlLiH_4$. Methyl methacrylate and para-methyl styrene (Aldrich) were distilled under reduced pressure at about 60°C; the purified monomers were sealed and stored in refrigerator before use. The compressed gas, sulfur dioxide (Matheson), was led through a P_2O_5 tower before introduced into reaction system. Hydroxyethyl acrylate (Aldrich) was used for polymerization without further purification.

Poly(1-hydroxy-2,6-methylene-phenylene) (PHMP), ortho, ortho'-coupled linear novolac, and para-substituted PHMPs: p-F-PHMP, p-Cl-PHMP, p-Br-PHMP, p-NO_2-PHMP, p-CH_3O-PHMP and p-CH_3-PHMP were synthesized according to the procedures described in reference (7). Varcum-2217 was supplied by BTL Specialty Resins Corp., polybisphenols (PBPh) through the courtesy of Mead Corp., which were fractionated by precipitation with a non-solvent; two PBPhs with distinct molecular weights were obtained. Poly(methyl acrylate) (PMA), poly(ethyl methacrylate) (PEMA), poly(butyl methacrylate) (PBMA) and poly(hexyl methacrylate) (PHMA) from Aldrich, and poly(methyl methacrylate) (PMMA) from Polyscience. The number average molecular weights and glass transition temperatures of above phenolic resins were determined on a Model 115 Perkin-Elmer Vapor Pressure Osmometer and a DuPont 910 Differential Scanning Calorimeter respectively. The results are tabulated in Table I.

Table I. Description of the phenolic resins

Polymer	Mn	Tg (°C)
PHMP	1000	92
p-F-PHMP	1000	113
p-Cl-PHMP	820	84
p-Br-PHMP	1300	108
p-NO_2-PHMP	900	125
p-OCH_3-PHMP	900	67
p-CH_3-PHMP	670	99
PBPh-1	1330	115
PBPh-2	4570	132
Varcum-2217	820	89

Syntheses. The polymerization reaction of poly(2-methyl pentene-1 sulfone) (PMPS) was carried out at -78°C. Purified 2-methyl pentene-1 (42 grams) and condensed SO_2 (about 125 grams) at a molar ratio of 1 to 4 were charged into the reaction system under atmospheric pressure, and the reaction was initiated by 2 milliliters of butyl hydroperoxide. The white polymer mass was purified by dissolving in acetone, then precipitating into methanol (8).

Copolymerization of MMA with hydroxyethyl acrylate (HEA) or para-methyl styrene (MST) was carried out in ethanol solution by free radical initiation (9-11). AIBN was used as the initiator at 0.3% by weight of the monomers. Each reaction was conducted at 70°C under nitrogen atmosphere, and terminated at a conversion about 10% to obtain the desired compositions. The copolymers were purified by the same procedure used for PMPS. The weight average molecular weights of MMA-HEA and MMA-MST copolymers range at 107,000 and 70,600, respectively, by GPC analysis with monodispersed polystyrenes as standards. The feed molar fractions, copolymer compositions, which were determined by ^{1}H-NMR for the MMA-HEA copolymer and UV spectroscopy for the MMA-MST copolymer, and reaction conversions for the copolymerization reactions are tabulated in Table II.

Table II. Characterization of MMA-copolymers

Copolymer	f_1	F_1	Conversion%
PMMA-c-HEA-1	0.05	0.02	0.14
PMMA-c-HEA-2	0.10	0.04	0.11
PMMA-c-HEA-3	0.20	0.10	0.10
PMMA-c-MST-1	0.05	0.11	0.13
PMMA-c-MST-2	0.12	0.21	0.15
PMMA-c-MST-3	0.21	0.31	0.10

Dissolution Measurement. Resist solutions in mixtures of isoamyl acetate/cyclohexanone/methyl isobutyl ketone (90:5:5 by volume) were filtered through 0.45 µm disc filters, then spin-coated onto silicon wafers at about 2000 rpm. The coated wafers were prebaked in a convection oven at 90°C for 1 hour, then stored in a desiccator. The basicities of the alkaline solutions were titrated by a standard HCl solution with a Fisher Accument pH meter, Model 805 MP. The film thickness is about 2 µm. Resist dissolution was measured by a He-Ne laser interferometer in a thermostated bath at the desired temperatures (12,13).

Water Diffusion. Water sorption of polymethacrylate/Varcum-2217 blends was measured by immersing dried film into water for various time periods. The wet films were pat-dried with tissue papers, followed by blowing nitrogen on the surface. The films before and after water immersion were carefully weighed with a 4-digit electric analytic balance (14).

Contact Angle Measurement. Contact angles of water, or glycerol on films of a series of methacrylate polymers were measured using a model 100-00 goniometer by Rame-Hart. A microslide coated with the blend film was sealed in a plastic box filled with a saturated atmosphere of the probe liquid. The size of the liquid droplet was controlled by a microsyringe. Contact angles were recorded after 2 minutes to allow for stabilization of the liquid drop.

Results and Discussion

Dependence on Base Concentration. The dissolution rates of substituted PHMPs at different alkali concentrations are displayed in Figure 1 for seven different novolac resins. In each case, there appears to be a limiting concentration C_o below which the rate of dissolution is too slow to be measured in the experimental time scale. The ascending portion of the curve can be represented by a power law dependence of the rate on concentration C, eq.(1),

$$R = A C^n \qquad (1)$$

as has been reported in the literature (2,5). The calculated values of n and A are listed in Table III. The magnitude of n and A ranges from 2.6 to 3.5 and 0.65 to 13630, respectively.

Table III. n and A values of PHMPs

Resin	n	A
p-NO$_2$-PHMP	2.6	13300
P-CH$_3$O-PHMP	3.4	13630
PHMP	3.4	5830
p-Cl-PHMP	2.6	540
p-F-PHMP	3.5	134
p-Br-PHMP	3.1	3
p-CH$_3$-PHMP	2.8	0.65

A careful examination of the data reveals that at least in some cases the rates deviate from the simple power law at low alkali concentrations. Such deviations are shown in Figure 2a for p-Cl-PHMP and PBPh-1. An excellent fit of the data over the entire concentration range was found by using $(C - C_o) = C_e$ in eq.(2) instead of C, where C_o is the limiting concentration,

$$R = A' C_e^{n'} \qquad (2)$$

Coincidently, the exponent of C_e is 2.1 for both resins (Figure 2b) (Table IV), but additional experiments are required to confirm the general applicability of eq.(2).

When a higher molecular weight polybisphenol resin, PBPh-2, was used, the exponent for C_e was determined to be 2.4 (Figure 2b) (Table IV). The change in the exponent is minor when the molecular weight of the resin increases by a factor of about 3. The limiting concentration, however, almost doubled. We are not certain of the significance of the different C_o values for the two PBPh samples. Possible differences in microstructures may contribute to our observation. The A' values vary also with the type of resins and their molecular weights (Table IV).

Effect of Added Salt. It is well known that the presence of salt promotes dissolution (1,3). For a fixed NaOH concentration, the dissolution rates of both p-Cl-PHMP and PBPh-1 were found to increase

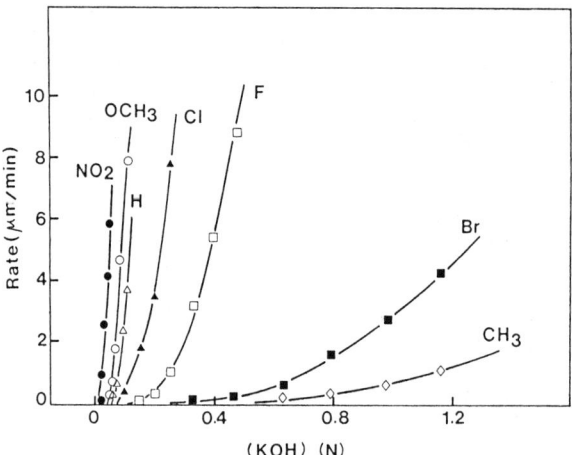

Figure 1. Dissolution of para-substituted PHMPs at different KOH concentrations.

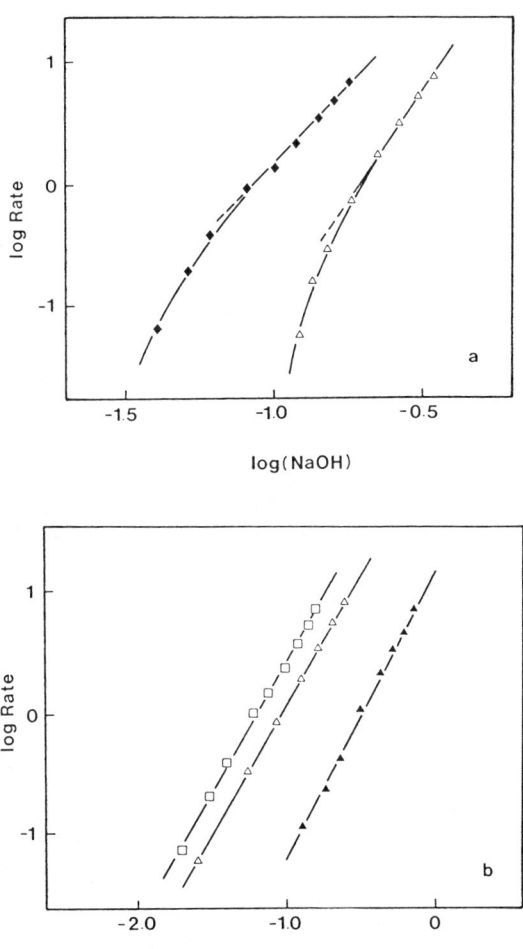

Figure 2. Dissolution of p-Cl-PHMP and PBPhs in NaOH solutions at 20.0°C. (a) log(Rate) plotted against log(NaOH), ◆ p-Cl-PHMP, △ PBPh-1; (b) log(Rate) plotted against log(△NaOH), □ p-Cl-PHMP, △ PBPh-1, ▲ PBPh-2.

Table IV. C_o, n' and A' values of p-Cl-PHMP and PBPhs

Resin	Mn	C_o	n'	A'
p-Cl-PHMP	820	0.020N	2.1	326
PBPh-1	1,330	0.095N	2.1	153
PBPh-2	4,570	0.17N	2.4	15

at first with increasing NaCl concentration and then reached plateau values at high salt concentrations (Figure 3). The rates can be represented by eq.(3),

$$\frac{R - R_o}{R_p - R_o} = \frac{B(NaCl)^m}{1 + B(NaCl)^m} \quad (3)$$

In eq.(3), R_o is the rate in the absence of NaCl and R_p is the plateau value. Again the values of m are identical for p-Cl-PHMP and PBPh-1, namely, 2.0.

When the dissolution rates at fixed NaCl concentrations are plotted against NaOH concentration, additional information is obtained:

(a) The C_o value, obtained by extrapolating the rates to zero, decreases with increasing salt concentration (Figure 4).

(b) The dependence of dissolution rate on alkali concentration still obeys a power law relation with C_e, but the exponent n decreases as salt concentration increases (Table V).

Table V. NaCl effect on dissolution kinetics

[NaCl] (N)	$[NaOH]_o$ (N)	n'
0	0.02	2.1
0.05	0.01	1.83
0.10	0.0087	1.74
0.15	0.0075	1.69
0.20	0.0063	1.66
0.25	0.0050	1.64
:	:	:
>0.40	0	1.5

Cation Size. In their early studies, Hinsberg ([1]) and Arcus ([3]) found that dissolution rates of resists decreased as the size of the cation of the base increased. Our results support their conclusion. In Figure 5, the dissolution rates of a PMPS(10%)/p-NO_2-PHMP film in different alkali solutions clearly show a decreasing trend with increasing cation size. In fact, the rate is inversely proportional to the cross-sectional area of the unhydrated cation (Figure 6). It is known in the diffusion of small molecules in polymers, the diffusion constant is inversely related to the size of the molecule ([15]). The observed dependence of dissolution rate on cation size is therefore suggestive of cation diffusion as a crucial step. It is

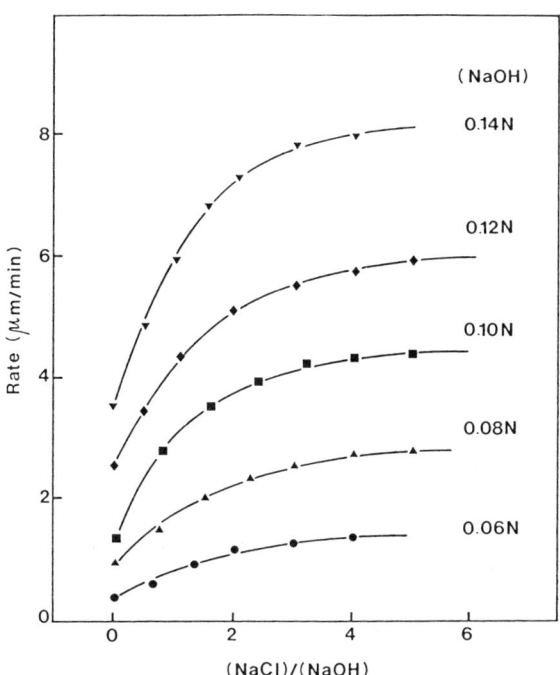

Figure 3. Effect of added salt on dissolution of p-Cl-PHMP at 20.0°C.

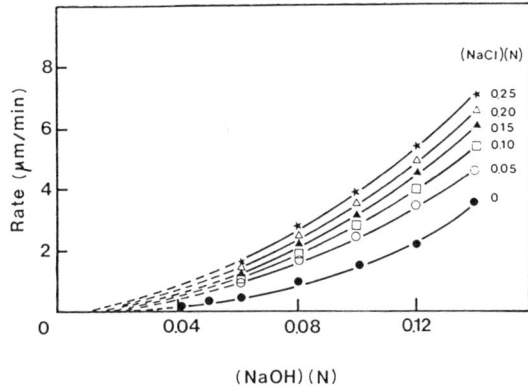

Figure 4. Effect of added NaCl on dissolution kinetics at fixed NaCl concentrations.

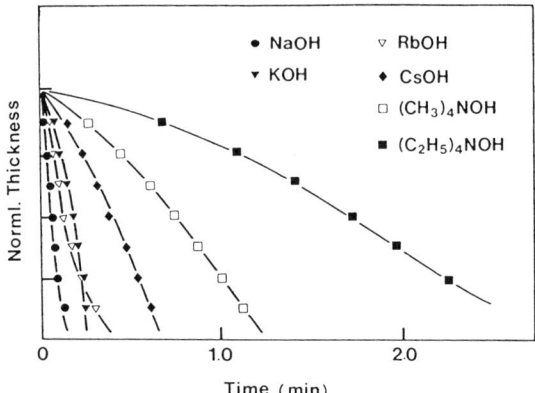

Figure 5. Dissolution of PMPS(10%)/p-NO_2-PHMP in different alkalis (0.08N, 27.5°C).

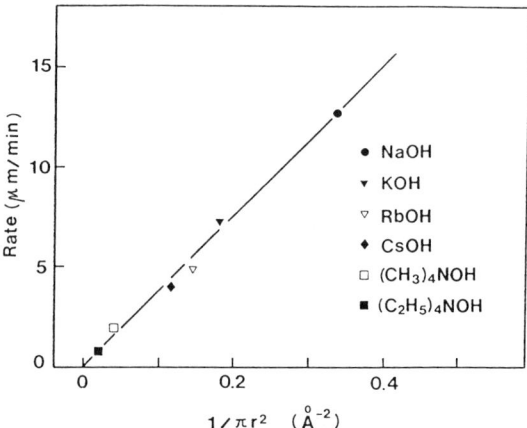

Figure 6. Dissolution rates of PMPS(10%)/p-NO_2-PHMP plotted as a function of the reciprocal of the cation cross-sectional areas.

conceded that a note of caution is due here in comparing results obtained at a single alkali concentration and a single temperature because we have not established that the power law parameters are the same for different bases. Therefore, additional experiments were conducted to determine the temperature coefficients of the rate processes from which the apparent activation energies of dissolution could be calculated.

<u>Influence of Chain Mobility</u>. The results of such measurements are displayed in Figure 7 for p-Cl-PHMP and poly(methyl acrylate)(15%)/Varcum-2217 blend. For PMA(15%)/Varcum blend, activation energy below Tg' was unable to be calculated because of insufficient data. As temperature increased from 5°C, the dissolution rate increased at first only slowly but then much more rapidly after a certain temperature was reached. The intersection (extrapolated) of the two sections of the curve can be interpreted as the glass transition temperature, Tg', of the medium as perceived by the diffusant (<u>16,17</u>). It is necessary to distinguish the observed Tg' from the Tg of the dry film because the former reflects the segment mobility of the partially solvated film as seen by the cation in the dissolution process. The larger cation requires, for its diffusion, the cooperative motion of a larger number of segments and hence "sees" a higher Tg'.

Both Tg' and the activation energy below Tg' increase with cation size (Table VI). The activation energy values for p-Cl-PHMP compare favorably with the results of ion conductivity measurements in cellulose acetate (<u>16,18</u>), shown in Figure 8. This relationship speaks strongly for cation diffusion as being involved in the rate determining step.

Table VI. Tg' and Ea' of Dissolution in Different Alkali Solutions

Cation	Ion Volume (A)	Tg' (°C) 1	Tg' (°C) 2	Ea' (kJ/mole) 1
Na^+	3.94	11.3	40.5	45.9
K^+	9.86	17.6	44.0	57.2
Rb^+	13.58	29.7	45.3	71.8
Cs^+	19.51	-	50.3	88.0

1. p-Cl-PHMP
2. PMA(15%)/Varcum-2217; insufficient data for calculating Ea'.

If cation diffusion is indeed involved in the rate determining step, the dissolution rates at temperatures above Tg' are expected to obey a WLF-type relationship (17). In the present context, the ratio of the dissolution rate at temperature T to that at Tg' replaces the shift factor,

$$\log \frac{R_T}{R_g} = \frac{C_1 (T - Tg')}{C_2 + (T - Tg')} \quad (4)$$

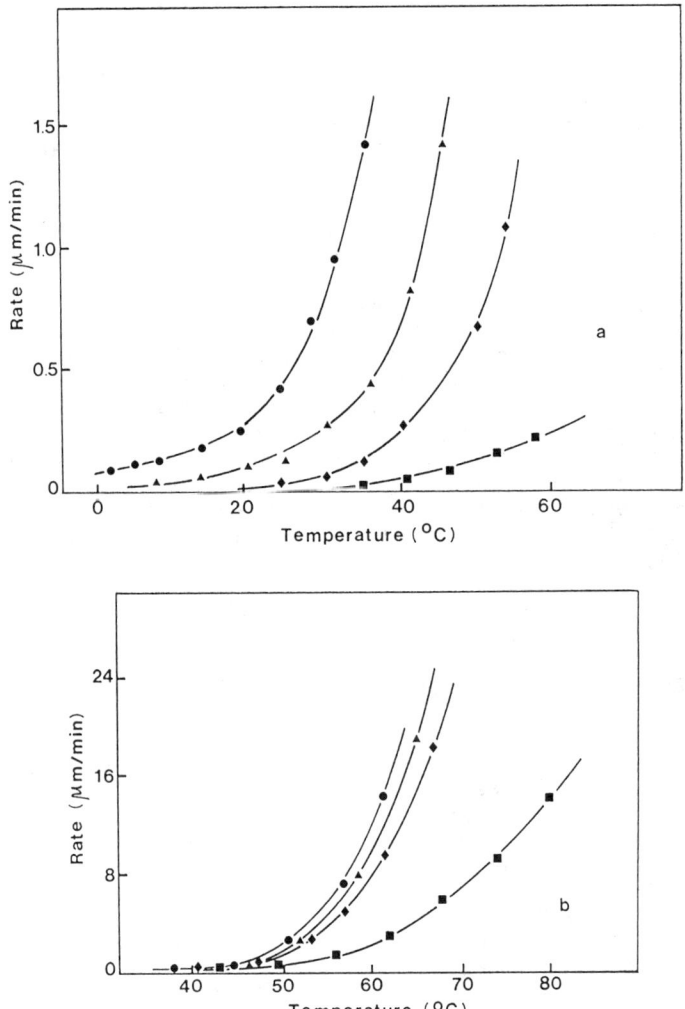

Figure 7. Dissolution of p-Cl-PHMP (0.06N) (a) and PMA(10%)/Varcum-2217 (0.8N) (b) in NaOH (●), KOH (▲), RbOH (♦) and CsOH (■) solutions at varied temperatures.

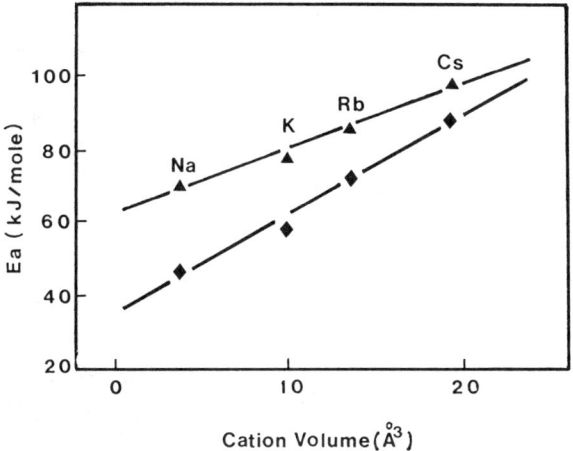

Figure 8. Activation energies of p-Cl-PHMP dissolution in different alkalis (♦) and ionic conductivities in cellulose acetate (▲).

According to eq.(4), a plot of $1/\log(R_T/R_g)$ v.s. $1/(T - Tg')$ should yield a straight line with a slope of C_2/C_1 and an intercept of $1/C_1$. Our results conform to eq.(4), as can be seen in Figure 9, and it is gratifying that the data for different cations fall on a single line. But the intercepts yield C_1 values of 21.7 and 4.3, respectively, rather than the WLF value of 17.4; the ratios of C_2/C_1 are 23.7 and 8.2 for the two films studied, in comparison with the value of 3.0 for the WLF equation (17). We note that in ionic conductivity measurements (19-22), the results have also been found to fit WLF-type equation but with different C_1 and C_2 values.

Polymer Inhibitors. In our study, poly(olefin sulfone), or polymethacrylates as polymeric inhibitors were especially of interest for investigating the nature of dissolution inhibition and establishing relationships between chemical structure and retardation effect. The 2-methyl pentene-1 sulfone polymers synthesized in our laboratory, with intrinsic viscosity from 45 to 70 cm³/g were found to be compatible with the substituted PHMPs and the Varcum resin. In order to understand the interaction between the two components, a model system consisting of para-substituted phenols and dipropyl sulfone in carbon tetrachloride/deutero-chloroform were analyzed by FT-IR and ¹H NMR techniques. In the presence of the sulfone compound, an apparent shift, about -170 to -180 cm⁻¹, of the O-H vibration frequencies of para-substituted phenols were observed. The magnitude of the shift is not sensitive to the ratio of phenol to sulfone as long as the solution is sufficiently dilute to minimize the self-association of phenol. A ratio of eight was used throughout this series of experiments. The interactions between the two functional groups were also found in the measurements of phenol/PMPS mixture (Table VII);

Table VII. IR and ¹H-NMR measurements of phenol-sulfone mixtures

Phenol	$\Delta \nu_{O-H}$ (cm⁻¹) sulfone	PMPS	$\Delta \delta$(ppm) sulfone	σ^+
p-F-phenol	180	184	-	-
p-Cl-phenol	182	187	0.87	0.04
p-Br-phenol	183	195	-	-
phenol	171	174	0.61	0.02
p-CH₃-phenol	175	171	0.55	0
p-CH₃O-phenol	171	170	0.51	-0.26
p-t-Bu-phenol	172	167	0.62	-0.28

the broadened O-H bands shifted toward lower frequencies were attributed to hydrogen bonding between the hydroxyl and the sulfonyl groups. But only a very slight decrease in the O=S=O stretching frequencies of about -8 cm⁻¹ was seen. The ¹H-NMR spectra indicated a chemical shift (0.5 - 0.8 ppm) of the resonance frequency of the hydroxyl proton toward lower field (Table VII). The data for the series of para-substituted phenols can be fitted by Hammett equation (eq.5) (23,24) (Figure 10):

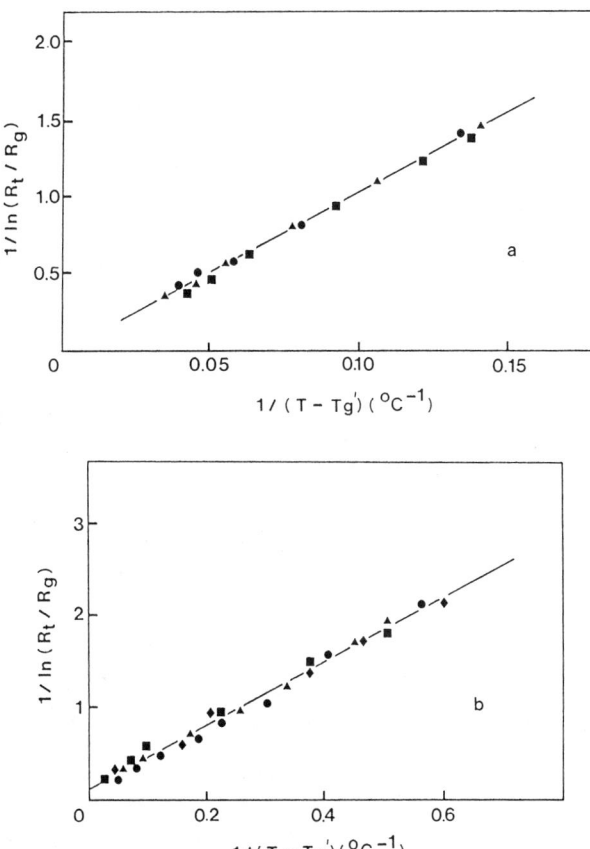

Figure 9. Dissolution of p-Cl-PHMP (a) and PMA(10%)/Varcum-2217 (b) at temperatures above Tg' fitted by the WLF equation.

$$\log \frac{\Delta \delta}{\Delta \delta_o} = \rho \sigma + h \qquad (5)$$

where $\Delta \delta$: O-H resonance shift of substituted phenol, in ppm,
       $\Delta \delta_o$: O-H resonance shift of unsubstituted phenol,
       ρ : reaction constant,
       σ : substituent constant,
       h : intercept.

It has been noticed that the resonance shift of p-CH_3O-phenol deviated from the correlation. This may be a consequence of the tendency that the hydroxyl groups can also form hydrogen bonding with CH_3O-group of another phenol molecule, resulting in a positive deviation from the Hammett's relation.

The carbonyl group is an effective hydrogen bond acceptor, as can be seen from the shifts in hydroxyl and carbonyl stretching frequencies (25,26). Judging from the magnitudes of the hydroxy frequency shifts, we conclude that the strength of interaction between phenol and carbonyl groups is only marginally stronger than that between phenol and sulfone groups. The inhibitory effects by PMMA and PMPS are shown in Figure 11.

In addition to the strength of interaction, the hydrophilic/hydrophobic characteristics of an inhibitor is also expected to have a direct bearing on its effectiveness. This is demonstrated amply by the results obtained from a series of methacrylate polymers containing different alkyl groups in the side chain (Figure 12). The decrease in dissolution rate as the hydrophobic character of the alkyl group increases from methyl to hexyl parallels the decrease in the diffusion rate and the equilibrium sorption of water in these polymers (Figure 13); their dissolution induction periods correlated with the surface wettability of the resist blends as well (Figure 14), where the induction period was estimated by extrapolating the steady state portion to intersect with the initial part of the curve. Additional evidence in support of this concept is obtained by copolymerizing methyl methacrylate with hydrophilic comonomer hydroxyethyl acrylate or hydrophobic para-methyl styrene. The abilities of these copolymers to alter dissolution are in full accord with expectations: the hydrophilic hydroxyethyl acrylate segments promote dissolution, whereas the para-methyl styrene segments slow down dissolution by virtue of their hydrophobicity (Figure 15).

<u>Model of Dissolution</u>. Based on the results described above, a model for the dissolution of phenolic resins in aqueous alkali solutions is proposed. The model is adapted from Ueberreiter's description for polymer dissolution in organic solvents (27). In Ueberreiter's model, the dissolution process takes place in several steps with the formation of (a) a liquid layer containing the dissolved polymer, (b) a gel layer, (c) a solid swollen layer, (d) an infiltration layer, and (e) the unattacked polymer. The critical step which controls the dissolution process is the gel layer. In adapting his model to our case, we need to take into account the dependence of solvation on phenolate ion formation. There is a partition of the cation and the hydroxide ion between the aqueous solution and the solid phase. The

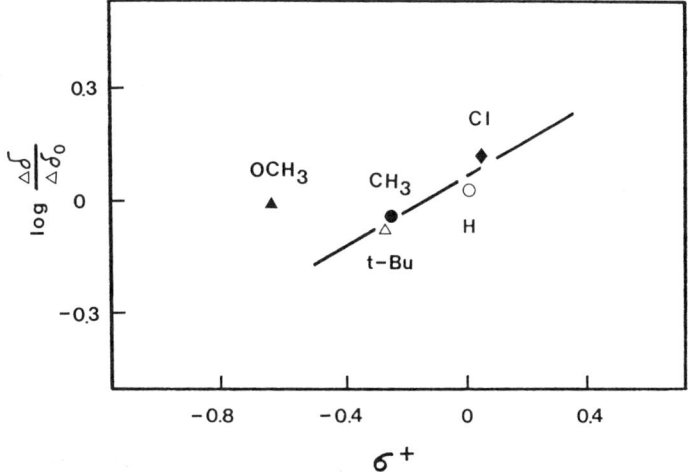

Figure 10. ^{1}H-NMR data of para-substituted phenol/sulfone mixtures fitted by Hammett equation.

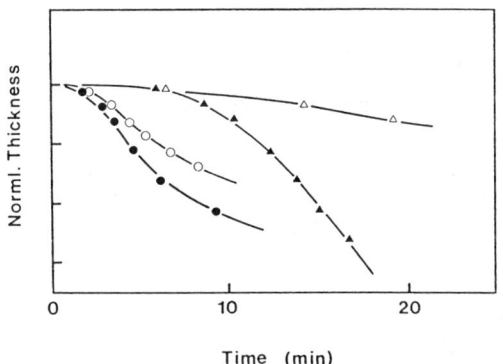

Figure 11. Dissolution of PMPS/Varcum-2217 and PMMA/Varcum-2217 blends in 1.2N KOH solutions at 24.5°C.

● PMPS(3%)/Varcum-2217   ▲ PMPS(10%)/Varcum-2217
○ PMMA(3%)/Varcum-2217   △ PMMA(10%)/Varcum-2217

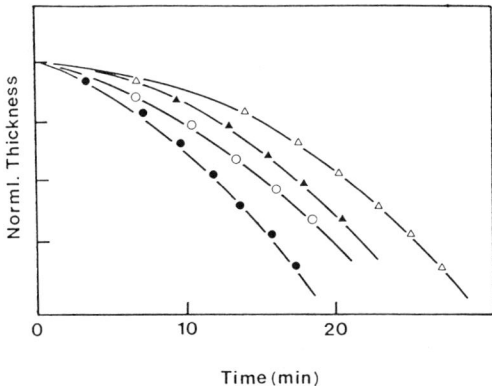

Figure 12. Dissolution of polymethacrylate(8%)/Varcum-2217 in 1.16N KOH solutions at 25.5°C.

- ● PMMA(8%)/Varcum-2217
- ○ PEMA(8%)/Varcum-2217
- ▲ PBMA(8%)/Varcum-2217
- △ PHMA(8%)/Varcum-2217

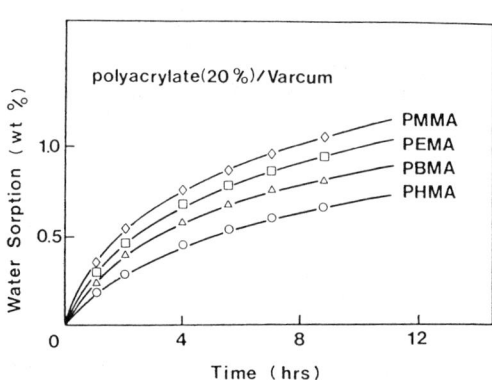

Figure 13. Water uptake of polymethacrylate(20%)/Varcum-2217 films.

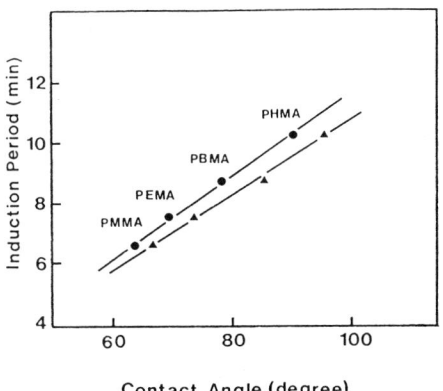

Figure 14. Induction periods of polymethacrylate(8%)/Varcum blends plotted against contact angles of the polymethacrylates.

● glycerol   ▲ water

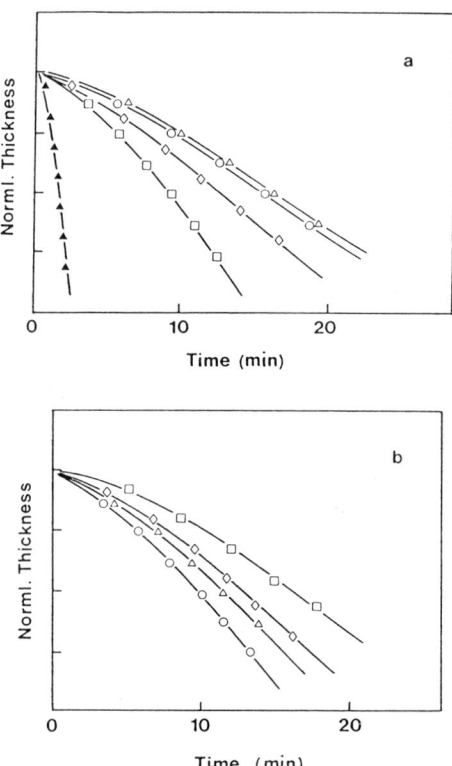

Figure 15. Dissolution of MMA-copolymer(10%)/Varcum-2217 blends in 1.0N KOH solutions at 30.0°C.

	(a)	(b)
Varcum-2217	▲	
10% PMMA	○	○
10% MMA-copolymers		

F_{HEA} = 0.02 △ F_{MST} = 0.11 △
           0.04 ◇   0.21 ◇
           0.10 □   0.31 □

hydroxyl ions, upon entering the solid phase, react with phenol groups to form phenolate ions and the extent of reaction is governed by equilibrium considerations. In the meantime, the cations diffuse to the site of the phenolate ion in order to maintain electroneutrality. Our model rests on two assumptions. (a) Once a sufficient number of phenolate ions are formed, solvation and additional reaction occur rapidly, leading to dissolution. The crucial step in the rate process is the formation of a critical number of phenolate ions per molecule; molecules with phenolate ions less than the critical amount will remain insoluble. This assumption leads naturally to the concept of a minimum alkali concentration for dissolution. At solution concentration less than C_o, the concentrations of ions in the solid are so small that only a few phenolic molecules attain the required degree of deprotonation for dissolution. The attendant low degree of solvation may result in a small diffusion constant of the cation. A discontinuous change in ionic transport with concentration is postulated here which has analogy in the penetration of solvent into glassy polymers. In the latter case, a discontinuity in the velocity of penetration (Case II) or the diffusion constant (Fickian) exists at the advancing front (28). We believe that the advancing front in our "gel" layer can be viewed in a similar way. (b) The diffusion of cations to form ion pairs with phenolate groups is involved in the rate determining step. The assumption is supported by the activation energies of dissolution at low temperatures and the conformity of the temperature dependence of the rate to a WLF-type equation above Tg'. The increase in the dissolution rate with alkali concentration then reflects changes in the cation diffusion constant (Huang, J. P.; Kwei, T. K.; Reiser, A. Macromolecules, in press.). The concentration dependence comes form the "opening" up of the secondary structure of the phenolic resin as the progressive formation of phenolate ions leads to higher degree of solvation and more facile diffusion paths. In this sense, the underlying reason of the dependence on alkali concentration is similar to, yet different from, the concentration dependence of solvent diffusion in polymers.

The effect of added salt on dissolution rate is not completely understood at present. A possible explanation of the plateau value is that the solubility of the salt in the solid reaches a limiting value at high concentrations. But additional studies are needed to arrive at a quantitative understanding of the observed decreases in C_o and n when salt is present.

The role of the dissolution inhibitor in our model rests on its ability to alter the path of water and ion transport in the solid. It is well known that water absorption bears a direct relation with the number of polar groups in the polymer and that ionic diffusion occur via "hopping" along hydrophilic sites. Therefore, the hydrophilic/ hydrophobic characteristics of the inhibitor exert a profound effect on dissolution rate.

Acknowledgments

Financial support of this work by IBM is gratefully appreciated.

Literature Cited

1. Hinsberg, W. D.; Guttierez, M. L. Proc. SPIE 1984, 57, 469.
2. Arcus, R. A. Proc. Kodak Microelectronic Seminar, 1985, p 25.
3. Arcus, R. A. Proc. SPIE 1986, 631, 124.
4. Templeton, M. K.; Szmanda, C. R.; Zampini, A. Proc. SPIE 1987, 771, 136.
5. Garaz, C. M.; Szmanda, C. R. Proc. SPIE 1988, 920, 321.
6. Hanabata, M.; Furuta, A.; Uemura, Y. Proc. SPIE 1987, 771, 85.
7. Pennacchia, J. Ph.D. Thesis, Polytechnic University, New York, 1986.
8. Bowden, M. J.; Thompson, L. F. J. Appl. Polym. Sci. 1973, 17, 3211.
9. Lewis, F. M.; Walling, C.; Mayo, F. R. J. Am. Chem. Soc. 1948, 70, 1519.
10. Walling, C.; Briggs, E. R.; Mayo, F. R. J. Am. Chem. Soc. 1948, 70, 1537.
11. Saini, G.; Leoni A.; Franco, S. Die Makromolekulare Chemie 1971, 146, 165.
12. Rodrigues, F.; Krasicky, P. D.; Groele, R. J. Solid State Technology 1985, 28, 125
13. Shiraishi, H. U.S. Patent 4 409 317, 1983
14. Turner, D. T. Polymer 1987, 28, 293
15. Stannett, V. In Diffusion in Polymer; Crank, J.; Park, G. S., Ed.; Academic Press: New York, 1968; Chapter 2.
16. Barker, R. E.; Thomas, C. R. J. Appl. Phys. 1964, 35, 87.
17. Kumius, C. A.; Kwei, T. K. In Diffusion in Polymer; Crank, J.; Park, G. S., Ed.; Academic Press: New York, 1968; Chapter 4.
18. Cotton, F. A.; Wilkinson, G. In Basic Inorganic Chemistry; John Wiley & Sons, Inc.: New York, 1976, p 222
19. Killis, A.; Nest, J-F. L.; Gandiniand, A.; Cheradame, H. Makromol. Chem. 1982, 183, 1037.
20. Wintersgill, M. C.; Fontanella, J. J.; Smith, M. K. Polymer 1987, 28, 633.
21. Watanabe, M.; Itoh, M.; Sanui, K.; Ogata, N. Macromolecules 1987, 20, 569.
22. Vincent, C. A. Prog. Solid St. Chem. 1987, 17, 145.
23. Swain, C. G.; Lupton, E. C., Jr. J. Am. Chem. Soc. 1968, 90, 4328.
24. DeRosa, T. F.; Pearce, E. M.; Charton, M. Macromolecules 1985, 18, 2277.
25. Pearce, E. M.; Kwei, T. K.; Min, B. Y. Macromol. Sci. 1984, A21, 1181.
26. Kwei, T. K.; Pearce, E. M.; Ren, F.; Chen, J. P. J. Polym. Sci. Poly. Phys. Ed., 1986, 24, 1597
27. Ueberreiter, K. In Diffusion in Polymer; Crank, J.; Park, G. S., Ed.; Academic Press: New York, 1968; Chapter 7.
28. Frisch, H. L.; Wang, T. T.; Kwei, T. K. J. Polym. Sci. Part A-2, 1969, 7, 879.

RECEIVED June 14, 1989

Chapter 23

Solvent Concentration Profile of Poly(methyl methacrylate) Dissolving in Methyl Ethyl Ketone

A Fluorescence-Quenching Study

William Limm[1], Mitchell A. Winnik[2], Barton A. Smith[3], and Deirdre T. Stanton[2]

[1]U.S. Army CRDEC, Attn: SMCCR—RSC—A (E3220), Aberdeen Proving Ground—Edgewood Arsenal, MD 21010—5423
[2]Lash Miller Chemistry Laboratory and Erindale College, University of Toronto, Toronto, Ontario M5S 1A1, Canada
[3]IBM Research Division, Almaden Research Center K91/801, San Jose, CA 95120—6099

A novel approach to determine the solvent concentration profile in a photoresist undergoing dissolution via fluorescence quenching and laser interferometry is introduced. Fluorescence arising from phenanthrene dye labels in a 1-μm-thick poly(methylmethacrylate) (PMMA) film is quenched by permeation of methyl ethyl ketone (MEK), a good solvent for PMMA. A steady-state MEK concentration profile has been estimated from quenching data with existing sorption and light scattering data. The profile contains all the features of Case II diffusion: the Fickian precursor, the solvent front, and the plateau region. However, the solvent front is not so steep as those observed in systems where penetrant diffusion is much slower. We account for these findings in detail.

Dissolution of polymers in organic solvents attracted much attention recently (1-11) due to the importance of the photoresist dissolution process in manufacturing integrated circuits (IC's). As their sophistication and circuit density increase, the understanding of fundamental aspects of photoresist dissolution becomes more critical. Currently, the state-of-the-art IC's have the minimum feature size of less than 1 μm. On this scale, photoresist swelling, which usually accompanies its dissolution, is often the limiting factor in obtaining higher circuit density. Therefore, a reduction of photoresist swelling upon exposure to developing solvent is of vital interest to those who are developing new photoresist materials.

Extent of polymer swelling is often gauged by gravimetric method - one monitors the weight-gain of a polymer sample when it is immersed in solvent. Thus, the solvent permeation rate (SPR) and, therefore, permeation mechanism can be determined from this method. However, it provides no information on how a polymer swells spatially. Furthermore, if solvent permeation is accompanied by polymer dissolution, as in the case of photoresist, the gravimetric method is inaccurate. A more useful approach is to determine the thickness of the gel layer during the polymer dissolution experiment since the gel layer is formed due to the swelling of bulk polymer. However, the determination of the gel layer thickness is not trivial. Krasicky et al.([4]) estimated the thickness of the transition layer, which encompasses all interphases between the bulk glassy polymer film and the bulk polymer solution, by measuring the change in optical interference intensity during and after the polymer dissolution process. To estimate the thickness of the gel layer from this method, a solvent concentration profile (SCP) across the transition layer has to be adopted. Conversely, if the correct SCP can be determined by some experimental method, the SCP would yield not only the gel layer thickness, but the manner of solvent diffusion into the polymer film as well.

Therefore, the best approach to investigate photoresist swelling is to determine, in-situ, the SCP in a polymer undergoing dissolution. Although Crank ([12]) proposed a descriptive SCP in 1953, firm experimental data started to appear only recently. Thomas and Windle's microdensitometry ([13]-[16]) and Kramer's Rutherford backscattering ([17]-[18]) produced SCP of several solvent-polymer combinations. However, these efforts were limited by the spatial resolution of their techniques (ca. 30 nm). In addition, these techniques have been applied to systems where the SPR's are on the order of 1 μm/hour or less. The SPR's are much greater for systems where solvent permeation is accompanied by polymer film dissolution. Therefore, the determination of SCP in such systems would require a technique that is quicker and less cumbersome for repeated measurements.

In our previous paper ([11]), we introduced a novel experimental method to study the mechanistic details of solvent permeation into thin polymer films. This method incorporates a fluorescence quenching technique ([19]-[20]) and laser interferometry ([6]). The former, in effect, monitors the movement of vanguard solvent molecules; the latter monitors the dissolution process. We took the time differences between these two techniques to estimate both the nascent and the steady-state transition layer thicknesses of PMMA film undergoing dissolution in 1:1 MEK-isoproanol solution. The steady-state thickness was in good agreement with the estimate of Krasicky et al. ([7]-[8]).

In this paper, to determine the steady state SCP across the transition layer, we analyze the fluorescence intensity decay of dye molecules covalently bound to the polymer chains. The decay is due to the permeation of solvent, which quenches dye fluorescence, into the polymer film. The diffusion coefficients of MEK, the quencher, were available from a light scattering experiment (21) at high MEK concentrations. For the low concentration regime, indirect experimental data were available (22). Once the diffusion coefficient is determined at a given concentration, the extent of fluorescence quenching can be predicted. Therefore, by working backward, one can determine the solvent diffusion coefficient and the solvent concentration in a polymer film from fluorescence quenching data. Consequently, if a polymer film dissolves in a solvent with a constant dissolution rate (DR)*, the solvent concentrations at different parts of the SCP can be determined. Finally, a SCP is constructed from these data.

Poly(methyl methacrylate) [PMMA] is an excellent polymer for studying photoresist dissolution because of its minimal swelling characteristic. For this work, PMMA molecules were labelled with phenanthrene (Phe) dye since its fluorescence is quenched by MEK. In addition, this dye has the advantage of forming few excimers (23-24) which results in self-quenching. Thus, the reduction in fluorescence intensity of PMMA-Phe* is virtually solely due to MEK quenching. Consequently, the permeation of MEK into a PMMA film can be monitored from fluorescence intensity decay.

The interferometry trace shows the change in the optical thickness of the polymer film with respect to time. Both the completion of the polymer film dissolution and the DR can be determined.

Experimental

The PMMA-Phe synthesis, characterization, film preparation, apparatus and experimental scheme are described elsewhere (11). Briefly, the PMMA chains, copolymerized from MMA and Phe-labelled monomers, were characterized via gel permeation chromatography (GPC): M_w = 411,000, M_n = 197,000 and M_w/M_n = 2.08. UV-absorption measurements indicated that ca. 1 % of all monomer units were Phe-labelled. The sample was dissolved in toluene and was spin-coated onto 1-inch diameter quartz disks. Then, the films (ca. 1 μm thick) were annealed at 160 C for 60 minutes under vacuum.

A PMMA-Phe film was seated in the flow cell and was continuously exposed to 290 nm radiation throughout the experiment (Figure 1). The Phe fluorescence was monitored at 365 nm maximum which did not shift appreciably with a change in MEK concentration in PMMA-Phe. Therefore, the decay in the Phe fluorescence intensity provides an accurate measure of MEK diffusion and its SCP.

The solvent pump was turned on at t = 0 sec. It takes ca. 20 sec for the solvent to reach the flow cell containing the PMMA-Phe sample. A significant reduction in fluorescence intensity signals the arrival of solvent at the PMMA-Phe surface.

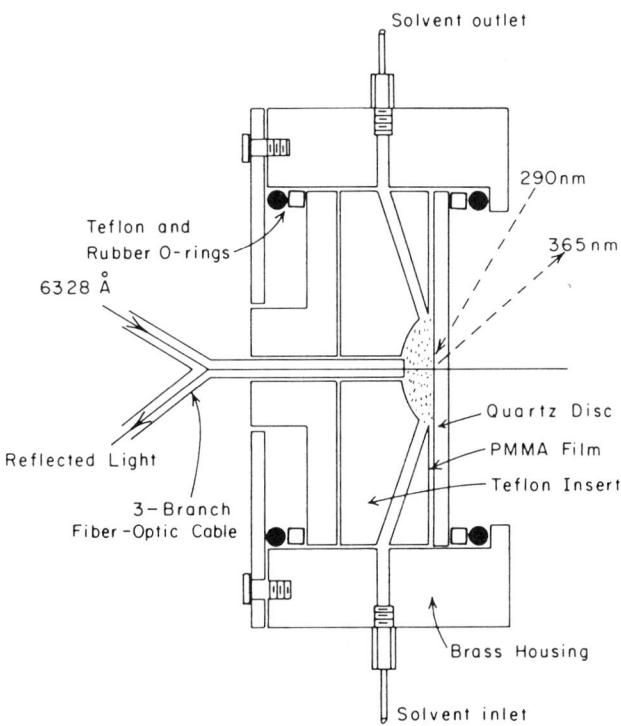

Figure 1. Flow Cell for Monitoring Solvent Permeation and PMMA Film Dissolution Simultaneously. The cell is placed in the sample chamber of a fluorescence spectrometer. (Reproduced with permission from Ref. 11. Copyright 1988 Wiley & Sons.)

Results

A typical time profile of the excited PMMA-Phe*
fluorescence intensity decay is shown in Figure 2. The
MEK permeation commences at 24 sec. The SPR increases
during the plasticization period until it becomes constant, the onset of the steady state. It is characterized
by a linear relationship between the amount of solvent
absorbed and time. It was determined from a linear regression analysis that the PMMA-Phe fluorescence intensity
starts to deviate from linearity at 197 sec. This indicates a decrease in the SPR and/or the unquenched PMMA-
Phe*. The decrease in SPR is unexpected at this film
thickness since the SPR in thicker PMMA-Phe films show no
anomaly at 1 μm. A more plausible explanation is the
reduction in available PMMA-Phe*, which is expected when
the front end of the SCP reaches the substrate.

The PMMA dissolution was monitored via laser interferometry. The resulting sinusoidal pattern (Figure 3)
can be transformed to illustrate the loss in PMMA film
thickness with respect to time (11). The dissolution commences at 27 sec and ends at 222 sec. By comparing the
latter with the fluorescence deviation at 197 sec, one
obtains the transition layer time-thickness is 25 sec.
Since the film is 1.0 μm thick, its overall DR is ca. 5.1
nm/sec. Then, the spatial thickness of the transition
layer is ca. 130 nm. Figure 3 also indicates that the DR
changes little during the course of dissolution. This
contrasts with the permeation process which possesses a
rather lengthy plasticization period as seen from the
curvature in the intensity versus time data at early times
in Figure 2.

Data Analysis

We wish to use the fluorescence decay data in Figure 2 to construct the SCP as it propagates through the
PMMA film, focusing on data at the final phase of the dissolution process. This analysis will involve the following three assumptions:
1) the SCP advances at a constant rate
2) the shape of the profile does not change and
3) the assumptions above hold even after the
   solvent molecules have reached the substrate.

In Figure 4 we present a picture of this model illustrating the propagation of the SCP through the film. According to assumptions 1) and 2), the SCP maintains its shape
as it advances at a constant rate. These assumption are
in accord both with the data (Figures 2 and 3) and with
the current level of understanding of Case II diffusion.
The third assumption is more troublesome but necessary for
evaluation of the data. As depicted in Figures 4b and 4c,
we ignore the fact that the frontier solvent molecules
accumulate as they reach the substrate, and assume that

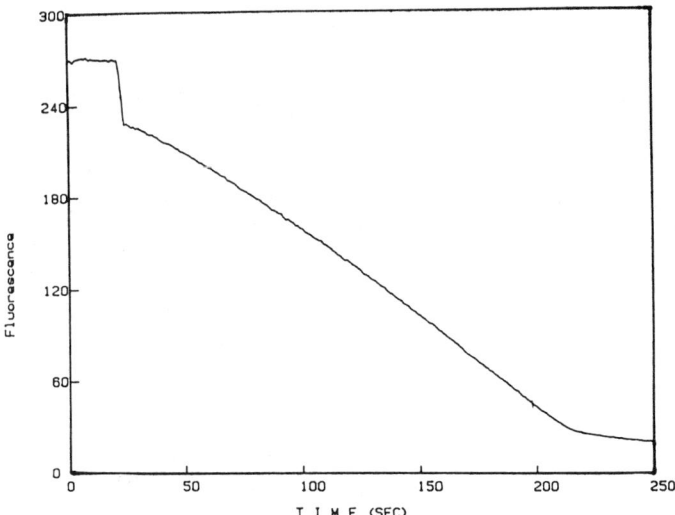

Figure 2. PMMA-Phe* Fluorescence Decay in MEK. Initial contact of MEK and PMMA-Phe* produces a sharp drop in intensity. This is followed by the plasticization period, the steady state and the termination.

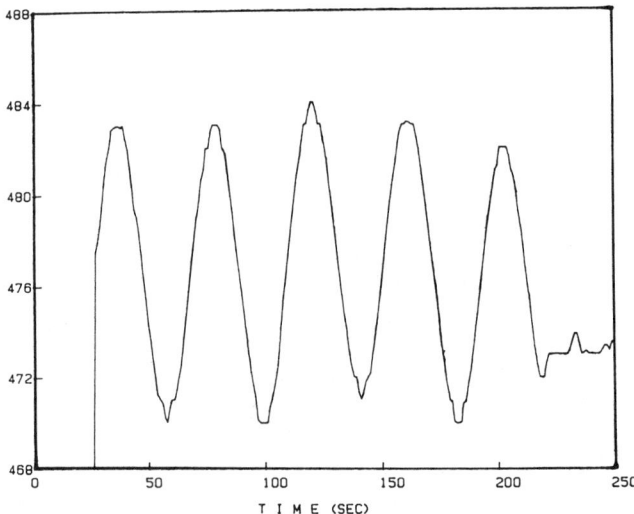

Figure 3. Interferometry Trace of a PMMA Film Dissolving in MEK.

the SCP maintains its shape. This is tantamount to assuming that the number of molecules already at the substrate surface is small compared to the number arriving at each increment in time.

This model allows us to use the fluorescence intensity obtained near the end of the dissolution process in Figure 2 to construct the SCP in the PMMA film. The strategy is to measure the incremental change in fluorescence intensity, ΔI, occurring in a one second interval in the curved region of Figure 2 and to relate this to the amount of residual fluorescence, I', in the Δt slice. The time profile of I' contains information about the SCP. The time profile can be transformed to the distance profile because the SCP propagates at a constant rate through the PMMA film.

In addition, one needs to know Io', the total (unquenched) fluorescence intensity corresponding to the Δt slices from which I' is determined. The ratio Io'/I' is related to a specific concentration of solvent (quencher) through independent knowledge of the fluorescence quenching process.

Let us consider what happens to the fluorescence intensity of a thin PMMA-Phe film at the PMMA-quartz interface. For the sake of simplicity, we focus on the region that lies between the substrate and 5.1 nm away from it, which we name the "Q-zone". 5.1 nm is the distance the steady state SCP travels in one second. In Figure 4a, it experiences no quenching. After one second, Figure 4b, quenching commences. The change in fluorescence intensity in that one second is completely due to quenching at the "Q-zone", which contains information about the MEK concen-tration at the <u>first</u> 5.1-nm of the SCP. After another second, we have the situation illustrated in Figure 4c. Similarly, the ΔI between Figures 4b and 4c represents the amount of quenching due to the <u>second</u> 5.1-nm of the SCP. Therefore, as the SCP propagates through the "Q-zone" (Figures 4c and 4d), the MEK concentration at each 5.1-nm portion of the SCP can be determined.

Fluorescence Quenching

Fluorescence quenching is described in terms of two mechanisms that show different dependencies on quencher concentration. In dynamic quenching, the quencher can diffuse at least a few nanometers on the time scale of the excited state lifetime (nanoseconds). In static quenching, mass diffusion is suppressed. Only those dye molecules which are accidentally close to a quencher will be affected. Those far from a quencher will fluoresce normally, unaware of the presence of quenchers in the system. These processes are described below for the specific case of PMMA-Phe* quenched by MEK.

In low viscosity media, the quenching of Phe* fluorescence by MEK is dynamic in nature and follows the Stern-Volmer equation (25):

$$I_o/I = 1 + k_q \tau^o c \tag{1}$$

where τ^o is the unquenched lifetime of the Phe, c is the molar concentration of MEK and k_q is the phenomenological second-order rate constant for the quenching process:

$$\text{Phe*} + \text{MEK} \xrightarrow{k_q} \text{Phe} + \text{MEK*} \tag{2}$$

Studies on the quenching of photoexcited 9-phenanthrylmethyl pivalate by MEK, a model for the PMMA-Phe/MEK system, provide a value of $k_q = 7.3 \times 10^8\ M^{-1}s^{-1}$ in cyclohexane (11). This value is nearly one order of magnitude lower than the diffusion-controlled rate. For reactions in which a diffusion step precedes a chemical step, the relationship between k_q and k_{diff} is given by:

$$\frac{1}{k_q} = \frac{1}{k_{diff}} + \frac{1}{k_{chem}} \tag{3}$$

where k_{chem} is the second-order rate constant for the quenching process under strictly chemical control ($k_{diff} \to \infty$). Since $k_{diff} = 5.0 \times 10^9\ M^{-1}s^{-1}$ in cyclohexane (11), we obtain $k_{chem} = 8.6 \times 10^8\ M^{-1}s^{-1}$.

Values of k_{diff} in PMMA-MEK mixtures can be calculated from the Smoluchowski expression:

$$k_{diff} = 4\pi N_A R_o D/1000, \tag{4}$$

where N_A is Avogadro's number, R_o is the capture radius for the quenching and D is the diffusion coefficient of MEK in PMMA-MEK mixtures. Values of D over the PMMA weight fraction range of 0 to 0.75 are available from light scattering experimentation (21). No data are available at higher PMMA concentrations for MEK, but D values in glassy PMMA have been reported for methyl acetate (MeAc) (22) which is isosteric with MEK. These values are plotted in Figure 5 and are combined by drawing a smooth line through both sets of data. With $R_o = 5\ A$ (11), we have all the data necessary to calculate k_{diff} and k_q values over the entire range of PMMA-MEK compositions.

MEK diffusion is too slow to make a significant contribution to quenching in mixtures rich in PMMA. Quenching occurs only if a MEK molecule is close enough to a Phe group at the moment it absorbs light. Static quenching is described by the Perrin equation (26-27):

$$\ln(I_o/I) = 4\pi N_A R_o [\text{MEK}]/3000 \tag{5}$$

where R_q, the radius of the active sphere, is 5 A for Phe* quenched by aliphatic ketones (11).

At an MEK concentration greater than 1 M, both the dynamic and the static quenching mechanisms have to be taken into account. Therefore, Frank and Vavilov's model of combined static and dynamic quenching model (28),

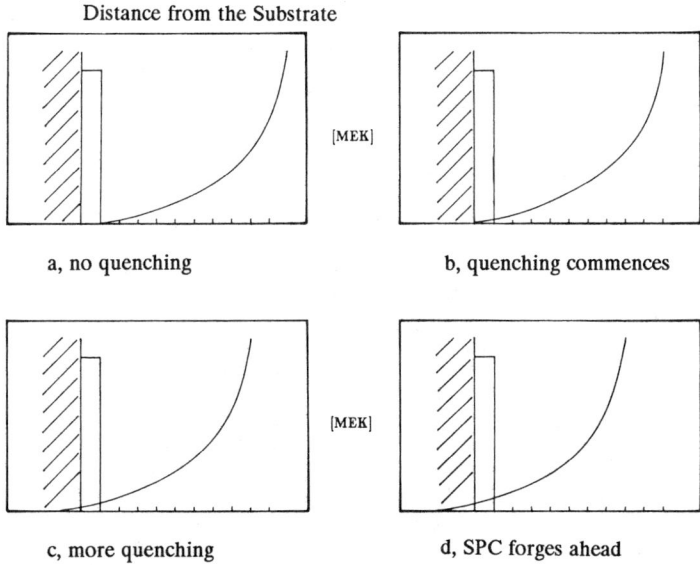

Figure 4. Propagation of Solvent Concentration Profile.

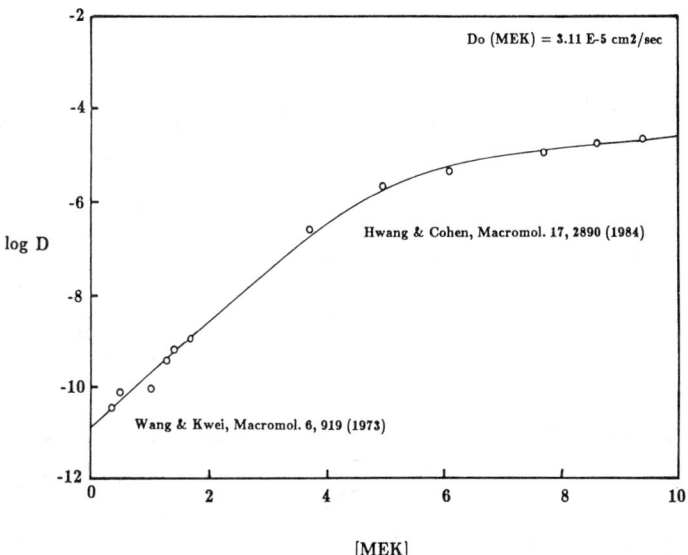

Figure 5. MEK Diffusion Coefficient as a Function of MEK Concentration.

$$I_o/I = (I_o/I)_{STATIC} \cdot (I_o/I)_{DYNAMIC} \quad (6)$$
$$= \exp(4\pi N_A R_o [MEK]/3000) \cdot (1 + k_q \tau^o [MEK]/1000)$$

is used to calculate the solvent concentration. At an MEK concentration greater than 7 M, quenching is too extensive to determine the residual fluorescence intensity.

Solvent Concentration Profile

The most complete model to date for describing Case II diffusion is that of Thomas and Windle (13-16). They envision the process as a coupled swelling-diffusion problem in which the swelling rate is treated as a linear viscoelastic deformation driven by osmotic pressure. This model leads to the idea of a precursor phase propagating ahead of a moving boundary, as we have depicted in Figure 4. While Thomas and Windle have used numerical methods to examine in detail the predictions of their model, this model is difficult to test with the data obtained here.

Consequently, we will follow the example of Mills et al. (29) who recently presented the first measurements of local solvent concentration using the Rutherford backscattering technique. They analyzed the case of 1,1,1-trichloroethane (TCE) diffusing into PMMA films in terms of a simpler model developed by Peterlin (30-31), in which the propagating solvent front is preceded by a Fickian precursor. The Peterlin model describes the front end of the steady state SCP as:

$$c(x) = c_o \exp(-vx/D) \quad (7)$$

where c_o is the limiting concentration of Fickian diffusion, v is the front velocity, x is the distance ahead of the moving front and D is the diffusion coefficient of penetrant.

We determine the SCP as it is terminated at the substrate in the following manner. First, we calculate $I_o' = dI/dt$ for the linear portion of the intensity decay. A linear least squares fit to the data in the time interval 170 - 190 sec produces a value of $dI/dt = 1150$ counts/sec. This represents the steady state quenching rate which is intimately related to the steady state SPR. Secondly, we calculate $I_t' = dI/dt$ at one-second intervals as the steady state ends with the arrival of the SCP at the substrate. For example, for the data in Figure 2, I_t's have been calculated commencing at 197 sec. Finally, we use these values of (I_o'/I_t') in conjunction with eq. (6) to calculate the concentration of MEK, $c(x)$, at each interval by a numerical method and, thereby, construct a histogram of the SCP (Figure 6).

Figure 7 illustrates the front end of the SCP as a semilogarithmic plot with respect to time. The plot is linear, in accord with eq. (7). Each second corresponds

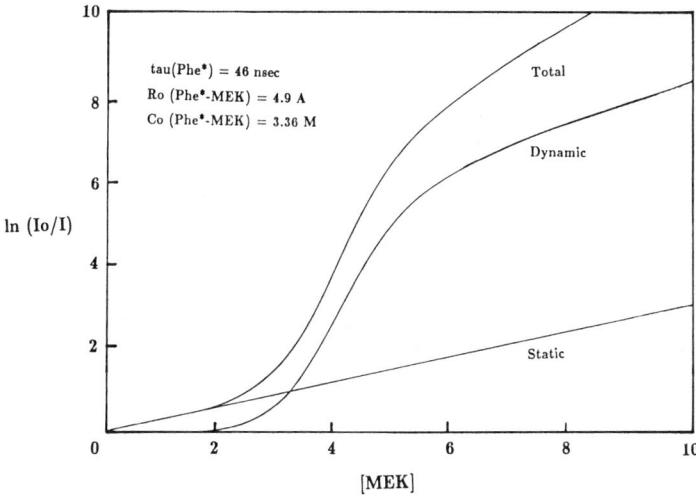

Figure 6. Calculated PMMA-Phe* Fluorescence Intensity from Static and Dynamic Quenching Theory as a Function of MEK Concentration.

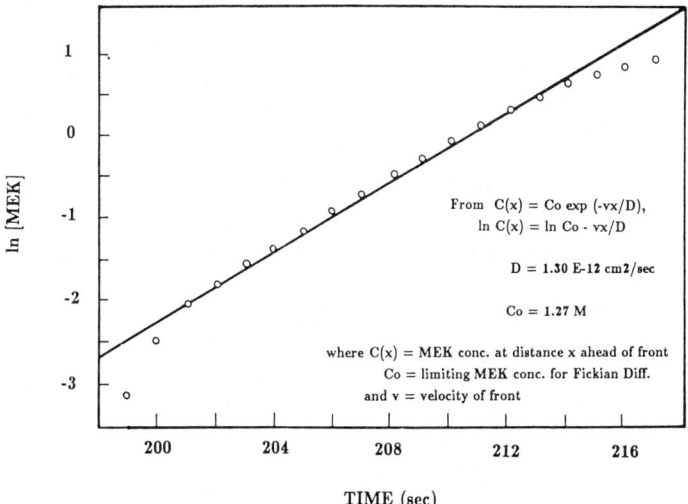

Figure 7. Fickian Precursor in the MEK Solvent Concentration Profile.

to a distance of 5.1 nm. The span of the plot in Figure 7 covers ca. 16 seconds corresponding to a permeation depth of at least 80 nm for the Fickian precursor ahead of the moving front. It is interesting to compare this value with that of the transition layer which we determined to be ca. 130 nm thick. This indicates that the Fickian precursor takes up the major portion of the transition layer for the PMMA-MEK system.

The reptation model for polymer diffusion would predict that the thickness of the gel phase reflects the dynamics of disentanglement. The important factors here are chain length, solvent quality and temperature since they affect the dimensions of the polymer coils in the gel phase. The precursor phase, on the other hand, depends upon solvency and temperature only through the osmotic force it can generate in the system and the viscoelastic response of the system in the region of the front. These factors should be independent of the PMMA molecular weight.

From the slope of the plot in Figure 7, we calculate $D = 1.3 \times 10^{-12}$ cm^2/sec and $c_o = 1.3$ M, which is ca. 0.11 in MEK mole fraction. By comparison, Wang and Kwei (22) reported $D = 5 \times 10^{-13}$ cm^2/sec for MeAc vapor at 30 C. In addition, Mills et al. (29) reported, for TCE diffusing into PMMA film, $D = 3 \times 10^{-12}$ cm^2/sec and $c_o = 0.09$ in TCE mole fraction from their Rutherford backscattering experiment. Therefore, our findings are in good agreement with other investigators' results.

In the high concentration regime, our SCP is different from a typical SCP observed in Case II diffusion. Specifically, our SCP lacks the sharp solvent front (Fig.8). The abrupt increase in solvent concentration normally observed is due to the long relaxation time of the polymer chain in response to solvent plasticization. Then, the absence of this feature points to a very rapid relaxation of PMMA chains by MEK. This is probably due to a good match in the solubility parameters of PMMA and MEK (= 9.3 for both).

The molar concentration of pure MEK is ca. 11.2 M. One might question why the concentration of MEK does not reach 11.2 M on the SCP. This is mostly due to the slow process of untangling PMMA chains. For the concentration of MEK to reach 11.2 M, the swollen polymer gel phase has to be untangled and removed from the vicinity of the quartz substrate. This is driven by the entropic force which works rather slowly in the absence of high solvent flow. For example, Mills et al. (29) report, for TCE diffusing into PMMA film, that the SCP of TCE stabilizes at a mole fraction of less than 0.2. By comparison, our results of [MEK] = 3.2 M corresponds to a mole fraction of ca. 0.3. This, again, reflects the better solubility of MEK in PMMA relative to TCE (δ = 9.6).

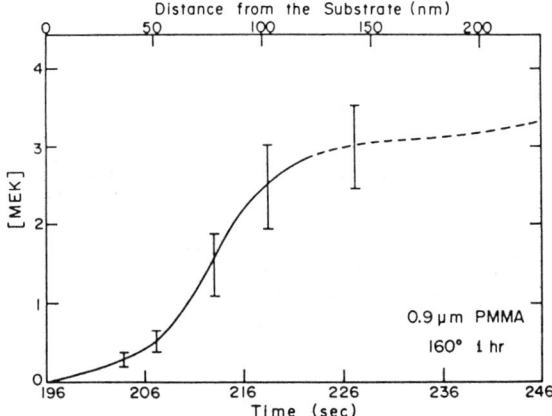

Figure 8. Estimation of the MEK Solvent Concentration Profile in a PMMA Film.

Summary

Experiments are reported on the dissolution rate and permeation rate for thin (1 μm) PMMA films exposed to liquid MEK. The films contain ca. 1 % of covalently-bound Phe, a fluorescent dye. By monitoring the dissolution rate by laser interferometry and the fluorescence quenching of Phe* by MEK, we can determine:
1) the thickness of the gel layer
2) the shape and the thickness of the Fickian precursor and
3) the diffusion coefficient of MEK in the glassy PMMA matrix.

Acknowledgments

The authors wish to thank the IBM SUR program, NSERC Canada and NRC Research Associateship program for their support of this work.

Literature Cited

1. Greeneich, J. S. J. Electrochem. Soc. 1974, 121, 1669.
2. Greeneich, J. S. J. Electrochem. Soc. 1975, 122, 970.
3. Ouano, A. C. Polym. Eng. Sci. 1978, 18, 306.
4. Ouano, A. C.; Carothers, J. A. Polym. Eng. Sci. 1980, 20, 160.
5. Ouano, A. C. In Polymers in Electronics Davidson, T., Ed.; ACS Symposium Series No. 242; American Chemical Society: Washington, DC, 1984; p. 79.
6. Rodriguez, F.; Krasicky P. D.; Groele, R. J. Solid State Tech. May 1985, p. 125.
7. Krasicky, P. D.; Groele, R. J.; Jubinsky, J. A.; Rodriguez, F.; Namaste, Y. M. N.; Obendorf, S. K. Polym. Eng. Sci. 1987, 27, 282.
8. Krasicky, P. D.; Groele, R. J.; Rodriguez, F. J. Appl. Polym. Sci. 1988, 35, 641.
9. Thompson, L. F.; Willson, C. G.; Bowden, M. J. Eds.; Introduction to Microlithography ACS Symposium Series No. 219; American Chemical Society: Washington, DC, 1983.
10. Papanu, J. S.; Manjkow, J.; Hess, D. W.; Soong, D. S.; Bell, A. T. In Advances in Resist Technology and Processing IV, Bowden, M. J., Ed.; SPIE 1987, 771, p. 93.
11. Limm, W.; Dimnik, G. D.; Stanton, D.; Winnik, M. A.; Smith, B. A. J. Appl. Polym. Sci. 1988, 35, 2099.
12. Crank, J. J. Polym. Sci. 1953, 11, 151.
13. Thomas, N. L.; Windle, A. H. Polymer 1978, 19, 255.
14. Thomas, N. L.; Windle, A. H. Polymer 1980, 21, 613.
15. Thomas, N. L.; Windle, A. H. Polymer 1981, 22, 627.
16. Thomas, N. L.; Windle, A. H. Polymer 1982, 23, 529.
17. Hui, C.-Y.; Wu, K.-C.; Lasky, R. C.; Kramer, E. J. J. Appl. Phys. 1987, 61, 5129.

18. Hui, C.-Y.; Wu, K.-C.; Lasky, R. C.; Kramer, E. J. J. Appl. Phys. 1987, 61, 5137.
19. Forster, Th. Faraday Soc. Disc. 1959, 27, 7.
20. Dexter, D. L. J. Chem. Phys. 1953, 21, 836.
21. Hwang, D.-H.; Cohen, C. Macromol. 1984, 17, 2890.
22. Wang, T. T.; Kwei, T. K. Macromol. 1973, 6, 919.
23. Birks, J. B.; Georghiou, S. J. Phys. B. 1968, 1, 958.
24. Stevens, B.; Dubois, J. T. Trans. Faraday Soc. 1966, 62, 1525.
25. Birks, J. B. Photophysics of Aromatic Molecules, p.443, Wiley-Interscience, New York, 1970.
26. Perrin, F. Compte. Rend. 1924, 178, 1978.
27. Perrin, F. Ann. Chem. Phys. 1932, 17, 283.
28. Frank, J. M.; Vavilov, S. I. Z. Physik 1931, 69, 100.
29. Mills, P. J.; Palmstrom, C. J.; Kramer, E. J. J. Mater. Sci. 1986, 21, 1479.
30. Peterlin, A. J. Polym. Sci. 1965, B3, 1083.
31. Peterlin, A. J. Res. NBS 1977, 81A, 243.

RECEIVED June 29, 1989

Chapter 24

Molecular Studies on Laser Ablation Processes of Polymeric Materials by Time-Resolved Luminescence Spectroscopy

Hiroshi Masuhara[1], Akira Itaya, and Hiroshi Fukumura

Department of Polymer Science and Engineering, Kyoto Institute of Technology, Matsugasaki, Kyoto 606, Japan

> By fluorescence analyses just upon laser ablation and of ablated surface, molecular aspects of ablation mechanism were elucidated and a characterization of ablated materials was performed. Laser fluence dependence of poly(N-vinylcarbazole) fluorescence indicates the importance of mutual interactions between excited singlet states. As the fluence was increased, a plasma-like emission was also observed, and then fluorescence due to diatomic radicals was superimposed. While the polymer fluorescence disappeared mostly during the pulse width, the radicals attained the maximum intensity at 100 ns after irradiation. Fluorescence spectra and their rise as well as decay curves of ablated surface and its nearby area were affected to a great extent by ablation. This phenomenon was clarified by probing fluorescence under a microscope.

There are a number of reports concerning laser ablation of polymeric materials in relation to microelectronics technology (1-8). Most of the polymers studied were polyimide, poly(methyl methacrylate), poly(ethylene terephthalate) and some photoresists. They have carbonyl as well as amino groups and hetero-atoms. In view of photophysics and photochemistry, this means that fluorescence lifetime is very short, intersystem crossing occurs with high quantum yield, and they are photochemically reactive (9).

On the other hand, there are other series of polymers having π-electronic chromophore such as N-carbazolyl and 1-pyrenyl groups, whose photophysical properties are quite different from the above polymers and whose laser chemistry is studied in detail. A relation among interchromophoric interaction, spectral shape and geometrical structure in the excited singlet, triplet, cationic and anionic

---

[1]Current address: ERATO, JRDC (1988-1993), 15 Morimoto-cho, Shimogamo, Sakyo-ku, Kyoto 606, Japan

states of these compounds has been elucidated in most detail by using fluorescence and transient absorption spectroscopic methods (10, 11). It is also confirmed that an increase of excitation intensity of laser pulse inevitably results in an efficient interaction between excited singlet states even in dilute solution. This process is one of the main factors leading to non-linear effects in fluorescence rise and decay curves, fluorescence yield, intersystem crossing yield, ionization, photochemical reactions, etc. (12) Therefore, laser ablation study on these polymers is expected to be very fruitful from mechanistic viewpoint, however, such a report is quite scarce (13-16).

Another advantage to examine these polymers is that characterizations of ablated materials can be made possible by fluorescence spectroscopy. Fluorescence is very sensitive, and such surrounding microenvironmental conditions around the π-chromophore as polarity and viscosity and chromophore aggregation can be probed. This indicates that a new characterization which could not be achieved by ESCA and FTIR is expected.

In the present work, we have examined poly(N-vinylcarbazole) (abbreviated hereafter as PVCz) and pyrene-doped poly(methyl methacrylate) (PMMA) films by using a time-resolved fluorescence spectroscopic method. Fluorescence spectra and their dynamic behavior of the former film were elucidated with a high intensity laser pulse and a streak camera, which makes it possible to measure dynamics just upon laser ablation. This method reveals molecular and electronic aspects of laser ablation phenomena (17). For the latter film a laser pulse with weak intensity was used for characterizing the ablated and masked areas. On the basis of these results, we demonstrate a high potential of fluorescence spectroscopy in molecular studies on laser ablation and consider its mechanism.

Experimental

Materials. PVCz (Takasago International Co. Ltd.) was purified by several reprecipitations from benzene-methanol solution. 1-Ethylpyrene (EPy) was recrystallized and sublimed in vacuo before use. PMMA was reprecipitated twice from tetrahydrofuran solution with methanol. PVCz films were prepared by spin-coating a 10 wt% anisole solution of the polymer on a quartz plate. PMMA and EPy were dissolved in chlorobenzene and cast on quartz or sapphire plates. Each film was dried under vacuum in several hours. Film thickness before and after ablation was measured with a Dektak 3030 or a Tencor Alpha 2000.

Instruments. A schematic diagram of the microcomputer-controlled system for the fluorescence measurement just upon ablation is shown in Figure 1. The excitation light source was a 351 nm laser (Lumonics He-400, 15 ns) and fresh surface of PVCz film was examined in air. The laser beam was focused onto a 2×1 mm^2 spot by using a quartz lens with a 250 mm focal length and an aperture. A copper mesh mask made photo-lithographically was used in a contact mode. The laser fluence was measured with a Gentec ED-200 power meter. The fluorescence just from the ablated area was led to a polychromator (Jobin-Yvon HR-320). Time-resolved fluorescence was measured by using a streak camera system (Hamamatsu C2830, M2493). The spectral data were averaged over 30 measurements.

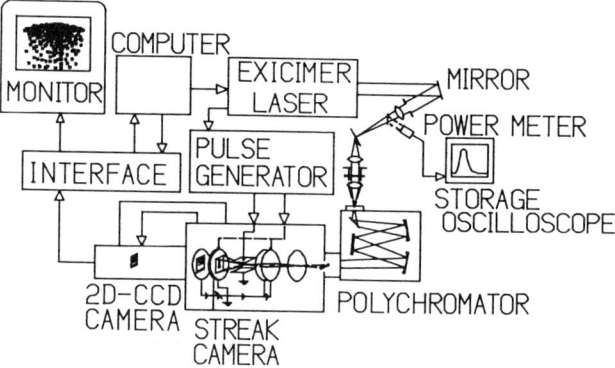

Figure 1. A schematic diagram of the streak camera system for laser ablation.

For EPy-doped PMMA film, a 308 nm excimer laser (Lumonics TE 430T-2, 6ns) was used as as exposure source. We used a time-correlated single photon counting system (18) for measuring fluorescence spectra and rise as well as decay curves of a small ablated area. The excitation was a frequency-doubled laser pulse (295 nm, 10ps) generated from a synchronously pumped cavity-dumped dye laser (Spectra Physics 375B) and a CW mode-locked YAG laser (Spectra Physics 3000). Decay curves under a fluorescence microscope were measured by the same system as used before (19).

Laser Ablation Dynamics of PVCz Film

Fluence-Dependent Luminescence Spectra. The surface of the present film was homogeneous and very smooth, while laser irradiation resulted in morphological changes. The irradiated volume was removed to some extent and a hole was left, showing laser ablation behavior. An etched depth brought about with one shot of excitation was plotted against the logarithm of laser fluence. If the laser pulse penetrates into the film according to the Lambert-Beer equation, $\log(I_0/I) = \varepsilon cd$, and the depth region where I is larger than the threshold is ablated, the etched depth should be proportional to the logarithm of the fluence. However, no linear relation was obtained. This is reasonable, since the energy of the 351 nm photon is lower than any bond energy of the polymer and a plurality of excited states and/or multi-photon absorption processes should be involved in bond cleavage. The ablation threshold was estimated to be a few tens of mJ/cm^2.

The fluorescence spectra measured just upon ablation are given in Figure 2A as a function of laser fluence. The contribution below 370 nm was suppressed, as a Hoya L37 filter was used in order to cut off the laser pulse. Fluorescence spectra of this polymer film consist of sandwich (max. 420 nm, lifetime 35 ns) and partial overlap (max. 370 nm, lifetime 16 ns) excimers (20). The latter excimer is produced from the initially excited monomer state, while the sandwich excimer from the partial overlap excimer and the monomer excited states. Since these processes compete with efficient interactions between identical and different excimers ($S_1 - S_1$ annihilation) (12), the sandwich excimer is quenched to a greater extent compared to the partial overlap one under a high excitation. Actually the fluence-dependent spectral change around the threshold can be interpreted in terms of $S_1 - S_1$ annihilation.

With increasing laser fluence, an additional tail in the long wavelength region was detected. Further increases above 1 J/cm^2 resulted in the structured bands superimposed on the broad spectra. The new tail might be a plasma-like emission and suggests that a new ablation process is involved. This type of emission has been confirmed also for polyimides, PMMA, graphite, and biological tissues (2, 21, 22). The Bremsstrahlung and recombination processes may respond to the continuum in problem. The structured bands at high fluence can be assigned to C_2 (Swan band) and CN radicals, indicating fragmentation of the polymer.

Temporal Characteristics of Luminescence. In order to reveal dynamics of these emissions, we adjusted the gate width of the streak

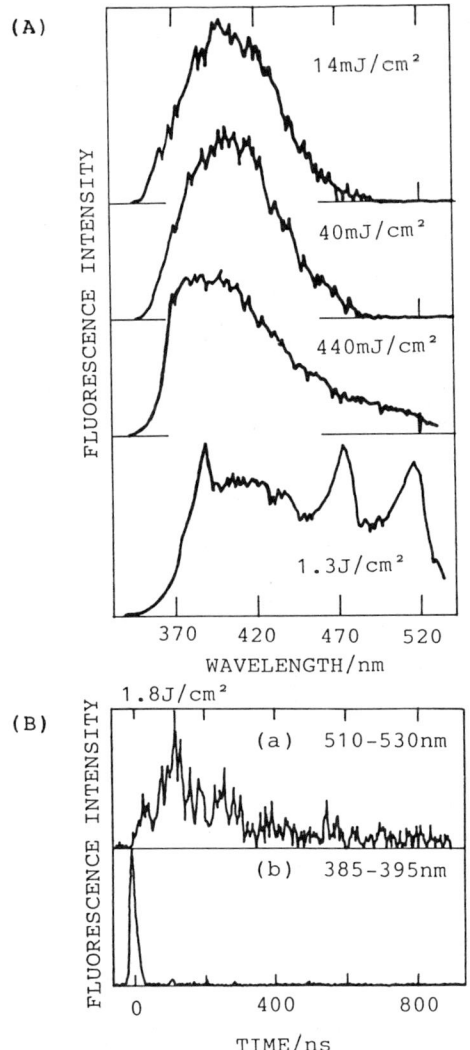

Figure 2. (A) Luminescence spectra of PVCz film just upon laser ablation. (B) Fluorescence rise and decay curves of C_2 radical (a) and partial overlap excimer of PVCz film (b).

camera to -15 ~ 72 ns, where the origin of the time axis was set to the maximum of the laser pulse. We were unable to detect any emission other than the two kinds of excimer, independent of laser intensity. Only the relative intensity of two excimer fluorescence was confirmed to change in the region of 3 mJ/cm^2 ~ 1.55 J/cm^2. The new emission in the long wavelength region and the structured bands became distinct at late stages after excitation. One of the typical examples is also given in Figure 2B. Compared to the partial overlap excimer, the maximum intensity of the radical emission was observed at 100 ns after irradiation.

Temporal characteristics at early stages were elucidated by measuring fluorescence intensity with the gate time of 1.74 ns as a function of the delay time. Compared to the laser pulse, the time where the maximum intensity is attained shifts to the early stage as the laser fluence becomes high. Of course, we could not find out any decay component with intrinsic fluorescence lifetime of 17 and 35 ns. It is concluded that an S_1 - S_1 annihilation occurs quite efficiently during the pulse width.

S_1 - S_1 Annihilation and Ablation Mechanism. The S_1 - S_1 annihilation process is responsible to laser ablation, which was supported by the following experiment. Total fluorescence intensity and the relative intensity of excimer emissions (-15 ~ 72 ns gate width) were plotted against the fluence in Figure 3. It is interesting that the relative contribution of excimers showed a similar change to that of total fluorescence intensity. This indicates that the S_1 - S_1 annihilation has an important role in the primary processes of laser ablation phenomena, since the relative contribution of excimers is determined by the degree of S_1 -S_1 annihilation, and the suppressed fluorescence intensity corresponds to the enhanced ablation.

Fluorescence intensity increases with the laser fluence, while its change was quite smooth even around the threshold. If a new process leading to ablation was involved in addition to the S_1 - S_1 annihilation, the relative fluorescence intensity of two excimers would not change furthermore above the threshold. Although the details are beyond our current knowledge, we conclude that the S_1 - S_1 annihilation is the origin of laser ablation in this fluence range.

The total fluorescence intensity saturated around a few hundreds of mJ/cm^2 which corresponds to the irradiation condition where the new plasma-like emission was observed. Above this value fluorescence intensity decreased, which is accompanied with the recovery of the relative intensity of excimer emissions. This means that a quite efficient deactivation channel of excitation intensity opens in this energy range, and the contribution of S_1 -S_1 annihilation is depressed. This suggests that fragmentation reactions to diatomic radicals are not induced by the annihilation process. Multi-photon absorption processes via the S_1 states and chemical intermediates should be involved, although no direct experimental result has as yet been obtained.

Characterization of Ablated PMMA Film with EPy

Fluorescence Spectra. Fluorescence spectra of PMMA films doped with a high concentration of EPy are composed of a structured monomer and a red-shifted broad excimer bands, and their relative contribution is modified by laser ablation. Figure 4 shows normalized fluorescence spectra of the ablated area of EPy-doped PMMA films irradiated with single and triple laser shots with the fluence of 4.3 J/cm^2. For comparison, the spectrum of the unirradiated film is also shown. An intensity ratio of the excimer fluorescence to the monomer one (I_E/I_M) decreased with the number of laser shots. It was also confirmed that this value of I_E/I_M decreases with increase of the laser fluence. This indicates that the high fluence leads to the larger change as compared with the low fluence, which is quite reasonable.

An effective thickness of the layer where the fluorescence is observed is assumed to be the depth where the excitation light intensity is 1/e of the initial value. The thickness was calculated to be 1.4 μm from an absorption coefficient of the film at 295 nm (excitation wavelength). Therefore, the observed fluorescence spectral change is due to that of aggregate states of EPy in the depth region of 1.4 μm from the ablated surface. Actually, it is well known in a PMMA matrix that the excimer band is due to the ground state dimer of the dopant (23).

Fluorescence Rise and Decay Curves. Both monomer and excimer fluorescence decay curves of the unirradiated film are non-exponential and the excimer fluorescence shows a slow rise component. This behavior is quite similar to the result reported for the PMMA film doped with pyrene. (23) A delay in the excimer formation process was interpreted as the time taken for the two molecules in the ground state dimer to form the excimer geometry. Dynamic data of the ablated area observed at 375 nm (monomer fluorescence) and 500 nm (excimer fluorescence) are shown in Figure 5. When the laser fluence increased, the monomer fluorescence decay became slower. The slow rise of the excimer fluorescence disappeared and the decay became faster.

One explanation is due to a change of EPy aggregation in the area left upon the laser ablation. Some of EPy dimers constitute the non-fluorescent quenching site, and others form the dimer which are converted to excimer more easily compared to before irradiation. In the fluorescence studies on vacuum-deposited films of ω-(1-pyrenyl)alkanoic acid, we reported an important role of aggregation of pyrenyl chromophores (24). During laser annealing of these films, their fluorescence spectral shape changed and finally its intensity decreased. This characteristic behavior was interpreted by introducing a non-fluorescent aggregate in the mechanism. We consider that the similar quenching sites are responsible for the present result. Another possible explanation is that some of EPy molecules in the left area are evaporated by the laser ablation, which seems to be consistent with the prolonged decay time of the monomer fluorescence. However, accelerated excimer formation process can not yet be explained.

Fluorescence Characterization of Ablated Polymeric Materials. In order to produce sharply etched patterns, the film was ablated with a photo-lithographically prepared mesh mask in the contact mode. The ablation was conducted with two laser shots with the laser fluence of 0.2 J/cm^2. The decay curves of the ablated film was measured by a

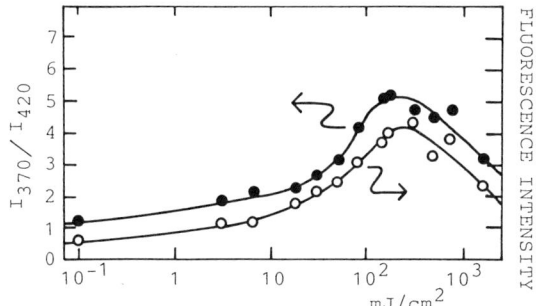

Figure 3. Fluence-dependences of relative intensity of two excimer emissions and fluorescence intensity.

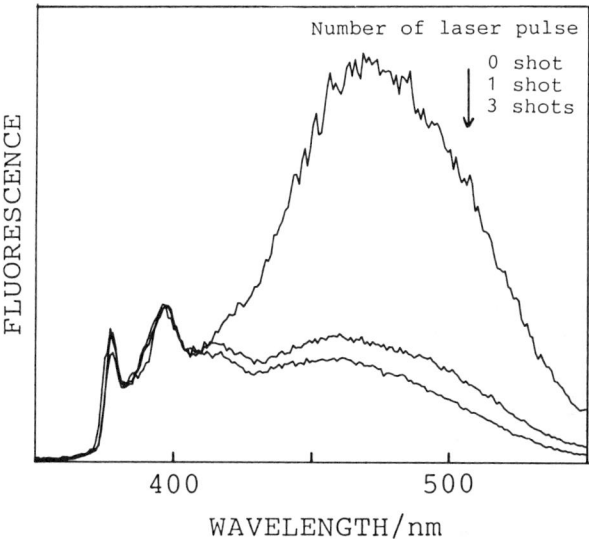

Figure 4. Fluorescence spectral change of EPy in PMMA induced by laser ablations.

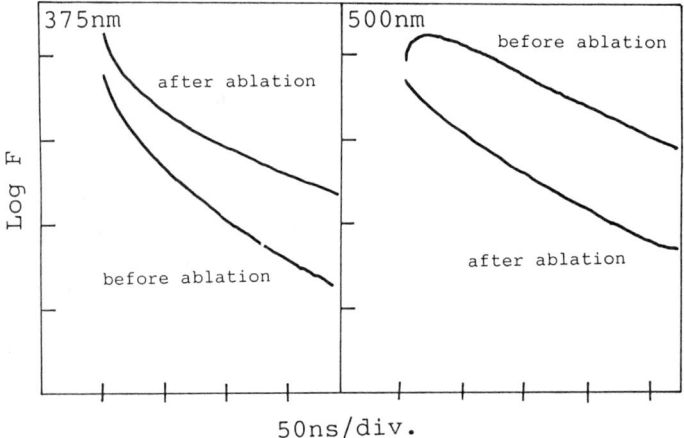

Figure 5. Fluorescence rise and decay curves of EPy in PMMA before and after laser ablation.

dynamic fluorescence microprobe apparatus with a two-dimensional resolution of 5 μm, which was constructed by us (19). The fluorescence behavior of the masked area was different from that of the ablated one. Furthermore, it is worth noting that the decay curve of the former area is not identical with that of the untreated film. We measured excimer fluorescence dynamics as a function of the horizontal position of the fluorescence microscope. The time when the intensity became one-twentieth of the initial one was used as a measure, because the decay curves were non-exponential. The values below 80 ns and around 100 ns correspond to the masked and the ablated areas, respectively. From these data, it was concluded that the width of the masked area was 37 μm, which is in agreement with the mesh dimension.

These results indicate that aggregation states of EPy in the region of about 20 μm around the ablated area were affected by the laser ablation with low fluence of 0.2 J/cm^2. Since the excitation energy cannot migrate up to 20 μm, another effect which causes this change of EPy aggregation may be transmitted through the PMMA matrix from the ablated area. When materials are ablated by the laser, a fast thermal elevation and a volume expansion occur simultaneously with the explosive desorption of the ablated materials. These expansions are transmitted through the PMMA matrix to unirradiated and unablated areas, and lead to different distributions of the dopant and to changes in the physico-chemical properties of the PMMA. These morphological changes have been probed here for the first time by the fluorescence characteristics of EPy aggregation.

Conclusion

The time-resolved fluorescence spectroscopic approach is very fruitful in elucidating electronic and molecular aspects of laser ablation of fluorescent polymers. While a plasma-like emission and fluorescence behavior of ablated radicals have been reported (21, 22), fluorescence dynamics of the polymer itself has been considered for the first time in the present work. A typical temporal characteristic of PVCz film at 1.3 J/cm^2 fluence was as follows. Only excimer emissions were observed during laser pulse, a broad plasma-like emission was detected later, and fragmented radicals became distinct. Ablation behavior can be interpreted in terms of photophysical and photochemical processes, including $S_1 - S_1$ annihilation.

It is confirmed that the polymer matrix around ablated area was also affected strongly by laser ablation. The change of the matrix properties are brought about over a few tens of μm. This type of information is basically important and indispensable for practical applications such as excimer laser lithography. The time-resolved fluorescence spectroscopy is one of the powerful characterization methods for ablated polymer matrix.

Acknowledgments

The authors wish to express their sincere thanks to Messrs. A. Kurahashi and S. Eura for their experimental efforts. Thanks are also due to Prof. I. Yamazaki and Dr. N. Tamai who helped us with the single photon counting measurements. The present work is partly supported by the Grant-in-Aid on Special Project Research for Photochemical Processes (63104007) and on Priority Area for Macromolecular Complexes (63612510), and the Grant-in-Aid for Scientific Research (63430003) from the Japanese Ministry of Education, Science, and Culture.

Literature Cited

1. Srinivasan, R.; Leigh, W.J. J. Am. Chem. Soc. 1982, 104, 6784.
2. Yeh, J. T. C. J. Vac. Sci. Tech. A 1986, 4, 653.
3. Srinivasan, R.; Braren, B.; Seeger, D.E.; Dreyfus, R.W. Macromolecules 1986, 19, 916.
4. Danielzik, D.; Fabricius, N.; Rowekamp, M.; Von der Linde, D. Appl. Phys. Lett. 1986, 48, 212.
5. Srinivasan, R.; Braren, B.; Dreyfus, R.W. J. Appl. Phys. 1987, 61, 372.
6. Larciprete, R.; Stuke, M. Appl. Phys. B 1987, 42, 181.
7. Koren, G. Appl. Phys. Lett. 1987, 50, 1030.
8. Estner, R.C.; Nogar, N.S. Appl. Phys. Lett. 1986, 49, 1175.
9. For example, Turro, N.J. Modern Molecular Photochemistry; Benjamin: New York, 1978.
10. Masuhara, H. Makromol. Chem. Suppl. 1985, 13, 75.
11. Masuhara, H. In Photophysical and Photochemical Tools in Polymer Science, Winnik, M. A. Ed.; Reidel: Dordrecht, 1986; p.65.
12. Masuhara, H.; In Photophysical and Photochemical Tools in Polymer Science, Winnik, M. A. Ed.; Reidel: Dordrecht, 1986; p.43
13. Masuhara, H.; Hiraoka, H.; Domen, N. Macromolecules 1987, 20, 450.
14. Masuhara, H.; Hiraoka, H.; Marinero, E.E. Chem. Phys. Lett. 1987, 135, 103.
15. Hiraoka, H.; Chuang, T.J.; Masuhara, H. J. Vac. Sci. Tech. B 1988, 6, 463.
16. Srinivasan, R.; Braren, B. Appl. Phys. A 1988, 45, 286.
17. Eura, S.; Itaya, A.; Masuhara, H. Polym. Preprints Japan 1988, 37, E591.
18. Yamazaki, I.; Kume, H.; Tamai, N.; Tsuchiya, H.; Oka, K. Rev. Sci. Instr. 1985, 56, 1185.
19. Itaya, A.; Kurahashi, A.; Masuhara, H.; Tamai, N.; Yamazaki, I. Chem. Lett. 1987, 1079.
20. Itaya, A.; Sakai, H.; Masuhara, H. Chem. Phys. Lett. 1987, 138, 231.
21. Koren, G.; Yeh, J.T.C. J. Appl. Phys. 1984, 56, 2120.
22. Gauthier, T.D.; Clarke, R.H.; Isner, J. M. J. Appl. Phys. 1988, 64, 2736.
23. Avis, P.; Porter, G. J. Chem. Soc., Faraday Trans. 2 1974, 70, 1057
24. Itaya, A.; Kawamura, T.; Masuhara, H.; Taniguchi, Y.; Mitsuya, M. Chem. Phys. Lett. 1987, 133, 235.

RECEIVED June 29, 1989

Chapter 25

Mechanism of Polymer Photoablation Explored with a Quartz Crystal Microbalance

Sylvain Lazare and Vincent Granier

Laboratoire de Photophysique et Photochimie Moléculaire, UA Centre National de la Recherche Scientifique 348, Université de Bordeaux I, 351 Cours de la Libération, F-33405 Talence, France

The amount of polymer ablated by absorption of the high intensity radiation of the excimer laser (ArF, KrF, XeF) was measured precisely with the aid of a quartz crystal microbalance, as a function of the laser intensity. From the work of Srinivasan and Dyer, it is known that ablation occurs in the time-scale of the laser pulse. Therefore with the aid of a new model of the "Moving Interface", we assumed that 1- the laser intensity (I_o) has to cross the plume of ablating products and is consequently attenuated by absorption (screening effect) and 2- the irradiated interface moves back only when $I > I_t$ and at a rate which is proportional to $I-I_t$ ($v = k(I-I_t)$). A fit of the experimental etch curve provides for each polymer a unique couple of k and ß, respectively ablation rate constant and screening coefficient, which characterize respectively the response of each polymer under the ablative conditions and the mean absorptivity of the ablation products. This is the first model attempting to evaluate the extent of the screening effect of the radiation and permitting to compare the ablation rates of different polymers.

The quartz crystal of the microbalance, placed at some distance of the ablating target polymer, is used to probe the rate of product deposition. The spatial distribution of the ablating products and deposition yields were determined as a function of polymer structure and experimental conditions. A high degree of aromaticity of the polymer facilitates the deposition whereas a highly oxygenated structure gives a low deposition yield.

The irradiation of a polymer surface with the high intensity, pulsed, far-UV radiation of the excimer laser causes spontaneous vaporization of the excited volume. This phenomenon was first described by Srinivasan (1) and called ablative photodecomposition. The attention of many researchers was drawn to the exceptional capabilities of photoablation (2). Etching is confined to the irradiated volume, which can be microscopic or even of submicron dimensions, on heat-sensitive substrates like polymers. In most experimental conditions, there is no macroscopic evidence of thermal damage, even when small volumes are excited with pulses of fluences as large as 20 J/cm^2. The best resolution is in general achieved at the shortest wavelength (193 nm). This is an appealing feature for many applications that require the careful etching of heat sensitive materials like organic polymers, VLSI electronic devices and living tissues.

The present challenge is to find the detailed mechanism of photoablation which has been the subject of many speculations in the past few years. There is no doubt that the polymer chains are degraded by absorption of the incident radiation, but the challenge is really to determine the extent of photochemical versus thermal character of the forces that produce decomposition and expellation of the products. No clear experimental evidence has been brought yet, but one may reasonably think of a mixture of both. The problem was addressed by various experimental approaches. One may mention in historical order of appearance, the determination of some stable ablation products ([3]), the spectroscopic recording of flame luminescence ([4]), the evaluation of the etch depth as a function of intensity ([5]), the laser induced fluorescence of C_2 in the plume ([6]), the surface modification after ablation ([7]), the thermal loading after ablation ([8]), the mass spectroscopic detection of some products ([9]), the fast recording of the transient pressure developed on the ablating substrate ([10]), the streak camera picturing of the plume ([11]), etc.. Most of the work has been carried out with common nanosecond pulses of the excimer laser, but more recently laser systems delivering femtosecond pulses have been built in several laboratories. A few reports ([12]) show that femtosecond ablation exhibits new attractive features due to the six order of magnitude greater intensity which allows coherent two photon absorption. This results in an easy ablation of materials at the wavelengths that are normally not absorbed when delivered at low intensity. Doping non absorbing polymers with absorbing molecules was shown ([13]) to improve ablation by ns pulses.

In this work, we have tried to relate the amount of polymer vaporized, to the instantaneous intensity of the laser pulse. Early measurements made by Srinivasan and Braren ([6]), showed that the amount of ablated material was a linear function of the logarithm of the total pulse energy. This was rationalized by a theory based on the static Beer's law ([14]). Recent and more extensive investigations ([15-17]) revealed new features of the etch curves that were not suspected before. This is mainly an important curvature which appears markedly for some polymers or wavelengths, as will be discussed below. This was accounted for by a dynamic theory ([18]) and we recently proposed an empirical model ([19]) which allows to extract from the experimental data, two parameters k and ß, featuring the behaviour of the polymer at a given wavelength. k is the ablation rate constant and ß is the screening coefficient that accounts for the attenuation of the laser beam from the gas ablation products. The latter is known as the screening effect, a phenomenon which was rather neglected in the past. Only one attempt from Koren ([20]), to evaluate the extend of a 248 nm beam attenuation by polyimide is recorded, but no data featuring each polymer was routinely measured, as with our model.

From couples of values k and ß, we can easily compare the different polymers and an attempt can be made to understand the influence of their chemical structure on their photoablative behaviour. Although the fitting of the experimental etch curves could not be approached perfectly at high fluence because of a presumed drastic change of mechanism, the ability to simulate these curves over a wide domain of fluence, is consistent with our dynamic model of the so-called "Moving Interface". With this model, photoablation appears more like a surface vaporization rather than an explosion. Furthermore ablation is the result of the competition between two conflicting sets of reactions, one being the photochemical or thermal dissociations of bonds and the other being recombinations of radical pairs or reformation of bonds. Most probably, these two types of events occur in the condensed phase as well as in the gas phase after ablation. During absorption, dissociation is predominant as long as the absorbed laser intensity is sufficient, but recombination becomes important as soon as the absorbed laser intensity falls under some threshold. After irradiation has ceased, heat relaxation occurs. It is clear that the temperature plays an important role by favouring the formation and diffusion of radicals issued from bond breaking. As long as the incident laser beam is absorbed either by the plume of ablation products or by the surface, a fraction of energy is converted into heat and the temperature increases. At the end of the irradiation phase which lasts only about 20 ns, the dissociation reactions are only activated by the thermal energy, which is diffusing to the bulk of the polymer (a process which takes a few microseconds for a polymer). For this reason ablation stops but further chemical reactions take place in the plume of ablation products (dissociation, oxidation, radical coupling and polymerization). When these products are brought into contact with a surface a new polymer film can deposit, with a rate that we have probed, in vacuum, with the quartz crystal microbalance. The space distribution and the yield of deposition are studied as a function of polymer structure and experimental conditions. It is shown that the tendency to form a film is favoured by low fluences, by an increase of the background pressure and by cooling the deposit substrate. All this information is useful in understanding the mechanism of ablation as well as in the development of its applications.

ETCH AND DEPOSITION RATE MEASUREMENTS

We first experimented with the Quartz Crystal Microbalance (QCM) in order to measure the ablation rate in 1987 (17). The only technique used before was the stylus profilometer which revealed enough accuracy for etch rate of the order of 0.1 μm, but was unable to probe the region of the ablation threshold where the etch rate is expressed in a few Å/pulse. Polymer surfaces are easily damaged by the probe tip and the meaning of these measurements are often questionable. Scanning electron microscopy (21) and more recently interferometry (22) were also used. The principle of the QCM was demonstrated in 1957 by Sauerbrey (23) and the technique was developed in thin film chemistry, analytical and physical chemistry (24). The equipment used in this work is described in previous publications (25). When connected to an appropriate oscillating circuit, the basic vibration frequency (F_0) of the crystal is 5 MHz. When a film covers one of the electrodes, a negative shift δF, proportional to its mass, is induced:

$$\delta F = -2.3 \times 10^6 F_0^2 (\delta m/A) \text{ with } \delta F \text{ in Hz, } F_0 \text{ in MHz and } \delta m/A \text{ in g/cm}^2$$

If the mass load on the electrode is not uniform, a calibration is then necessary to account for the radial sensitivity of the vibrating device (Lazare, S.; Granier, V., unpublished results). The maximum of sensitivity is obtained at the centre of the electrode. This allows, for instance, etching over surface areas as small as a 2 mm diameter disc, with a minimum detectable mass of one nanogram. The calibration is performed in this case by using a fluence at which the ablation rate is known, in order to determine the sensitivity factor.

Although quartz crystals are used in liquid media (26) the limitations with solid films are twofold. This is on one hand the maximum thickness of the film and on the other hand the elastic modulus of the material. Both limits are related so that polymers having a low elastic modulus are accepted only in small thicknesses. A larger thickness dampens the acoustic vibration to such an extent that resonance is no longer possible. In good cases, thicknesses up to 10 μm are measurable. Etching results in positive shifts ($\delta F > 0$) whereas deposition gives negative shifts ($\delta F < 0$).

AT quartz crystals, as those used in this work, are designed to display the smallest temperature dependence at room temperature, but frequency shifts due to temperature variation must be eliminated in order to get only the mass variation contribution. Although this is not a major problem, low laser repetition rates are systematically used, therefore avoiding any possible heat accumulation.

The excimer laser was a Lambda Physik EMG 200E, equipped with a stable resonator and for the ArF, KrF and XeF radiations respectively at 193, 248 and 351 nm. The fluence is varied either by lens focusing or by attenuation with silica plates. The central part of the beam is usually used. For a given fluence, a record of frequency versus number of pulses is achieved and the average etch rate (etch depth per pulse) is obtained by linear regression. As previously reported (25), for low fluence it is commonplace that the etch rate varies by an incubation process. The initial etch depth is characteristic of the virgin polymer. A precise determination of the ablation threshold can be achieved by using the linear variation of etch depth with low fluence values. Fig. 1 shows etch curves in the above threshold region for several polymers. The various intercepts give a convenient and precise way of measuring threshold fluences.

The plot of etch depth versus logarithm of fluence or etch curves for different polymers and wavelengths are displayed in Fig.2. A quasilinear variation with fluence is denoted by a gradual curvature on the log plot, for fluences immediately above the threshold. As seen below, this is due to a moderate screening effect. Most of the incident radiation goes to the solid polymer to perform ablation. As the fluence increases, a portion of straight line is reached for polymers displaying a strong screening effect. For those having low screening the etch plot is curved all over. For some conditions and polymers, environmental effects on etch rate have been seen. One is due to the presence of oxygen which can modify the etch rate by attacking the excited surface. This was observed for polycarbonate and polystyrene at 248 nm (25). The second is due to the background pressure which can favour the products redepositing on the ablated surface. Then a difference between vacuum and atmospheric pressure etching is observed (27). This was strongly visible for polystyrene and poly(α-methyl styrene).

The deposition rate measurements were done in a vacuum chamber (10^{-5} torr). The polymer was irradiated through a silica window and the products deposition rates were probed by a series of 6 quartz crystals placed on a quarter of a circle (radius 4 cm) centred on the target. The fluence was varied with the aid of a spherical lens placed outside the chamber. In this configuration, each mass sensor gives one experimental point of the curve giving the deposition thickness as a function of ejection angle. The integration of this curve over the half space faced by the irradiated polymer surface leads to the total amount of material deposited. Knowing the etched volume, the yield of deposition can be inferred precisely (results are given below).

SCREENING EFFECT AND ABLATION RATE CONSTANT

The screening effect by the gaseous ablation products is a consequence of the high rate of ablation which takes place within the irradiation of the laser pulse (20 ns), and tends to reduce the ablation depth. With our "Moving Interface" model the absorption of the products is characterized by a coefficient ß, assumed constant. The optical thickness of the plume of products is equal to ßx, x being the etch depth at time t. The attenuation of the incident intensity is therefore $I_o(t)e^{-\beta x}$. The fitting of the experimental etch depth versus fluence (19), by integration of the kinetic law of the "Moving Interface" (2b) $v = k(I_o(t)e^{-\beta x(t)} - I_t)$ over the period of time during which the intensity is greater that I_t, provides the two parameters k and ß. k is the ablation rate constant and represents the speed of the interface ablating under an intensity equal to $1MW/cm^2 + I_t$. k may also represent the mass of ablated polymer per unit of time. These two definitions are proportional and the linear speed (19) is obtained by dividing the mass rate (ng.ns^{-1}.MW^{-1}.cm^2) by the density of the polymer when available.

The theoretical curves are the solid lines displayed in Fig.2 and it can be seen that the agreement with the experimental data is good over a large range of fluence from the threshold region up to a fluence of several joules per cm^2. At high fluence a sharp increase of the experimental etch depth is not reproduced by the model. This point is discussed in the next paragraph. The screening coefficients ß and ablation rate constants

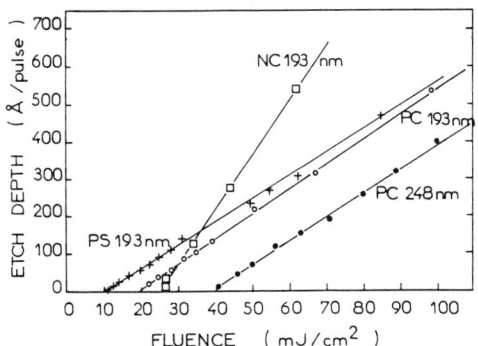

Figure 1. Etch rate showing the linear dependence on fluence at low intensity. Key: +, PS, polystyrene at 193 nm; ○, PC, polycarbonate at 193 nm; ●, PC at 248 nm; and □, NC, dinitrocellulose at 193 nm.

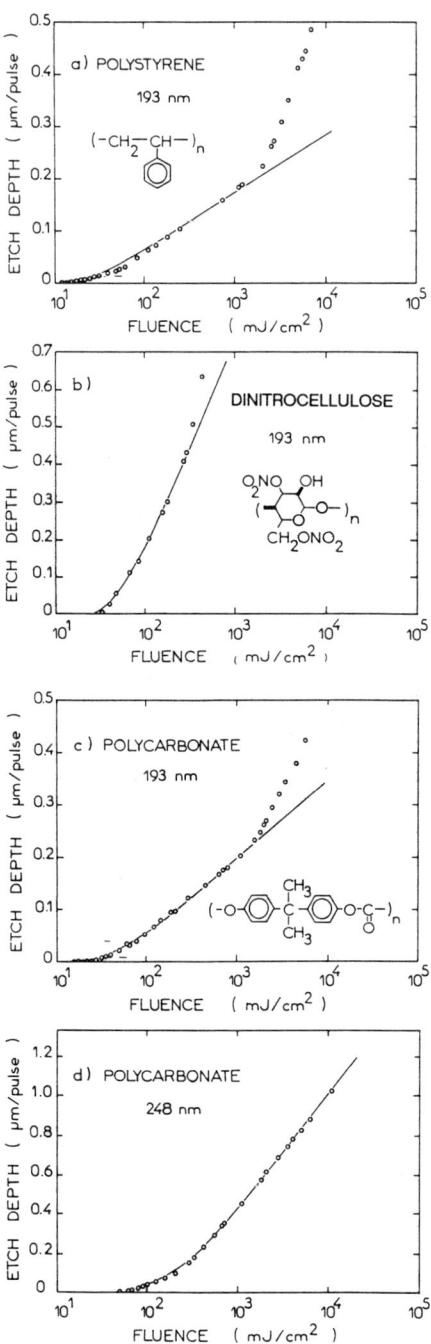

Figure 2. Etch depth vs fluence: a, PS, 193 nm; b, dinitrocellulose, 193 nm; c, polycarbonate, 193 nm; and d, polycarbonate, 248 nm.

k for a selected set of polymers are given in Table I. Accuracy of the fit (better than 1%) and uncertainty on etch depth have been reported elsewhere (19). The ablation parameters ß and k show a strong dependence on both wavelength and polymer structure (discussed below). In general, the shortening of the wavelength gives higher k and ß. For instance, this is particularly obvious for polyphenylquinoxaline (19) which has been studied at several wavelengths (see Table I). ß, the mean high intensity absorption coefficient of the products, is consistently higher at shorter wavelength. For the same polymer, variation of k with wavelength denotes a change of the efficiency of the corresponding photon (quantum yield). It is generally found that a shorter wavelength has a better ablation efficiency. This is in agreement with the idea that the corresponding high energy photon has more power to directly break a bond. Here a more detailed modelling will be developed to fully explore the significance of this strong wavelength dependence of k.

Table I. Ablation parameters of various polymers

	193 nm				248 nm			
POL.[a]	I_t^b	α^c	β^d	k^e	I_t^b	α^c	β^d	k^e
PET	1.4	3.0	1.16	303	1.96	1.60	0.46	348
PBT	1.61	2.80	1.04	313	3.92	1.70	0.38	540
PPQ	1.26	0.65	2.4	152	2.24	0.37	0.54	166
PI[f]	1.12	4.2	1.86	254	3.78	2.2	0.51	264
BPPC	1.33	5.5	1.42	216	2.73	0.1	0.36	164
PS	0.77	8.0	1.69	155	2.8	0.06	0.31	136
P4tBS	1.0	8.4	2.84	338				
PαMS	1.05	8.0	1.21	245				
NC	1.75	1.8	0.38	1137				
PES	3.4	0.14	0.0	2348				

PPQ at 351 nm, $I_t^b = 3.4$, $\alpha^c = 0.27$, $\beta^d = 0.17$, $k^e = 113$

PI[f] at 308 nm, $I_t^b = 3.5$, $\alpha^c = 1.0$, $\beta^d = 0.105$, $k^e = 130$

[a] PET = poly(ethylene terephthalate), PBT = poly(butylene terephthalate), PPQ = polyphenylquinoxaline, PI = polyimide, BPPC = polycarbonate, PS = polystyrene, P4tBS = poly(4-tert-butylstyrene), PαMS = poly(α-methylstyrene), NC = dinitrocellulose, PES = poly(ethylene succinate)

[b] Threshold intensity (MW/cm^2)
[c] Low intensity absorption coefficient (10^5 cm^{-1})
[d] Ablation products attenuation coefficient (10^5 cm^{-1})
[e] Ablation rate constant (ng.ns^{-1}.MW^{-1}.cm^2)
[f] Experimental data taken from ref 16

HIGH FLUENCE MECHANISM

At very high fluence (>3-5 J/cm^2), some polymers display a sharp increase of the etch depth which is not predicted by the model that describes the low fluence behaviour. This is clearly demonstrated by the 193 nm etch curves of polycarbonate and polystyrene (Fig.2). This phenomenon was already observed with the ablation of polyimide (28) and was attributed to the appearance of a dominant thermal mechanism. This is not so clear to us, since a plasma assisted etching may also operate, like in the case of metals or semiconductors (29). It has been observed in these conditions of high fluence, that polymers that strongly deviate from the model, exhibit constantly a large ß coefficient. Conversely, polymers with negligible ß like dinitrocellulose do not deviate from the theoretical curve of our model. Large ß coefficients give rise to a strong attenuation of the incident laser beam and therefore to a large deposition of energy (2b) in the gas phase products. A plasma regime may ensue, giving rise to a new interaction mechanism with the polymer surface. In order to develop this point of view more experimental work has to be performed.

ROLE OF POLYMER STRUCTURE

From the work of Srinivasan and Braren (5) we know that the kinetics of organic materials ablation depends on their chemical structure. This dependence should not be confused with the absorptivity of the polymer, difficult to evaluate at laser intensity because it is the combination of several non-linear effects. In principle, two polymers of different chemical structures having similar laser beam absorptivities are not expected to have identical ablation kinetics parameters. Polystyrenelike polymers (PS, PαMS, P4tBS), whose repeat unit bears the same chromophore display quite different ablation rate constants (Table I), revealing that the degradation pathways are strongly influenced by a slight change of structure.

The definition of the ablation rate constant k allows an easy comparison of the different polymer structures, since it represents the amount of material vaporized, by the same quantity of energy above the threshold (1MW/cm^2). In order to go into a more detailed analysis, several working hypotheses are being examined, like for instance: bond energies, aromatic contents, oxygen content, thermal stability of the polymer. For aromatic polymers, the results (ß and k) are plotted in Fig.3 as a function of aromatic content (mass % of aromatic carbon atoms). A nearly linear correlation is found and shows that the ablation rate constant k (Fig.3a) decreases linearly with the aromaticity. The presence of aromatic bonds, having larger energy than single bonds, tends to decrease k probably because they must absorb several photons to break up. This multiphoton process takes a longer time than the one photon breaking of a single bond. As the solid becomes richer in aromatic bonds, its ablation is then made more difficult and this results in smaller values of k. As a result, single bonds are first and preferentially broken and most of the aromatic moieties must desorb probably untouched. This effect is revealed by the coefficient ß which increases linearly as in Fig.3b as the polymer becomes richer in aromatic rings. The exception of poly-t-butylstyrene is however to be noted and has not received any explanation yet.

Conversely, it is believed that non aromatic polymers NC and PES, ablate much faster than aromatic ones, as shown by the k values of Table I, because single bonds are statistically broken with one photon only. Besides, the higher ablation speed cannot be accounted for by a presumed higher absorptivity, since it is suggested to be lower by low intensity absorption coefficients displayed in Table I.

The thermal stability of the polymer is an interesting criterion for one to develop because several authors have considered photoablation as proceeding through a purely thermal degradation. Polyimide and polyphenylquinoxaline, two polymers of the thermostable family, are stable up to a temperature of approximately 500°. As seen in Figure 3a, they display quite

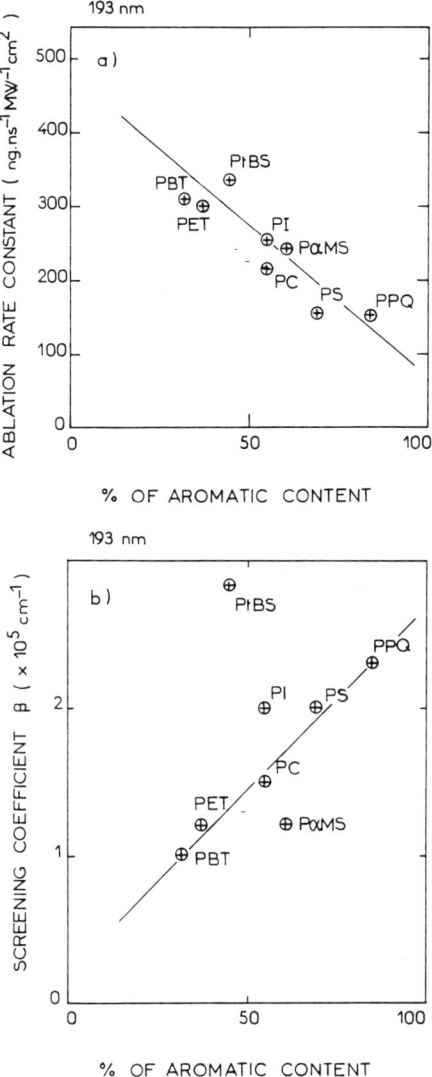

Figure 3. Ablation parameters at 193 nm for various polymers: a, rate constant k; and b, screening coefficient β. See Table I for acronyms.

different ablation rate constants. Polyphenylquinoxaline is the slowest polymer to ablate and this indeed is in good agreement with the concept of thermal ablation. But polyimide is much faster to ablate although having a similar thermal stability. It is even much faster than polystyrene which is known to be very thermolabile. This shows that the concept of thermal degradation mechanism of ablation cannot alone account for these ablation data.

As mentioned above the presence of oxygen in the molecular structure of the polymer is thought to facilitate ablation, since it should readily give very stable products like H_2O, CO_2 and CO (3),(9). This only led to a poor correlation between k and the oxygen content of the polymer. Poly(ethylene succinate) whose repeat unit ($-CH_2-O_2C-CH_2-CH_2-CO_2-CH_2-$) is highly oxygenated and contains two potential CO_2, displays the highest ablation rate constant known today (k = 2350 $ng.ns^{-1}.MW^{-1}.cm^2$). Surprisingly, dinitrocellulose that contains more oxygen (57%), has a lower rate constant (1137 $ng.ns^{-1}.MW^{-1}.cm^2$), probably this is because some of these oxygen are linked to nitrogen rather than carbon as in poly(ethylene succinate). Much effort has to be focussed on the understanding of these intriguing differences.

PRODUCTS DEPOSITON RATE MEASUREMENTS

The deposition reaction of the ablation products when they hit a surface, was early recognized (30) to be one of the obstacles to the development of photoablation for microlithographic purpose. However some new horizons in thin film synthesis by pulsed laser evaporation, pushed many groups (31) to investigate the features of these new reactions. This may in fact give information on the ablation mechanism. Fig.4 shows the spatial distribution probed with the microbalance of the deposition obtained for the ablation of polyphenylquinoxaline at 248 nm. The curve gives the mass of material (accumulation of 600 pulses) deposited as a function of the angle Θ between the normal to quartz surface and the normal to the ablated polymer surface. It can be fitted with a mathematical expression like $t = t_o \cos^n \Theta$ where t_o is the thickness at $\Theta = 0$ and it was shown that n is a parameter that describes the directivity of the distribution and it is dependent on fluence, background pressure, polymer structure and wavelength. By integration of $t(\Theta)$ over the half space into which the ablation products are desorbed one may know the total amount of products deposited and this can be compared to the mass of ablated polymer, known from the ablation curves, to give the yield of the deposition reaction. A few examples of yields (193 nm) are displayed in Fig.5 as a function of ablation depth. Fig.5 really shows that the polymer structure strongly influences the yield of deposition. Aromatic polymers tends to deposit more easily than others like poly(ethylene succinate) which are more likely to give non reactive products like CO_2. For all polymers the yield decreases by increasing the ablation rate. Higher ablation rates are necessarily accompanied by an increase of the kinetic energy of the products and a more intense attenuation of the laser radiation, which leads to smaller fragments, unfavorable to an efficient deposition. An increase of the background pressure results also in better yields.

The variations of yield with the experimental conditions provide a rapid information on the reactivity and nature of the ablation products, which are difficult to study by others means. Products are formed in a large distribution. They are expelled from the surface with a very high kinetic energy, they absorb many photons and are fragmented into smaller ones during their flight in the laser beam. It is clear that stable molecules eventually formed during ablation like H_2O, CO_2, CO, H_2 etc.. account for the amount of the material that does not deposit. Therefore the structures of ablated and deposited polymers are in general different. However, aromatic moieties are commonly found in the deposited material, showing that they easily survive to the laser photolysis.

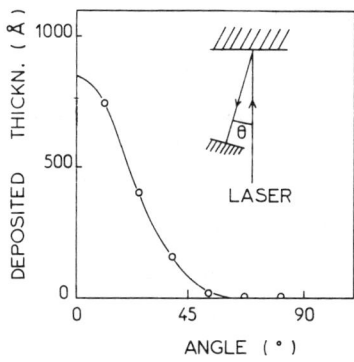

Figure 4. Deposited thickness measured with the microbalance as a function of ejection angle. Polyphenylquinoxaline at 248 nm, 200 mJ/cm^2, 600 pulses, pressure = 10^{-5} torr.

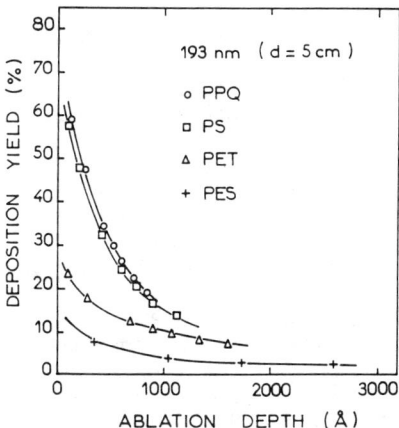

Figure 5. Yield of deposition as a function of ablation depth for various polymers. *See* Table I for acronyms.

CONCLUSIONS

The photoablation behaviour of a number of polymers has been described with the aid of the moving interface model. The kinetics of ablation is characterized by the rate constant k and a laser beam attenuation by the desorbing products is quantified by the screening coefficient ß. The polymer structure strongly influences the ablation parameters and some general trends are inferred. The deposition rates and yields of the ablation products can also be precisely measured with the quartz crystal microbalance. The yields usually depend on fluence, wavelength, polymer structure and background pressure.

ACKNOWLEDGEMENTS

The authors thank Bernard Sillon for a gift of polyphenylquinoxaline and Jean Claude Wittmann for a gift of poly(butylene terephthalate).

LITERATURE CITED

1. Srinivasan, R.; Mayne-Banton, V. Appl. Phys. Lett. 1982, 41, 576.
2. For recent reviews see
   a) Srinivasan, R. Science 1986, 234, 559.
   b) Lazare, S.; Granier, V. Laser Chem. 1989, 10, 25.
3. Srinivasan, R.; Leigh, W. J. J. Am. Chem. Soc. 1982, 104, 6784.
4. Davis, G. M.; Gower, M. C.; Fotakis, C.; Efthimiopoulos, T.; Argyrakis, P. Appl. Phys. 1985, A 36, 27.
5. Srinivasan, R.; Braren, B. J. Polym. Sci. Polym. Chem. Edit. 1984, 22, 2601.
6. Srinivasan, R.; Braren, B.; Seeger, D. E.; Dreyfus, R. W. Macromolecules 1986, 19, 916.
   Srinivasan, R.; Braren, B.; Dreyfus, R. W.; Hadel, L.; Seeger, D. E. J. Opt. Soc. Am. 1986, B3, 785.
   Srinivasan, R.; Braren, B.; Dreyfus, R. W. J. Appl. Phys. 1986, 61, 372.
7. Lazare, S.; Hoh, P. D.; Baker, J. M.; Srinivasan, R. J. Am. Chem. Soc. 1984, 106, 4288.
   Srinivasan, R.; Lazare, S. Polymer 1985, 26, 1287.
   Lazare, S.; Srinivasan, R. J. Phys. Chem. 1986, 90, 2124.
8. Gorodetsky, G.; Kazyaka, T. G.; Melcher, R. L.; Srinivasan, R. Appl. Phys. Lett. 1985, 46, 828.
   Dreyfus, R. W.; McDonald, F. A.; von Gutfeld, R. J. J. Vac. Sci. Technol. 1987, B5, 1521.
   Dyer, P. E.; Sidhu, J. J. Appl. Phys. 1985, 57, 1420.
9. Danielzik, B.; Fabricius, N.; Röwekamp, R.; Von der Linde, D. Appl. Phys. Lett. 1986, 48, 212.
   Estler, R. C.; Nogar, N. S. Appl. Phys. Lett. 1986, 49, 1175.
   Larciprete, R.; Stuke, M. Appl. Phys. 1987, B 42, 181.
   Feldmann, D.; Kutzner, J.; Laukemper, J.; MacRobert, S.; Welge, K. H. Appl. Phys. 1987, B 44, 81.
10. Dyer, P. E.; Srinivasan, R. Appl. Phys. Lett. 1986, 48, 445.
    Srinivasan, R.; Dyer, P. E.; Braren, B. Lasers Surg. Med. 1987, 6, 514.
11. Dyer, P. E.; Sidhu, J. J. Appl. Phys. 1988, 64, 4657.
12. Nikolaus, B. Conf. Las. Elect. Opt. Techn. Dig. Ser. 1986, 5 (OSA Washington).
    Srinivasan, R.; Sutcliffe, E.; Braren, B. Appl. Phys. Lett. 1987, 51, 1285.
    Küper, S.; Stuke, M. Appl.Phys. 1987, B44, 199; Appl. Phys. Lett. 1988, 54, 4.

13. Srinivasan, R.; Braren, B.; Dreyfus, R. W.; Hadel, L.; Seeger, D.E. J. Opt. Soc. Am. 1986, B3, 785.
    Masuhara, H.; Hiraoka, H.; Domen, K. Macromolecules 1987, 20, 452.
    Masuhara, H.; Hiraoka, H.; Marinero, E. E. Chem. Phys. Lett. 1987, 135, 103.
    Srinivasan, R.; Braren, B. Appl. Phys. 1988, A45, 289.
    Hiraoka, H.; Chuang, T. J.; Masuhara, H. J. Vac. Sci. Technol. 1988, B6, 463.
    Chuang, T. J.; Hiraoka, H.; Mödl, A. Appl.Phys. 1988, A45, 277.
14. Jellinek, H. H. G.; Srinivasan, R. J. Phys. Chem. 1984, 88, 3048.
15. Srinivasan, V.; Smrtic, M. A.; Babu, S. V. J. Appl. Phys. 1986, 59, 3861.
16. Srinivasan, R.; Braren, B.; Dreyfus, R. W. J. Appl. Phys. 1987, 61, 372.
17. Lazare, S.; Soulignac, J. C.; Fragnaud, P. Appl. Phys. Lett. 1987, 50, 624.
18. Sutcliffe, E.; Srinivasan, R. J. Appl. Phys. 1986, 60, 3315.
19. Lazare, S.; Granier, V. Appl. Phys. Lett. 1989, 54, 862.
20. Koren, G. Appl. Phys. Lett. 1987, 50, 1030.
21. Andrew, J. E.; Dyer, P. E.; Foster, D.; Key, P. H. Appl. Phys. Lett. 1983, 43, 717.
22. Hansen, S. G.; Robitaille, T. E. Appl. Phys. Lett. 1987, 62, 1394.
23. Sauerbrey, G. Z. Phys. Verhandl. 1957, 8, 113.
24. In Applications of Piezoelectric Quartz Crystal Microbalances; Lu, C.; Czanderna, A., Eds.; Elsevier: New York, 1984.
25. Lazare, S.; Granier, V. J. Appl. Phys. 1988, 63, 2110.
26. Konash, P. L.; Bastiaans, G. J. Anal. Chem. 1980, 52, 1929.
    Nomura, T.; Okuhara, M. Anal. Chim. Acta 1982, 142, 281.
27. Koren, G.; Oppenheim, U. P. Appl. Phys. 1987, B42, 41.
28. Srinivasan, V.; Smrtic, M. A.; Babu, S. V. J. Appl. Phys. 1986, 59, 3861.
29. Poprawe, R.; Beyer, E.; Herziger, G. Inst. Phys. Conf. Ser. 1984, 72, 67.
    Herziger, G.; Krentz, E. W. Spring. Ser. Chem. Phys. 1984, 39, 90.
    Shinn, G. B.; Steigerwald, F.; Stiegler, H.; Sauerbrey, R.; Tittel, F. K.; Wilson, W. L. J. Vac. Sci. Technol. 1986, B4, 1273.
30. von Gutfeld, R. J.; Srinivasan, R. Appl. Phys. Lett. 1987, 51, 15.
31. Dijkkamp, D.; Venkatesan, T.; Wu , X. D.; Shaheen, S. A.; Jisrawi, N.; Min-Lee, Y. H.; McLean, W. L.; Croft, M. Appl. Phys. Lett. 1987, 51, 619.
    Hansen, S. G., Robitaille, T. E. Appl. Phys. Lett. 1988, 52, 81; J. Appl. Phys. 1988, 64, 2122.

RECEIVED July 13, 1989

Chapter 26

Mechanism of UV- and VUV-Induced Etching of Poly(methyl methacrylate)

Evidence for an Energy-Dependent Reaction

Nobuo Ueno[1], Tsuneo Mitsuhata[1], Kazuyuki Sugita[1], and Kenichiro Tanaka[2]

[1]Department of Image Science and Technology, Faculty of Engineering, Chiba University, Yayoicho, Chiba 260, Japan
[2]Photon Factory, National Laboratory for High-Energy Physics, Tsukuba, Ibaraki 305, Japan

> We performed a mass spectroscopic study of the vaporized species generated during ultraviolet (UV) and vacuum ultraviolet (VUV) irradiation of poly(methyl methacrylate) (PMMA), and found a remarkable difference between the spectra measured during UV and VUV irradiation. In the case of VUV irradiation, mass peaks were observed, which can be ascribed to products of the direct main-chain scission of PMMA. The results are explained by an energy-dependent and site-selective photochemical reactions in PMMA. The present finding indicates that the use of energy dependent photochemical reactions is potentially useful in the photoetching of polymer resists. The effects of temperature and oxygen gas on the photoetching of PMMA are also described in this article.

Among a variety of etching techniques in microlithography, photo-induced etching is becoming important as a non-destructive method. High energy photons, i.e. synchrotron radiation including soft X-ray, were used for the etching of various materials, and found to be useful in microlithography (1-3).

In general, the absorption coefficients of solids consisting of light elements tend to increase with photon energy and reach a maximum in the photon energy range of about 10 - 30 eV [vacuum ultraviolet (VUV) region]. For photo-induced etching, large penetration depth of the incident light into materials is not required. Hence, we can realize more rapid etching by an effective use of VUV light which is absorbed at the surface region. Further, we expect a different photochemical reaction by VUV irradiation, since VUV can excite solids to higher energy states. In fact, Ueno et al. (4) showed the efficacy of VUV in photoetching of Si (100) and PMMA [poly(methyl methacrylate)] using a conventional light source.

In the case of PMMA, for example, VUV excites electrons localized along the polymer backbone as well as those in the side chain, while UV excites only the π electrons in the side chain. The absorption coefficient in the VUV region is more than several hundreds times larger than that in the UV region (see Figure 1)(5). Thus, we can expect more effective reactions which produce volatile species by VUV-induced main-chain scission. The use of such energy-selective or site-selective reactions (6) would become more important not only for the photoetching applied in the fabrication of semiconductor substrate but also for resist technology (7,8) in microlithography.

We present here the results of a mass spectroscopic study of the vaporized species produced during UV and VUV irradiation of PMMA. A difference was observed between the mass spectra of the vaporized species produced by the two methods of irradiation, indicating direct main-chain scission induced by VUV absorption of PMMA. Further, the effects of heating and introduction of oxygen on UV etching will be described.

Experimental

We used two experimental setups as shown in Figure 2. Mass analysis of species resulting from UV irradiation was performed with method A (Figure 2a), including the studies on the effects of heating and introduction of oxygen on the etching reaction. The inner surface of an etch tube was coated with PMMA (thickness < 0.1 mm) and irradiated with the UV light of a commercial D_2 lamp (Original Hanau D200F) after filtering by a quartz window. Thus the film was exposed to UV light ($h\nu$ < 6.8 eV) in high vacuum. The vaporized species produced by this exposure were introduced into the main vacuum chamber of a mass spectrometer and analyzed by a QP mass filter (ULVAC MSQ-150A) after ionization by 70 eV e^- bombardment. A light shutter was placed between the light source and the window. The film could be heated and its temperature was measured with a thermocouple attached to the etch tube. For the etching experiment under oxygen, we introduced oxygen gas (>99.99%) into the etch tube at a pressure of 10^{-2} Torr. Method B (Figure 2b) was used for VUV etching of PMMA. The light source for this setup was a capillary-gas-discharge lamp (9). The lamp was mounted on the main chamber with a differential pumping unit and a light shutter. In this experiment, hydrogen gas was used for the discharge, since it can emit intense VUV light in the wavelength region of 85 - 167 nm (7.4 - 14.5 eV)(10). The light emitted from the discharge lamp was collimated by two glass tubings of an inner diameter of 1 mm, and irradiated the surface area of a specimen of about 4 mm in diameter. As shown in Figure 1, these photons can excite the main chain of PMMA. A PMMA film (thickness< 1 μm) spin-coated on a Si wafer was placed on the rotatable sample holder. Both heater and thermocouple were mounted on the sample holder. For a direct comparison of etching products produced by UV and VUV irradiation, we also used this setup for UV irradiation by simply exchanging the capillary-discharge lamp for the D_2 lamp. The light power density at the sample position was measured with a thermopile to be 33 mW/cm^2 for UV and less than 0.02 mW/cm^2 for VUV. The temperature rises of the PMMA films resulting from the absorption of photons were negligible. Although the intensity of VUV was much weaker than that of UV, the vaporized species by VUV irradiation to PMMA were detected. This can be ascribed to the large absorption coefficient of PMMA in the VUV region (see Figure 1). For example, the penetration of VUV (Lyman α, λ=121.6 nm) into a PMMA film is compared in Figure 3 with that of UV (λ=213 nm).

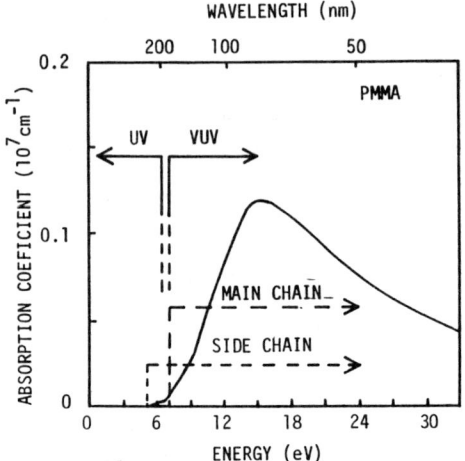

Figure 1. Absorption coefficient of PMMA. The energy regions are illustrated for main-chain and side-chain excitations. (Reproduced with permission from ref. 5. Copyright 1978 American Institute of Physics.)

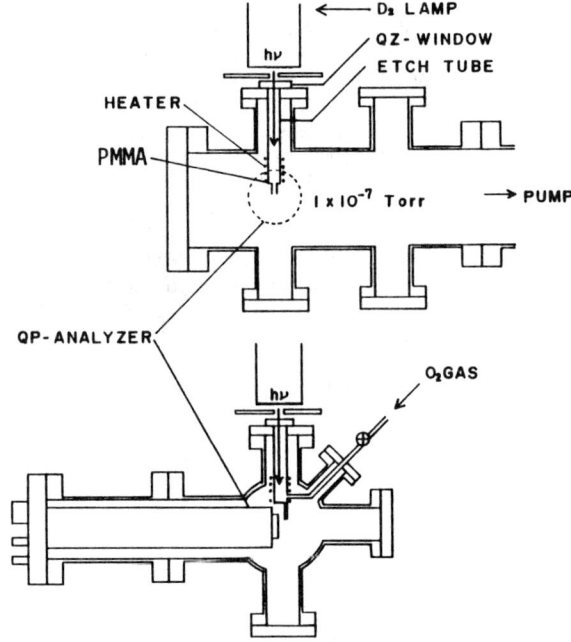

(a) METHOD A

Figure 2. Experimental setups for the mass analysis of vapor products a, with etch tube. Continued on next page.

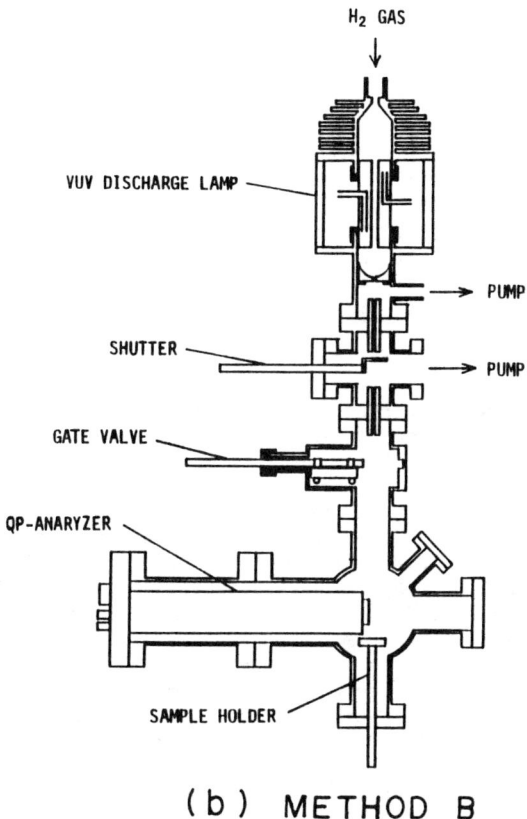

(b) METHOD B

Figure 2. Continued. b, With rotatable sample holder.

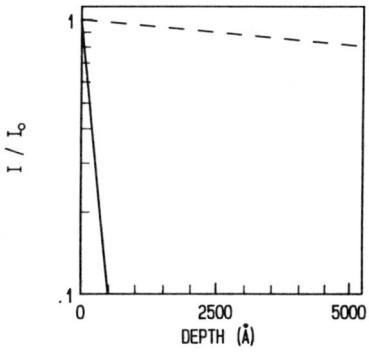

Figure 3. Attenuation of UV and VUV light in PMMA film.
---: UV (λ =213 nm). ———: VUV (λ =121.6 nm). The intensity of light before and after absorption is I_0 and I, respectively.

Results and Discussion

UV Etching. A typical mass spectrum of the vaporized UV etching products is shown in Figure 4, together with a background spectrum obtained without UV irradiation. The comparison clearly shows that UV irradiation causes an increase in intensity for various mass peaks. For example, the intensity of the peaks of m/e=15, 31, 59 and m/e=41, 69 increased drastically by UV irradiation. The former three are due to side-chain scission caused by UV absorption at the C=O unit, while the latter two are due to main-chain scission initiated by side-chain scission (11). The structure and mass numbers of typical vaporized species are shown in Table I. From here on, we use the spectral intensity after the background is subtracted.

Figure 5 shows the time dependence of the spectral intensity for m/e=31, 41, 69 and 100 during irradiation. The intensity for the species of larger mass numbers increases more slowly with time than that for the smaller species. This means that the species of m/e=41 and 69 are not dominated by the fragments of MMA monomer (m/e=100) due to the e^- bombardment in the ionization chamber but by the etching products vaporized from the PMMA film, since the fragments produced by the e^- bombardment of MMA should show similar time dependence to that of MMA^+. The difference in the time dependence of the spectral intensity for the species with different mass numbers tended to be smaller when PMMA was irradiated at higher temperature. The tendency can be ascribed to the more rapid increase of the diffusion constants for species of larger mass number in PMMA with increase in temperature. Such a consideration is supported by experimental results that the etch rate increased with the film thickness and with temperature in the photoetching of the positive-working resist (12-15), where the diffusion of decomposed species in the film plays an important role. In passing, we note that the intensity of MMA monomer ion was about 10^3 times larger at 90°C than at 50°C, while the intensity of peaks of lower mass numbers due to side-chain scission did not show a comparable increase. In Figure 6, the drastic increase of the species due to main-chain scission as a result of heating is shown; the increase of the species generated by side-chain scission is also shown for comparison. We consider that the drastic increase in intensity of MMA^+ is a result of effective unzipping reaction in vacuum occurring even at 90°C which is sufficiently lower than the ceiling temperature Tc=164°C at 1 atm (16), since the vaporization of MMA reduces the concentration of MMA in the film followed by a reduction of Tc.

Generally, the dominant mechanism of the UV etching of degradable polymer resist in the presence of oxygen is believed to be a reaction of polymer molecule with oxygen atoms produced by photodissociation of molecular oxygen (17). Therefore, it is interesting to compare the mass spectra of the etching products by UV-irradiation to PMMA in vacuum and in oxygen gas. The results are shown in Figure 7, where the intensity variations of the mass peaks for m/e=44 (CO_2^+) and m/e=41 ($C_3H_5^+$) are compared. A remarkable increase in intensity of mass peak of m/e=44 (CO_2^+) was observed upon UV irradiation in oxygen, while other species did not show comparable increases. That is, the integrated carbon number in all of the

26. UENO ET AL. *UV- and VUV-Induced Etching of Poly(methyl methacrylate)* 429

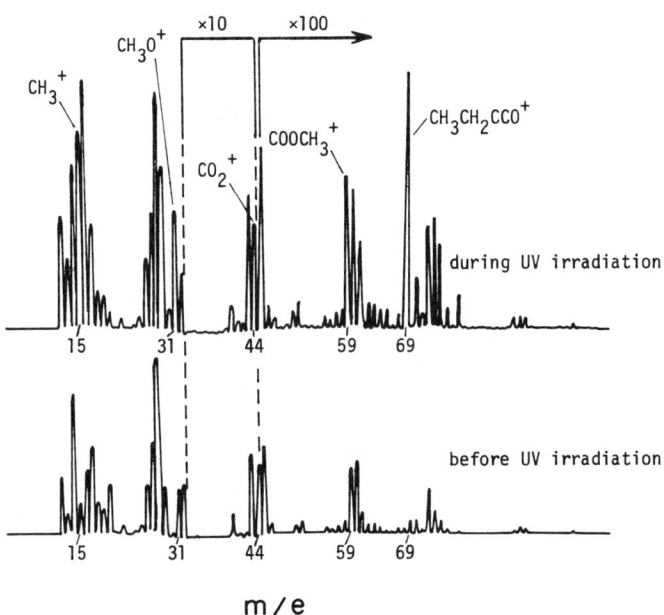

Figure 4. Typical mass spectra before (lower) and during (upper) UV irradiation of PMMA at 50°C without O_2 gas.

Table I. Molecular structure and mass numbers of selected products by photoetching of PMMA

15	16	28	29	31	41	44
CH_3	O	CO	CHO	$\begin{array}{c}O\\|\\CH_3\end{array}$	$\begin{array}{c}CH_3\\|\\CH_2-C\end{array}$	CO_2

55	59	69	83	100
$\begin{array}{c}CH_3\\|\\CH_2-C-CH_2\end{array}$	$\begin{array}{c}C=O\\|\\O\\|\\CH_3\end{array}$	$\begin{array}{c}CH_3\\|\\CH_2-C\\|\\C=O\end{array}$	$\begin{array}{c}CH_3\\|\\CH_2-C-CH_2\\|\\C=O\end{array}$	$\begin{array}{c}CH_3\\|\\CH_2-C\\|\\C=O\\|\\O\\|\\CH_3\end{array}$

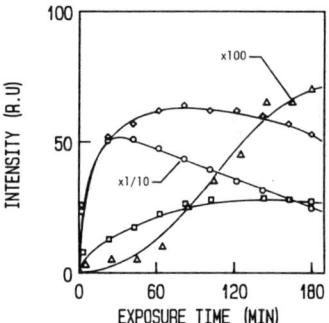

Figure 5. Intensity variations of mass peaks of m/e=31(○), 41(◊), 69(□), and 100(△) as a function of UV-exposure time of PMMA at 50°C without O_2.

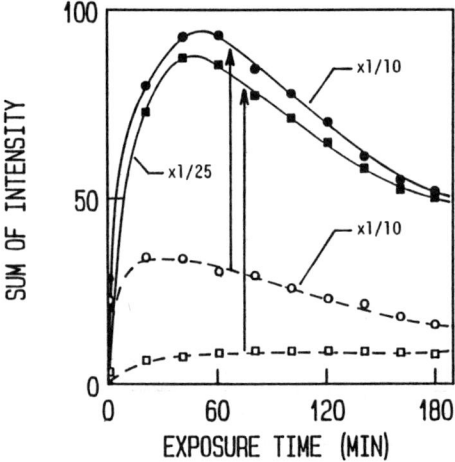

Figure 6. Temperature dependence of sum of peak intensity for m/e=41, 69, 100 (products by main-chain scission + unzipping) (□;50°C, ■;90°C) and for m/e=15, 28, 29, 31, 44, 59 (products by side-chain scission)(○;50°C, ●;90°C).

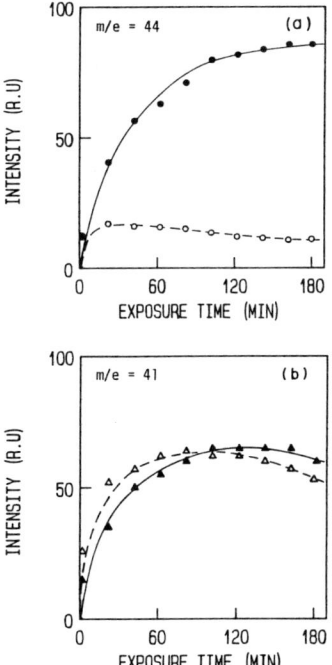

Figure 7. Intensity variations of mass peaks of m/e=44 (CO_2^+) (a) and m/e=41 ($C_3H_5^+$) (b) as a function of UV-exposure time for PMMA with and without O_2 gas. ●,▲:With O_2 gas. ○,△:Without O_2 gas. Temperature of the sample was 50°C.

vaporized species did not increase appreciably by introducing oxygen gas during UV irradiation to PMMA. We conclude that reactions of PMMA with oxygen during UV irradiation do not play major role in the etching reaction under the pressure range used in our study.

From the above results, the main UV-etching reactions of PMMA are considered as follows,

$$\mathrm{-CH_2-\underset{\underset{\underset{CH_3}{|}}{\overset{\overset{CH_3}{|}}{C}}-\overset{|}{C=O}}{|}}\;\xrightarrow{UV}\; \mathrm{-CH_2-\underset{\underset{\underset{CH_3}{|}}{\overset{\overset{CH_3}{|}}{\dot{C}}-}}{|}C=O}\; +\; \mathrm{\underset{\underset{CH_3}{|}}{\overset{\overset{\dot{C}=O}{|}}{|}}} \;\longrightarrow\; \mathrm{-CH_2-\underset{\underset{\underset{CH_3}{|}}{\overset{\overset{CH_3}{|}}{\dot{C}}-}}{|}C=O}\; +\; \mathrm{CH_2=\underset{\underset{\underset{CH_3}{|}}{\overset{CH_3}{|}}}{\dot{C}}\cdot} \quad (1)$$

$$\longrightarrow\; \mathrm{-CH_2-\underset{\underset{\dot{C}=O}{|}}{\overset{\overset{CH_3}{|}}{C}-}}\; +\; \mathrm{\dot{O}-CH_3}\;\longrightarrow\; \mathrm{-CH_2-\underset{\underset{\underset{CH_3}{|}}{\overset{\overset{CH_3}{|}}{\dot{C}}-}}{|}C=O}\; +\; \mathrm{CH_2=\underset{C=O}{\overset{CH_3}{|}}{C}} \quad (2)$$

with O, O_2

$$\longrightarrow\; \mathrm{-CH_2-\underset{\underset{\dot{O}}{|}}{\overset{\overset{CH_3}{|}}{C}-}C=O}\;\longrightarrow\; \mathrm{-CH_2-\underset{\dot{O}}{\overset{\overset{CH_3}{|}}{\dot{C}}-}}\; +\; \mathrm{\dot{C}=O} \quad (3)$$

<u>VUV Etching.</u> As in the case of UV etching, we found that the mass spectral pattern changed with VUV-irradiation time. The results are shown in Figure 8, where the contribution of the background is subtracted. Further, the comparison of mass spectra during VUV and UV irradiation without oxygen gas is shown in Figure 9. A clear difference was observed between the mass spectra observed for UV and VUV irradiation. The main difference is that the species of m/e=55 and 83 were clearly observed for VUV irradiation. Such intense peaks of m/e=55 and 83 were never observed in the decomposition of PMMA by UV irradiation and heating. The structure of these species is shown in Table I. We ascribe the appearance of these species to the results of direct main-chain scission by the electronic excitation of the main chain by VUV absorption in the following ways,

26. UENO ET AL. *UV- and VUV-Induced Etching of Poly(methyl methacrylate)* 433

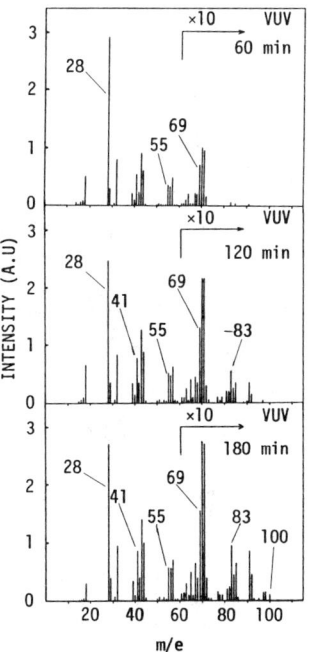

Figure 8. Time dependence of mass spectra of vaporized products during VUV irradiation to PMMA at 30°C. The time after the switch-on of the irradiation is indicated in the figure. The intensity is 10 times enlarged above m/e=60.

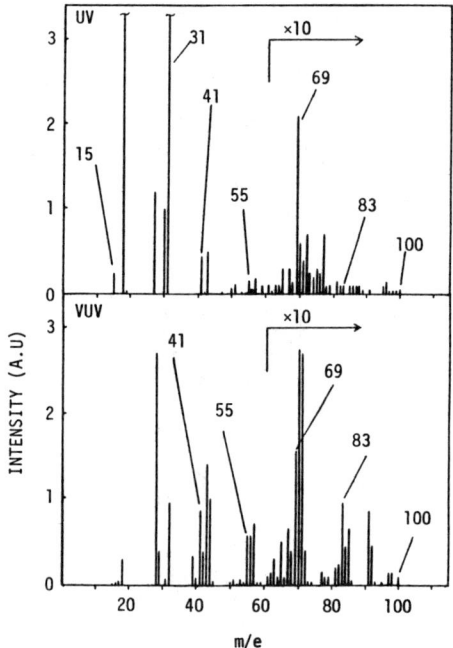

Figure 9. Comparison of mass spectra of vaporized products by VUV (lower) and UV (upper) irradiation of PMMA. The intensity is 10 times enlarged above m/e=60.

$$\begin{array}{c}\text{VUV}\text{VUV}\\\diagdown\diagdown\\\text{CH}_3\text{CH}_3\text{CH}_3\\|||\\-\text{CH}_2-\text{C}-\text{CH}_2-\text{C}-\text{CH}_2-\text{C}-\\|||\\\text{C=O}\text{C=O}\text{C=O}\\|||\\\text{O}\text{O}\text{O}\\|||\\\text{CH}_3\text{CH}_3\text{CH}_3\end{array}$$

$$\longrightarrow\quad\begin{array}{c}\text{CH}_3\\|\\-\text{CH}_2-\text{C}\cdot\\|\\\text{C=O}\\|\\\text{O}\\|\\\text{CH}_3\end{array}\;+\;\begin{array}{c}\text{CH}_3\\|\\\text{CH}_2=\text{C}\\|\\\text{C=O}\\|\\\text{O}\\|\\\text{CH}_3\end{array}\;+\;\begin{array}{c}\text{CH}_3\\|\\\text{CH}_2=\text{C}\\|\\\text{C=O}\\|\\\text{O}\\|\\\text{CH}_3\end{array}\quad(4)$$

$$\longrightarrow\quad\begin{array}{c}\text{CH}_3\\|\\-\text{CH}_2-\text{C}\cdot\\|\\\text{C=O}\\|\\\text{O}\\|\\\text{CH}_3\end{array}\;+\;\begin{array}{c}\text{CH}_3\\|\\\text{CH}_2-\text{C}-\text{CH}_2\\|\\\text{C=O}\\|\\\text{O}\\|\\\text{CH}_3\end{array}\;+\;\begin{array}{c}\text{CH}_3\\|\\\cdot\text{C}\cdot\\|\\\text{C=O}\\|\\\text{O}\\|\\\text{CH}_3\end{array}\quad(5)$$

Acknowledgments

The present work was partly supported by Grants-in-Aid for Scientific Research from the Ministry of Education, Science and Culture (No. 61470018 and No. 62470100).

Literature Cited

1. Urisu T.; Kyuragi H. J. Vac. Sci. Technol. 1987, B5 1436.
2. Kyuragi H.; Urisu T. Appl. Phys. Lett. 1987, 50 1254.
3. Yamada H.; Hori M.; Morita S.; Hattori S. J. Electrochem. Soc. Solid-State Science and Technology 1988, 135 967.
4. Ueno N.; Mitsuhata T.; Sugita K.; Tanaka K. Jpn. J. Appl. Phys. 1988, 27 1723.
5. Ritsko J. J.; Brillson L. J.; Bigelow R. W.; Fabish T. J. J. Chem. Phys. 1978, 69 3931.
6. Eberhardt W.; Sham T. K.; Carr R.; Krummacher S.; Strongin M.; Weng S. L.; Wesner D. Phys. Rev. Lett. 1983, 50 1038.
7. Ueno N.; Doi Y.; Sugita K.; Sasaki S.; Nagata S. J. Appl. Polym. Sci. 1987, 34 1677.
8. Mochiji K.; Kimura T.; Obayashi H. Appl. Phys. Lett. 1985, 46 387.
9. Ueno N.; Ikegami A.; Hayasi Y.; Kiyono S. Jpn. J. Appl. Phys. 1977, 16 1655.
10. Samson J. A. R. Techniques of Vacuum Ultraviolet Spectroscopy; John Wiley and Sons.: New York, London, and Sydney, 1967.
11. Hiraoka H. IBM J. Res. Develop. 1977, 21 121.
12. Ueno N.; Sugita K. Jpn. J. Appl. Phys. 1986, 25 1455.
13. Lanagan M.; Lindsey S.; Viswanathan N. S. Jpn. J. Appl. Phys. 1983, 22 L67.

14. Ueno N.; Konishi S.; Tanimoto K.; Sugita K. Jpn. J. Appl. Phys. 1981, 20 L709.
15. Sugita K.; Ueno N.; Konishi S.; Suzuki Y. Photogr. Sci. Eng. 1983, 27 146.
16. Ivin K. J. Trans. Faraday Soc. 1955, 51 1267.
17. Vig J. R. In Surface Contamination, Genesis, Detection and Control ; Mittel K. L. Ed.; Plenum Press, New York and London, 1979; Vol. 1.

RECEIVED June 29, 1989

INDEXES

Author Index

Baiocchi, Frank A., 188
Ban, Hiroshi, 174
Berry, A. K., 85
Bogan, L. E., Jr., 85
Clecak, N., 114
Dems, B. C., 233
Downing, J., 114
Ebina, Mayumi, 56
Endo, Masayuki, 266
Fahey, J., 99
Frank, C. W., 346
Fréchet, Jean M. J., 73,99
Fukumura, Hiroshi, 397
Granier, Vincent, 408
Graziano, K. A., 85
Hashimoto, Michiaki, 316
Hayase, Shuzi, 132
Horiguchi, Rumiko, 132
Houlihan, Francis M., 38
Hoyle, C. E., 277
Huang, J. P., 361
Hutchens, D. E., 277
Igarashi, K., 99
Iizawa, T., 99
Imamura, Saburo, 174
Inaki, Yoshiaki, 300
Itaya, Akira, 397
Ito, Hiroshi, 56
Iwayanagi, Takao, 316
Jurek, M. J., 157
Jurgensen, Charles W., 209
Karatsu, T., 114
Klingensmith, K. A., 114
Kosbar, L. L., 346
Krasicky, P. D., 233
Kuan, S. W. J., 346
Kwei, T. K., 361
Lazare, Sylvain, 408
Limm, William, 382
MacDonald, Scott A., 26
Martin, P. S., 346
Masuhara, Hiroshi, 397
Matuszczak, Stephen, 73

Matuszczak, Stephen, 73
McKean, D. R., 26
McKinley, A. J., 114
Michl, J., 114
Miller, R. D., 114
Mitsuhata, Tsuneo, 421
Moghaddam, Minoo Jalili, 300
Nalamasu, Omkaram, 188
Nishikubo, T., 99
Nomura, Noboru, 266
Onishi, Yasunobu, 132
Pearce, E. M., 361
Pease, R. F. W., 346
Reck, Berndt, 73
Reichmanis, Elsa, 1,38,157
Reiser, A., 361
Rodriguez, F., 233
Sasago, Masaru, 266
Schaedeli, U., 26
Sheats, James R., 329
Smith, Barton A., 382
Sooriyakumaran, R., 114
Stanton, Deirdre T., 382
Stillwagon, L. E., 251
Stöver, Harald D. H., 73
Sugita, Kazuyuki, 421
Takemoto, Kiichi, 300
Tanaka, Akinobu, 174
Tanaka, Kenichiro, 421
Tani, Yoshiyuki, 266
Tarascon, Regine G., 38
Taylor, Gary N., 188,251
Thackeray, J. W., 85
Thames, S. F., 277
Thompson, Larry F., 1,38
Uchino, Shou-ichi, 316
Ueda, Mitsuru, 56
Ueno, Nobuo, 421
Ushirogouchi, Toru, 132
Wallraff, G., 114
Willson, C. Grant, 73,99
Winnik, Mitchell A., 382

Affiliation Index

AT&T Bell Laboratories, 1,38,157,188,209,251
Chiba University, 421
Cornell University, 73,99,233
Hewlett Packard Laboratories, 329
Hitachi Ltd., 316
IBM Almaden Research Center, 26,56,73,99,114,382
Kanagawa University, 99
Kyoto Institute of Technology, 397
Matsushita Electric Industrial Company Ltd., 266
National Laboratory for High-Energy Physics, 421
NTT Basic Research Laboratories, 174
NTT LSI Laboratories, 174
Osaka University, 300
Polytechnic University, 361
Rohm and Haas Company, 85
Shipley Company, Inc., 85
Stanford University, 346
Toshiba Corporation, 132
U.S. Army, 382
Université de Bordeaux I, 408
University of Southern Mississippi, 277
University of Texas at Austin, 114
University of Toronto, 382
Yamagata University, 56

Subject Index

A

Ablation rate constant
 definition, 418
 determination, 414,417t
Ablative photodecomposition
 description, 411
 See also Photoablation
Acetylated m-cresol–novolac copolymers, preparation, 193
Acetylated poly(phenylsilsesquioxane)
 characterization, 176–177
 molecular structure determination, 176
 ^{29}Si-NMR spectra, 176,178f
 synthesis, 176
Acid analysis, merocyanine dye methods, 30,31f,33f
Acid-catalyzed photoresist films
 acid diffusion, 35
 acid generation, 30,32,33f,34t
 advantages, 28
 catalytic chain length, 34,35t
 development of classes of cationic photoinitiators, 28
 experimental procedure, 35–36
 generation mechanism from irradiation of triphenylsulfonium salts, 28–29
 merocyanine dye method for acid analysis, 30,31f,33f
 use of *tert*-butoxycarbonyl functionality as protecting groups, 28,31
Acid diffusion, determination for acid-catalyzed photoresist films, 35

Acid generation in photoresist films
 acid photogeneration vs. dose, 32,33f
 acid present after irradiation, 32,34t
 acid present before irradiation, 32
 quantum yield, 32,34
Acid hardening resin resists
 cross-linking activation energy determination, 87,89
 cross-linking chemistry, 87
 determination of acid generated, 87–88
 effect of postexposure bake temperature and time, 87
Activation energy of cross-linking, determination for novolac-based negative resists, 90,93,94f,95t
Active polyformals
 first-order dependence of acidolysis, 106,108f
 preparation under phase-transfer catalysis, 104,106t,107–108f
 scanning electron micrograph of self-developed image, 109,110f
 schematics of imaging process, 106,108f,109
 thermogravimetric analysis, 106,107f
 UV absorbance vs. time during acidolysis, 106,107f
Alkali-developable silicon-containing positive photoresists, development, 175
Alkoxysilanes, polymerization, 134,136,142t
Allylic ethers and polyethers, thermolytic cleavage, 103,104,105f

A

Anthracenes
  photobleaching chemistry of polymers, 332–346
  rate equation, 339
Antireflective coatings, description, 13

B

Benzylic ethers and polyethers, thermolytic cleavage, 103,104,105f
Bilevel processing, description, 18
Bis(pyrimidine) derivatives
  photopolymerization, 308,312f
  structure, 306,307f
  wavelength dependence of photopolymerization, 308,310f
Bleaching contrast
  definition, 334
  influencing factors, 334,339
  relationship to ratio of oxidation rate constant, 339–340
Block copolymers
  applications, 158–159
  synthetic procedures, 159
Bombardment, role in pattern transfer regime, 215–218,220f
Bombardment-induced kinetics
  etching rate, 211–212
  model, 211
Bombardment-induced pattern transfer, models, 230–231
Brönsted acid, generation mechanism from triphenylsulfonium salts in acid-catalyzed photoresist films, 28–36
tert-Butoxycarbonyl-containing resists
  characterization, 28,30
  use as protecting groups, 28,31
p-tert-Butylphenyl glycidyl ether, evaluation as planarizing layers, 263

C

Catalytic chain length
  determination for acid-catalyzed photoresist films, 34,35t
  experimental procedure, 36
Chemical amplification
  approaches, 100–101
  chemical reactions, 11,12f
  description, 74
  design of negative resists, 11
  development of high-resolution, dual-tone resist materials, 11,12f
  development of nonionic acid precursors based on nitrobenzyl ester photochemistry, 11,14f

Chemical amplification—*Continued*
  examples, 74
  sensitivity improvement in resists, 11
  three-component aqueous-base developable, positive-tone resists, 11,13,14f
Chemically amplified deep-UV resists
  characteristic curve for sensitivity measurement, 78,80f
  contrast measurements, 78t,79–80f
  cross-linking process via electrophilic aromatic substitution, 78,79f
  cross-linking via nonpolymeric multifunctional latent electrophile, 78–80f
  design, 75,76
  developmental approaches, 74
  gel permeation chromatogram of copolymer, 75,77f,78
  imaging experiments, 81,82f
  mechanistic and model studies, 81,83f
  preparation of copolymers containing electrophilic and nucleophilic groups, 75,76f
  preparation of copolymers of vinylbenzylacetate and 4-(*tert*-butoxycarbonyl)styrene, 84
  removal of *tert*-butoxycarbonyl protecting groups, 84
  resist modeling experiments, 84
  scanning electron micrographs, 81,82f
  sensitivity measurements, 78t,79–80f
  use of difunctional cross-linker, 78,80f,81
  UV spectrum, 78,79f
Chemically amplified negative resists, description, 86
Chemically amplified resists
  effect of polymer structure on performance, 43–54
  experimental materials, 41
  GC–MS studies, 42–43
  gel permeation chromatographic studies, 41
  lithographic characterization, 51,52–53t,54
  lithographic evaluation, 43
  performance criteria, 40t
  photochemically initiated depolymerization, 44–50
  polymer molecular properties, 41,42t
  process considerations, 50–51
  thermally initiated depolymerization, 44,45t
Chloromethylated polystyrene, cross-linking mechanism, 7
Circuit densities, increase via improvements in technologies, 27
Contact printing, use for device production, 1,3t

INDEX

Contrast-enhanced lithography
  categories of photobleachable dyes, 319–320
  description, 319
Contrast enhancement materials, 13,15f
Contrast for resist, determination, 4,5f
Conventional photolithography, technological alternatives, 3
Conventional positive photoresist, 9,10f
Copolymer approach to sensitive deep-UV resist system design
  CF_4 etching of copolymers and reference polymers, 68,72f
  cross section of negative images, 68,69f
  cross section of positive and negative images, 68,71f
  deep-UV contrast curves, 65,69f
  differential scanning calorimetry, 59,62f
  dry etch resistance, 68,72f
  experimental materials, 58,60
  experimental measurements, 58–59
  GC–MS, 59,63f
  IR spectra, 65,67f
  lithographic evaluation, 59
  resist imaging, 65
  scanning electron micrographs of negative images, 68,71f
  scanning electron micrographs of positive and negative images, 68,70f
  structure of two-component copolymer resist, 68,72f,73
  thermal gravimetric analysis, 59,61f,65,66f
  UV spectra, 65,67f
o-Cresol–novolac, evaulation as planarizing layers, 260–262t,263

D

Deep–mid-UV systems, 3–4
Deep-UV photolithography, 27–28
Deep-UV resist
  performance criteria, 40t
  properties, 58
  system design, 58–72
Deep-UV sensitive dissolution inhibitors, examples, 9
Developer, selection, 6
Diazonaphthoquinone–novolac resists, 58
Diazonium salts
  application to photoacid generators, 327,329,330f
  application to two-layer resist system, 322,327,328f
  chemical structures, 320,323t
  contrast-enhanced lithographic dye, 320–321

Diazonium salts—*Continued*
  exposure curves of positive photoresist with and without diazonium salt–contrast-enhanced lithography, 322,326f
  IR spectrum of photoproduct, 322,324f
  lithographic evaluation, 320–321
  negative two-layer resist formation, 321
  NMR spectrum of photoproduct, 322,324f
  photoacid generator formation, 321
  photobleaching curve, 322–325f
  scanning electron microscopic photographs of resist patterns, 322,326f
  synthesis, 320–322
  thermal stability, 322,323f
  use as photosensitive material in photobleachable two-layer resist system, 320
  UV absorption spectra, 322,325f
Dissolution inhibition resists, 8–10
Dissolution of novolac resins
  activation energies, 373t,375f
  characterization of copolymers, 366t
  contact angle measurement, 366
  dependence on base concentration, 367t,368–369f,370t
  description of resins, 365t
  dissolution measurement, 366
  effect of added salt, 367,370t,371f
  effect of cation size, 370,372f,373
  effect of chain mobility, 373–377
  effect of polymer inhibitors, 376–382
  effect of temperature, 373,374f
  experimental materials, 365
  fitting by WLF equation, 373,376,377f
  induction periods of blends vs. contact angles, 378,381f
  IR and ^{1}H-NMR measurements, 376t
  model, 378,383
  purification of materials, 365
  synthesis of resins, 365–366
  water diffusion, 366
  water uptake of polymer films, 378,380f
Dynamic random access memory device, circuit density vs. size, 1

E

Electron-beam lithography
  limitations, 349
  use in device production, 3t
Etching resistance of resist, definition, 4

F

Film acid analysis, experimental procedure, 36

Fluorescence spectroscopy, examination of laser-ablated polymers, 401

G

Gas-phase functionalization
 description, 189–190
 influencing factors, 190
 scheme for single-layer resists, 190,191f
Gas-phase functionalized resist schemes
 absorption spectrum, 205,207f
 effect of polymer structure on $TiCl_4$ incorporation, 198
 lithographic sensitivity vs. amount and position of chlorine in polymers, 205t
 O_2 reactive-ion etching behavior of films vs. Ti on film surface, 195,197t
 photooxidative imaging scheme, 202,204f
 Rutherford backscattering spectra, 200,201f,202,203f
 scanning electron micrograph, 205,206f
 sensitivity curves, 202,205,206–207f
 structures and functionalization nature of polymers, 202,203f
 Ti concentrations vs. mole percent acetyl groups, 198,200t
 Ti incorporation vs. background pressure, 195t
 Ti layer thickness vs. $TiCl_4$ treatment time, 195,196f
GC–MS, chemically amplified resists, 42–43
Gel permeation chromatography, chemically amplified resists, 43

H

High-molecular-weight polysilanes, lithographic potential, 123–130
High-resolution, dual-tone resist materials, use of chemical amplification principle, 11,12f
High-resolution lithography, exposure systems, 349
Hydroxy-3-(aminoethanethiosulfuric acid) propyl phenyl ether
 concentration vs. photolysis time, 296,298f
 photolysis, 296,298f,299t
 quantum yields for disappearance, 296,299t
 synthetic scheme, 283,286
5-Hydroxy-1,1,3,3-tetramethylindan, formation, 46,49–50

I

Imaging via thermolytic main-chain cleavage
 mechanism of cleavage of polyesters and polycarbonates, 101–103
 polycarbonates, polyesters, and polyethers, 101
 preparation of active polyformals under phase-transfer catalysis, 104,106t,107–108f
 procedure for polycondensation under phase-transfer conditions, 109
 testing of polyformal as dry-developing imaging system, 106,108f,109,110f
 thermolytic cleavage of allylic and benzylic ethers and polyethers, 103,104,105f
Integrated circuits, patterning with photolithography, 39

K

Kinetics, polymer etching in oxygen glow discharge, 210–231
KrF excimer laser lithography
 development of negative deep-UV resist, 270–278
 lack of high-resolution resist, 269–270

L

Langmuir–Blodgett polymer films, See Ultrathin Langmuir–Blodgett polymer films
Langmuir–Blodgett technique, preparation of ultrathin polymer films, 350
Laser ablation of polymeric materials
 ablation mechanism, 405
 experimental materials, 401
 fluence dependence of relative intensity of excimer emissions, 405,407f
 fluence-dependent luminescence spectra, 403,404f
 fluorescence characterization, 406,409
 fluorescence rise and decay curves, 406,408f
 fluorescence spectra, 405–406,407f
 instrumentation, 401,402f,403
 polymers studied, 400–401
 Si–Si annihilation, 405
 temporal characteristics of luminescence, 403,405
Laser interferometry, 234–250
Liquid monomer films, evaluation as planarizing layers, 256–257

INDEX

Lithographic characterization of chemically
amplified resists
absorbance, 53t
contrast, 53t,54f
differential solubility, 52
performance, 51,52t
scanning electron micrographs, 53,54f
sensitivity, 53t
Lithographic evaluation
chemically amplified resists, 43
novolac–dimethyl siloxane block copolymers
Auger depth profiles, 165,168f,169
effect of block lengths on resolution
capabilities, 170t
effect of phenolic component's chemical
structure on lithographic
properties, 163
experimental procedures, 159–163
materials characterization, 160
molecular weight determination, 162t,163
O_2 reactive-ion etching, 169t,170
quantitative nature of silylamine–phenol
reaction, 160,162
SEM photographs, 170–173
structure of copolymers, 160,161f
TEM, 165,166–168f
thermal characterization of copolymers,
163,164–165t

M

Mass spectroscopy of UV- and vacuum
UV-irradiated poly(methyl methacrylate)
attenuation of UV and VUV light in film,
425,427f
comparison of mass spectra of vaporized
products, 432,434f
experimental procedure, 425,427f
experimental setups, 425,426f,428f
intensity variations of mass peaks,
428,430f,431f,432
mass spectra of vaporized UV etching
products, 428,429f
molecular structure and mass numbers of
photobleaching products, 428,429t
temperature dependence of sum of peak
intensity, 428,430f
time dependence of mass spectra of
VUV-irradiated products, 422,433f
Merocyanine dye
absorption spectra, 30,31f
effect of bleaching on addition of
fractional equivalents of TFA, 30,33f
Methylated melamine, structure, 89
Microelectronics technology
evolution, 1
power delay product vs. year and cost per
bit vs. year, 1,2f

Microlithography
developments in technology, 3
evolution, 1
photolithographic trends, 1,3t
size vs. year of commercialization for
devices, 1,2f
Monomers, evaluation as planarizing layers,
262–263t,264
Multilayer lithography, role of polymer
etching kinetics, 210–211
Multilevel resist chemistry
discussion, 13–19
process sequences involving planarization
techniques, 16,17f
Multilevel resist technology, description, 6

N

Near-surface imaging, introduction, 189
Negative deep-UV resist for KrF excimer
laser lithography
chemical structures, 270,272f
development, 270
dissolution characteristics, 273,277f
dissolution kinetics, 273,276,277f
effect of alkaline concentration on
contrast and sensitivity, 276,277f
effect of concentration on contrast,
271,274f
effect of concentration on exposure
characteristics, 271,272f
Fourier-transform IR spectra, 273,274f
lithographic evaluation, 271,276,278f
optimization of composition,
271,272f,274f
preparation, 270
scanning electron microscopic photography,
276,278f
spectroscopic characterization,
273,274–275f
UV spectra, 273,275f
Negative resists
chemistry, 6
classes of inherently cross-linking
polymers, 7,10f
cyclized poly(cis-1,4-isoprene)
resist, 6–7
developer selection, 6
Negative tone resists, three-component
systems, 87
Nonionic acid precursors based on
nitrobenzyl ester photochemistry, 11,14f
Novolac resins
dissolution model, 378,383
factors influencing dissolution, 364–382
properties, 134
role in microelectronic industry as
positive photoresists, 364

Novolac-based negative resists
  acid concentration for cross-linking, 96t,97
  acid concentration for cross-linking of resist films, 89–92
  Arrhenius plots, 93,94f
  cross-linking activation energy determination, 90,93,94f,95t
  cross-linking kinetics, 93,95
  effect of acid on lithographically useful cross-linking, 96t,97
  experimental procedures, 89
  lithographic characteristics, 95–96,97f
  lithographic potential curve, 90,91f
  scanning electron micrographs, 96,97f
Novolac-based resist, characteristics, 177t
Novolac–dimethyl siloxane block copolymers, lithographic evaluation, 159–173
Numerical aperture steppers, improvements in size, 3

O

Oligosilane derivatives, absorption in UV spectral region, 116
Optical lithography
  advantages as patterning method, 332
  limitations, 349
  pattern transfer processes, 333
  photochemical image enhancement, 333–346
  resolution aspects, 332–333
Organic materials as planarizing layers
  comparison of planarization of films of positive photoresist, o-cresol–novolac, and poly(α-methylstyrene), 259,260t,261
  drawing of film profile over hole, 254,255f
  experimental procedures, 254,256f,257t,258
  factors influencing planarization, 257
  film shrinkage, 263t,264
  T_g measurement, 257
  T_g of films, 263t,264
  O_2 reactive-ion etching rate measurement, 257
  planarization achieved by thick films, 261t,262t
  planarization achieved by unbaked and baked films, 257,258,259t
  planarization determination, 254,256
  property requirements, 254
  resin information, 256t
  schematic drawing of topographic substrate, 254,255f
Organic polymers
  bombardment-induced yields in pattern transfer regime, 218–219,220f,221

Organic polymers—*Continued*
  carbon atom yield vs. etching rate at constant plasma density, 219,220f,221
  chemical etching regime, 212
  etching rate vs. bombardment energy flux, 216,217f
  ion beam studies, 212–214
  O_2 reactive-ion etching rate vs. pressure, 216,217f,218
  O_2 reactive-ion etching rate vs. self-bias voltage, 218,220f
  role of bombardment in pattern transfer regime, 215–218,220
  role of O atoms in pattern transfer regime, 214
Organosilicon moieties, use in defining submicrometer features in bilevel resist structures, 158
Organosilicon polymers
  anomalous transport regime, 229–230
  diffusion-controlled regime, 224–225
  etching rates in steady-state regime, 225–226,227f
  etching regimes, 221–222
  kinetics of oxide growth, 222–223
  selectivity for etching organic vs. silyl novolac vs. bombardment energy, 226,227f
  silicon atom yield of novolac vs. bombardment energy, 226,227f
  steady-state oxide thickness, 226,228–229
  transient regime, 223–224
  use as imaging layers in bilayer resist processes, 133
Organosilicon resists, examples, 18,19f
Oxygen reactive-ion etching of thin polymer films, quantitative analysis of laser interferometer wave form, 234–250

P

Photoablation
  ablation parameters for polymers, 417t,418,419f,420
  ablation rate constant, 414,417t
  amount of polymer vaporized vs. instantaneous intensity of laser pulse, 412
  capabilities, 411–412
  deposited thickness vs. ejection angle, 420,421f
  determination of extent of photochemical vs. thermal character of forces, 412
  effect of oxygen in molecular structure of polymer, 420
  effect of polymer chemical structure on photoablative behavior, 412–413

INDEX

Photoablation—*Continued*
  effect of thermal stability, 418,420
  etch and deposition rate measurements by quartz crystal microbalance, 413–414,415–416f
  high fluence mechanism, 418
  product deposition rate measurements, 420,421f
  role of polymer structure, 418,419f,420
  screening effect by gaseous ablation products, 414,417
  yield of deposition vs. ablation depth, 418
  See also Ablative photodecomposition
Photoacid generators
  application of diazonium salts, 327,329–330f
  exposure curve, 327,329f
  scanning electron microscopic photograph, 327,330f
Photobleachable resist system
  application of diazonium salts, 322,327,328f
  exposure curve, 322
  scanning electron microscopic photograph, 327,328f
Photobleaching chemistry of polymers containing anthracenes
  bleaching kinetics, 334–341
  deep-UV exposure, 343–344,345–346f
  experimental materials, 334
  experimental procedures, 334,336–338f
  kinetic behavior of copolymers, 336–337f,340–341
  percent thickness remaining vs. incident energy, 334,337f
  photoreactions under inert gas, 341–342
  second-order system as linear image transfer agent, 336f,340
  sensitivity, 343
  sensitizer concentration effects, 343,345f
  transmittance vs. incident energy, 334–338f,345–346
Photochemical image enhancement, 333
Photochemically initiated depolymerization
  chemically amplified resists, 44–50
  gel permeation chromatograms, 46,47f
Photochemistry of polysilanes
  electron spin resonance spectra, 121–122,124f
  1,1-elimination, 122
  formation of silyl radicals, 121
  mechanism, 119
  quantum yields for polymer scission and cross-linking, 122–123,124t
  source of hydrogen atoms in abstraction reaction, 121
  trapping products from irradiation, 118–119,120t

Photochemistry of polysilanes—*Continued*
  wavelength-dependence model, 121
  yield of adduct vs. irradiation wavelength, 119,120f,121
Photoinduced etching, advantages, 424
Photolithography, improvements, 39–40
Photooxidation of polymers
  analytical methods, 194
  exposures, 193
  O_2 reactive-ion etching conditions, 194
  schematic of gas–solid reaction cell, 194,196f
  $TiCl_4$ treatment procedure, 194
Photoresist dissolution, importance in integrated circuit manufacture, 385
Photoresist films, acid generation, 30,32,33f,34t
Planarization
  determination, 254,256
  influencing factors, 257
  unbaked and baked films, 258,259t
  unbaked films, 257,258t
Planarization processes, schematic drawing, 252,253f,254
Planarization techniques for multilevel resist schemes
  approaches to good local and global planarization, 16
  process sequences, 16,17f
  transferring of defined image through planarizing layer to substrate, 16,18
Plasma processes, effect on etching kinetics, 211
Polyamide(s), resolution patterns, 314,317f
Polyamide isomers containing thymine photodimer
  photoreversal, 308,311,312f
  structure, 306,307f
Poly[γ-(amino-β-sulfonic acid) ether]—*Continued*
  concentration dependence on apparent viscosity, 285,287f
  Fourier-transform IR spectra film, 292,296,297f
  isothermal weight loss of film, 290,293f
  kinematic viscosity vs. time, 285,288f
  percent transmission vs. time, 285,289f
  photolysis of films, 292,296,297f
  synthetic scheme, 283–284
  thermal polymerization of films, 290,291f,293f
Poly[γ-(amino-β-thiosulfate) ether]
  decomposition and cross-linking of thiosulfate moiety, 292,294
  elemental analysis, 281
  experimental materials, 281
  Fourier-transform IR spectra of film, 290–293,295
  images produced, 300–301f
  instrumentation, 281–282

Poly[γ-(amino-β-thiosulfate) ether]—
  Continued
  photolysis of films, 283,292–295
  preliminary examination of imaging
    characteristics, 296,300–301f
  preparative photolysis, 282
  properties, 280–281
  quantitative photolysis, 282–283
  rheological behavior, 285,287–289f
  synthetic procedures, 282
  synthetic scheme, 283–284
  titration, 285,286f
  weight loss of films vs. temperature,
    290,291f
Poly[4-(*tert*-butoxycarbonyloxy)-
    α-methylstyrene]
  photochemically initiated
    depolymerization, 44–50
  solid-state depolymerization mechanism,
    46,48,50
  thermally initiated depolymerization,
    41,44,45t
Poly[4-(*tert*-butoxycarbonyloxy)styrene]
  photochemically initiated evaluation,
    44–50
  thermally initiated depolymerization, 44
Poly[4-(*tert*-butoxycarbonyloxy)styrene–
    sulfone]
  composition and reaction conditions for
    preparation, 41t
  molecular properties, 41,42t
  photochemically initiated
    depolymerization, 44–50
  thermally initiated depolymerization, 44
Polycarbamates, main-chain cleavage, 101
Poly(dimethylsilane), discovery, 115–116
Polyesters, main-chain cleavage, 101
Polyethers, main-chain cleavage, 101
Polyformal-based imaging systems
  experimental procedures, 109–111
  imaging experiments, 111
  imaging via thermolytic main-chain
    cleavage, 101–110
  UV monitoring of acidolysis of polyether,
    109,111
Polymer containing thymine photodimers,
  photolysis, 304
Polymer etching kinetics in oxygen glow
  discharge
  bombardment-induced pattern transfer
    models, 230–231
  bombardment-induced process, 211–212
  organic polymers, 212–221
  organosilicon polymers, 221–230
  problems with plasmas, 211
  role in pattern transfer step in
    multilayer lithography, 210–211
Polymer(s) in microlithography
  multilevel resist chemistry, 13–19

Polymer(s) in microlithography—*Continued*
  resist properties, 4
  single-level resist chemistry, 6–14
Polymer refractive index, determination,
  235–240
Polymer structure, role in photoablation,
  418,419f,420
Polymer swelling
  determination by solvent concentration
    profile, 385
  measurement, 385–386
Polymethacrylates, thermal and
  acid-catalyzed deesterification, 59–60
Polymethacrylates containing 6-cyanouracil,
  resolution pattern, 314,316f
Polymethacrylates containing pendant thymine
  bases
  lithographic evaluation, 311,313t
  quantum yields and maximum
    photodimerization conversion, 313t
Polymethacrylates containing pyrimidine
  bases
  quantum yields, 306,308t
  structure, 304,305f
Polymethacrylates containing thymine
  derivatives
  quantum yields, 308,309f
  structure, 304,305f
Polymethacrylates containing thymine
  photodimer
  photoreversal, 311,315f
  photosensitivity spectra, 313–314,315f
  structure, 306,309f
Poly(methyl methacrylate)
  absorption coefficient, 424–425,426f
  acid-catalyzed deprotection, 57–58
  high-resolution characteristics, 8
  mechanism of UV- and VUV-induced etching,
    424–425
  positive resist, 8
  preparation, 193
  sensitivity enhancement, 57
  solvent concentration profile, 387–397
  UV-etching reactions, 432
  VUV-etching reactions, 432,435
Poly(α-methylstyrene)
  chain scission and depolymerization, 46
  evaluation as planarizing layers,
    261t,262
Poly(olefin sulfones), chain scission
  positive resists, 8
Poly(phthalaldehyde), preparation and
  photocleavage, 100,102f
Polysilane(s)
  λ_{max} of copolymer, 143,144f
  absorption spectrum and response to
    irradiation, 116,117f
  factors influencing spectral
    characteristics, 116

INDEX

Polysilane(s)—*Continued*
microlithographic applications, 134
photochemistry, 118–124,137,143–146
potential reactions, 133
relative molecular structure in polar
  solvents, 143,144f
^{29}Si-NMR chemical shifts vs. temperature,
  145,146f
solubility measurement, 136
spectral properties, 137,143–146
UV absorption changes during cooling,
  143,145,146f
Polysilane derivatives
applications, 115
discovery, 115–116
formula, 115
preparation by modified Wurtz
  coupling, 116
Polysilane lithography
bilayer imaging, 126,129f
O_2 reactive-ion etching, 126,127f
spectral bleaching of film, 123,125f,126
stepwise lithographic contrast curves,
  126,128
Polysilane photophysics
calculated first singlet excitation
  energy, 116,117f,118
degree of polarization, 118
nature of photoexcited state, 118
Polysilanes substituted with carboxylic
  acids
cross section of resist pattern, 150,151f
mask size vs. resist pattern size,
  150,152f
polymer structure, 136,138f
resist properties, 150,151–152f
sensitivity curve, 150,151f
solubility, 145,147,148f
synthesis, 134,140t
thermal stability, 147,149f
UV absorption change during photolysis,
  147,149f
Polysilanes substituted with phenol group
molecular weight and quantum yield of
  copolymers, 137,138t
polymer structure, 136,138f
potential reactions, 134,135f
solubility, 137
structure of stable polysilane, 136,139f
synthesis, 136–137,139f
thermal stability, 147,149f
Polysiloxanes, O_2 reactive-ion etching
  durability, 134
Polysiloxanes with phenolic group bonded
  to Si
resist properties, 153,154f,155
solubility in base and softening point,
  153,154t
structure, 134,135f

Polysiloxanes with phenolic group bonded
  to Si—*Continued*
synthesis, 150,152f,153
synthesis of monomers, 134,141t
Polystyrene
incorporation of chloromethyl groups, 7
use as resist, 7
Poly(styrene–sulfone), chain scission and
  depolymerization, 46
Positive photoresists, composition, 350
Positive resists
definition, 8
dissolution inhibition resists, 8–9
mechanism of action, 8
poly(methyl methacrylate), 8
poly(olefin sulfones), 8
production of negative-tone images, 9,12f
Process parameters for chemically amplified
  resists
influencing factors, 50–51
postexposure thermal treatment process
  step, 50
temperature of postexposure bake, 51
time between exposure and postexposure, 51
time of postexposure bake, 51
1:1 Projection, use in device production, 3t
5:1 Projection, use in device production, 3t
Pyrimidine bases, photochemical
  dimerization, 303–304
Pyrimidine-containing polymers, reversible
  photoreaction, 304
Pyrimidine derivatives as lithographic
  materials
experimental materials, 304–308
instrumentation, 306
lithographic sensitivity,
  311,313t,314,315f
photodimerization, 306–308
photoreversal of photodimers, 308–312
resolution evaluation, 314,316–317f
Pyrimidine photodimers, photoreversal,
  308,311,313f,315f

Q

Quantitative analysis of laser
  interferometer wave form
advantages, 250
amplitude reduction factor and O_2
  reactive-ion etching rate, 245,246t
amplitude reduction factor and self-bias
  potential vs. power density, 243,244f
background, 235–240
basic film model, 235,236f
calculation of polymer refractive index,
  242,243t
data analysis, 243
end-point behavior, 246,249f

Quantitative analysis of laser
interferometer wave form—*Continued*
experimental procedure, 240,241f,242
influencing factors, 250
O$_2$ reactive-ion etching interferogram,
237,241f
polymer refractive index determination,
235–240
reactive-ion etching–laser interferometer,
240,241f,242
scanning electron micrographs of etched
films, 246,247,248f
surface roughness of etched films,
243,245t
wave form dependence on concentration,
235,236f
Quantum yield, determination for acid
generation, 32,34
Quartz crystal microbalance
etch depth vs. fluence, 414,416f
limitations with solid films, 413
linear dependence on fluence at low
intensity, 414,415f
measurement of ablation rate, 413–414
theory, 413–414

R

Reflectance
definition, 235,237
shape of R vs. time curve, 235,236f
Resins, evaluation as planarizing layers,
257–262
Resist
contrast curves, 4,5f
etching resistance, 4,6
resolution and sensitivity, 4
Resolution of resist, definition, 4

S

Screening effect
description, 412
gaseous ablation products, 414–417
Sensitivity for negative and positive
resists, defined, 4
Silicon-based positive photoresist
absorption intensity relationship,
182,184f,185
application to near-UV lithography, 177
characteristics, 177t
electron beam lithography, 177,179
excimer laser lithography, 179,183f
experimental materials, 185,187
image reversal chemistry, 182–186
IR absorption changes of diazo groups,
182,183f

Silicon-based positive photoresist—*Continued*
IR absorption changes with monofunctional
sensitizer, 185,186f
lithographic procedure, 187
measurements, 187
preparation, 177
scanning electron micrographs, 177–181
sensitivity curves, 171,179,180–181f
UV absorption changes, 182,184f
X-ray lithography, 179,181f
Silicon-containing resists, 175
Single-layer resist(s), gas-phase
functionalization scheme, 190,191f
Single-level resist chemistry, 6–14
Solution quantum yield
determination for acid generation, 32,34
experimental procedure, 36
Solvent concentration profile
determination by fluorescence quenching,
386–397
estimation of gel layer thickness, 386
Solvent concentration profile of poly(methyl
methacrylate)
data analysis, 389,391,393f
estimation of methyl ethyl ketone solvent
concentration profile in film,
396,397f
experimental procedure, 387,388f
Fickian precursor in methyl ethyl ketone
solvent concentration profile,
394,395f,396
flow cell, 387,388f
fluorescence intensity vs. methyl ethyl
ketone concentration, 394,395f
fluorescence quenching, 391–392,393f,394
interferometric trace, 389,390f
methyl ethyl ketone diffusion coefficient
vs. concentration, 392,393f,394
model, 394,395f,396,397f
propagation, 389,391,393f
time profile, 389,390f
Submicrometer features in bilevel resist
structures, definition, 158
Substrate planarization
need in etchback processing, 252
schematic drawing, 252,253f,254

T

Thermally initiated depolymerization,
chemically amplified resists, 44,45t
Thiosulfate-functionalized polymer,
characterization, 280–301
Three-component, aqueous-base developable,
positive-tone resists, use of chemical
amplification principle, 11,13,14f
TiCl$_4$
effect on etching selectivities, 190

INDEX

$TiCl_4$—*Continued*
  use in organic-on-organic bilayer resist scheme, 190,192f
Time-resolved luminescence spectroscopy, molecular studies on laser ablation processes of polymeric materials, 401–409
Trilevel processing, description, 16,18

U

Ultrathin Langmuir–Blodgett polymer films
  electron beam exposure, 352
  electron beam lithography, 352,353f,354
  emission spectrum, 354,357f
  excitation spectra, 354,358f,359t
  experimental materials, 351
  film preparation, 351–352
  fluorescence measurements, 352–362
  optical exposure, 352
  optical lithography, 352,354,355–356f
  ratio of excimer to monomer fluorescence intensities, 354,359f
  schematic representation, 354,360f,361
  schematic representation of dyes, 361,362f
  substrate preparation, 351
Ultrathin polymer films
  Langmuir–Blodgett preparation technique, 350

Ultrathin polymer films—*Continued*
  obstacles in investigation, 350–351
  preparation on wafer surface, 350
Ultrathin resists, technological advantages, 349–350
UV-induced etching of poly(methyl methacrylate), mechanism, 424–435

V

Vacuum UV (VUV)-induced etching of poly(methyl methacrylate), mechanism, 424–435

X

X-ray lithography, use in device production, 3t

Z

Zwitterionic water-soluble thiosulfate polymer, synthesis and characterization, 280–301

*Production: Becki K. Weiss and Paula M. Bérard
Indexing: Deborah H. Steiner
Acquisition: Cheryl Shanks*

*Elements typeset by Hot Type Ltd., Washington, DC
Printed and bound by Maple Press, York, PA*

*Paper meets minimum requirements of American National Standard
for Information Sciences—Permanence of Paper for Printed Library
Materials, ANSI Z39.48–1984* ∞

Other ACS Books

Chemical Structure Software for Personal Computers
Edited by Daniel E. Meyer, Wendy A. Warr, and Richard A. Love
ACS Professional Reference Book; 107 pp;
clothbound, ISBN 0–8412–1538–3; paperback, ISBN 0–8412–1539–1

Personal Computers for Scientists: A Byte at a Time
By Glenn I. Ouchi
276 pp; clothbound, ISBN 0–8412–1000–4; paperback, ISBN 0–8412–1001–2

Biotechnology and Materials Science: Chemistry for the Future
Edited by Mary L. Good
160 pp; clothbound, ISBN 0–8412–1472–7; paperback, ISBN 0–8412–1473–5

Polymeric Materials: Chemistry for the Future
By Joseph Alper and Gordon L. Nelson
110 pp; clothbound, ISBN 0–8412–1622–3; paperback, ISBN 0–8412–1613–4

The Language of Biotechnology: A Dictionary of Terms
By John M. Walker and Michael Cox
ACS Professional Reference Book; 256 pp;
clothbound, ISBN 0–8412–1489–1; paperback, ISBN 0–8412–1490–5

Cancer: The Outlaw Cell, Second Edition
Edited by Richard E. LaFond
274 pp; clothbound, ISBN 0–8412–1419–0; paperback, ISBN 0–8412–1420–4

Practical Statistics for the Physical Sciences
By Larry L. Havlicek
ACS Professional Reference Book; 198 pp; clothbound; ISBN 0–8412–1453–0

The Basics of Technical Communicating
By B. Edward Cain
ACS Professional Reference Book; 198 pp;
clothbound, ISBN 0–8412–1451–4; paperback, ISBN 0–8412–1452–2

The ACS Style Guide: A Manual for Authors and Editors
Edited by Janet S. Dodd
264 pp; clothbound, ISBN 0–8412–0917–0; paperback, ISBN 0–8412–0943–X

Chemistry and Crime: From Sherlock Holmes to Today's Courtroom
Edited by Samuel M. Gerber
135 pp; clothbound, ISBN 0–8412–0784–4; paperback, ISBN 0–8412–0785–2

For further information and a free catalog of ACS books, contact:
American Chemical Society
Distribution Office, Department 225
1155 16th Street, NW, Washington, DC 20036
Telephone 800–227–5558